# Discoidin Domain Receptors in Health and Disease

Rafael Fridman • Paul H. Huang
Editors

# Discoidin Domain Receptors in Health and Disease

Springer

*Editors*
Rafael Fridman
Department of Pathology
Wayne State University School of Medicine and Karmanos Cancer Institute
Detroit, MI, USA

Paul H. Huang
Division of Cancer Biology
The Institute of Cancer Research
London, UK

ISBN 978-1-4939-6381-2 ISBN 978-1-4939-6383-6 (eBook)
DOI 10.1007/978-1-4939-6383-6

Library of Congress Control Number: 2016947091



Printed on acid-free paper

This Springer imprint is published by Springer Nature
The registered company is Springer Science+Business Media LLC New York

# Preface

Sensing and responding to dynamic changes in the tissue microenvironment is pivotal for the robust functioning of biological systems. Cells have thus evolved sophisticated mechanisms to interact with their microenvironment, which include a diverse array of receptor systems that recognize and respond to external stimuli. The Discoidin Domain Receptors (DDRs), comprising DDR1 and DDR2, are one major class of cell surface sensors and the only tyrosine kinases that bind to collagen, the major component of the extracellular matrix (ECM). As such, DDRs are unique Receptor Tyrosine Kinases (RTKs), which activate signal transduction pathways through their intrinsic tyrosine kinase activity. Cognate binding to collagen triggers a series of signaling events that ultimately regulates a host of fundamental cellular functions. As a consequence, it is perhaps unsurprising that dysregulations in DDR function, including alterations in receptor expression levels, structural mutations, and aberrant signaling, are key factors in a number of major diseases of unmet medical need such as cancer, fibrosis, atherosclerosis, and arthritis. These findings raised considerable interest on DDRs as potential therapeutic targets in various disease conditions. In spite of these encouraging trends, the biological functions of DDRs and the mechanisms of receptor regulation and signaling in human pathologies remain ill defined. Therefore, much research is still needed to fully unveil and understand the action of these unique collagen receptors. With this goal in mind, we set to recruit an international group of scientist experts in the field of DDRs to contribute their knowledge of DDR structure, biology, and regulation and their emerging roles in health and disease. Their scholarly and generous contribution has been assembled in the first book dedicated to the DDR field.

In the first section of the book, we focus on the biological aspects of DDR function and their importance in physiology. This includes an in-depth analysis of the binding properties and cell adhesion function of these receptors and how these receptors work in concert with the integrins to perform their biological roles (Leitinger). Given their capacity to bind collagen, the DDRs are critical for organizing collagens into higher-order structures, essential for maintaining the structural integrity of the collagenous matrix. The extensive work in characterizing this process, known as fibrillogenesis, which is also a hallmark of pathological fibrosis, is comprehensively covered in the second chapter (Argarwal). Unlike other RTKs, which bind soluble ligands, the DDRs are unique in their ability to bind to a component of the ECM. Understanding the molecular basis of this interaction and how collagen binding ultimately leads to receptor activation requires the use of crystallographic and NMR-based methods. The third chapter summarizes our remarkable advances in the structural biology of the DDRs and how this information can be harnessed to develop novel agonists and antagonists for these receptors (Nishida and Shimada). We have learnt much of the physiological role of the DDRs from the use of genetically engineered model organisms. The last two chapters in this section provide insights into how genetic manipulations in mouse models (Kano et al.) and invertebrates (Unsoeld et al.) have shed light on the mechanisms by which this class of receptors regulates development and maintains key physiological functions.

The second section of the book delves into our understanding of the impact of the DDRs on cancer initiation and progression. In the era of genomic medicine, large-scale molecular profiling studies in combination with in-depth molecular pathology and functional analyses have shown that the DDRs are implicated in a wide range of cancer types. This includes the two major cancer killers worldwide, lung and breast cancers, which together account for more than a quarter of all cancer-related deaths (summarized by Longmore and Grither). Chapters in this section illuminate on the pathophysiological, genomic, and functional evidence that DDRs play a role in the etiology of the disease and novel approaches by which we can exploit this dependency for the development of new cancer therapeutics. These include chapters on the dysregulation of DDRs in lung cancer (Zhang et al.), breast cancer (Fernandez-Valdivia and Fridman), and cancers of the reproductive system (Marco and Gill). To effectively treat these cancers, strategies to pharmacologically target DDRs are required, and in this area, there have been significant advances in recent years in the development of small-molecule inhibitors of the DDRs (Lu and Ding, Ruiz-Castro et al.). The DDRs display a distinctive activation profile of slow and sustained phosphorylation upon collagen binding, which eventually leads to the propagation of complex downstream signaling networks. It is thus also necessary to characterize the oncogenic signaling networks propagated by the DDRs to fully understand their role in cancer progression (Iwai et al. and Ruiz-Castro et al.). Enhanced cell migration, deregulated differentiation, and loss of adherens junctions are key characteristics of oncogenesis and metastasis. The last chapter in this section details the extensive mechanistic studies that have been performed to demonstrate how DDR1, in concert with the integrins (another set of collagen receptors), plays an important role in maintaining the stability of adherens junctions (Yeh and Tang).

The book also includes chapters on the expression and dysregulation of the DDRs in diseases other than cancer. Indeed, DDRs have been implicated in the pathogenesis of arthritis, atherosclerosis, and fibrosis. These prominent aging-associated disorders are major public health issues for which there is an urgent need to develop new medicines and therapeutic strategies to improve quality of life and reduce burden on society. Defects in DDR2 expression and function have been linked to the development of osteoarthritis, a joint disease that afflicts 25% of adults aged over 65 years. Studies from mouse models of osteoarthritis suggest that targeting DDR2 has clinical utility in slowing the process of progressive joint failure (Xu et al.). Fibrosis, the pathological deposition of fibrillar collagen, causes significant impairment in organ function and is a major health problem. However, there are no effective therapies to diminish the damage of tissue fibrosis. Recent evidence from DDR-deficient mice indicates a key role for DDRs in fibrosis. Two chapters in this section summarize the current knowledge on the roles of DDRs in kidney (Chatziantoniou et al.) and liver (Olaso et al.) fibrosis and provide support for exploring DDRs as therapeutic targets in fibrotic conditions. Atherosclerosis, the accumulation of plaque in the coronary arteries, is the biggest cause of deaths in the developed world. DDR1 has been implicated in the promotion of atherosclerotic plagues through multiple mechanisms including inflammation and calcification. Thus, inhibiting this receptor may prove beneficial for reducing atherosclerosis in patients (Bendeck). DDRs have been implicated in heart development. Understanding the biology of the DDRs in cardiac development and function will facilitate a deeper insight into the role of these receptors in cardiovascular disease (Goldsmith et al.).

This volume is only possible due to the time and energy generously given by our contributing authors, for which we are extremely grateful. We would also like to thank the staff at our publisher Springer, in particular Joy Bramble, Jeffrey Taub, and Fiona Sarne who have kept us on track and provided all the much-needed administrative support. We believe that this volume will serve as a scholarly reference and guide for scientists and clinicians interested in the emerging topic of the DDRs and hope that moving forward it will provide the foundation for new discoveries in this exciting field.

Detroit, MI, USA — Rafael Fridman
London, UK — Paul H. Huang

# Contents

# Contributors

**Gunjan Agarwal** Department of Biomedical Engineering, The Ohio State University, Columbus, OH, USA

**Dina Allam** UCL School of Public Policy, London, UK

**Beatriz Arteta, Ph.D.** Department of Cell Biology and Histology, School of Medicine and Enfermery, University of the Basque Country, Leioa, Spain

**Iker Badiola, Ph.D.** Department of Cell Biology and Histology, School of Medicine and Enfermery, University of the Basque Country, Leioa, Spain

**Michelle P. Bendeck** Department of Laboratory Medicine and Pathobiology, University of Toronto, Toronto, ON, Canada

**Aitor Benedicto, Ph.D.** Department of Cell Biology and Histology, School of Medicine and Enfermery, University of the Basque Country, Leioa, Spain

**Gerold Bepler, M.D., Ph.D.** Karmanos Cancer Institute, Wayne State University, Detroit, MI, USA

**Aliccia Bollig-Fischer** Wayne State University School of Medicine and Karmanos Cancer Institute, Detroit, MI, USA

**Christos Chatziantoniou, Ph.D.** Inserm UMR S 1155 and Sorbonne University, Pierre and Marie Curie University (University of Paris 6), Paris, France

**Ke Ding** School of Pharmacy, Jinan University, Guangzhou, China

Guangzhou Institutes of Biomedicine and Health, Chinese Academy of Sciences, Guangzhou, China

**Aude Dorison, Ph.D.** Inserm UMR S 1155 and Sorbonne University, Pierre and Marie Curie University (University of Paris 6), Paris, France

**Jean-Claude Dussaule, Ph.D.** Inserm UMR S 1155 and Sorbonne University, Pierre and Marie Curie University (University of Paris 6), Paris, France

**Sandamali A. Ekanayaka** Department of Pathology, Wayne State University, Detroit, MI, USA

**Rodrigo Fernandez-Valdivia** Department of Pathology, Wayne State University School of Medicine and Karmanos Cancer Institute, Detroit, MI, USA

**Rafael Fridman** Department of Pathology, Wayne State University School of Medicine and Karmanos Cancer Institute, Detroit, MI, USA

**Paul R. Gill, Ph.D** Unidad de Microbiología Molecular, Instituto de Investigaciones Biológicas Clemente Estable, IIBCE, Montevideo, Uruguay

El Grupo Esmeralda Uruguay, Sierras de Rocha, IIBCE, Montevideo, Uruguay

**Jack G. Goldsmith** Department of Cell Biology and Anatomy, University of South Carolina, Columbia, SC, USA

**Edie C. Goldsmith** Department of Cell Biology and Anatomy, University of South Carolina, Columbia, SC, USA

**Whitney R. Grither** Departments of Medicine (Oncology), The ICCE Institute, Washington University, St. Louis, MO, USA

Departments of Biochemistry, The ICCE Institute, Washington University, St. Louis, MO, USA

**Alexander Hartman** Department of Cell Biology and Anatomy, University of South Carolina, Columbia, SC, USA

**Adam Hoffman** Department of Cell Biology and Anatomy, University of South Carolina, Columbia, SC, USA

**Paul H. Huang** Division of Cancer Biology, Institute of Cancer Research, London, UK

**Harald Hutter** Department of Biological Sciences, Simon Fraser University, Burnaby, BC, Canada

**Leo K. Iwai, M.Sc., Ph.D.** Laboratório Especial de Toxinologia Aplicada (LETA), Center of Toxins, Immune-response and Cell Signaling (CeTICS), Instituto Butantan, São Paulo, Brazil

**Gabor Jarai, Ph.D.** Novartis Institutes for Biomedical Research, Respiratory Diseases, Horsham, UK

Fibrotic Diseases, Disease Sciences and Biology, Bristol-Myers Squibb, Pennington, NJ, USA

**David E. Justus** Department of Cell Biology and Anatomy, University of South Carolina, Columbia, SC, USA

**Kiyoshi Kano** Laboratory of Developmental Biology, Joint Faculty of Veterinary Medicine, Yamaguchi University, Yamaguchi, Japan

Laboratory of Veterinary Anatomy, Joint Faculty of Veterinary Medicine, Yamaguchi University, Yamaguchi, Japan

**Yasuo Kiso** Laboratory of Veterinary Anatomy, Joint Faculty of Veterinary Medicine, Yamaguchi University, Yamaguchi, Japan

**Celina G. Kleer** Department of Pathology and Comprehensive Cancer Center, University of Michigan, Ann Arbor, MI, USA

**Ken Takeshi Kusakabe** Laboratory of Veterinary Anatomy, Joint Faculty of Veterinary Medicine, Yamaguchi University, Yamaguchi, Japan

**Peter L. Lee** Department of Developmental Biology, Harvard School of Dental Medicine, Boston, MA, USA

Faculty of Medicine, Harvard Medical School, Boston, MA, USA

**Birgit Leitinger** Molecular Medicine Section, NHLI, Imperial College London, London, UK

**Yefu Li, M.D., Ph.D.** Department of Developmental Biology, Harvard School of Dental Medicine, Boston, MA, USA

Faculty of Medicine, Harvard Medical School, Boston, MA, USA

**Gregory D. Longmore** Departments of Medicine (Oncology), The ICCE Institute, Washington University, St. Louis, MO, USA

Departments of Cell Biology and Physiology, The ICCE Institute, Washington University, St. Louis, MO, USA

**Xiaoyun Lu** School of Pharmacy, Jinan University, Guangzhou, China

Guangzhou Institutes of Biomedicine and Health, Chinese Academy of Sciences, Guangzhou, China

**Agnes Malysa** Karmanos Cancer Institute, Wayne State University, Detroit, MI, USA

**Marta Marco, Ph.D.** Area de Biologia Tumoral, Departamento de Bioquímica Clínica, Facultad de Quimica, Universidad de la República, Montevideo, Uruguay

Laboratorio de Biotecnología, Polo Tecnológico de Pando, Facultad de Química, Universidad de la República, Montevideo, Uruguay

**Joana Marquez, Ph.D.** Department of Cell Biology and Histology, School of Medicine and Enfermery, University of the Basque Country, Leioa, Spain

**Ekaterina Mironova** Department of Cell Biology and Anatomy, University of South Carolina, Columbia, SC, USA

**Noritaka Nishida** Graduate School of Pharmaceutical Sciences, The University of Tokyo, Tokyo, Japan

**Elvira Olaso, Ph.D.** Department of Cell Biology and Histology, School of Medicine and Enfermery, University of the Basque Country, Leioa, Spain

**Leo S. Payne, Ph.D.** School of Biological Sciences, The University of Auckland, Auckland, New Zealand

**Jay D. Potts** Department of Cell Biology and Anatomy, University of South Carolina, Columbia, SC, USA

**Pedro A. Ruiz-Castro** Novartis Institutes for Biomedical Research, Respiratory Diseases, Horsham, UK

Sanford-Burnham Medical Research Institute, La Jolla, CA, USA

Division of Gastroenterology and Hepatology, Novartis Institutes for Biomedical Research, Zurich, Switzerland

**Duncan Shaw, Ph.D.** Novartis Institutes for Biomedical Research, Respiratory Diseases, Horsham, UK

Global Discovery Chemistry, Novartis Institutes for Biomedical Research, Cambridge, MA, USA

**Ichio Shimada** Graduate School of Pharmaceutical Sciences, The University of Tokyo, Tokyo, Japan

**Ming-Jer Tang, M.D., Ph.D.** Department of Physiology, National Cheng Kung University Medicine College, Tainan, Taiwan

**Jesse Taylor** Department of Biological Sciences, Simon Fraser University, Burnaby, BC, Canada

**Thomas Unsoeld** Department of Biological Sciences, Simon Fraser University, Burnaby, BC, Canada

**Lin Xu, M.D., Ph.D.** Department of Developmental Biology, Harvard School of Dental Medicine, Boston, MA, USA

Faculty of Medicine, Harvard Medical School, Boston, MA, USA

**Yi-Chun Yeh** Department of Physiology, National Cheng Kung University Medicine College, Tainan, Taiwan

**Yingtao Zhang** Karmanos Cancer Institute, Wayne State University, Detroit, MI, USA

# Part I
# Discoidin Domain Receptor Biology and Roles in Physiology

# Chapter 1
# DDRs: Binding Properties, Cell Adhesion and Modulation of Integrin Function

**Birgit Leitinger**

## 1.1 Introduction

The search for new receptors with tyrosine kinase function led to the isolation of discoidin domain receptor (DDR) cDNAs by several groups in the 1990s. The receptors were named based on the presence of a discoidin (DS) homology domain in their extracellular regions but remained orphan receptors until 1997 when two independent studies demonstrated that several different types of collagen, major constituents of all extracellular matrices (ECMs), are functional DDR ligands [1, 2]. This was a surprising finding, since most receptor tyrosine kinases (RTKs) were known to interact with peptide-like growth factors or hormones such as epidermal growth factor (EGF) or insulin, and cell–ECM interactions were thought to be mediated mainly by the integrin family of cell surface receptors. The discovery of a direct interaction between RTKs and the matrix and the realization of shared ligands between integrins and RTKs changed the perceptions in the field of cell–matrix interactions.

The two initial studies hinted at a unique receptor activation mechanism for the DDRs. Not only was the ligand an insoluble protein of the ECM, but DDR activation, as assessed by receptor autophosphorylation, was found to occur with protracted kinetics [1, 2]. While conventional RTKs are activated by their ligands within seconds to minutes, DDR autophosphorylation takes place in the time frame of hours [1, 2]. It was speculated that the slow receptor activation may be a result of initial cellular interactions with integrins before activating the lower affinity DDRs, or that collagen added to cells may need to change to a higher order structure before being capable to bind to the DDRs [1].

Like conventional RTKs, the DDRs regulate many fundamental cellular functions, including cell differentiation, proliferation, metabolism and survival [3]. On the

B. Leitinger (✉)
Molecular Medicine Section, NHLI, Imperial College London, London, UK
e-mail: b.leitinger@imperial.ac.uk



R. Fridman, P.H. Huang (eds.), *Discoidin Domain Receptors in Health and Disease*, DOI 10.1007/978-1-4939-6383-6_1

other hand, the DDRs also have overlapping functions with matrix-binding integrins and regulate cell adhesion, migration and invasion. Additionally, they can directly impact on ECM structure and composition through their ability to control matrix metalloproteinase expression and activity. This chapter provides a discussion on the DDR–ligand interactions and considers the interplay between the DDRs and integrins and their role in regulating cell adhesion.

## 1.2 DDR Binding Properties

### *1.2.1 DDR Domain Organization*

As members of the RTK family, the DDRs are single-span type I transmembrane proteins with a cytoplasmic kinase domain (Fig. 1.1). Many other RTKs have structural domains in their extracellular regions that are widely distributed, including fibronectin domains, immunoglobulin-like domains or cysteine-rich domains [4]. The DDRs, on the other hand, have a unique structural arrangement of two globular domains: an amino-terminal DS domain, tightly linked to a discoidin-like (DS-like) domain [5]. The two globular domains are followed by a juxtamembrane region that is likely unstructured, and at least for DDR1, highly flexible [6]. The transmembrane domain connects the extracellular region to the cytoplasmic region, which contains an unusually large juxtamembrane region followed by the carboxy-terminal kinase domain. While the globular domains are highly conserved between DDR1 and DDR2 (59 % sequence identity in the DS domains, 51 % identity in the DS-like domains and 68 % identity in the kinase domains), the juxtamembrane regions, in particular the extracellular juxtamembrane regions, are poorly conserved [3].

As a result of alternative splicing, there are five isoforms of DDR1, all with common extracellular and transmembrane regions but distinct cytoplasmic sequences [7, 8]. While DDR1a-c are functional receptors, DDR1d and DDR1e are truncated proteins with non-functional kinase domains. The most common DDR1 isoforms are DDR1a and DDR1b, which are both widely expressed in different tissues. The longest isoform, DDR1c, contains 919 amino acids and differs from DDR1b by the insertion of six amino acids in its kinase domain. DDR1a lacks 37 amino acids in the intracellular juxtamembrane region, with respect to DDR1b (Fig. 1.1). The biological significance of the six amino acid insert in DDR1c is unknown. Only one isoform of DDR2 is known. Both DDRs contain a number of predicted N- and O-glycosylation sites in their DS-like and juxtamembrane regions (Fig. 1.1).

### *1.2.2 Collagen Structure*

The collagen family consists of 28 different types in vertebrates, where they are the most abundant proteins in the various organisms (making up ~30 % of total body mass) [9]. The functions of the collagens are twofold: to perform key structural

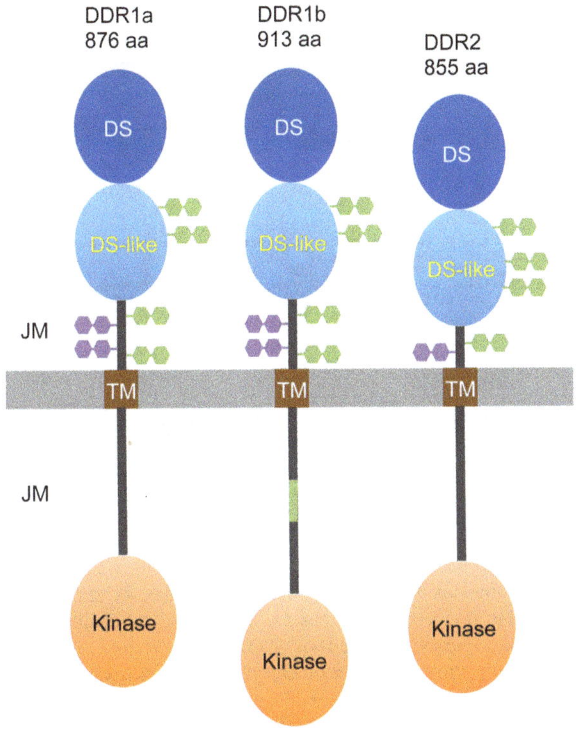


**Fig. 1.1** Domain organization of the DDRs. The extracellular regions are composed of an amino-terminal DS domain, a DS-like domain and a juxtamembrane (JM) region. In DDR1, this JM region contains ~50 amino acids, whereas the DDR2 JM region has 30 amino acids. The transmembrane (TM) domains are followed by a large intracellular JM region (up to 170 amino acids in DDR1, ~140 amino acids in DDR2) and a carboxy-terminal kinase domain. The plasma membrane is represented by a *grey bar*. Predicted glycosylation sites are indicated (*green*, N-glycosylation and *purple*, O-glycosylation). The two main isoforms of DDR1 are shown: DDR1a and DDR1b, which differ in their cytoplasmic regions. The DDR1b region with additional 37 amino acids, relative to DDR1a, is highlighted in *green*

roles in ECMs and to mediate cellular interactions. All ECMs contain a characteristic spectrum of different types of collagens, as well as other key ECM molecules, and the particular molecular composition defines the biomechanical and physical properties of tissues. Collagens impart stiffness to tissues and are well known to provide tensile strength to tendon, cartilage, bone and skin [10]. In these tissues, collagen is present as elongated fibrils. However, other forms of collagen exist, such as in basement membranes, where collagen forms two-dimensional networks.

The hallmark of all collagens is a characteristic triple-helical structure, whereby three polypeptide chains wrap around one another with a one residue stagger to result in a very tight, right-handed triple helix that resembles a stiff cable [11] (Fig. 1.2a). The collagen polypeptide chains are termed α chains. Different types of collagen vary in the number of distinct α chains they incorporate. Some collagens are composed of two (e.g. two α1 and one α2 chains) or three different α chains (e.g. α1, α2 and α3). These collagens are termed heterotrimeric collagens, while

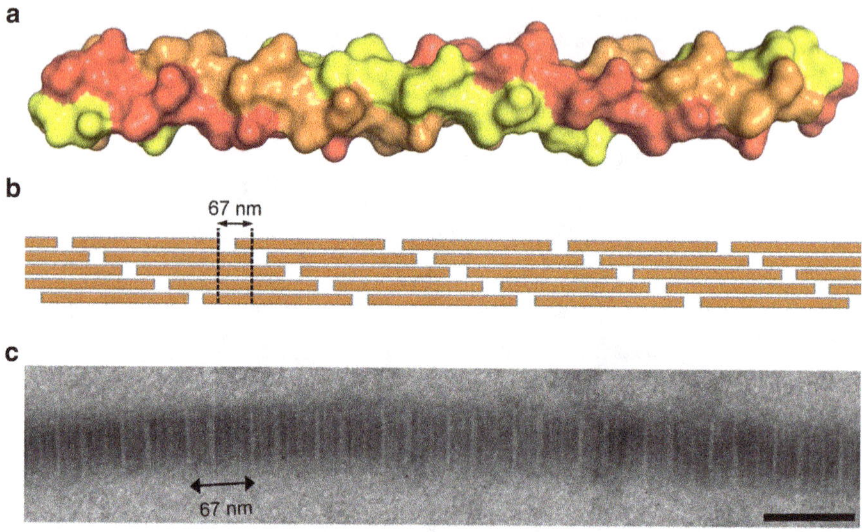


**Fig. 1.2** Collagen structure. (**a**) Surface representation of a collagen triple helix, represented by the idealized collagen-mimetic peptide $(POG)_{11}$. The three peptide α chains are wound around one another with a one residue stagger between chains. *Yellow*, leading chain; *red*, middle chain and *orange*, trailing chain. The Figure was created with PyMOL (http://www.pymol.org). (**b**) Schematic representation of individual triple helices arranged in a collagen fibril. Triple helices associate laterally into a quarter-staggered array, staggered by 67 nm. The length of the triple helices are not drawn to scale. (**c**) Electron micrograph of a heterotypic collagen fibril isolated from articular cartilage. Note periodic banding pattern of 67 nm. Scale bar, 100 nm. Panel C courtesy of Dr. Uwe Hansen, University Hospital Münster, Germany

homotrimeric collagens are made up of three identical α chains. All collagens are designated with roman numerals, for example collagen I, collagen II, collagen IV, etc. Collagens have a very characteristic amino acid composition, with repeating glycine-X-Y motifs. Amino acids in the X and Y positions can be any amino acid but are frequently proline and 4-hydroxyproline, respectively [12]. Glycine, as the smallest amino acid, is required in every third position, because of the structural constraints imparted by the tight packing of the α chains within the triple helix.

In tissues, collagens are not found in the form of individual triple helices but assemble into supramolecular structures, in which the triple-helical molecules form higher-order assemblies such as fibrils and fibres or sheet-like networks. Collagens are grouped into subfamilies, such as fibril-forming, beaded filaments, network-forming collagens or fibril-associated collagens with interrupted triple helices [10]. Prominent fibril-forming collagens are the most abundant ones: collagens type I–III. Network forming collagens include the basement membrane collagen type IV. All collagens are initially synthesized as proteins in which the triple-helical regions are flanked by non-collagenous regions. In the fibril-forming collagens, these non-collagenous domains are proteolytically cleaved once the collagen has been secreted [13]. Fibril-forming collagens contain one major triple-helical region with uninterrupted glycine-X-Y repeats of about 1000 amino acids. Upon removal of the non-collagenous regions, individual triple helices self-assemble laterally into the macromolecular structures known as fibrils or fibres that are essential constituents of many tissues (Fig. 1.2b). The precise staggering of individual triple helices within fibrils leads to their unique banding in electron micrographs (Fig. 1.2c). Some collagens that do not form fibrils co-associate with collagen fibrils; others form different kinds of supramolecular assemblies, such as the network of collagen IV that results from self-assembly in which non-collagenous domains mediate key interactions.

### 1.2.3 Ligand-Binding Specificity of the DDRs

The DDRs, like collagen-binding integrins, belong to a family of transmembrane receptors that bind only to native, triple-helical collagens [14]. The two initial studies that identified the DDRs as collagen receptors demonstrated that DDR activation, as assessed by receptor phosphorylation, was only induced by collagen in its native conformation but not by heat-denatured collagen, such as gelatin [1, 2]. Studies with recombinant DDR ectodomains showed the DDRs to interact directly with collagen and to display high affinity towards collagen in solid-phase binding assays, as long as the recombinant proteins were present in dimerized or multimerized form [15–17].

Both receptors are activated by a number of different collagen types and thus display a broad ligand binding specificity. Fibrillar collagens are ligands for both DDRs [1, 2], but non-fibrillar collagens are recognized with distinct preferences. For example, the basement membrane collagen type IV is a ligand only for DDR1 but not for DDR2 [1, 2, 16]. Furthermore, collagen type II seems to be preferentially recognized by DDR2 [18]. Additionally DDR2, not DDR1, is a receptor for collagen type X, which is exclusively found in the hypertrophic zone of the growth plate [19]. DDR1 can also interact with collagen type VIII [20], but it is not known whether DDR2 can bind to this collagen. The collagen-binding preferences of the DDRs are in agreement with their expression patterns. DDR1, but not DDR2, is expressed by epithelial cells [7], which are in contact with basement membranes. Although the roles that DDR1 plays as a collagen IV receptor are not well established, DDR1 functions in some organs are likely to be dependent on its interactions with collagen IV in basement membranes. For example, DDR1 functions in kidney, where it is expressed in glomerular epithelial cells, are presumably due to DDR1 interacting with collagen IV in the glomerular basement membrane [21–23]. DDR2, on the other hand, is found in articular cartilage and bone cells, where it can interact with collagens type II and X. DDR2 has well-established roles in the growth of long bones where it controls several aspects such as chondrocyte maturation [24] and osteoblast differentiation [25]. However, a detailed understanding about which collagen ligand(s) controls the various processes in bone growth is still missing.

### 1.2.4 DDR Binding Sites in Collagen

Collagens can interact with many different proteins, including other matrix molecules or cellular receptors. Binding sites for collagen interaction partners are best characterized for the fibrillar collagens, which contain over 50 biologically active sites [26]. Mapping binding sites in fibril-forming collagens is challenging because of their tissue insolubility and large size. Early studies used collagen fragments obtained by cyanogen bromide digestion or rotary shadowing electron microscopy to obtain the rough locations of some binding sites within fibril-forming collagens. Later work has taken advantage of libraries of overlapping synthetic collagen-mimetic peptides, the so-called Collagen Toolkit peptides, which allow comprehensive analysis of binding sites within the fibrillar collagens type II and type III [27].

DDR binding sites have been mapped in fibrillar collagens and some binding motifs have been characterized in detail. Similar to collagen-binding integrins, the DDRs bind to distinct amino acids motifs rather than recognizing general structural features of the triple helix. The first study to address DDR binding sites in collagen used atomic force microscopy to visualize DDR2 binding to collagen type I but did not reveal specific binding motifs [15]. Early mapping data used recombinant variants of collagen II and localized a DDR2 binding motif to the second quarter of the collagen II collagenous domain (amino acids 235–468 of the triple helix) [18]. A more detailed understanding of DDR binding motifs in collagens was obtained with the use of the Collagen Toolkit peptides, which allowed comprehensive screening for DDR binding sites in the homotrimeric collagens type II and type III. Studies with these peptides mapped DDR2 binding to several sites in both collagens, whereas DDR1 seemed to be restricted to one main binding site that is shared with DDR2 [28, 29]. Through the use of further synthetic peptides this shared binding site was characterized as a six amino acid motif, GVMGFO (O is 4-hydroxyproline), which functions as a high-affinity interaction site for both DDR1 and DDR2 [28, 29]. GVMGFO is found in the fibrillar collagens types I–III but not in collagen IV. Thus, DDR1 is likely to bind to a different type of motif in non-fibrillar collagens. The region encompassing the GVMGFO motif is also a functional interaction site for two unrelated secreted proteins: von Willebrand factor [30] and SPARC (Secreted protein, acidic, cysteine rich) [31]. It is thus conceivable that the DDRs compete for binding to this site with von Willebrand factor or SPARC.

As discussed above, collagens form supramolecular assemblies in tissues that contain multiple binding sites for interaction partners. It is currently not clear in what form collagen functions as a physiological DDR ligand. For DDR activation experiments, fibril-forming collagens are commonly added as isolated triple helices to cells. During the incubation period at 37 °C, these triple helices may produce fibrils in cell culture medium, and it is therefore not clear in what form of assembly (fibril or single triple helices) collagen activates the DDRs. However, it is clear that DDR activation does not require collagen to be present in its fibrillar state. This conclusion comes from experiments with DDR ligands that are unable to assemble into fibrils. For example, GVMGFO-containing triple-helical peptides induced

DDR phosphorylation with the same kinetics as full-length, tissue-derived collagen [28, 29]. Furthermore, pro-collagen type II, with intact N- and C-terminal pro-peptides, was able to induce DDR phosphorylation similarly to collagen type II [18]. However, some DDR2-mediated functions, such as cell cycle arrest of cancer cells or down-regulation of focal adhesion kinase in smooth muscle cells were only observed with polymerized collagen type I but not with collagen I in the form of isolated triple helices [32, 33]. It is possible that single triple helices and collagen fibrils have different DDR binding modes and thus transmit different signals into cells. However, it remains to be established whether collagen fibrils are functional DDR ligands.

### 1.2.5 Molecular Basis of Collagen Binding

Most RTKs are monomeric when not ligand-bound, but the DDRs form constitutive dimers on the cell surface [6, 34, 35]. While we do not understand the molecular mechanism of how collagen binding to the DDR ectodomain induces intracellular kinase activation [36], the binding mode of the GVMGFO motif with the DDRs is understood at atomic level detail [37]. The structures of the DDR DS and DS-like domains are known from nuclear magnetic resonance (NMR) and X-ray crystallography [5, 37, 38]. DS domains are contained in a variety of unrelated proteins, usually found as single or tandem repeats [39, 40]. DS domains function as interaction modules and mediate a number of different biological functions in secreted proteins (such as blood coagulation factors V and VIII) or transmembrane proteins (such as neuropilin). DS domains interact with diverse ligands, ranging from lipids to carbohydrates and proteins.

A crystal structure of a DDR1 ectodomain region comprising the DS and DS-like domains was obtained in complex with an anti-DDR1 antibody fragment [5]. Figure 1.3 shows the structure of the DS/DS-like domain tandem. The DDR DS domains contain about 160 amino acids, and, in common with all DS domains, adopt a β-barrel structure consisting of two anti-parallel sheets with a total of eight β-strands [5, 37, 38]. The bottom of the barrel forms a flat surface, whereas the top of the barrel contains five protruding loops. The structure is stabilized by two intra-molecular, conserved disulfide bridges. Cys31 (DDR1) or Cys30 (DDR2) links the DS domain amino terminus to Cys185 near the carboxy terminus. Cys74 (DDR1) or Cys73 (DDR2) forms a disulfide bond with Cys177, which links loops 2 and 6.

The DS-like domain of DDR1 contains about 180 amino acids and adopts an eight stranded β-barrel structure, similar to the preceding DS domain (Fig. 1.3), despite very low amino acid sequence conservation [5]. Compared with the DS domain, there are five additional strands in the DS-like domain that protrude between the β1 and β2 strands. These comprise a calcium-ion binding site and are glycosylated at two asparagine residues. The DS-like domain contains a conserved disulfide bridge that connects the β4 and β7 strands but does not have a disulfide bond between the domain's amino- and carboxy-terminal ends, as seen in the DDR DS domains.

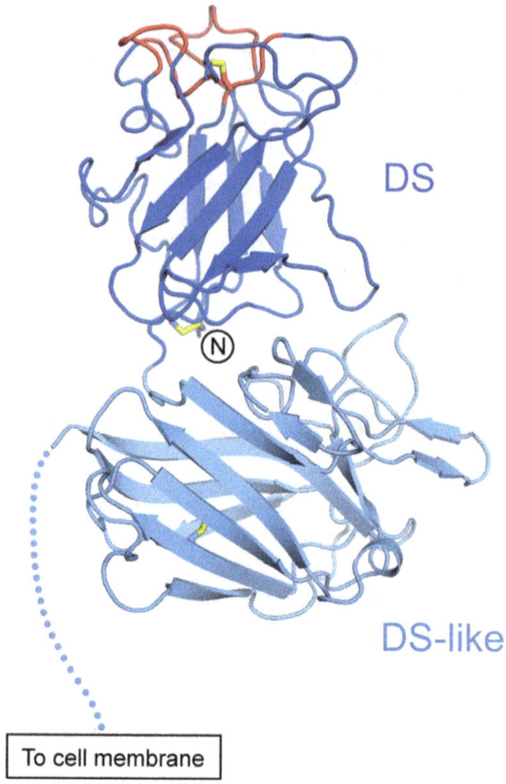


**Fig. 1.3** Crystal structure of the globular domains of DDR1. Cartoon drawing of the structure of the DS and DS-like domains. Disulfide bonds are shown in *yellow*; the amino terminus is indicated (N). The *dashed line* represents the flexible extracellular juxtamembrane region. The Figure was prepared using the coordinates of PDB entry 4AG4 [5]

The collagen-binding sites of the DDRs are entirely contained in their DS domains, as demonstrated with experiments using recombinant extracellular proteins that mimicked the receptors' collagen-binding specificities [16, 38]. Initial mapping was done by loop swapping mutagenesis that defined three spatially adjacent surface-exposed loops as important for the DDR–collagen interaction [16, 41]. These loops are highly conserved between DDR1 and DDR2. A subsequent study that used NMR to determine the solution structure of the DDR2 DS domain and transferred cross-saturation experiments to map the collagen-binding site, confirmed the loop region as the interaction site for collagen [38]. The binding site can be described as a trench that is created by the five protruding loops at the top of the DS domain.

The identification of the GVMGFO motif as a DDR ligand enabled a crystal structure determination of a complex between the DDR2 DS domain and a triple-helical peptide encompassing this motif [37] (Fig. 1.4). The crystal structure confirmed the previously defined collagen footprint [38] and revealed an amphiphilic-binding pocket that accommodates the apolar GVMGFO. The two essential amino acids of the GVMGFO motif (M and F; ref [28]) are inserted into this specificity pocket. An important feature of the binding mode is that these key collagen residues are provided by two distinct collagen α chains, explaining why the

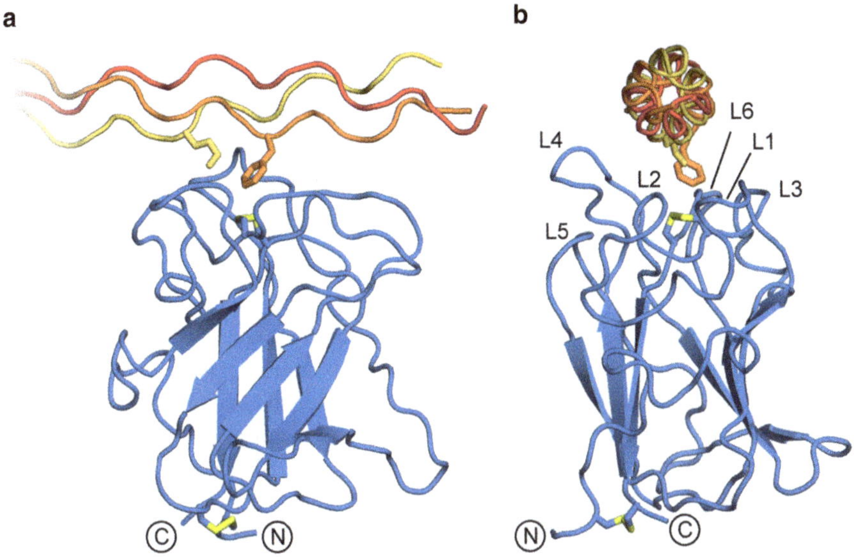


**Fig. 1.4** Crystal structure of the DDR2 DS domain in complex with a collagen-mimetic peptide. (**a**) Cartoon representation of the DS domain (*blue*) and the collagen peptide (*yellow*, leading chain; *orange*, middle chain and *red*, trailing chain). Disulfide bonds are shown in *yellow*. The side chains of key collagen residues (Met21 of the leading chain and Phe23 of the middle chain) are shown. The amino and carboxy termini of the DS domain are indicated (N, C). (**b**) Orthogonal view of the complex related to (**a**) by a 90 °C rotation around a vertical axis, with the collagen peptide viewed from amino to carboxy terminus. Loops at the top of the DS domains are labelled

DDR2–collagen interaction requires a triple-helical conformation and does not occur with denatured collagen. One wall and the floor of the collagen-binding trench are made up by apolar residues (Trp52, Thr56, Asn175 and Cys73-Cys177), whereas the other wall is characterized by a salt bridge (Arg105-Glu113). These key ligand-binding residues are strictly conserved in DDR1, consistent with both receptors binding to fibril-forming collagens. However, at the periphery of the GVMGFO peptide-binding interface, there are several residues that are not conserved in DDR1, and these residues are responsible for the distinct collagen-binding preference of DDR1. When these residues in DDR2 were substituted for those of DDR1, a chimeric receptor was created that was able to bind to collagen IV [29]. Thus, specific regions within the DS domains allow discriminating between fibril-forming and non-fibrillar collagen types.

As mentioned above, it is not clear which ligands for DDR2 regulate various aspects of bone growth. DDR2-dependent functions in human skeletal growth were realized through molecular analysis of a rare, autosomal recessive genetic disorder termed spondylo-meta-epiphyseal dysplasia with short limbs (SMED-SL) and abnormal calcifications [42]. Affected individuals are characterized by disproportionate short stature, short limbs with broad fingers, bone abnormalities and premature calcifications. While most of the identified *DDR2* mutations responsible for the disorder result in trafficking defects with DDR2 protein retained in the endoplasmic reticulum [43, 44], one disease mutation did not affect DDR2 trafficking but resulted in a cell surface receptor that failed to bind collagen [44]. In this case, the missense mutation responsible for the disorder (pE113K) affected a key collagen-binding residue: Glu113 that forms a salt bridge with Arg105 in the wild-type receptor. In the disease mutant, the glutamate is replaced by a lysine, thus replacing a negatively charged residue with a positively charged amino acid. Consequently, the salt bridge that is essential to maintain the binding pocket is lost, resulting in loss of collagen binding of the mutant protein [44]. The fact that a residue involved in binding to the GVMGFO motif is affected is consistent with DDR2 interactions with collagen II being essential for proper bone development.

As mentioned in Sect. 1.2.4, the GVMGFO-containing DDR binding region in collagen overlaps with the binding sites for von Willebrand factor and SPARC. While the amino acid requirements for the von Willebrand factor interactions with collagen are somewhat different to those of the DDRs [30], SPARC has an essentially identical collagen-binding mode to the DDRs [31, 45], with the M and F of the GVMGFO motif provided by two separate α chains and inserted into an amphiphilic-binding pocket on SPARC. Recent findings show that recessive osteogenesis imperfecta, a bone fragility disorder, which is often caused by mutations in the genes encoding collagen type I chains, can also be caused by missense mutations in *SPARC* [46]. Interestingly, like in the DDR2 disease mutant (pE113K), the disease mutations result in disruption of a salt bridge, formed by Arg166 and Glu263, that is essential for the structure through which SPARC binds to the GVMGFO motif.

## 1.3 DDR Functions in Cell Adhesion and Migration

Both DDRs have been shown to regulate cell adhesion and migration, at least in cell culture models. In many studies, pro-adhesive and pro-migratory functions have been ascribed to the DDRs. This is particularly evident in immune cells, where both receptors seem to be important players that can help activated leukocytes to migrate into infectious or inflammatory tissue sites. For example, DDR1 mediates cell migration of monocytic cells and T cells in three-dimensional collagen matrices [47, 48]. Similarly, DDR2, which is expressed on circulating neutrophils, is required for migration in three-dimensional collagen matrices [49]. Integrins, which are key mediators of leukocyte interactions with the ECM, are not involved in leukocyte migration in three-dimensional collagen [50]. It is thus possible that the DDRs are the main mediators of leukocyte cell migration.

Pro-migratory roles have also been found in a number of cancer cell lines (for a comprehensive review, see [51]) and epithelial cells (e.g. [52, 53]). How the DDRs enhance cell migration is poorly understood, but two studies have shown that DDR1-induced signalling has an effect on the tyrosine kinase Syk (Spleen tyrosine kinase). Syk can inhibit cell migration, and DDR1 signalling can block this inhibitory effect in both megakaryocytes and epithelial cells [52, 54] (Fig. 1.5). In carcinoma

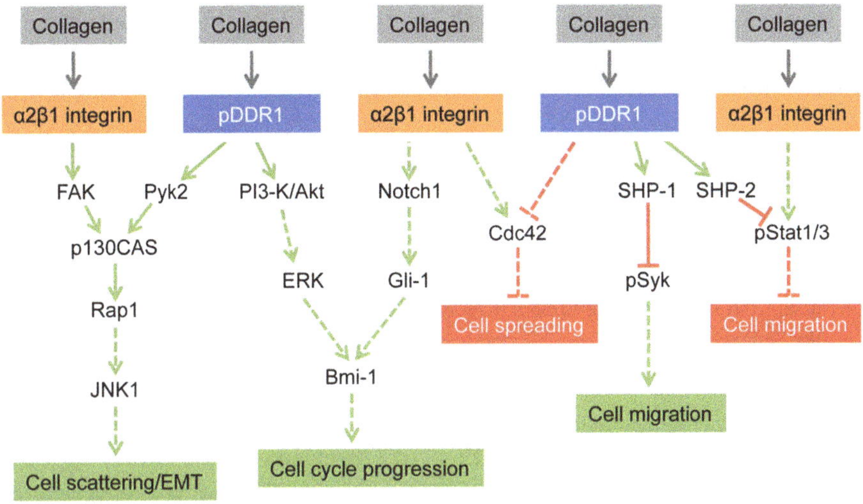


**Fig. 1.5** Selected intracellular events activated by collagen binding to DDR1 or integrin α2β1, illustrating cross-talk between DDR1 and α2β1. Cellular outcomes are shown that depend on the cellular context. In pancreatic cancer cells DDR1 and α2β1 cooperate with one another, leading to cell scattering/EMT. Similarly, in mouse embryonic stem cells, the two receptors cooperatively signal to allow cell cycle progression. In MDCK cells, DDR1 antagonizes α2β1 functions, such as cell spreading or cell migration. In other cell types, including megakaryocytes or smooth muscle cells, DDR1 promotes cell migration. *Solid lines* depict direct signalling effector molecules and *dashed lines* indicate indirect steps. Cellular outcomes marked in *green* are promoted by DDR1 activation, cellular outcomes in *red* are processes that are suppressed by DDR1 activation

cells, DDR1 is required for collective cell migration by interacting with the cell polarity regulators Par3 (Partitioning defective protein 3) and Par6 at cell–cell contacts, which decreases actomyosin contractility [55].

In terms of cell adhesion, a number of studies found adhesion-promoting roles for the DDRs. Thus, in glioma and pituitary adenoma cells, DDR1 enhanced cell adhesion to collagen type I [56, 57]. Similarly, in the leukaemic cell line THP-1, DDR1 promoted cell adhesion to collagen type I independently of integrin function [48]. In human embryonic kidney cells, overexpression of either DDR1 or DDR2 enhanced cell adhesion to collagen [58]. In this model system, however, integrin function was enhanced by the DDRs (see below). DDR2 was observed to positively regulate fibroblast adhesion to three-dimensional collagen [59]. Additionally, a number of studies using DDR1 knockout mice observed reduced adhesion to collagen in the absence of DDR1, which is consistent with a pro-adhesive function of DDR1. For example, lack of DDR1 in smooth muscle cells decreased adhesion to collagens types I and VIII without affecting adhesion to fibronectin or vitronectin [20]. Furthermore, macrophages from mice lacking DDR1 had reduced adhesion to collagen type IV [60], whereas mesangial cells from these mice showed impaired adhesion to collagen I [61]. Intriguingly, knockdown of DDR1 in melanocytes affected only adhesion to collagen IV but not to collagen I [62].

In contrast to the examples mentioned above, where DDR1 enhances cell adhesion and cell migration, a negative regulatory role was found in MDCK cells, a cell line often used as a model system to understand epithelial cell function. Here, DDR1 was shown to decrease cell adhesion to collagen type I, without affecting adhesion to fibronectin or vitronectin [63]. Furthermore, DDR1 negatively regulated cell spreading, as well as cell migration on collagen I [63–65]. While it is clear that both DDRs can control cell adhesion and migration in cell culture models, it is not clear how these functions relate to physiological processes in healthy adult tissues. Given that the DDRs can both enhance and block cell adhesion and cell migration, the physiological roles that the DDRs play in these processes may be dependent on the cellular and tissue context.

## 1.4 Cooperation of the DDRs with Integrins

RTK-induced signalling pathways are highly interconnected with other signalling pathways, such as those triggered by other RTKs or different classes of cell surface receptors. In this way, cells can fine-tune their signalling outcome. As mentioned in the introduction, the DDRs are unique amongst RTKs in that they interact with the ECM. Integrins are the most prominent cellular receptors for ECM molecules and are well known to regulate cell adhesion [66]. Integrins are heterodimeric transmembrane proteins, composed of non-covalently associated $\alpha$ and $\beta$ subunits. Eighteen $\alpha$ subunits and 8 $\beta$ subunits combine in different ways to form 24 distinct integrins in higher vertebrates. Both subunits have large extracellular regions composed of several subdomains, a single transmembrane helix and short cytoplasmic tails with which they link to the cytoskeleton. While not all integrin ligands are ECM molecules, many ECM proteins are functional integrin ligands, including the key ECM components fibronectin, laminin and collagens. Four of the 24 integrins are collagen receptors [14]. These receptors share a common $\beta$ subunit, the $\beta1$ subunit, which associates with many different $\alpha$ subunits. The collagen-binding subfamily consists of $\alpha1\beta1$, $\alpha2\beta1$, $\alpha10\beta1$ and $\alpha11\beta1$. These integrins are expressed in many different tissues, with $\alpha1\beta1$ (mostly in mesenchymal cells) and $\alpha2\beta1$ (predominantly in epithelial cells, as well as in platelets) the most widely distributed. The expression of $\alpha11\beta1$ is mostly restricted to fibroblasts in regions of highly organized interstitial collagen networks, whereas $\alpha10\beta1$ is prominent in chondrocytes.

Like the DDRs, collagen-binding integrins bind to a number of different collagens with distinct binding preferences, such as $\alpha1\beta1$ preferring collagen type IV over fibril-forming collagens or $\alpha2\beta1$ preferring fibril-forming collagens over type IV collagen [14]. Like the DDRs, collagen-binding integrins recognize collagen by binding specific amino acid motifs in their triple-helical regions. The use of synthetic triple-helical peptides was key to the identification of these motifs. GFOGER was the first integrin-binding motif to be identified for both $\alpha1\beta1$ and $\alpha2\beta1$ [67–69]. Subsequently, a series of motifs with the consensus GxOGER were identified for $\alpha2\beta1$, with the identity of x determining the affinity for integrin binding [70, 71].

While α2β1 has high affinity for GFOGER, α1β1 prefers another sequence, GLOGEN [72].

DDR activation by collagen is independent of β1 integrins [73], but integrins and DDRs can modulate each other's functions. Cross-talk between DDR1 and collagen-binding integrins was observed in several studies. Depending on the cell type, DDRs and integrins can either cooperate with one another in cell adhesion and signal transduction or inhibit each other's functions. Cooperation of DDR1 with α2β1 is required in pancreatic cancer cells, which undergo epithelial-mesenchymal transition upon interaction with collagen I. In these cells, both receptors coordinate to activate c-jun N-terminal kinase (JNK), which in turn up-regulates N-cadherin expression and promotes cell scattering [74]. Integrin binding to collagen I activates focal adhesion kinase (FAK), whereas DDR1 binding to collagen I engages the FAK homologue Pyk2 (Proline rich tyrosine kinase 2), with both signals being coordinated through complex formation of DDR1, α2β1, FAK and Pyk2 with the scaffold protein p130CAS (p130 Crk-associated substrate) (Fig. 1.5). The signal gets to JNK through the small GTPase Rap1 (Ras-related protein 1) [74].

Another example of DDR1 cooperating with α2β1 is provided by mouse embryonic stem cells where DDR1 and α2β1 promote self-renewal through regulation of the cell cycle, with the signalling pathways converging at the gene regulatory plycomb protein Bmi-1 (B-lymphoma Moloney murine leukaemia virus insertion region-1) [75]. Here, collagen binding to α2β1 activates Notch and Gli-I (Glioma-associated oncogene homolog 1), whereas DDR1 binding to collagen I results in phosphoinositide 3 (PI3) kinase/Akt activation, possibly through the small GTPase Ras (Fig. 1.5). In contrast, opposing effects of DDR1 and β1 integrins were found in human adipose stromal cells in three-dimensional collagen matrices. Here, DDR1 activates transcription of stromal aromatase while β1 integrins play an inhibitory role [76]. However, in contrast to the cooperation examples given above, where DDR1 and integrin signalling intersect at a common downstream signalling molecule, in adipose cells DDR1 and β1 integrins seem to act independently from one another.

The DDRs can also affect the activity of integrins. In MDCK cells, several studies observed a negative regulation of α2β1 by DDR1. Cell interactions with collagen type I resulted in DDR1-induced suppression of cell migration [63, 64], which was due to DDR1 interactions with the phosphatase SHP-2 (SH2 domain containing phosphatase 2), which in turn suppressed α2β1-mediated collagen-induced phosphorylation of the transcription activators Stat1 (Signal transducer and activator of transcription 1) and Stat3 [63] (Fig. 1.5). Furthermore, DDR1 inhibited α2β1-induced cell spreading in MDCK cells, through inhibiting the function of the small Rho-family GTPase Cdc42 (Cell division cycle 42) [65].

As mentioned above, while DDR1 can negatively regulate cell adhesion and migration in MDCK cells, in many cell types DDR1 has a pro-adhesive function. In agreement with this effect, DDR1 was shown to positively regulate integrin functions. For example, overexpression of DDR1 or DDR2 in human embryonic kidney cells enhanced integrin activation of both α1β1 and α2β1 without affecting integrin expression levels [58]. While the molecular mechanism behind this effect has not yet been explored, it was shown that DDR-mediated signalling increased integrin activation, such that cells were able to bind more effectively to medium affinity integrin-binding sites for α1β1 (GLOGEN sequence) and α2β1 (GMOGER sequence). This suggests that DDR signalling can regulate the affinity state of collagen-binding integrins. Another study also found DDR1 regulation of β1 integrin function. Overexpression of DDR1 in NIH 3T3 cells induced increased activation of α2β1 and the fibronectin-binding integrin α5β1 in focal adhesions, as well as enhanced focal adhesion maturation [77]. Here, DDR1 overexpression was accompanied by enhanced surface expression of α2β1 and α5β1, suggesting that DDR1 may regulate surface transport of β1 integrins.

In conclusion, The DDRs mediate cross-talk to integrins in several ways. There is both positive and negative regulation by the DDRs of integrin-mediated cellular functions. DDR signalling can directly affect the expression or activity of integrins. In addition, DDR-triggered signalling can converge with integrin-induced pathways. However, it is not known whether integrin signalling affects DDR activation. Furthermore, most of the known DDR-modulated integrin functions affect the collagen-binding integrins, and it is not clear whether the DDRs can additionally regulate other families of matrix-binding integrins.

## 1.5 Conclusions

Since the DDRs were identified as collagen receptors, we have made good progress in understanding how they interact with collagens. While it remains to be seen what kind of amino acid motifs function as DDR ligands in non-fibrillar collagens, the nature of the DDR interaction motifs in fibril-forming collagens is known and we have a good understanding of the structural basis of ligand recognition. However, we only have a poor understanding of the nature of DDR ligands in vivo. It is not clear whether collagen in the form of fibrils or fibres activates the DDRs. We also do not know whether different ligands, or different physical states of collagen, induce distinct signalling pathways. Moreover, despite crystallographic characterization of the DDR–collagen interaction, we lack insight into the molecular mechanism of transmembrane signalling and do not know how the information of ligand binding by the DS domains is transmitted across the cell membrane to induce kinase activation. We also do not understand how unwanted DDR activation is controlled or how DDR signalling is switched off.

As discussed in this chapter, the DDRs are at the interface between RTKs and matrix receptors, and we have some understanding of how DDR-mediated signalling and integrin-mediated signalling can intersect. However, we have little knowledge of how DDR signalling interfaces with signalling pathways triggered by other receptor families or how cells use receptor cross-talk to fine-tune cellular behaviour. We also do not understand whether integrins or other cell surface receptors can modulate DDR activity. Since collagen-independent functions of the DDRs have been described [3], it is possible that other receptors, including other RTKs, can induce DDR activation. Alternatively, as yet undefined, non-collagenous DDR ligands may be responsible for collagen-independent DDR functions.

## References

1. Shrivastava A, Radziejewski C, Campbell E, Kovac L, McGlynn M, Ryan TE, Davis S, Goldfarb MP, Glass DJ, Lemke G, Yancopoulos GD (1997) An orphan receptor tyrosine kinase family whose members serve as nonintegrin collagen receptors. Mol Cell 1(1):25–34
2. Vogel W, Gish GD, Alves F, Pawson T (1997) The discoidin domain receptor tyrosine kinases are activated by collagen. Mol Cell 1(1):13–23
3. Leitinger B (2014) Discoidin domain receptor functions in physiological and pathological conditions. Int Rev Cell Mol Biol 310:39–87
4. Lemmon MA, Schlessinger J (2010) Cell signaling by receptor tyrosine kinases. Cell 141(7):1117–1134
5. Carafoli F, Mayer MC, Shiraishi K, Pecheva MA, Chan LY, Nan R, Leitinger B, Hohenester E (2012) Structure of the discoidin domain receptor 1 extracellular region bound to an inhibitory Fab fragment reveals features important for signaling. Structure 20(4):688–697. doi:10.1016/j.str.2012.02.011
6. Xu H, Abe T, Liu JK, Zalivina I, Hohenester E, Leitinger B (2014) Normal activation of discoidin domain receptor 1 mutants with disulfide cross-links, insertions, or deletions in the extracellular juxtamembrane region: mechanistic implications. J Biol Chem 289(19):13565–13574. doi:10.1074/jbc.M113.536144
7. Alves F, Vogel W, Mossie K, Millauer B, Hofler H, Ullrich A (1995) Distinct structural characteristics of discoidin I subfamily receptor tyrosine kinases and complementary expression in human cancer. Oncogene 10(3):609–618
8. Alves F, Saupe S, Ledwon M, Schaub F, Hiddemann W, Vogel WF (2001) Identification of two novel, kinase-deficient variants of discoidin domain receptor 1: differential expression in human colon cancer cell lines. FASEB J 15(7):1321–1323
9. Ricard-Blum S (2011) The collagen family. Cold Spring Harb Perspect Biol 3(1):a004978. doi:10.1101/cshperspect.a004978
10. Kadler KE, Baldock C, Bella J, Boot-Handford RP (2007) Collagens at a glance. J Cell Sci 120(Pt 12):1955–1958
11. Shoulders MD, Raines RT (2009) Collagen structure and stability. Annu Rev Biochem 78:929–958
12. Brodsky B, Persikov AV (2005) Molecular structure of the collagen triple helix. Adv Protein Chem 70:301–339
13. Myllyharju J, Kivirikko KI (2004) Collagens, modifying enzymes and their mutations in humans, flies and worms. Trends Genet 20(1):33–43
14. Leitinger B (2011) Transmembrane collagen receptors. Annu Rev Cell Dev Biol 27:265–290. doi:10.1146/annurev-cellbio-092910-154013
15. Agarwal G, Kovac L, Radziejewski C, Samuelsson SJ (2002) Binding of discoidin domain receptor 2 to collagen I: an atomic force microscopy investigation. Biochemistry 41(37):11091–11098
16. Leitinger B (2003) Molecular analysis of collagen binding by the human discoidin domain receptors, DDR1 and DDR2. Identification of collagen binding sites in DDR2. J Biol Chem 278(19):16761–16769. doi:10.1074/jbc.M301370200
17. Agarwal G, Mihai C, Iscru DF (2007) Interaction of discoidin domain receptor 1 with collagen type 1. J Mol Biol 367(2):443–455
18. Leitinger B, Steplewski A, Fertala A (2004) The D2 period of collagen II contains a specific binding site for the human discoidin domain receptor, DDR2. J Mol Biol 344(4):993–1003. doi:10.1016/j.jmb.2004.09.089
19. Leitinger B, Kwan AP (2006) The discoidin domain receptor DDR2 is a receptor for type X collagen. Matrix Biol 25(6):355–364. doi:10.1016/j.matbio.2006.05.006
20. Hou G, Vogel W, Bendeck MP (2001) The discoidin domain receptor tyrosine kinase DDR1 in arterial wound repair. J Clin Invest 107(6):727–735
21. Gross O, Girgert R, Beirowski B, Kretzler M, Kang HG, Kruegel J, Miosge N, Busse AC, Segerer S, Vogel WF, Muller GA, Weber M (2010) Loss of collagen-receptor DDR1 delays renal fibrosis in hereditary type IV collagen disease. Matrix Biol 29(5):346–356

22. Kerroch M, Guerrot D, Vandermeersch S, Placier S, Mesnard L, Jouanneau C, Rondeau E, Ronco P, Boffa JJ, Chatziantoniou C, Dussaule JC (2012) Genetic inhibition of discoidin domain receptor 1 protects mice against crescentic glomerulonephritis. FASEB J 26(10):4079–4091. doi:10.1096/fj.11-194902
23. Rubel D, Frese J, Martin M, Leibnitz A, Girgert R, Miosge N, Eckes B, Muller GA, Gross O (2014) Collagen receptors integrin alpha2beta1 and discoidin domain receptor 1 regulate maturation of the glomerular basement membrane and loss of integrin alpha2beta1 delays kidney fibrosis in COL4A3 knockout mice. Matrix Biol 34:13–21. doi:10.1016/j.matbio.2014.01.006
24. Zhang Y, Su J, Yu J, Bu X, Ren T, Liu X, Yao L (2011) An essential role of discoidin domain receptor 2 (DDR2) in osteoblast differentiation and chondrocyte maturation via modulation of Runx2 activation. J Bone Miner Res 26(3):604–617
25. Lin KL, Chou CH, Hsieh SC, Hwa SY, Lee MT, Wang FF (2010) Transcriptional upregulation of DDR2 by ATF4 facilitates osteoblastic differentiation through p38 MAPK-mediated Runx2 activation. J Bone Miner Res 25(11):2489–2503
26. Sweeney SM, Orgel JP, Fertala A, McAuliffe JD, Turner KR, Di Lullo GA, Chen S, Antipova O, Perumal S, Ala-Kokko L, Forlino A, Cabral WA, Barnes AM, Marini JC, San Antonio JD (2008) Candidate cell and matrix interaction domains on the collagen fibril, the predominant protein of vertebrates. J Biol Chem 283(30):21187–21197
27. Farndale RW, Lisman T, Bihan D, Hamaia S, Smerling CS, Pugh N, Konitsiotis A, Leitinger B, de Groot PG, Jarvis GE, Raynal N (2008) Cell-collagen interactions: the use of peptide Toolkits to investigate collagen-receptor interactions. Biochem Soc Trans 36(Pt 2):241–250. doi:10.1042/BST0360241
28. Konitsiotis AD, Raynal N, Bihan D, Hohenester E, Farndale RW, Leitinger B (2008) Characterization of high affinity binding motifs for the discoidin domain receptor DDR2 in collagen. J Biol Chem 283(11):6861–6868. doi:10.1074/jbc.M709290200
29. Xu H, Raynal N, Stathopoulos S, Myllyharju J, Farndale RW, Leitinger B (2011) Collagen binding specificity of the discoidin domain receptors: binding sites on collagens II and III and molecular determinants for collagen IV recognition by DDR1. Matrix Biol 30(1):16–26. doi:10.1016/j.matbio.2010.10.004
30. Lisman T, Raynal N, Groeneveld D, Maddox B, Peachey AR, Huizinga EG, de Groot PG, Farndale RW (2006) A single high-affinity binding site for von Willebrand factor in collagen III, identified using synthetic triple-helical peptides. Blood 108(12):3753–3756
31. Giudici C, Raynal N, Wiedemann H, Cabral WA, Marini JC, Timpl R, Bachinger HP, Farndale RW, Sasaki T, Tenni R (2008) Mapping of SPARC/BM-40/osteonectin-binding sites on fibrillar collagens. J Biol Chem 283(28):19551–19560
32. Wall SJ, Werner E, Werb Z, DeClerck YA (2005) Discoidin domain receptor 2 mediates tumor cell cycle arrest induced by fibrillar collagen. J Biol Chem 280(48):40187–40194
33. Bhadriraju K, Chung KH, Spurlin TA, Haynes RJ, Elliott JT, Plant AL (2009) The relative roles of collagen adhesive receptor DDR2 activation and matrix stiffness on the downregulation of focal adhesion kinase in vascular smooth muscle cells. Biomaterials 30(35):6687–6694
34. Mihai C, Chotani M, Elton TS, Agarwal G (2009) Mapping of DDR1 distribution and oligomerization on the cell surface by FRET microscopy. J Mol Biol 385(2):432–445
35. Noordeen NA, Carafoli F, Hohenester E, Horton MA, Leitinger B (2006) A transmembrane leucine zipper is required for activation of the dimeric receptor tyrosine kinase DDR1. J Biol Chem 281(32):22744–22751. doi:10.1074/jbc.M603233200
36. Carafoli F, Hohenester E (2013) Collagen recognition and transmembrane signalling by discoidin domain receptors. Biochim Biophys Acta 1834(10):2187–2194. doi:10.1016/j.bbapap.2012.10.014
37. Carafoli F, Bihan D, Stathopoulos S, Konitsiotis AD, Kvansakul M, Farndale RW, Leitinger B, Hohenester E (2009) Crystallographic insight into collagen recognition by discoidin domain receptor 2. Structure 17(12):1573–1581. doi:10.1016/j.str.2009.10.012
38. Ichikawa O, Osawa M, Nishida N, Goshima N, Nomura N, Shimada I (2007) Structural basis of the collagen-binding mode of discoidin domain receptor 2. EMBO J 26(18):4168–4176

39. Baumgartner S, Hofmann K, Chiquet-Ehrismann R, Bucher P (1998) The discoidin domain family revisited: new members from prokaryotes and a homology-based fold prediction. Protein Sci 7(7):1626–1631
40. Kiedzierska A, Smietana K, Czepczynska H, Otlewski J (2007) Structural similarities and functional diversity of eukaryotic discoidin-like domains. Biochim Biophys Acta 1774(9):1069–1078. doi:10.1016/j.bbapap.2007.07.007
41. Abdulhussein R, McFadden C, Fuentes-Prior P, Vogel WF (2004) Exploring the collagen-binding site of the DDR1 tyrosine kinase receptor. J Biol Chem 279(30):31462–31470
42. Bargal R, Cormier-Daire V, Ben-Neriah Z, Le Merrer M, Sosna J, Melki J, Zangen DH, Smithson SF, Borochowitz Z, Belostotsky R, Raas-Rothschild A (2009) Mutations in DDR2 gene cause SMED with short limbs and abnormal calcifications. Am J Hum Genet 84(1):80–84
43. Al-Kindi A, Kizhakkedath P, Xu H, John A, Sayegh AA, Ganesh A, Al-Awadi M, Al-Anbouri L, Al-Gazali L, Leitinger B, Ali BR (2014) A novel mutation in DDR2 causing spondylo-meta-epiphyseal dysplasia with short limbs and abnormal calcifications (SMED-SL) results in defective intra-cellular trafficking. BMC Med Genet 15:42. doi:10.1186/1471-2350-15-42
44. Ali BR, Xu H, Akawi NA, John A, Karuvantevida NS, Langer R, Al-Gazali L, Leitinger B (2010) Trafficking defects and loss of ligand binding are the underlying causes of all reported DDR2 missense mutations found in SMED-SL patients. Hum Mol Genet 19(11):2239–2250. doi:10.1093/hmg/ddq103
45. Hohenester E, Sasaki T, Giudici C, Farndale RW, Bachinger HP (2008) Structural basis of sequence-specific collagen recognition by SPARC. Proc Natl Acad Sci U S A 105(47): 18273–18277
46. Mendoza-Londono R, Fahiminiya S, Majewski J, Care4Rare Canada Consortium, Tetreault M, Nadaf J, Kannu P, Sochett E, Howard A, Stimec J, Dupuis L, Roschger P, Klaushofer K, Palomo T, Ouellet J, Al-Jallad H, Mort JS, Moffatt P, Boudko S, Bachinger HP, Rauch F (2015) Recessive osteogenesis imperfecta caused by missense mutations in SPARC. Am J Hum Genet 96(6):979–985. doi:10.1016/j.ajhg.2015.04.021
47. Hachehouche LN, Chetoui N, Aoudjit F (2010) Implication of discoidin domain receptor 1 in T cell migration in three-dimensional collagen. Mol Immunol 47(9):1866–1869
48. Kamohara H, Yamashiro S, Galligan C, Yoshimura T (2001) Discoidin domain receptor 1 isoform-a (DDR1a) promotes migration of leukocytes in three-dimensional collagen lattices. FASEB J 15(14):2724–2726
49. Afonso PV, McCann CP, Kapnick SM, Parent CA (2013) Discoidin domain receptor 2 regulates neutrophil chemotaxis in 3D collagen matrices. Blood 121(9):1644–1650. doi:10.1182/blood-2012-08-451575
50. Friedl P, Weigelin B (2008) Interstitial leukocyte migration and immune function. Nat Immunol 9(9):960–969. doi:10.1038/ni.f.212
51. Valiathan RR, Marco M, Leitinger B, Kleer CG, Fridman R (2012) Discoidin domain receptor tyrosine kinases: new players in cancer progression. Cancer Metastasis Rev 31(1-2):295–321. doi:10.1007/s10555-012-9346-z
52. Neuhaus B, Buhren S, Bock B, Alves F, Vogel WF, Kiefer F (2011) Migration inhibition of mammary epithelial cells by Syk is blocked in the presence of DDR1 receptors. Cell Mol Life Sci 68(22):3757–3770. doi:10.1007/s00018-011-0676-8
53. Shitomi Y, Thogersen IB, Ito N, Leitinger B, Enghild JJ, Itoh Y (2015) ADAM10 controls collagen signaling and cell migration on collagen by shedding the ectodomain of discoidin domain receptor 1 (DDR1). Mol Biol Cell 26(4):659–673. doi:10.1091/mbc.E14-10-1463
54. Abbonante V, Gruppi C, Rubel D, Gross O, Moratti R, Balduini A (2013) Discoidin domain receptor 1 protein is a novel modulator of megakaryocyte-collagen interactions. J Biol Chem 288(23):16738–16746. doi:10.1074/jbc.M112.431528
55. Higashiyama S, Nanba D, Nakayama H, Inoue H, Fukuda S (2011) Ectodomain shedding and remnant peptide signalling of EGFRs and their ligands. J Biochem 150(1):15–22. doi:10.1093/jb/mvr068
56. Yoshida D, Teramoto A (2007) Enhancement of pituitary adenoma cell invasion and adhesion is mediated by discoidin domain receptor-1. J Neurooncol 82:29–40

57. Ram R, Lorente G, Nikolich K, Urfer R, Foehr E, Nagavarapu U (2006) Discoidin domain receptor-1a (DDR1a) promotes glioma cell invasion and adhesion in association with matrix metalloproteinase-2. J Neurooncol 76(3):239–248
58. Xu H, Bihan D, Chang F, Huang PH, Farndale RW, Leitinger B (2012) Discoidin domain receptors promote alpha1beta1- and alpha2beta1-integrin mediated cell adhesion to collagen by enhancing integrin activation. PLoS One 7(12):e52209. doi:10.1371/journal.pone.0052209
59. Kim D, You E, Min NY, Lee KH, Kim HK, Rhee S (2013) Discoidin domain receptor 2 regulates the adhesion of fibroblasts to 3D collagen matrices. Int J Mol Med 31(5):1113–1118. doi:10.3892/ijmm.2013.1320
60. Franco C, Ahmad PJ, Hou G, Wong E, Bendeck MP (2010) Increased cell and matrix accumulation during atherogenesis in mice with vessel wall-specific deletion of discoidin domain receptor 1. Circ Res 106(11):1775–1783. doi:10.1161/CIRCRESAHA.109.213637
61. Curat CA, Vogel WF (2002) Discoidin domain receptor 1 controls growth and adhesion of mesangial cells. J Am Soc Nephrol 13(11):2648–2656
62. Fukunaga-Kalabis M, Martinez G, Liu ZJ, Kalabis J, Mrass P, Weninger W, Firth SM, Planque N, Perbal B, Herlyn M (2006) CCN3 controls 3D spatial localization of melanocytes in the human skin through DDR1. J Cell Biol 175(4):563–569
63. Wang CZ, Su HW, Hsu YC, Shen MR, Tang MJ (2006) A discoidin domain receptor 1/SHP-2 signaling complex inhibits {alpha}2beta1-integrin-mediated signal transducers and activators of transcription 1/3 activation and cell migration. Mol Biol Cell 17(6):2839–2852
64. Wang CZ, Hsu YM, Tang MJ (2005) Function of discoidin domain receptor I in HGF-induced branching tubulogenesis of MDCK cells in collagen gel. J Cell Physiol 203(1):295–304
65. Yeh YC, Wang CZ, Tang MJ (2009) Discoidin domain receptor 1 activation suppresses alpha-2beta1 integrin-dependent cell spreading through inhibition of Cdc42 activity. J Cell Physiol 218(1):146–156. doi:10.1002/jcp.21578
66. Hynes R (2002) Integrins. Bidirectional, allosteric signaling machines. Cell 110(6):673–687
67. Knight CG, Morton LF, Onley DJ, Peachey AR, Messent AJ, Smethurst PA, Tuckwell DS, Farndale RW, Barnes MJ (1998) Identification in collagen type I of an integrin a2 b1-binding site containing an essential GER sequence. J Biol Chem 273(50):33287–33294
68. Knight CG, Morton LF, Peachey AR, Tuckwell DS, Farndale RW, Barnes MJ (2000) The collagen-binding A-domains of Integrins a1b1 and a2b1 recognize the same specific amino acid sequence, GFOGER, in native (triple-helical) collagens. J Biol Chem 275(1):35–40
69. Xu Y, Gurusiddappa S, Rich RL, Owens RT, Keene DR, Mayne R, Hook A, Hook M (2000) Multiple binding sites in collagen type I for the integrins a1b1 and a2b1. J Biol Chem 275(50):38981–38989
70. Raynal N, Hamaia SW, Siljander PR, Maddox B, Peachey AR, Fernandez R, Foley LJ, Slatter DA, Jarvis GE, Farndale RW (2006) Use of synthetic peptides to locate novel integrin alpha2beta1-binding motifs in human collagen III. J Biol Chem 281(7):3821–3831
71. Siljander PR, Hamaia S, Peachey AR, Slatter DA, Smethurst PA, Ouwehand WH, Knight CG, Farndale RW (2004) Integrin activation state determines selectivity for novel recognition sites in fibrillar collagens. J Biol Chem 279(46):47763–47772
72. Hamaia SW, Pugh N, Raynal N, Nemoz B, Stone R, Gullberg D, Bihan D, Farndale RW (2012) Mapping of potent and specific binding motifs, GLOGEN and GVOGEA, for integrin alpha-1beta1 using collagen toolkits II and III. J Biol Chem 287(31):26019–26028. doi:10.1074/jbc.M112.353144
73. Vogel W, Brakebusch C, Fassler R, Alves F, Ruggiero F, Pawson T (2000) Discoidin domain receptor 1 is activated independently of b1 integrin. J Biol Chem 275(8):5779–5784
74. Shintani Y, Fukumoto Y, Chaika N, Svoboda R, Wheelock MJ, Johnson KR (2008) Collagen I-mediated up-regulation of N-cadherin requires cooperative signals from integrins and discoidin domain receptor 1. J Cell Biol 180(6):1277–1289
75. Suh HN, Han HJ (2011) Collagen I regulates the self-renewal of mouse embryonic stem cells through alpha2beta1 integrin- and DDR1-dependent Bmi-1. J Cell Physiol 226(12):3422–3432. doi:10.1002/jcp.22697

76. Ghosh S, Ashcraft K, Jahid MJ, April C, Ghajar CM, Ruan J, Wang H, Foster M, Hughes DC, Ramirez AG, Huang T, Fan JB, Hu Y, Li R (2013) Regulation of adipose oestrogen output by mechanical stress. Nat Commun 4:1821. doi:10.1038/ncomms2794
77. Staudinger LA, Spano SJ, Lee W, Coelho N, Rajshankar D, Bendeck MP, Moriarty T, McCulloch CA (2013) Interactions between the discoidin domain receptor 1 and b1 integrin regulate attachment to collagen. Biol Open 2(11):1148–1159

# Chapter 2
# DDRs and Collagen Fibrillogenesis

**Gunjan Agarwal**

## 2.1 Introduction

Collagen type I in its mature fibrillar state is the major component of the extracellular matrix (ECM) in a variety of vertebrate tissues. Collagen fibrils not only impart mechanical strength to the tissue but also interact with cells through cell surface receptors and soluble proteins, which is integral to cell adhesion, differentiation, proliferation, migration, and survival. The assembly of collagen fibrils (fibrillogenesis) is a complex process regulated in part by a variety of collagen-binding proteins (CBPs), which directly interact with the collagen molecules and/or fibrils. Several CBPs belonging to the family of proteoglycans (e.g., decorin, lumican, and fibromodulin) or glycoproteins (e.g., SPARC and thrombospondin-2), just to mention a few have been shown to influence collagen fibrillogenesis (Table 2.1). All these CBPs occur as cell-secreted, soluble proteins in the ECM. Discoidin domain receptors (DDR1 and DDR2) are unique as they belong to the family of collagen receptors on the cell surface. The collagen-binding extracellular domain (ECD) of DDRs exists both as a cell surface anchored and as a soluble glycoprotein in the ECM. Thus, understanding the modulation of collagen fibrillogenesis by DDRs can help provide novel insights into cell–matrix interactions, ECM remodeling, and receptor function.

In this chapter, we describe the role of DDRs in collagen fibrillogenesis. We begin with Sect. 2.1, which summarizes the experimental evidence on the role of DDRs in collagen fibrillogenesis. In the most comprehensive Sect. 2.2, we describe the various stages of fibrillogenesis of collagen type I with special attention to the putative role(s) of DDRs at each stage. In Sect. 2.3, we discuss the

G. Agarwal (✉)
Department of Biomedical Engineering, The Ohio State University, Columbus, OH, USA
e-mail: agarwal.60@osu.edu

 
R. Fridman, P.H. Huang (eds.), *Discoidin Domain Receptors in Health and Disease*, DOI 10.1007/978-1-4939-6383-6_2

**Table 2.1** Regulation of collagen fibrillogenesis in the skin of mice lacking a collagen-binding protein (CBP) as indicated

| CBP | Collagen deposition | Fibril diameter | Tensile strength | Reference |
|---|---|---|---|---|
| *Small leucine-rich proteoglycans (SLRPs)* | | | | |
| Decorin[a] | – | Irregular | Reduced | [1] |
| Lumican[a] | – | Increased | Reduced | [2] |
| Fibromodulin | – | Irregular and decreased | Reduced | [3, 4] |
| Mimecan/osteoglycin | | Increased | Reduced | [5] |
| *Matrix proteins* | | | | |
| Dermatopontin | Reduced | Increased | Reduced | [6] |
| Osteopontin | – | Decreased | No change | [7] |
| SPARC[a] | Reduced | Decreased | Reduced | [8, 9] |
| Thrombospondin2 | Reduced | Irregular and increased | Reduced | [10] |
| Periostin | Reduced | Decreased | Reduced | [11] |

Comparisons of collagen fibril diameter and skin tensile strength have been made between the CBP knockout mice relative to their wild-type littermates

[a]Consistent results were observed in studies using recombinant proteins in vitro

functional consequences of the modulation of collagen fibrillogenesis by DDRs in health and disease. Finally, Sect. 2.4 summarizes the role of DDRs in modulating assembly of other collagen types.

## 2.2 DDRs and Collagen Fibrillogenesis

The first evidence for plausible role of DDRs in collagen fibrillogenesis arose when Agarwal and coworkers examined the binding pattern(s) of recombinant ECD of DDRs on monomeric collagen type I using atomic force microscopy (AFM) [12–14]. As shown in Fig. 2.1a while DDR2 ECD bound to isolated molecules of pepsin-digested collagen type I (~290 nm in length), DDR1 ECD was observed to bind at overlapping collagen molecules (>300 nm in length). Binding of DDR1 ECD to overlapping collagen monomers suggested its possible role in modulating collagen fibrillogenesis. In vitro collagen turbidity assays involving recombinant DDR ECDs and exogenous addition of fluorescently labeled collagen to cells expressing full-length DDRs confirmed that DDRs delay and/or disrupt the process of collagen fibrillogenesis (Fig. 2.1b, c). The ultrastructure of collagen fibrils formed in vitro in the presence of DDR ECDs showed alterations when examined using transmission electron microscopy (TEM) (Fig. 2.1d). The characteristic D-periodic banding pattern was weakened or disrupted in collagen fibrils formed in the presence of recombinant DDR ECDs in vitro.

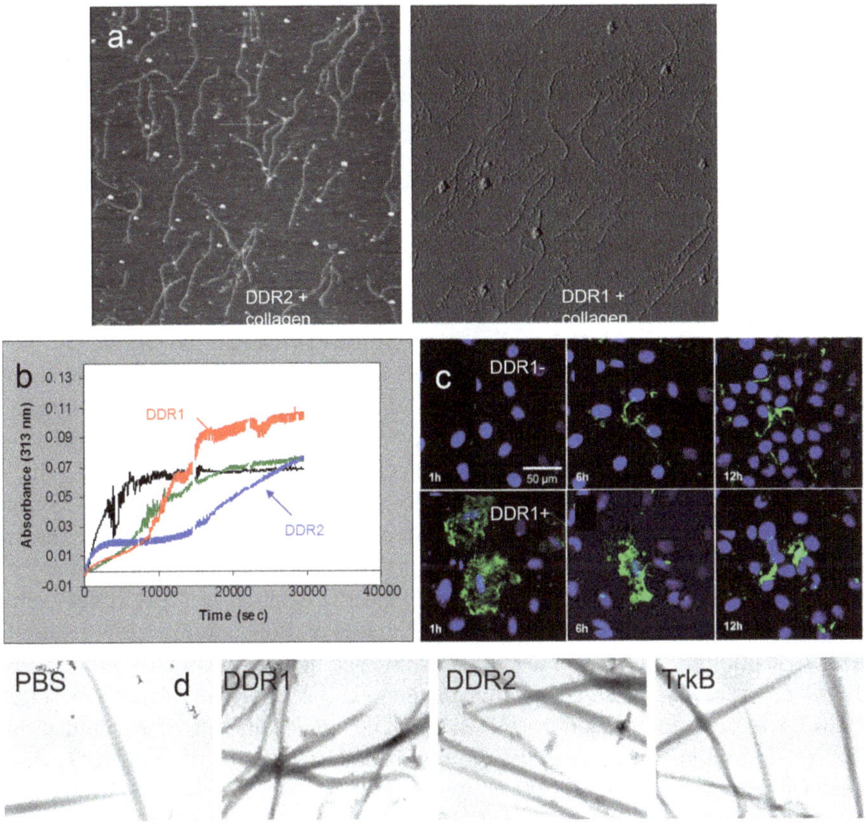


**Fig. 2.1** Inhibition of in vitro collagen fibrillogenesis by DDRs [12–14]. (**a**) Binding pattern of DDR2 and DDR1 on pepsin-digested bovine dermal collagen. (**b**) Collagen turbidity measurements in the presence of DDRs. (**c**) Fibrillogenesis of fluorescently labeled collagen when added to cells expressing DDRs. (**d**) Ultrastructure of collagen fibrils formed in the presence of DDRs

To further examine if DDRs affect fibrillogenesis of collagen endogenously secreted by the cells, stable cell lines expressing DDR ECDs as a cell surface anchored or a soluble protein were employed [15, 16]. The intracellular kinase domain (KD) of DDRs was especially removed in these DDR ECD constructs to eliminate the effect of DDR-mediated cell-signaling events (including regulation of MMPs) in ECM remodeling. The preosteoblastic cells E1 MC3T3 used in this study endogenously secrete and assemble collagen in their ECM. Hydroxyproline assay revealed that the amount of collagen deposited in the ECM of cells expressing DDR ECDs was reduced compared to control cells, consistent with the role of DDRs in delaying collagen fibrillogenesis. Ultrastructural examination by TEM showed collagen fibrils with weakened or disrupted D-periodicity in the ECM of DDR ECD expressing cells, similar to that observed in in vitro experiments. Interestingly, even though not observed in vitro, collagen fibrils formed by DDR ECD expressing cells were smaller in diameter compared to control cells. Taken together, these results elucidate that by binding to collagen, DDR ECDs inhibit the kinetics of collagen fibrillogenesis and alter the ultrastructure of the collagen fibrils thus formed.

Consistent results were obtained during examination of collagen content and structure in the aorta of DDR1 knockout (KO) mice [17]. Lack of DDR1 resulted in increased collagen deposition and a higher percentage of fibril with larger diameters and a more pronounced D-periodicity. The difference in contrast of D-periodicity was attributed to increased 'depth' of D-periods in the collagen fibrils present in the DDR1 KO mice as compared to their wild-type littermates, ascertained using AFM. No significant changes in the length of D-periods were observed in the collagen fibrils in in vitro or in vivo studies as a function of DDRs.

Additional reports have supported the role for DDRs in inhibiting accumulation of collagen. An increased collagen deposition has been reported in the mammary gland [18] and inner ear [19] of DDR1 KO mice. DDR1 deletion was found to decrease atherosclerotic plaque size by accelerating matrix accumulation and reducing inflammation in low-density lipoprotein receptor-deficient mice [20]. Since both macrophages as well as epithelial and smooth muscle cells (SMCs) express DDR1, in order to identify the cell type responsible for matrix accumulation in the DDR1 KO mice, chimeric mice with tissue-specific deletion of DDR1 mice were generated. In this seminal study, DDR1 deficiency in bone marrow cells reduced leukocyte infiltration [21], whereas vessel-wall-specific deletion of DDR1 increased matrix deposition in the atherosclerotic plaques [22]. While limited reports exist for DDR2, it appears to have similar effects as DDR1 in limiting matrix accumulation in vivo. In a model of chronic liver injury, hepatic stellate cells from DDR2 KO mice led to increased fibrosis [23].

Conflicting reports exist on the role of DDRs in inhibiting fibrosis in vivo, especially when accompanied by inflammation. Loss of DDR1 expression in the kidney delayed and blunted renal inflammation and fibrosis in hereditary type IV collagen disease [24] in Ang II-induced hypertension [25] and in unilateral ureteral obstruction (UUO) model [26]. Along similar lines, reduced inflammation and fibrosis has been observed in DDR1 KO mice upon bleomycin-induced lung injury, which was associated with limited myofibroblast expansion [27]. In a rat model of carotid injury, the neointimal area and collagen deposition was significantly reduced by siRNA-mediated knockdown of DDR2 [28]. In another study, expression of DDR2 was enhanced in an alcohol-induced liver fibrosis rat model [29]. These confounding observations can partly be explained by considering the feedback loop between fibrosis and inflammation. Factors produced during the injury and acute/chronic inflammatory phase are thought to facilitate the fibrogenic phase by promoting fibroblast activation [30]. On the other hand, factors such as collagen type I, produced during the fibrogenic phase, can regulate termination of the injury/inflammatory phase [31]. Hence, reduced fibrosis observed in DDR1 or DDR2 KO mice may be a consequence of blunted inflammation resulting in inadequate activation and proliferation of fibroblasts compared to their wild-type littermates. In addition, an increased collagen secretion by DDR-deficient cells may further blunt inflammation and the subsequent fibrogenic phase.

In clinical studies, it is not clear if DDRs are a cause or an effect in regulating fibrosis. DDR2 was significantly downregulated in the right atrial appendage in patients with chronic atrial fibrillation as compared to the control group [32]. In contrast, both DDR1 and DDR2 were found to be upregulated in nodules of lymphangioleiomyomatosis as compared to normal controls and were expressed in lesions of atherosclerosis [33]. Epithelial cell-enriched samples from livers of patients with primary biliary cirrhosis displayed an elevated expression of DDR2 compared to nondiseased livers [34].

Besides binding of DDR ECDs to collagen, other factors can also contribute to modulation of collagen fibrillogenesis by DDRs in vivo (1) DDRs are expressed as both full-length and kinase inactive isoforms which may differentially regulate the receptor function. For instance, DDR-mediated signaling can regulate cell proliferation and inflammation, which often accompanies the pathology of fibrosis [35]; shedded DDR ECD may serve to inhibit collagen fibrillogenesis as well as block ligand-binding of full-length receptor. (2) DDRs are expressed in a multitude of cell types, not all of which are responsible for collagen production and regulation, and (3) ECM remodeling in vivo comprises of a variety of collagen intermediates such as monomers, microfibrils, fibrils, and mature collagen fibers. DDRs may differentially regulate the occurrence of these various collagen intermediates. Thus, a closer look at the role of DDRs in regulating collagen fibrillogenesis is required to understand the multifaceted role of these collagen receptors. The following section outlines the various stages of fibrillogenesis of collagen type I with special emphasis on the role of DDRs in each stage.

## 2.3 Fibrillogenesis of Collagen Type I

The collagen molecule contains all the necessary information to assemble into fibrils. Collagen fibrils formed in vitro are microscopically identical to fibrils formed in vivo with a few exceptions (1) their lateral order is generally not as good, (2) they are not cross-linked and lack tensile strength, and (3) the multitude of CBPs are absent in in vitro generated collagen fibrils. A number of in vivo, in vitro, and in silico models have been employed to understand collagen fibrillogenesis. In particular, two major advances made in the past decade have provided novel insights into collagen fibrillogenesis (a) the binding sites of several collagen-binding proteins have been mapped onto the collagen triple helix and (b) the collagen fibril structure and intermolecular packing has been elucidated using X-ray diffraction. This section revisits collagen fibrillogenesis by including these recent insights and examining the role of DDRs at every stage.

### *2.3.1 Collagen Biosynthesis*

The biosynthesis of collagen type I (Col1) occurs with the synthesis of Col1$\alpha_1$ and Col1$\alpha_2$ polypeptide chains. The collagen $\alpha$-chains are characterized by the frequently occurring Gly-X-Y motif where X and Y are often prolines and hydroxyprolines, respectively. Noncollagenous domains (the N and C-propeptide) flank the two ends of the collagen α-chains. Collagen is secreted from the cells only as the triple-helical molecule and not as individual α-chains. However, mRNA expression level of individual α-chains is frequently used as a measure of collagen synthesis.

Collagen expression appears to be differentially regulated by DDR1 and DDR2. Expression and activation of DDR1 suppresses collagen synthesis in several cell types, whereas DDR2 has little effect on collagen expression. Studies using primary smooth muscle cells (SMC) from DDR1 KO mice have revealed that mRNA for collagen type I (and III) was significantly increased in DDR1$^{-/-}$ cells compared to DDR1$^{+/+}$ cells [20, 22]. Consistent with these results, overexpression by transient transcription of DDR1 in human SMC was found to induce a significant decrease in collagen type I mRNA, which was accompanied by a similar decrease at the protein level [33]. Chondrocytes isolated from the temporomandibular joint of DDR1$^{-/-}$ mice also expressed high levels of collagen type I [36]. Genes coding for collagen type I and III were suppressed by collagen-induced activation of DDR1 in both human and mouse fibroblasts [37].

In contrast to a negative correlation between collagen and DDR1 expression, both SMCs [38] and cardiac fibroblasts [39] isolated from DDR2 KO mice showed no significant differences in collagen type I expression compared to DDR2$^{+/+}$ cells. In another study employing dermal fibroblasts, collagen type I mRNA and protein expression was actually reduced significantly in DDR2$^{-/-}$ as compared to DDR2$^{+/-}$ cells [40]. Collagen-induced activation of DDR2 did not affect the expression of collagen types I and III in cultured fibroblasts [37]. One contradictory evidence is reported where overexpression of DDR2 in SMCs led to decrease in collagen expression, similar to that observed for DDR1 [33]. Thus taken together, DDR1 and DDR2 differentially regulate collagen synthesis. While expression and activation of DDR1 suppresses collagen synthesis, DDR2 does not significantly impact collagen synthesis.

A critical step in formation of the collagen triple helix is posttranslational modifications resulting in hydroxylation of several lysines and prolines in the collagen α-chains, mediated by three lysyl hydroxylases and two prolyl hydroxylases, respectively [41] (Fig. 2.2). Activity of prolyl-4-hydroxylase (P4H) is crucial for the stability of the triple helix [42], whereas 3-hydroxyproline resulting from P3H activity aids in intermolecular recognition and fibril assembly [43]. Depletion of lysyl hydroxylases LH1 [44], LH2 [45], and LH3 [46] modulate the extent of collagen secretion and deposition in the ECM as well as changes in the collagen fibril ultrastructure. Very little is understood about regulation of these enzymes by DDRs. Decrease in levels of LH1 [47] and of an isoform of P4H [39] have been reported in cells lacking DDR2. Glycosylation of certain hydroxylysine is the final step before the three α-chains assemble into a triple helix. The role of collagen glycosylation in DDR–collagen interaction is discussed in Sect. 2.3.7.

After posttranslational modifications, the two Col1$\alpha_1$ and one Col1$\alpha_2$ chains assemble together to form the heterotrimeric right-handed triple helical *procollagen* molecule. The assembly of the collagen triple helix is initiated by the noncollagenous C-propeptides and proceeds in a zipper-like fashion from the C to the N-terminus [48]. The collagen triple helix is stabilized by peptide bonds (which are

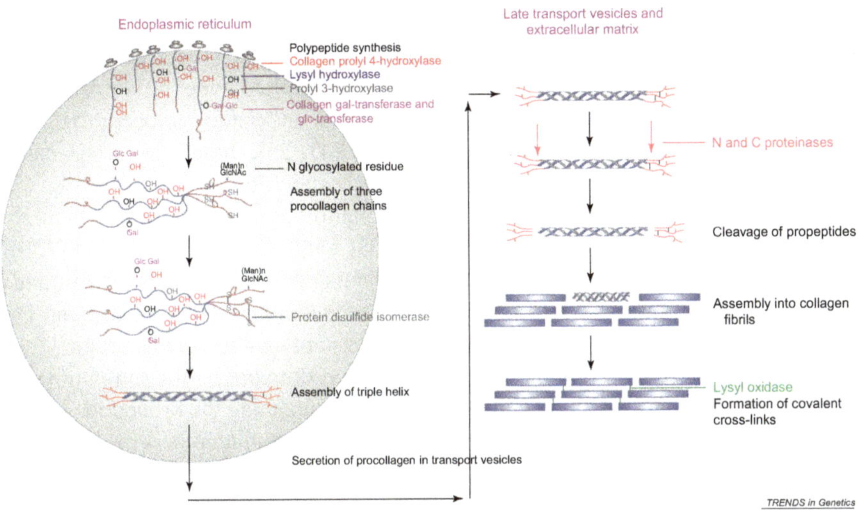


**Fig. 2.2** The main steps in biosynthesis of a collagen type I [41]. The three polypeptide α-chains are synthesized on membrane-bound ribosomes and secreted into the lumen of the endoplasmic reticulum, where they undergo: cleavage of the signal peptides, hydroxylation of certain proline and lysine residues, glycosylation of some of the hydroxyl-lysine residues, and of certain asparagine residues in the N and C propeptides. Thereafter, the C propeptides on the three α-chains associate by specific recognition sequences and mediate the assembly of the triple helix. The triple helix is stabilized by formation of intramolecular and intermolecular disulfide bonds. The procollagen molecules thus formed are transported from the endoplasmic reticulum through the Golgi stacks. They begin to aggregate laterally during transport to form secretory vesicles. The subsequent steps are cleavage of the N and C propeptides and spontaneous self-assembly of the resulting collagen molecules. Formation of covalent cross-links initiated by oxidation of certain lysine and hydroxylysine residues helps to stabilize the collagen microfibril

all trans), hydrogen bonds, interchain cross-linking, side-chain interactions, and disulphide bonds. In addition, preferential binding of chaperones, such as the heat-shock protein (HSP47) to its folded, native conformation further stabilize the collagen triple helix [49, 50]. As a result of the formation of the collagen triple helix, new recognition sites are encoded which modulate intermolecular assembly as well as interactions of the collagen molecule with various CBPs.

## 2.3.2 *Propeptide Cleavage*

Collagen is secreted from the cells in the form of *procollagen*, which consists of the collagen triple helix (a trimer of three α-chains) along with globular propeptide units flanking its ends (Fig. 2.2). For collagen fibrillogenesis to begin, the propeptides must be cleaved off from the procollagen molecule, especially the C-propeptide [51]. Retention of the N-terminal propeptide (resulting in pN-collagen) does not completely inhibit fibrillogenesis but results in abnormal collagen fibrils [52]. Mutations near the propeptide cleavage site(s) account for the altered collagen fibril structure witnessed in Ehlers–Danlos syndrome (EDS) phenotype [53] or osteogenesis imperfecta (OI) [54]. Cleavage of propeptides is understood to occur extracellularly in the proximity of the plasma membrane [41, 55], though in certain cases intracellular propeptide cleavage has also been reported [56, 57].

The C-propeptide is processed by the tolloid family of metalloproteases, namely the bone morphogenetic protein 1 (BMP-1) [58] and tolloid-like-1 (TLL-1) [59]. The processing of N-propeptides involves members of the a-disintegrin-and-metalloproteinase-with-thrombospondin-like-motifs family (ADAMTS 2, 3 and 14) [60, 61]. Both the BMP-1 and ADAMTS proteases are synthesized as proenzymes which are activated only after the removal of their prodomains by the action of furin [62]. Though there are no reports on DDRs directly impacting the propeptide cleavage in procollagen, it is interesting to note that the furin and members of the metalloprotease family are involved in ligand-induced shedding of the DDR1 ECD [63–66]. Thus, shedding of DDR1 ECD as a soluble protein in the ECM may occur in concert with the propeptide cleavage of newly synthesized collagen. Shedding of DDR2 ECD is not well characterized.

### 2.3.3 Interaction of DDRs with Tropocollagen

Upon the cleavage of the propeptides, the triple helical collagen monomer, now known as *tropocollagen* is generated. Tropocollagen also consists of nonhelical segments called *telopeptides*, which remain at the N and C terminus and are critical to initiate collagen fibrillogenesis. The triple helical region of the collagen monomer consists of 1014 aa and is estimated to be ~300 nm in length with 0.286 nm/residue [67, 68]. Over 50 known binding partners have been identified for the tropocollagen molecule, including several proteoglycans and matrix proteins [69].

DDRs bind to the collagen triple helix and not to the individual α-chains, propeptides, or telopeptide regions of the collagen molecule. Collagenase treated, thermally denatured collagen or CNBr fragments of collagen fail to activate DDRs [70, 71]. While it remains to be examined if DDRs bind to the procollagen intracellularly, several studies have established that DDRs do bind to the collagen triple helix in neutral pH conditions as found in the ECM. Single-molecule AFM studies have shown that DDR2 binds to at least three preferred sites along the collagen I triple helix [12]. However, unlike DDR2, DDR1 did not bind to isolated collagen molecules but bound only to overlapping collagen molecules >300 nm in length [13]. Using triple-helical homotrimeric collagen type II and III toolkit peptides, five putative DDR2 binding sites [72] and one DDR1 binding site [73, 74] on the collagen type II or III triple helix have been identified.

The five DDR2 binding sites have been mapped onto the collagen type I triple helix through sequence alignment and identified according to their toolkit peptide and AFM mapping as summarized in Table 2.2. In evaluating these sites, it should be

**Table 2.2** Location of DDR2 binding sites on tropocollagen I and their functional significance

| Binding site | Toolkit peptide | Distance from N-telo | Sequence on col1 α-chains | Significance |
|---|---|---|---|---|
| DDR2[1] | III-5 | 18–26 nm | Amino acid location: 64–90<br>$\alpha_1$(I): GERGPPGPQGARGLP GTAGLPGMKGHR<br>$\alpha_2$(I): GERGVVGPQGARGF PGTPGLPGFKGIRG | Cross-linking and glycosylation at K87; fibromodulin binding |
| DDR2[2] | II-13 | **62–70 nm** | Amino acid location: 217–243<br>$\alpha_1$(I): GAKGANGAPGIAG APGFPGARGPSGPQ<br>$\alpha_2$(I): GAKGAAGLPGVAG APGLPGPRGIPGPP | Collagen glycosylation at K219; KSPG region on microfibril |
| DDR2[3]<br>DDR1 | II-22/<br>II-23 | **113–116 nm** | Amino acid location: 394–405<br>$\alpha_1$(I): GPPGARGQAGVMGFPGPK<br>$\alpha_2$(I): GPVGARGEPGNIGFPGPK | Binding site for SPARC, vWF; DSPG region on microfibril |
| DDR2[4] | III-39 | – | Not conserved | – |
| DDR2[5] | II-44/<br>III-44 | **222–227 nm** | Amino acid location: 775–792<br>$\alpha_1$(I): GIAGQRGVVGLPGQRGER GFPGLPGPS<br>$\alpha_2$(I): GLLGAPGILGLPGSRG ERGLPGVAGAV | Collagenase cleavage site; MMP1 binding; fibrillogenesis inhibition; fibromodulin binding; KSPG region |

The five DDR2 binding sites as determined using toolkit peptides are numbered as DDR2[1] to DDR2[5] in the order in which they occur from the N to C terminus of tropocollagen I. Positions of the DDR2 binding sites from the start of the triple helical region on collagen type 1 are indicated in columns 3 and 4. The bold numbers in Column 3 indicate the distances which are consistent with AFM mapping of 65 ± 8 and 105 ± 8 nm from one end of *a* ~ 290 nm long pepsin-digested collagen type 1

noted that the Col1$\alpha_1$ chain shows higher sequence similarity to the Col2$\alpha_1$ chain than to the Col1$\alpha_2$ or Col3$\alpha_1$ chains [75]. Mapping of the DDR2 binding motifs onto the tropocollagen I molecule is consistent with respect to results obtained using single-molecule AFM studies on collagen type I monomers and peptide toolkits representative of collagens types II and III. Single-molecule AFM-mapped DDR2 binding site(s) to 35 ± 8, 65 ± 8 and 105 ± 8 nm from one end of the ~290 nm long pepsin-digested monomeric collagen type I [12], with the latter two binding sites being more frequent. These distances match the locations of three of the five DDR2 binding sites from either the N or the C-terminus of tropocollagen (Table 2.2). Both high-resolution microscopy studies [12] as well as X-ray diffraction studies [76] have shown that binding of DDR2 does not induce a bend in the collagen triple helix.

DDR2[3], the primary collagen-binding site for DDR2 (and DDR1) comprises of the GVMGFO motif. Receptor phosphorylation could only be induced by triple helical peptides corresponding to this motif [73, 74]. The GVMGFO motif present in collagens types II and III is conserved in the Col1$\alpha_1$ chain. DDR2 binding was primarily governed by this sequence on the Col1$\alpha_1$ chain in the triple helical peptide [72]. However, the sequence of the Col1$\alpha_2$ chain that aligns with the GVMGFO motif on Col1$\alpha_1$ chain was also recognized by DDR1 with similar affinity, indicating that both $\alpha_1$ and $\alpha_2$ chains of collagen may contribute to binding of DDR1 [73, 74]. The $\alpha_2$ chain is more hydrophobic than the Col1$\alpha_1$ chain and plays an important role in driving collagen fibrillogenesis [77]. Thus, the differences in binding of DDR1 vs. DDR2 to the collagen triple helix may contribute to their relative role(s) in modulating collagen fibrillogenesis.

### *2.3.4 Collagen Microfibril*

Collagen fibrillogenesis is a multistep process primarily driven by self-assembly of the tropocollagen molecules and consists of several intermediates [78]. Besides self-assembly of tropocollagen, additional factors can regulate collagen fibrillogenesis. These include the temperature, pH, and ionic species in the surrounding environment [79], presence of proteoglycans, CBPs, and other collagen types (e.g., types III and V) [80]. In vivo, cell surface receptors, cross-linking enzymes, and collagenases also work in concert to modulate collagen fibrillogenesis.

The self-assembly of tropocollagen is initiated by the N-(17 aa) and C-(26 aa) terminal telopeptides [52]. The N-telopeptide is elongated, whereas the C-telopeptide is folded back onto the triple helix. The telopeptides also contain sites for intra- and intermicrofibrillar cross-linking. In the widely accepted Hodge–Petruska scheme for packing of collagen molecules, each tropocollagen molecule is divided into five domains; four of length 1D, and the fifth of ~0.46D, where D is 67 nm. It is understood that in the very first stage, the collagen monomers self-assemble into ~570 nm long, 4D-staggered collagen dimers. Such dimers have been reported in both neutral [81] and acidic buffer conditions [82]. The critical concentration for de novo collagen polymerization has been estimated to be ~0.5 μg/mL at 37 °C and much higher for collagen monomers extracted from tissues [51].

Alongside the dimer formation, the tropocollagen molecules self-assemble in an entropy driven process into the basic unit of the collagen fibril called the *microfibril* ~4 nm in diameter (Fig. 2.3) [51]. Isolated microfibrils have not been found in vivo but are understood to be held together in collagen fibrils and fibers by intermicrofibrillar cross-links. Ample evidence exists for the existence of microfibrils as a structural unit within the collagen fibril. In situ X-ray diffraction studies [84], high-resolution microscopy of collagen fibrils in cornea [85] and bone [86], micromechanical testing of collagen fibrils from connective tissues [87], and in silico modeling of the collagen fibril mechanics [88] convincingly elucidate the existence and structure of the collagen microfibril.

Electron micrographs of collagen fibers typically show a light-dark periodic banding pattern of ~67 nm. This D-periodicity is also present in the collagen microfibril and arises due to a 'gap' and an 'overlap' zone (Fig. 2.3). Each D-period in a microfibril contains the complete collagen sequence from the laterally staggered arrangement

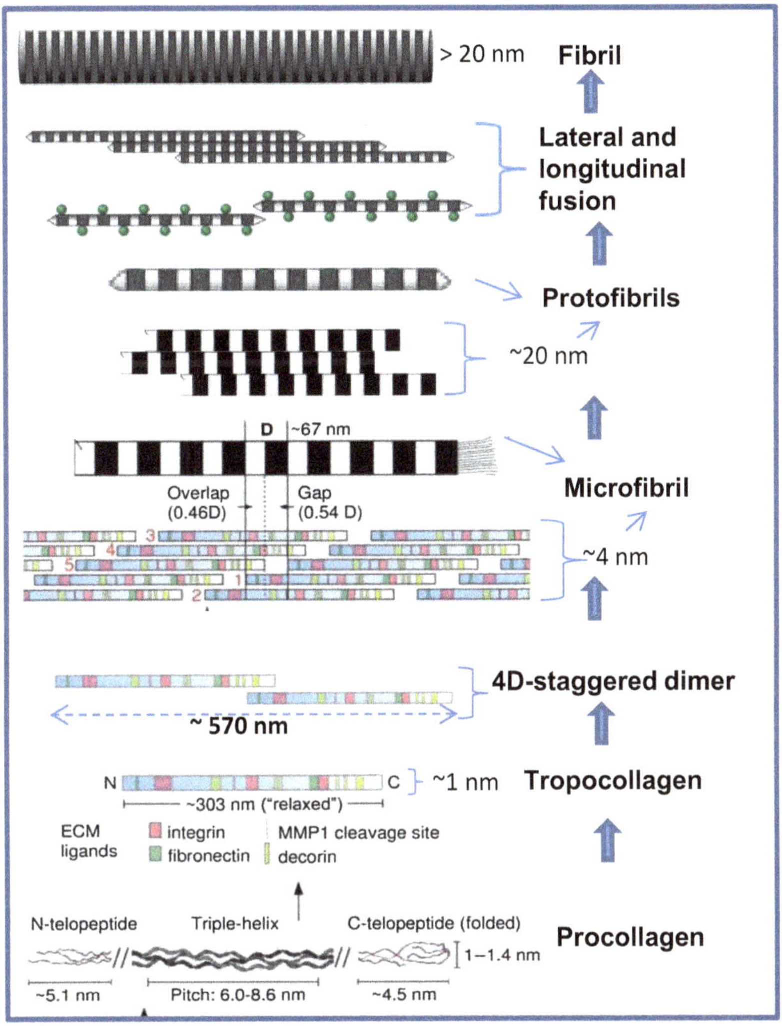


**Fig. 2.3** The main steps in fibrillogenesis of collagen type I. Procollagen type I secreted from the cells is cleaved to form tropocollagen, which self-assembles into various intermediates. The first step is formation of the 4D-staggered dimer followed by the 5-mer microfibril. The microfibril is the basic building block of the collagen fibril. The microfibrils and protofibrils undergo lateral and longitudinally fusion to form the collagen fibril. Interaction of other collagen types and CBPs (like DDRs) with the tropocollagen, microfibril, or protofibril can impact the fibrillogenesis process. Adapted from [51, 81, 83, 84]

of 5 mers (M1 through M5). In situ fiber diffraction studies have elucidated that in the overlap region, the segments from all five monomers follow a common tilt [84]. In a 3D space, the five mers are understood to pack in a quasi-hexagonal lattice with a distance of ~1.3 nm from its nearest neighbor [89]. The gap region is more disorganized with the segments from the 5-mers following different paths with respect to each other and as compared to the overlap region and includes two large cavities. The collagen microfibril is understood to be a dynamic structure, as studies using solid-state NMR have demonstrated that fibrillogenesis increases the heterogeneity of conformations of side chain amino acids, impacting 40 % of the imino acids [90].

The formation of microfibril results in certain regions of the triple helix (on different monomers) to come in close proximity to each other (Fig. 2.4). In particular, three major ligand-binding regions (MLBR) and a cell and a matrix interaction domain have been identified on the collagen microfibril [69]. Two of the DDR2 binding sites lie in one of the MLBR of the microfibril. DDR2[3], the major DDR1/DDR2 binding site consisting of the GVMGFO motif lies in the matrix interaction region in the gap region of the collagen microfibril. It is interesting to note that assembly of tropocollagen into the microfibril results in four of the five DDR2 binding sites to lie in the proteoglycan interaction regions and at least two of them are in line with the cross-linking sites on the collagen microfibril (Table 2.2 and Fig. 2.4). However, it remains to be elucidated if DDRs remain bound to collagen monomers and/or microfibrils during and/or after the process of collagen fibrillogenesis and for how long.

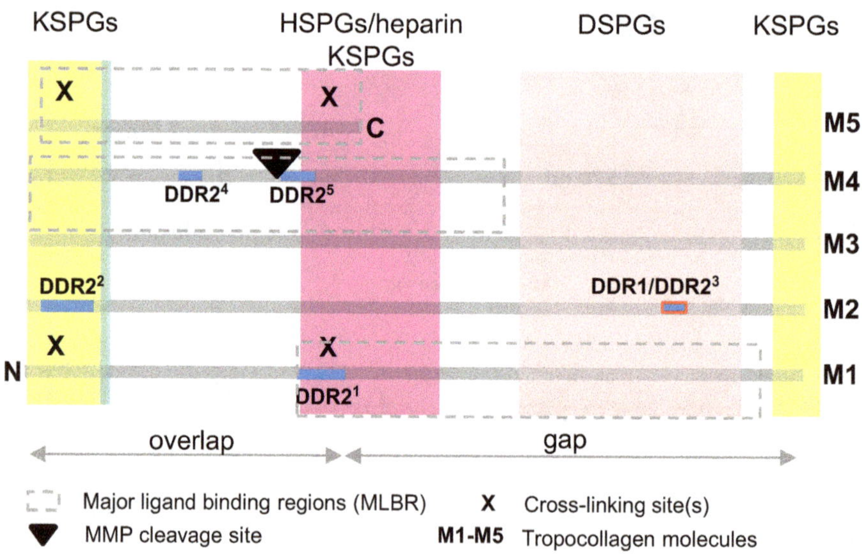


**Fig. 2.4** Schematic diagram of the collagen D-period present in the 5-mer collagen microfibril. The five collagen molecules are indicated as M1 through M5. A single D-period contains the entire collagen sequence. Approximate location(s) of the five DDR2 and one DDR1 binding sites is mapped onto this diagram. Adapted from [69, 89]

Do DDRs affect the assembly of the microfibril? A closer look at the DDR binding site(s) on tropocollagen can provide insight into the possible mechanism(s) by which DDRs affect the microfibril formation. At least one of the binding sites, namely DDR2$^{5}$ overlaps with the 50 % fibrillogenesis inhibition site. This site on the collagen triple helix binds to the C-telopeptide of the adjacent collagen monomer during microfibril assembly and gives rise to the characteristic one-quarter D-stagger [91] (Fig. 2.5a). Binding of DDR2 to this site may inhibit collagen fibrillogenesis by preventing C-telopeptide binding and thus sequester out collagen monomers from microfibril formation.

The primary DDR1/DDR2 binding sites, namely DDR2$^{3}$ overlaps with that of vWF-A3 and SPARC [93]. While the binding site of vWF-A3 is shifted toward the N-terminus, SPARC binding is centered at the GVMGFO albeit with a larger footprint than DDR2 (750 vs. 530 Å$^{2}$). Crystal structure analysis of triple helical collagen III peptides has provided novel insights into collagen I chain registry and protein binding. Intriguingly, DDR2, SPARC, and VWF-A3 each bind to a different face of the collagen triple helix (Fig. 2.5b). DDR2 interacts with the leading and middle chains, SPARC with the middle and trailing chains, and A3 predominantly with the leading and trailing chains [92]. While no reports exist on modulation of collagen fibrillogenesis by vWF, SPARC is understood to enhance procollagen processing and fibril formation [8, 9]. It is possible that DDRs may impede collagen fibrillogenesis by competing against SPARC for collagen binding.

Of the remaining binding site(s), DDR2$^{1}$ binding site overlaps with K87 residue, which is critical for collagen hydroxylation, glycosylation, and cross-linking (see the following sections). The sequence encompassed by DDR2$^{2}$ only seems to mildly (5 %) affect collagen fibrillogenesis in vitro [91]. Additional mechanisms may exist via which DDR1 inhibits collagen fibrillogenesis. For instance, it has been shown that DDR1 ECD clusters [94] upon collagen binding which could provide steric hindrance to the self-assembly of collagen. Alternatively, DDR1 binding to overlapping collagen molecules may sequester out collagen by locking 2 mers in a conformation incompatible with the microfibril formation.

Thus, binding of DDRs may inhibit the formation of the collagen microfibril by two mechanisms (1) binding to and sequestering tropocollagen molecules out of the fibrillogenesis process, thereby reducing the effective concentration of collagen undergoing fibrillogenesis or (2) by binding to tropocollagen and perturbing the arrangement of collagen molecules in the microfibril and beyond. The observations that DDRs delay the kinetics of fibrillogenesis and disrupt the ultrastructure of the collagen fibrils thus formed indicate that both the mechanisms may be involved.

### 2.3.5 Cell Surface

Collagen fibrillogenesis is understood to be a process tightly controlled by the cells and localized near the plasma membrane [95]. Functional involvement of the cytoskeleton, rearrangement of intracellular organelles, and formation of deep recesses

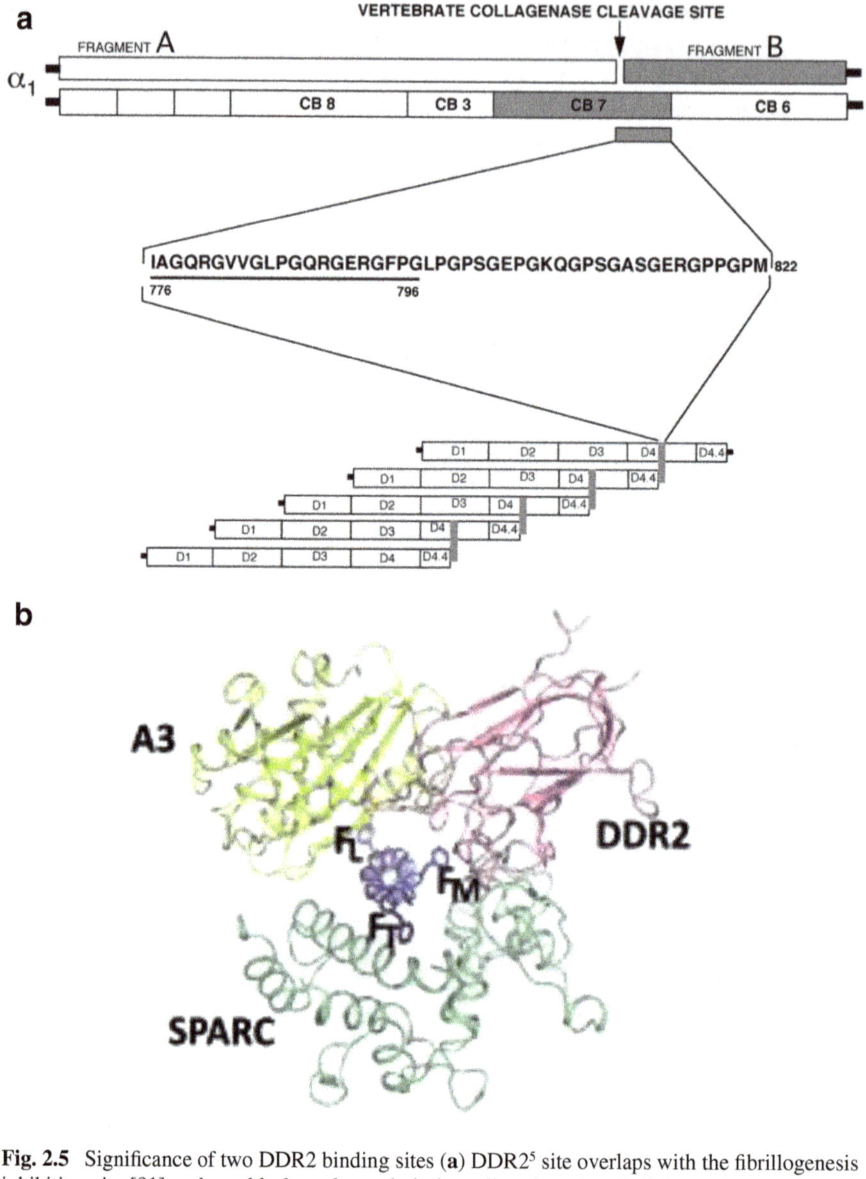


**Fig. 2.5** Significance of two DDR2 binding sites (**a**) DDR2[5] site overlaps with the fibrillogenesis inhibition site [91] and would play a key role in impeding the microfibril formation and (**b**) A3, DDR2[3], and SPARC bind to the same location but to a different surface of the collagen III helix [92]

like structures in particular in fibroblasts [78] are critical for organizing collagen secretion and assembly. The role of membrane receptors (e.g., integrins), collagen V, and fibronectin cannot be ignored in initiating and regulating the microfibril formation. Secretion of procollagen itself is thought to be regulated in part by fibronectin as the two have been found colocalized in the secretory pathway in cultured fibroblasts [96]. DDR-mediated cell signaling is known to affect cytoskeletal reorganization [97], integrin activation [98, 99], and cell–matrix adhesion [100] thus making them active players in cell–matrix interactions.

Cells expressing full-length DDR1b [13] or kinase dead DDR2 [15] on their surface inhibited collagen fibrillogenesis, suggesting a direct role of cell surface DDRs independent of signal transduction. In addition, as discussed above, interaction of DDRs with procollagen on the cell surface may offer competition to other proteins such as SPARC, which shares the DDR2[3] binding site on collagen and has been shown to modulate procollagen processing and collagen fibrillogenesis in the pericellular region [9]. It remains to be investigated if DDRs compete against SPARC for collagen binding or vice versa.

Additional events like receptor endocytosis, recycling, or shedding may also impact collagen fibrillogenesis on the cell surface. In particular, DDR1b is rapidly oligomerized and endocytosed by the cells upon collagen stimulation [101] thus making it unavailable to modulate collagen fibrillogenesis on the cell surface. One possible scenario to explain how DDR1b could still inhibit collagen fibrillogenesis would be that the receptor mediates endocytosis of collagen. In addition, it has been shown that DDR1 ECD is shed as soluble protein in the ECM [63, 66], which could preserve the capacity to inhibit fibrillogenesis. Further studies are needed to elucidate these mechanisms.

### 2.3.6 Collagen Fibril

The next stage of collagen fibrillogenesis involves coalescing of microfibrils near the cell surface to form D-periodic collagen protofibrils (diameter ~20 nm and length 4–12 μm) with tapered ends [83]. The mature fibril is assembled by linear or lateral fusion of micro and protofibrils [78]. Evidence for lateral fusion can be found in ultrastructural TEM and X-ray diffraction studies of tendon collagen fibrils which reveal a lateral spacing of ~4 nm consistent with microfibril units [102]. Time-lapse AFM imaging of in vitro collagen fibrillogenesis also shows that fibrils grow laterally in steps of ~4 nm, indicating incorporation of microfibrils through self-assembly [103]. The fibrils thus formed can further undergo further lateral fusion to form fibrils with nearly perfect registry of their respective D-periods [103, 104]. Collagen fibrils range from 15 to 500 nm in diameter depending upon the tissue type [105].

The longitudinal growth of collagen fibrils is dependent on the polarity of the fibrils, that is, if they are unipolar (C–N) or bipolar (N–N) [104]. Collagen fibrils can grow in length by (1) addition of collagen monomers to their pointed N-terminus ends for both unipolar or bipolar fibrils [106]. Monomers attach to the pointed ends in a systematic manner such that each mer is staggered in integral multiples of D with its nearest neighbors as in a microfibril. (2) Longitudinal fusion of unipolar fibrils from their C-terminus to form bipolar fibrils [103, 104]. Bipolar fibrils can further elongate from both their N-terminus ends by addition of monomers. Fibril length has been estimated to be 20–40 μm in the developing tissue and in the mm range for the adult tissue [107].

In vivo, the lateral growth of the collagen fibril is controlled by the binding of small leucine-rich proteoglycans (SLRPs) as well as other CBPs (Fig. 2.3) [108]. In the cornea, small diameter fibrils with constant interfibrillar spacing are associated with a high keratan sulfate proteoglycan (KSPG) content, while the larger diameter fibrils in the sclera are characterized by an absence of KSPG and presence of chondroitin sulfate (CSPG), dermatan sulfate (DSPG), and hyaluronic acid [109]. The proteoglycan content is significantly lower at the fibril tips which permits longitudinal fusion of fibrils [104].

What is the role of DDRs in regulating fusion of collagen fibrils? Insights into this question can be gained from collagen turbidity measurements, which measure the rise in absorbance as monomeric collagen assembles into fibrils. Based on analysis of the various phases of collagen turbidity measurements [110], we can say that the long lag time observed for DDR2 ECD containing samples indicates that DDR2 does inhibit the lateral fusion of fibrils in vitro [14]. This feature is also supported by the observations that DDR2 ECD expressing cells resulted in collagen fibrils with reduced diameters [15, 16]. DDR1 ECD on the other hand did not have a significant effect on collagen turbidity lag times [13]. However like DDR2, DDR1 ECD also decreased the fibril diameter when overexpressed as a cell-secreted protein [16]; and a trend toward larger fibril diameters was observed in the DDR1 KO mice [17].

Further insights into the role of DDR2 in regulating fibril fusion can be gained by analyzing collagen fibrils endogenously assembled by cells expressing DDR2 ECD as a membrane anchored vs. soluble protein. In these experiments, the membrane-anchored DDR2 ECD enabled a slow lateral growth of collagen fibril diameter with time [15], whereas the soluble DDR2 (and DDR1) ECD completely inhibited increase in fibril diameter [16]. These observations suggest that while cell surface receptors primarily mediate the initial micro- and protofibril assembly in pericellular regions, the soluble matricellular proteins control the later stages of fibril fusion and growth. It is interesting to note that two DDR2 binding sites lie in the KSPG binding region of the collagen microfibril [69] (Fig. 2.4), a proteoglycan known to control the fibril diameter. Thus, while there is evidence that DDRs regulate the fibril diameter, the mechanisms behind this aspect remain to be elucidated.

Another feature of collagen fibrils formed in the presence of purified recombinant DDRs or DDRs expressed in cultured cells was that they exhibited disrupted or weakened D-periodicity. Once again, the disrupting effect was more pronounced when DDR1 or DDR2 ECD were present as soluble vs. membrane anchored proteins. This disruption in the collagen ultrastructure both in the pericellular and extracellular regions suggests that DDRs may disrupt the initial assembly of microfibrils as well as the fusion of micro and protofibrils into fibrils. Two conflicting models exist regarding the orientation of the microfibril in the collagen fibril namely (1) the Herr and Farndale [75] and (2) Perumal and Orgel [111]. The two models primarily differ with regard to which of the telopeptides (N or C) remains buried inside the 3D fibril. Depending on the microfibril orientation, the DDR2 binding site(s) may be more or less exposed on the fibril surface and/or may require proteolytic treatment of collagen or fibril bends, kinks, or imperfections to be fully exposed for ligand binding. Further investigations are required to understand how DDRs modulate the intrafibrillar structure of collagen fibrils, fibril fusion, and if they remain bound to collagen fibrils.

### 2.3.7 Role of Glycosylation

Collagen glycosylation is another important factor in fibrillogenesis. Individual α-chains in the tropocollagen molecule possess *O*-linked carbohydrate units attached to certain hydroxyl lysine residues. These carbohydrate moieties contain either a single galactose residue, α-D-galactopyranosyl hydroxylysine or the disaccharide α-D-glucopyranosyl-(1 → 2)-β-D-galactopyranosyl hydroxylysine, abbreviated as Gal-Hyl and Glc-Gal-Hyl, respectively. Both enzymatic as well as nonenzymatic glycosylation of collagen have been shown to impact collagen fibrillogenesis. An increase in enzymatically glycosylated hydroxlysine residues was found associated with a decrease in fibril diameter of both in vivo and in vitro formed fibrils [112]. In another study, when collagen glycosylation (Glc-Gal-Hyl content) was reduced by suppression of lysyl hydroxylase 3 (LH3) in cells [113], an accelerated collagen fibrillogenesis with larger fibril diameters was observed compared to the controls. Modification in the galactosyltransferase activity such as due to mutations in the B4GALT7 gene can lead to defective collagen fibrillogenesis with an altered collagen ultrastructure [114]. Consistent with these observations, in vitro glycosylation of collagen by glycosyltransferase inhibited the rate of collagen fibrillogenesis while deglycosylation of collagen enhanced it [115]. Presence of monosaccharides such as D(+)glucose and D(+)mannose have been found to delay collagen fibril formation in vitro [116]. Interactions of glycosylated CBPs as a function of collagen glycosylation are much less understood. In one study, inhibition of collagen fibrillogenesis by the decorin proteoglycan was proportional to the degree of collagen glycosylation [117].

DDRs are a family of glycosylated proteins and DDR–collagen interactions are understood to be glycosylation dependent. Deglycosylation of collagen by periodate treatment has been reported to inhibit DDR2 activation [71, 118, 119]. On the other hand, the nonglycosylated toolkit peptides used to identify DDR binding site on collagen preserve the capacity to bind to and activate DDRs. Recently mass spectrometric analysis has helped identify five glycosylation sites in type I collagen ($\alpha_{1,2}$-87, $\alpha_{1,2}$-174, and $\alpha_2$-219, all localized near the N-terminal of the collagen triple helix). The predominant glycosylation site was $\alpha_1$-87, which is also one of the major helical cross-linking sites [120]. Two of these glycosylation sites, $\alpha_{1,2}$-87 and $\alpha_2$-219 overlap with DDR2 binding sites (Table 2.2). Thus, collagen glycosylation may play a major role in DDR2–collagen interaction and the ability of DDR2 to modulate collagen fibrillogenesis.

Receptor glycosylation is another mechanism which can impact modulation of collagen fibrillogenesis by DDRs. DDRs are glycosylated in the DS-like domain and in the transmembrane domain [121] and consist of both *N*- and O-glycosylation sites. In particular, N-glycosylation of Asn 213 in DDR2 and Asn 211 in DDR1 control the expression and stability of the DDR proteins [122]. However, although mutation at Asn211 does not affect binding of DDRs to collagen binding, N-glycosylation of Asn211 was shown to be a negative repressor of receptor dimerization and phosphorylation in the absence of ligand [123]. Recent evidence thus seems to suggest that glycosylation sites in the receptor ECD may regulate the oligomeric state of the receptor. This is especially important, as dimeric and oligomeric DDR ECDs are known to exhibit a higher affinity for collagen than monomeric DDRs [94, 124].

### 2.3.8 Collagen Cross-Linking

The formation of covalent intermolecular cross-links is the final modification in collagen fibrillogenesis and is critical for the stability of assembled collagen fibrils [125]. At the molecular level, lysine-derived cross-links mediated by lysyl oxidase (LOX) play the major role in intermolecular bonding. Assembly of the 4D-staggered tropocollagen molecules and formation of collagen fibrils recruits LOX [126] and results in oxidative deamination of telopeptide lysines or hydroxylysines. The resulting aldehyde or aldol condensation leads to formation of divalent inter- or intramolecular cross-links followed by the mature trivalent cross-links, involving pyrrole and pyridinoline. The four major sites of LOX-mediated cross-linking are (1) a residue of allysine or hydroxylysine at position $9^{N}$ in the N-telopeptide, (2) hydroxylysine at position 87, (3) 930 in the triple helical region, and (4) residue of allysine or hydroxylysine at position $16^{C}$ in the C-telopeptide. Intermolecular cross-links between $9^{N}$ and 930 and between $16^{C}$ and 87 stabilize a 4-D stagger relationship between collagen molecules that is important for the microfibril structure. Lateral networks of cross-links across the fibril have also been found related by 4D and 0D staggers. LOX activity is understood to primarily affect collagen organization and deposition [127–129]. Recent studies have elucidated how LOX activity may also be crucial to collagen fibrillogenesis. Inhibition of LOX in tendon-like constructs not only inhibited the formation of cross-links in collagen, but also resulted in structurally abnormal collagen fibrils with irregular profiles and widely dispersed diameters, representative of connective tissue disorders [130]. It is important to note that LOX activity mainly dictates the quantity of cross-links, whereas the quality of cross-links formed is governed by the extent of lysine hydroxylation mediated by LH1, LH2, and LH3 (see Sect. 2.3.1).

Not much is known about DDRs and their role in collagen cross-linking. However, it is interesting to note that one of the DDR2 binding sites, that is, DDR$2^{1}$ corresponding to toolkit peptide (III-5) overlaps with K87, a key site for collagen cross-linking. Mapping the DDR2 binding sites on the collagen microfibril (Fig. 2.4) reveals how DDR$2^{1}$ and DDR$2^{5}$ are aligned or adjacent to the cross-linking sites involving the C-telopeptide on the collagen microfibril. These sites have also been recently shown to be important for fibromodulin binding and its modulation of LOX activity [131]. A third DDR2 binding site, namely DDR$2^{2}$, is aligned with the N-telopeptide cross-linking site on the microfibril. These observations suggest that DDR2 binding may either directly or indirectly modulate LOX-mediated collagen cross-linking. DDR2 signaling is also known to regulate the expression of cross-linking enzymes. Binding and activation of DDR2 by collagen mediated increase in LOX [132], whereas the expression of LOX and LH1 was found decreased in the skin of DDR2 KO mice [47]. Since LOX activity is understood to occur primarily after the onset of collagen fibrillogenesis [126], it is tempting to speculate that DDR2 may regulate collagen cross-linking in ECM remodeling.

Besides LOX, other extracellular cross-linking enzymes, such as tissue transglutaminases (TG2), also play a role in cross-linking of the collagen fibril through transamidase activity involving γ-carboxamide group of peptide-bound glutamine and a primary amine [133]. The specific sites of TG2-mediated cross-linking for collagen are not yet elucidated. However, TG2 activity has been shown to influence collagen fibrillogenesis by impacting the fibril diameter [129] and fibril structure [134]. No reports exit on regulation of TG2 expression or activity by DDRs.

## 2.4 Functional Consequences

Modulation of collagen fibrillogenesis can impact both the quantity and quality of the fibrils formed, which in turn can impact the functional properties of the ECM. For instance, the skin mechanics in mice lacking certain CBPs is affected by the amount of collagen deposition (Table 2.1). The fibrillogenesis process can also impact the packing of collagen triple helices and its manifestation on the fibril morphology and physicochemical properties. Much less is understood on how CBPs impact these parameters.

Studies on the evolution of collagen demonstrate that the ~67 nm length of D-periods in the collagen fibril are preserved across various organisms and tissues and are constrained to alterations [135]. In diseases characterized by collagen mutations, the mean length of D-periods does not differ although differences in the distribution of D-period lengths have been reported [136, 137]. Our observations that the mean length of D-periods in the collagen fibrils is unaltered as a function of DDRs support the paradigm of preservation of the lateral assembly of collagen molecules during fibrillogenesis.

Unlike the length of D-periods, the depth of D-periods shows significant variations among tissues and/or species types. D-depths measured using AFM range from ~2.5 nm in cornea to 5.5 nm in the sclera [138]. In our studies, the depth of D-periods in collagen fibrils was reduced in the presence of DDRs [17], indicating that other factors besides lateral alignment of collagen molecules are affected in collagen fibrillogenesis. Recent X-ray diffraction data on triple helices formed by (Gly-Pro-Hyp) peptides [139], in situ X-ray diffraction studies [140] and numerical models [141] have shown that the packing of collagen triple helices in a fibril forms a quasi-hexagonal lattice and the degree of disorder in packing can be affected in certain conditions [142].

The following sections discuss how DDRs mediated modulation of collagen fibrillogenesis impacts functional properties of the ECM.

### *2.4.1 Cell–Matrix Interactions*

Cells communicate with the ECM via cell surface receptors and secreted proteins, which bind to collagen(s) among other ECM components. The binding of CBPs can be dependent on (1) specific motifs present on the collagen triple helix, (2) the various intermediates formed during collagen fibrillogenesis, or (3) the accessibility of the binding motif on the collagen fibril formed. Using peptide toolkits and crystal structure studies, the past decade has elucidated the binding sites and properties of several CBPs (summarized in [143]). While these studies are useful in understanding how CBPs interact with the monomeric collagen, much less is known about how CBPs interact with the collagen fibril and/or its intermediates. The resulting cell–matrix interactions can be especially affected by how CBPs interact with native vs. structurally altered collagen fibril. Impact of an altered collagen fibril structure on cell–matrix interactions has been observed in angiogenesis [144] and in platelet-collagen adhesion [145, 146].

DDRs can influence cell–matrix interactions via three mechanisms (1) competition of other CBPs with DDR binding sites on collagen, (2) regulation of the quality and quantity of collagen intermediates by DDRs, and (3) modulation of the collagen fibril structure by DDRs. The overlap of the DDR2$^3$ site with that for binding of vWF and SPARC suggest that DDRs may compete against these proteins for collagen binding and vice versa. It is also interesting to note that DDR2$^5$ overlaps with the binding site of the soluble collagenase MMP1 [147], which is known to preferentially cleave collagen monomers instead of fibrils [148] (Fig. 2.5a). DDR binding may thus either aid or inhibit the binding and activity of this enzyme, which, along with inhibition of collagen fibrillogenesis may contribute to the quality and quantity of collagen intermediates present at sites of active remodeling. Regarding the effect of the collagen fibril structure, it is interesting to note that cell proliferation was enhanced on the ECM secreted by DDR1$^{-/-}$ cells as compared to in vitro polymerized collagen [22]. Thus, DDRs can have a multifaceted role in regulating cell–matrix interactions.

### 2.4.2 Matrix Mineralization

In recent years, in vitro mineralization studies of collagen fibrils have provided novel insights into the formation of intrafibrillar mineral deposits [149]. The structural and compositional properties of collagen mineralized in vitro are similar to those mineralized in osteoblast cultures or occurring in vivo in long bones [150]. Collagen mineralization involves stabilization of amorphous calcium phosphate (ACP) and its infiltration in the collagen fibers. This process is mediated by inhibitors of hydroxyapatite crystallization, such as polyaspartic acid (pAsp) or acidic CBPs. The positive net charge near the C-terminal end of the collagen molecules attracts the negatively charged ACP–pAsp complexes, and promotes the infiltration of ACP into the fibrils [151]. The primary site of infiltration is located in the ~9 nm wide 'a-band' region spanning the interface of the gap and overlap regions in the collagen fibril. After infiltration of ACP, clusters of charged amino acids, both in the gap and the overlap regions, form nucleation sites controlling the conversion of ACP into arrays of oriented apatite crystals aligned parallel to the long axis of the fibril. Computer simulations have further demonstrated how charged amino acids are crucial in mediating this intrafibrillar nucleation of calcium phosphate [152]. Collagen cross-linking and the microfibril subunits also play an important role in regulating hydroxyapatite crystal deposition [153].

Certain acidic CBPs secreted in the ECM mimic the role of pAsp in matrix mineralization by inhibiting extrafibrillar mineralization and mediating infiltration of ACP into the collagen fibrils. The CBPs known to mediate ACP infiltration are osteopontin [154], phosphorylated dentin phosphophoryn (DPP) [155], fetuin, and the C-terminal fragment of the dentin matrix protein 1 (C-DMP1) [156]. While there is some evidence that these CBPs also modulate collagen fibrillogenesis [157, 158], a clear connection between collagen fibril structure and mineralization is yet to be established. It is interesting to note that CBPs such as decorin [159] and thrombospondin 1 [160], which modulate collagen fibrillogenesis and the collagen fibril structure, also modulate matrix mineralization.

DDRs are also understood to regulate matrix mineralization though the mechanisms involved are not completely understood. Osteoblast cultures expressing DDR1 or DDR2 ECD as a membrane-anchored or secreted protein revealed an enhanced matrix mineralization compared to controls [16]. Knockdown of full-length DDR2 in osteoblasts suppressed osteogenic marker gene expression and matrix mineralization [161]. Consistent with these observations, in vivo studies revealed lower mineralization in the fibula bone of DDR1 KO mice [18]. Atherosclerotic plaques in the vessel wall of DDR1 KO mice exhibited reduced calcification [162]. In SMED patients (with a phenotype similar to the DDR2 knockout mice), an aberrant mineralization can be correlated to mutations in DDR2 [163], which result in intracellular confinement of the receptor or its failure to bind collagen [164, 165]. These observations suggest that the extracellular expression of DDR ECDs and their ability to bind collagen are important mediators of matrix mineralization. At the same time, the role of DDR-mediated cell-signaling events in matrix mineralization cannot be ruled out. For instance, DDR1$^{+/+}$ vascular smooth muscle cells exhibit a higher alkaline phosphatase activity compared to DDR1$^{-/-}$ cells [162].

### *2.4.3 Matrix Mechanics*

Mechanical properties of collagen fibrils are dependent on several factors including collagen concentration, intrafibrillar packing, and cross-linking. While the collagen fibrils become more rigid with increasing collagen concentration, fibril diameter has little impact on its mechanical property as elucidated by studies conducted on in vitro polymerized collagen [166, 167] and skin mechanics of mice lacking CBPs (Table 2.1). Intrafibrillar packing and cross-linking are the dominant parameters in collagen mechanics. It has been elucidated how intermolecular spacing between the tropocollagen molecules within the collagen fibril in osteogenesis imperfecta (OI) is increased due to OI mutations, resulting in lower levels of cross-links and a loss in mechanical strength of the collagen fibril [168]. Slipping of microfibrils within a collagen fibril due to changes in cross-linking contribute significantly to its viscoelastic properties [87].

The multifaceted ability of DDRs to modulate collagen fibrillogenesis is indicative of their potential role in regulating collagen mechanics. It is interesting to note that collagen fibrils formed in vitro in the presence of recombinant DDR2 ECD were more flexible (reduced persistence length) than the controls [166]. On the contrary, cutaneous wounds in DDR2$^{-/-}$ mice showed reduced tensile strength which correlated with a significant reduction in collagen content and defective collagen cross-linking [47]. Further studies are required to elucidate the mechanisms and extent to which DDRs modulate the mechanical properties of the ECM, which in turn may dictate cell–matrix interactions.

## 2.5 Other Collagen Types

Besides the fibril-forming collagen types I–III, DDRs also bind to nonfibril-forming collagens such as the network-forming collagen type IV (DDR1 only) [73, 74] and the nonfibrillar collagen type X (DDR2 only) [169]. However, much less is understood about the impact of DDRs on the assembly of other collagen types. Insights into interaction of DDRs with the various collagen types have been crucial to enhance our understanding of DDRs with collagen type I and vice versa. In particular as discussed in Sect. 2.3.3, the binding sites of DDRs on the collagen triple helix have been accomplished using the homotrimeric col II and col III peptide toolkits and match well with the mapping of DDR2 binding sites(s) on collagen type I triple helix. In evaluating the interaction of DDR1 with collagen type IV, it is important to note that none of the six Col4$\alpha$ chains contain the amino acid sequence, GVMGFO, required for binding of DDR1 to collagens II and III [170]. In the collagen IV heterotrimer, SPARC recognizes GFP or GLP [171] while the vWF binding motifs occur on the Col4$\alpha_2$ (RGQPGVPGVPGMKGD), $\alpha_1$, $\alpha_4$ (RGQPGEMGD), and, possibly, the $\alpha_3$ (RGQPGRKGL) chains. (DDR2 does not bind to collagen IV.) It is unclear whether DDR1 also binds to these motifs on Col4$\alpha$ chains.

Similar to its effect on collagen type I, there are indications that DDRs may also regulate the assembly and degradation of other collagen types. Collagen type II present in the articular cartilage is a ligand for DDRs, especially DDR2. Activation of DDR2 in chondrocytes by collagen type II results in upregulation of the receptor [172]. The rate of cartilage degeneration was attenuated in mice heterozygous for DDR2 as compared to the WT controls or to mice with type XI collagen haploinsufficiency [173]. The network-forming collagen type IV is the main component of basement membrane (BM). Ultrastructural electron microscopy investigations have shown excess synthesis and thickening of the glomerular [174] and strial [19] BM in the kidney and inner ear of the DDR1 KO mice. It remains to be investigated if the differences in the BM composition are due to differences in secretion and/or assembly of collagen type IV in the DDR1 KO mice.

Another aspect common to interaction of DDRs with various collagen types can be observed in ECM remodeling. Studies using primary cell culture [175] and conditional transgenic mice [73, 74] have elucidated how the pericellular matrix surrounding the chondrocytes in physiological conditions prevents DDR2–collagen II interaction. Biochemical or biomechanical disruption of the pericellular matrix occurring in ECM remodeling accompanying pathologies such as osteoarthritis is required for activation of DDR2 by the underlying collagen type II. These observations are paralleled in tissues consisting primarily of collagen type I wherein DDR1 is upregulated in regions of ECM remodeling. DDR1 expression was increased and colocalized with collagen in carotids after vascular injury in control animals [176] and upregulated within glomeruli of remnant rat kidneys [177]. ECM remodeling in pathologies, such as lymphangioleiomyomatosis, is accompanied by an increase in DDR1 expression in the affected nodules [33]. Thus, interaction of DDRs with various collagen types may be more prominent in active ECM remodeling.

## 2.6 Conclusions

DDR1 and DDR2 have multifaceted roles in collagen regulation including collagen synthesis, fibrillogenesis, and degradation. Both the binding of DDR1 and DDR2 ECDs to collagen as well as their intracellular signaling events can contribute to collagen regulation. While DDR1 and DDR2 share similarities, they also differ in the mechanisms and extent to which they regulate collagen. The expression and activation of DDRs and their various isoforms along with the cell types involved should be carefully analyzed to comprehend their functional role(s) in health and disease. A number of reagents and tools are now available for DDR research, including expression constructs, recombinant proteins, poly- and monoclonal antibodies, intra- and extracellular inhibitors, and the knockout mice. The upcoming years thus hold the promise to significantly enhance our understanding of these intriguing receptors.

**Acknowledgment** This work was supported in part by the NSF CMMI award 1201111 to GA.

## References

1. Danielson KG et al (1997) Targeted disruption of decorin leads to abnormal collagen fibril morphology and skin fragility. J Cell Biol 136(3):729–743
2. Chakravarti S et al (1998) Lumican regulates collagen fibril assembly: skin fragility and corneal opacity in the absence of lumican. J Cell Biol 141(5):1277–1286
3. Jepsen KJ et al (2002) A syndrome of joint laxity and impaired tendon integrity in lumican- and fibromodulin-deficient mice. J Biol Chem 277(38):35532–35540. http://www.ncbi.nlm.nih.gov/pubmed/12089156. Accessed 26 May 2015
4. Svensson L et al (1999) Fibromodulin-null mice have abnormal collagen fibrils, tissue organization, and altered lumican deposition in tendon. J Biol Chem 274(14):9636–9647. http://www.ncbi.nlm.nih.gov/pubmed/10092650. Accessed 26 May 2015
5. Tasheva ES et al (2002) Mimecan/osteoglycin-deficient mice have collagen fibril abnormalities. Mol Vis 8:407–415. http://www.ncbi.nlm.nih.gov/pubmed/12432342. Accessed 14 Mar 2016

6. Takeda U et al (2002) Targeted disruption of dermatopontin causes abnormal collagen fibrillogenesis. J Invest Dermatol 119(3):678–683. http://www.ncbi.nlm.nih.gov/pubmed/12230512. Accessed 14 Mar 2016
7. Liaw L et al (1998) Altered wound healing in mice lacking a functional osteopontin gene (spp1). J Clin Invest 101(7):1468–1478. http://www.pubmedcentral.nih.gov/articlerender.fcgi?artid=508725&tool=pmcentrez&rendertype=abstract. Accessed 2 Nov 2015
8. Bradshaw AD et al (2003) SPARC-null mice display abnormalities in the dermis characterized by decreased collagen fibril diameter and reduced tensile strength. J Invest Dermatol 120(6):949–955. http://www.ncbi.nlm.nih.gov/pubmed/12787119. Accessed 17 Nov 2014
9. Rentz TJ et al (2007) SPARC regulates processing of procollagen I and collagen fibrillogenesis in dermal fibroblasts. J Biol Chem 282(30):22062–22071. http://www.ncbi.nlm.nih.gov/pubmed/17522057. Accessed 3 June 2014
10. Kyriakides TR et al (1998) Mice that lack thrombospondin 2 display connective tissue abnormalities that are associated with disordered collagen fibrillogenesis, an increased vascular density, and a bleeding diathesis. J Cell Biol 140(2):419–430. http://www.pubmedcentral.nih.gov/articlerender.fcgi?artid=2132586&tool=pmcentrez&rendertype=abstract. Accessed 26 May 2015
11. Norris RA et al (2007) Periostin regulates collagen fibrillogenesis and the biomechanical properties of connective tissues. J Cell Biochem 101(3):695–711. http://www.pubmedcentral.nih.gov/articlerender.fcgi?artid=3393091&tool=pmcentrez&rendertype=abstract. Accessed 29 May 2015
12. Agarwal G et al (2002) Binding of discoidin domain receptor 2 to collagen I: an atomic force microscopy investigation. Biochemistry 41(37):11091–11098
13. Agarwal G, Mihai C, Iscru DF (2007) Interaction of discoidin domain receptor 1 with collagen type 1. J Mol Biol 367(2):443–455
14. Mihai C et al (2006) Discoidin domain receptor 2 inhibits fibrillogenesis of collagen type 1. J Mol Biol 361(5):864–876
15. Blissett AR et al (2009) Regulation of collagen fibrillogenesis by cell-surface expression of kinase dead DDR2. J Mol Biol 385:902–911
16. Flynn LA et al (2010) Inhibition of collagen fibrillogenesis by cells expressing soluble extracellular domains of DDR1 and DDR2. J Mol Biol 395(3):533–543
17. Tonniges J et al Collagen fibril ultrastructure in mice lacking discoidin domain receptor 1. Microsc Microanal 2016;22(3):599–611
18. Vogel WF et al (2001) Discoidin domain receptor 1 tyrosine kinase has an essential role in mammary gland development. Mol Cell Biol 21(8):2906–2917
19. Meyer zum Gottesberge AM et al (2008) Inner ear defects and hearing loss in mice lacking the collagen receptor DDR1. Lab Invest 88(1):27–37. http://www.ncbi.nlm.nih.gov/pubmed/18026164. Accessed 22 Feb 2015
20. Franco C et al (2008) Discoidin domain receptor 1 (ddr1) deletion decreases atherosclerosis by accelerating matrix accumulation and reducing inflammation in low-density lipoprotein receptor-deficient mice. Circ Res 102(10):1202–1211
21. Franco C et al (2009) Discoidin domain receptor 1 on bone marrow-derived cells promotes macrophage accumulation during atherogenesis. Circ Res 105(11):1141–1148
22. Franco C et al (2010) Increased cell and matrix accumulation during atherogenesis in mice with vessel wall-specific deletion of discoidin domain receptor 1. Circ Res 106(11):1775–1783
23. Olaso E, Lin H-C et al (2011) Impaired dermal wound healing in discoidin domain receptor 2-deficient mice associated with defective extracellular matrix remodeling. Fibrogenesis Tissue Repair 4(1):5. http://www.pubmedcentral.nih.gov/articlerender.fcgi?artid=3044108&tool=pmcentrez&rendertype=abstract. Accessed 8 Nov 2014
24. Gross O et al (2010) Loss of collagen-receptor DDR1 delays renal fibrosis in hereditary type IV collagen disease. Matrix Biol 29(5):346–356. http://www.ncbi.nlm.nih.gov/pubmed/20307660. Accessed 2 June 2015

25. Flamant M et al (2006) Discoidin domain receptor 1 null mice are protected against hypertension-induced renal disease. J Am Soc Nephrol 17(12):3374–3381. http://www.ncbi.nlm.nih.gov/pubmed/17093065. Accessed 2 June 2015
26. Guerrot D et al (2011) Discoidin domain receptor 1 is a major mediator of inflammation and fibrosis in obstructive nephropathy. Am J Pathol 179(1):83–91. http://www.pubmedcentral.nih.gov/articlerender.fcgi?artid=3123843&tool=pmcentrez&rendertype=abstract. Accessed 2 June 2015
27. Avivi-Green C, Singal M, Vogel WF (2006) Discoidin domain receptor 1-deficient mice are resistant to bleomycin-induced lung fibrosis. Am J Respir Crit Care Med 174(4):420–427
28. Shyu K-G et al (2008) RNA interference for discoidin domain receptor 2 attenuates neointimal formation in balloon injured rat carotid artery. Arterioscler Thromb Vasc Biol 28(8):1447–1453. http://www.ncbi.nlm.nih.gov/pubmed/18497308. Accessed 17 Nov 2014
29. Zhang X-H et al (2010) Expression of discoidin domain receptors (DDR2) in alcoholic liver fibrosis in rats. Arch Med Res 41(8):586–592. http://www.ncbi.nlm.nih.gov/pubmed/21199726. Accessed 17 Nov 2014
30. Wynn TA (2011) Integrating mechanisms of pulmonary fibrosis. J Exp Med 208(7):1339–1350. http://jem.rupress.org/content/208/7/1339. Accessed 8 Oct 2014
31. Yang J et al (2013) Activated alveolar epithelial cells initiate fibrosis through secretion of mesenchymal proteins. Am J Pathol 183(5):1559–1570. http://www.pubmedcentral.nih.gov/articlerender.fcgi?artid=3814683&tool=pmcentrez&rendertype=abstract. Accessed 5 Nov 2014
32. Zhang P et al (2013) Protein analysis of atrial fibrosis via label-free proteomics in chronic atrial fibrillation patients with mitral valve disease. PloS One 8(4):e60210. http://www.pubmedcentral.nih.gov/articlerender.fcgi?artid=3617171&tool=pmcentrez&rendertype=abstract. Accessed 17 Nov 2014
33. Ferri N, Carragher NO, Raines EW (2004) Role of discoidin domain receptors 1 and 2 in human smooth muscle cell-mediated collagen remodeling: potential implications in atherosclerosis and lymphangioleiomyomatosis. Am J Pathol 164(5):1575–1585
34. Mao TK et al (2002). Elevated expression of tyrosine kinase DDR2 in primary biliary cirrhosis. Autoimmunity 35(8):521–529. http://www.ncbi.nlm.nih.gov/pubmed/12765478. Accessed 17 Nov 2014
35. Borza CM, Pozzi A (2014) Discoidin domain receptors in disease. Matrix Biol 34:185–192. http://www.ncbi.nlm.nih.gov/pubmed/24361528. Accessed 17 Nov 2014
36. Schminke B et al (2013) A discoidin domain receptor 1 knock-out mouse as a novel model for osteoarthritis of the temporomandibular joint. Cell Mol Life Sci 71(6):1081–1096. http://www.ncbi.nlm.nih.gov/pubmed/23912900
37. Faraci E et al (2003) An extracellular matrix-specific microarray allowed the identification of target genes downstream of discoidin domain receptors. Matrix Biol 22(4):373–381. http://www.ncbi.nlm.nih.gov/pubmed/12935821. Accessed 8 May 2015
38. Hou G, Wang D, Bendeck MP (2012) Deletion of discoidin domain receptor 2 does not affect smooth muscle cell adhesion, migration, or proliferation in response to type I collagen. Cardiovasc Pathol 21:214–218. http://www.ncbi.nlm.nih.gov/pubmed/21865059
39. Cowling RT et al (2014) Discoidin domain receptor 2 (DDR2) germline gene deletion leads to altered heart structure and function in the mouse. Am J Physiol Heart Circ Physiol 307(5):H773–H781. http://ajpheart.physiology.org/content/early/2014/06/30/ajpheart.00142.2014.full-text.pdf+html
40. Olaso E et al (2002) Discoidin domain receptor 2 regulates fibroblast proliferation and migration through the extracellular matrix in association with transcriptional activation of matrix metalloproteinase-2. J Biol Chem 277(5):3606–3613. http://www.ncbi.nlm.nih.gov/pubmed/11723120. Accessed 2 June 2015
41. Myllyharju J, Kivirikko KI (2004) Collagens, modifying enzymes and their mutations in humans, flies and worms. Trends Genet 20(1):33–43. http://www.cell.com/article/S0168952503003196/fulltext. Accessed 5 Sept 2014

42. Berg RA, Prockop DJ (1973) The thermal transition of a non-hydroxylated form of collagen. Evidence for a role for hydroxyproline in stabilizing the triple-helix of collagen. Biochem Biophys Res Commun 52(1):115–120. http://www.ncbi.nlm.nih.gov/pubmed/4712181. Accessed 14 Mar 2016
43. Weis MA et al (2010) Location of 3-hydroxyproline residues in collagen types I, II, III, and V/XI implies a role in fibril supramolecular assembly. J Biol Chem 285(4):2580–2590. http://www.pubmedcentral.nih.gov/articlerender.fcgi?artid=2807315&tool=pmcentrez&rendertype=abstract. Accessed 14 Mar 2016
44. Takaluoma K et al (2007) Tissue-specific changes in the hydroxylysine content and cross-links of collagens and alterations in fibril morphology in lysyl hydroxylase 1 knock-out mice. J Biol Chem 282(9):6588–6596. http://www.ncbi.nlm.nih.gov/pubmed/17197443. Accessed 9 Nov 2014
45. Gilkes DM et al (2013) Hypoxia-inducible factor 1 (HIF-1) promotes extracellular matrix remodeling under hypoxic conditions by inducing P4HA1, P4HA2, and PLOD2 expression in fibroblasts. J Biol Chem 288(15):10819–10829. http://www.pubmedcentral.nih.gov/articlerender.fcgi?artid=3624462&tool=pmcentrez&rendertype=abstract. Accessed 9 Nov 2014
46. Risteli M et al (2009) Reduction of lysyl hydroxylase 3 causes deleterious changes in the deposition and organization of extracellular matrix. J Biol Chem 284(41):28204–28211. http://www.pubmedcentral.nih.gov/articlerender.fcgi?artid=2788872&tool=pmcentrez&rendertype=abstract. Accessed 9 Nov 2014
47. Olaso E, Arteta B et al (2011) Loss of discoidin domain receptor 2 promotes hepatic fibrosis after chronic carbon tetrachloride through altered paracrine interactions between hepatic stellate cells and liver-associated macrophages. Am J Pathol 179(6):2894–2904. http://www.pubmedcentral.nih.gov/articlerender.fcgi?artid=3260862&tool=pmcentrez&rendertype=abstract. Accessed 2 June 2015
48. Khoshnoodi J et al (2006) Molecular recognition in the assembly of collagens: terminal non-collagenous domains are key recognition modules in the formation of triple helical protomers. J Biol Chem 281(50):38117–38121. http://www.ncbi.nlm.nih.gov/pubmed/17082192. Accessed 14 Mar 2016
49. Ishida Y et al (2006) Type I collagen in Hsp47-null cells is aggregated in endoplasmic reticulum and deficient in N-propeptide processing and fibrillogenesis. Mol Biol Cell 17(5):2346–2355. http://www.pubmedcentral.nih.gov/articlerender.fcgi?artid=1446091&tool=pmcentrez&rendertype=abstract. Accessed 14 Mar 2016
50. Makareeva E, Leikin S (2007) Procollagen triple helix assembly: an unconventional chaperone-assisted folding paradigm. PloS One 2(10):e1029. http://www.pubmedcentral.nih.gov/articlerender.fcgi?artid=2000351&tool=pmcentrez&rendertype=abstract. Accessed 20 Nov 2014
51. Kadler KE, Hojima Y, Prockop DJ (1987) Assembly of collagen fibrils de novo by cleavage of the type I pC-collagen with procollagen C-proteinase. Assay of critical concentration demonstrates that collagen self-assembly is a classical example of an entropy-driven process. J Biol Chem 262(32):15696–15701. http://www.ncbi.nlm.nih.gov/pubmed/3316206. Accessed 17 Nov 2014
52. Hulmes DJS (2002) Building collagen molecules, fibrils, and suprafibrillar structures. J Struct Biol 137(1-2):2–10. http://www.ncbi.nlm.nih.gov/pubmed/12064927. Accessed 19 Aug 2014
53. Malfait F et al (2013) Helical mutations in type I collagen that affect the processing of the amino-propeptide result in an Osteogenesis Imperfecta/Ehlers-Danlos Syndrome overlap syndrome. Orphanet J Rare Dis 8:78. http://www.pubmedcentral.nih.gov/articlerender.fcgi?artid=3662563&tool=pmcentrez&rendertype=abstract. Accessed 17 Nov 2014
54. Lindahl K et al (2011) COL1 C-propeptide cleavage site mutations cause high bone mass osteogenesis imperfecta. Hum Mutat 32(6):598–609. http://www.pubmedcentral.nih.gov/articlerender.fcgi?artid=3103631&tool=pmcentrez&rendertype=abstract. Accessed 17 Nov 2014

55. Birk DE et al (1990) Collagen fibrillogenesis in situ. Discontinuous segmental assembly in extracellular compartments. Ann N Y Acad Sci 580:176–194. http://www.ncbi.nlm.nih.gov/pubmed/2337297. Accessed 20 Nov 2014
56. Canty-Laird EG, Lu Y, Kadler KE (2012) Stepwise proteolytic activation of type I procollagen to collagen within the secretory pathway of tendon fibroblasts in situ. Biochem J 441(2):707–717. http://www.pubmedcentral.nih.gov/articlerender.fcgi?artid=3430002&tool=pmcentrez&rendertype=abstract. Accessed 17 Nov 2014
57. Humphries SM et al (2008) Active negative control of collagen fibrillogenesis in vivo. Intracellular cleavage of the type I procollagen propeptides in tendon fibroblasts without intracellular fibrils. J Biol Chem 283(18):12129–12135. http://www.ncbi.nlm.nih.gov/pubmed/18285337. Accessed 1 June 2014
58. Kessler E et al (1996) Bone morphogenetic protein-1: the type I procollagen C-proteinase. Science 271(5247):360–362. http://www.ncbi.nlm.nih.gov/pubmed/8553073. Accessed 17 Nov 2014
59. Scott IC et al (1999) Mammalian BMP-1/Tolloid-related metalloproteinases, including novel family member mammalian Tolloid-like 2, have differential enzymatic activities and distributions of expression relevant to patterning and skeletogenesis. Dev Biol 213(2):283–300. http://www.ncbi.nlm.nih.gov/pubmed/10479448. Accessed 5 Nov 2014
60. Colige A et al (1997) cDNA cloning and expression of bovine procollagen I N-proteinase: a new member of the superfamily of zinc-metalloproteinases with binding sites for cells and other matrix components. Proc Natl Acad Sci U S A 94(6):2374–2379. http://www.pubmedcentral.nih.gov/articlerender.fcgi?artid=20095&tool=pmcentrez&rendertype=abstract. Accessed 17 Nov 2014
61. Colige A et al (2002) Cloning and characterization of ADAMTS-14, a novel ADAMTS displaying high homology with ADAMTS-2 and ADAMTS-3. J Biol Chem 277(8):5756–5766. http://www.ncbi.nlm.nih.gov/pubmed/11741898. Accessed 17 Nov 2014
62. Thomas G (2002) Furin at the cutting edge: from protein traffic to embryogenesis and disease. Nat Rev Mol Cell Biol 3(10):753–766. http://www.pubmedcentral.nih.gov/articlerender.fcgi?artid=1964754&tool=pmcentrez&rendertype=abstract. Accessed 1 Oct 2014
63. Fu H-L et al (2013) Shedding of discoidin domain receptor 1 by membrane-type matrix metalloproteinases. J Biol Chem 288(17):12114–12129. http://www.pubmedcentral.nih.gov/articlerender.fcgi?artid=3636896&tool=pmcentrez&rendertype=abstract. Accessed 1 June 2014
64. Shitomi Y et al (2015) ADAM10 controls collagen signaling and cell migration on collagen by shedding the ectodomain of discoidin domain receptor 1 (DDR1). Mol Biol Cell 26(4):659–673. http://www.ncbi.nlm.nih.gov/pubmed/25540428. Accessed 1 Apr 2015
65. Slack BE, Siniaia MS, Blusztajn JK (2006) Collagen type I selectively activates ectodomain shedding of the discoidin domain receptor 1: involvement of Src tyrosine kinase. J Cell Biochem 98(3):672–84. http://www.pubmedcentral.nih.gov/articlerender.fcgi?artid=2593136&tool=pmcentrez&rendertype=abstract. Accessed 17 Nov 2014
66. Vogel WF (2002) Ligand-induced shedding of discoidin domain receptor 1. FEBS Lett 514(2-3):175–180. http://www.ncbi.nlm.nih.gov/pubmed/11943146. Accessed 1 June 2014
67. Kadler K (1995) Extracellular matrix 1: fibril-forming collagens. Protein Profile 2(5):491–619. http://www.ncbi.nlm.nih.gov/pubmed/7584473. Accessed 2 June 2014
68. Keene DR et al (2000) Decorin binds near the C terminus of type I collagen. J Biol Chem 275(29):21801–21804. http://www.ncbi.nlm.nih.gov/pubmed/10823816. Accessed 2 June 2014
69. Sweeney SM et al (2008) Candidate cell and matrix interaction domains on the collagen fibril, the predominant protein of vertebrates. J Biol Chem 283(30):21187–21197. http://www.pubmedcentral.nih.gov/articlerender.fcgi?artid=2475701&tool=pmcentrez&rendertype=abstract. Accessed 28 May 2014
70. Shrivastava A et al (1997) An orphan receptor tyrosine kinase family whose members serve as nonintegrin collagen receptors. Mol Cell 1(1):25–34

71. Vogel W et al (1997) The discoidin domain receptor tyrosine kinases are activated by collagen. Mol Cell 1(1):13–23
72. Konitsiotis AD et al (2008) Characterization of high affinity binding motifs for the discoidin domain receptor DDR2 in collagen. J Biol Chem 283(11):6861–6868. http://www.ncbi.nlm.nih.gov/pubmed/18201965. Accessed 27 Feb 2014
73. Xu H et al (2011) Collagen binding specificity of the discoidin domain receptors: binding sites on collagens II and III and molecular determinants for collagen IV recognition by DDR1. Matrix Biol 30(1):16–26
74. Xu L et al (2011) Intact pericellular matrix of articular cartilage is required for unactivated discoidin domain receptor 2 in the mouse model. Am J Pathol 179(3):1338–1346. http://www.pubmedcentral.nih.gov/articlerender.fcgi?artid=3157196&tool=pmcentrez&rendertype=abstract. Accessed 22 Feb 2015
75. Herr AB, Farndale RW (2009) Structural insights into the interactions between platelet receptors and fibrillar collagen. J Biol Chem 284(30):19781–19785. http://www.pubmedcentral.nih.gov/articlerender.fcgi?artid=2740402&tool=pmcentrez&rendertype=abstract. Accessed 12 Mar 2014
76. Carafoli F et al (2009) Crystallographic insight into collagen recognition by discoidin domain receptor 2. Structure 17(12):1573–1581. http://www.pubmedcentral.nih.gov/articlerender.fcgi?artid=2807035&tool=pmcentrez&rendertype=abstract. Accessed 28 May 2014
77. McBride DJ et al (1992) Self-assembly into fibrils of a homotrimer of type I collagen. Matrix 12(4):256–263. http://www.ncbi.nlm.nih.gov/pubmed/1435509. Accessed 23 Nov 2014
78. Trelstad RL, Birk DE, Silver FH (1982) Collagen fibrillogenesis in tissues, in a solution and from modeling: a synthesis. J Invest Dermatol 79(Suppl 1):109s–112s. http://www.ncbi.nlm.nih.gov/pubmed/7086181. Accessed 25 Nov 2014
79. Li Y, Douglas EP (2013) Effects of various salts on structural polymorphism of reconstituted type I collagen fibrils. Colloids Surf B Biointerfaces 112:42–50. http://www.ncbi.nlm.nih.gov/pubmed/23948153. Accessed 25 May 2014
80. Wenstrup RJ et al (2004) Type V collagen controls the initiation of collagen fibril assembly. J Biol Chem 279(51):53331–53337. http://www.ncbi.nlm.nih.gov/pubmed/15383546. Accessed 16 Nov 2014
81. Ward NP, Hulmes DJ, Chapman JA (1986) Collagen self-assembly in vitro: electron microscopy of initial aggregates formed during the lag phase. J Mol Biol 190(1):107–112. http://www.ncbi.nlm.nih.gov/pubmed/3783694. Accessed 27 Nov 2014
82. Kobayashi K, Ito T, Hoshino T (1985) Electron microscopic demonstration of acid-labile, 4D-staggered intermolecular association of collagen formed in vitro. Coll Relat Res 5(3):253–260. http://www.ncbi.nlm.nih.gov/pubmed/4042602. Accessed 27 Nov 2014
83. Birk DE, Bruckner P (2011) Collagens, suprastructures, and collagen fibril assembly. In: Mecham RP (ed) The extracellular matrix: an overview. Springer, Berlin, pp 77–115
84. Orgel JPRO et al (2006) Microfibrillar structure of type I collagen in situ. Proc Natl Acad Sci U S A 103(24):9001–9005. http://www.pubmedcentral.nih.gov/articlerender.fcgi?artid=1473175&tool=pmcentrez&rendertype=abstract. Accessed 22 Nov 2014
85. Holmes DF et al (2001) Corneal collagen fibril structure in three dimensions: structural insights into fibril assembly, mechanical properties, and tissue organization. Proc Natl Acad Sci U S A 98(13):7307–7312. http://www.pubmedcentral.nih.gov/articlerender.fcgi?artid=34664&tool=pmcentrez&rendertype=abstract. Accessed 22 Nov 2014
86. Granke M et al (2013) Microfibril orientation dominates the microelastic properties of human bone tissue at the lamellar length scale. PloS One 8(3):e58043. http://www.pubmedcentral.nih.gov/articlerender.fcgi?artid=3589472&tool=pmcentrez&rendertype=abstract. Accessed 22 Nov 2014
87. Yang L et al (2012) Micromechanical analysis of native and cross-linked collagen type I fibrils supports the existence of microfibrils. J Mech Behav Biomed Mater 6:148–158. http://www.ncbi.nlm.nih.gov/pubmed/22301184. Accessed 19 Nov 2014

88. Gautieri A et al (2011) Hierarchical structure and nanomechanics of collagen microfibrils from the atomistic scale up. Nano Lett 11(2):757–766. http://www.ncbi.nlm.nih.gov/pubmed/21207932. Accessed 22 Nov 2014
89. Orgel JPRO, San Antonio JD, Antipova O (2011) Molecular and structural mapping of collagen fibril interactions. Connect Tissue Res 52(1):2–17. http://www.ncbi.nlm.nih.gov/pubmed/21182410. Accessed 1 Dec 2013
90. De Sa Peixoto P et al (2013) Solid-state NMR study reveals collagen I structural modifications of amino acid side chains upon fibrillogenesis. J Biol Chem 288(11):7528–7535. http://www.pubmedcentral.nih.gov/articlerender.fcgi?artid=3597793&tool=pmcentrez&rendertype=abstract. Accessed 16 May 2014
91. Prockop DJ (1998) Inhibition of the self-assembly of collagen I into fibrils with synthetic peptides. Demonstration that assembly is driven by specific binding sites on the monomers. J Biol Chem 273(25):15598–15604. http://www.jbc.org/content/273/25/15598.full#F9. Accessed 21 Nov 2014
92. Brondijk THC et al (2012) Implications for collagen I chain registry from the structure of the collagen von Willebrand factor A3 domain complex. Proc Natl Acad Sci U S A 109(14):5253–5258. http://www.pubmedcentral.nih.gov/articlerender.fcgi?artid=3325688&tool=pmcentrez&rendertype=abstract. Accessed 30 May 2014
93. Farndale RW et al (2008) Cell-collagen interactions: the use of peptide Toolkits to investigate collagen-receptor interactions. Biochem Soc Trans 36(Pt 2):241–250. http://www.ncbi.nlm.nih.gov/pubmed/18363567. Accessed 1 June 2014
94. Yeung D et al (2013) Oligomerization of DDR1 ECD affects receptor-ligand binding. J Struct Biol:1–6. http://www.ncbi.nlm.nih.gov/pubmed/23810922
95. Kadler KE, Hill A, Canty-Laird EG (2008) Collagen fibrillogenesis: fibronectin, integrins, and minor collagens as organizers and nucleators. Curr Opin Cell Biol 20(5):495–501. http://www.pubmedcentral.nih.gov/articlerender.fcgi?artid=2577133&tool=pmcentrez&rendertype=abstract. Accessed 3 June 2014
96. Ledger PW, Uchida N, Tanzer ML (1980) Immunocytochemical localization of procollagen and fibronectin in human fibroblasts: effects of the monovalent ionophore, monensin. J Cell Biol 87(3 Pt 1):663–71. http://www.pubmedcentral.nih.gov/articlerender.fcgi?artid=2110792&tool=pmcentrez&rendertype=abstract. Accessed 16 Nov 2014
97. Huang Y et al (2009) The collagen receptor DDR1 regulates cell spreading and motility by associating with myosin IIA. J Cell Sci 122(Pt 10):1637–1646. http://www.ncbi.nlm.nih.gov/pubmed/19401332. Accessed 14 Nov 2014
98. Xu H et al (2012) Discoidin domain receptors promote α1β1- and α2β1-integrin mediated cell adhesion to collagen by enhancing integrin activation. PloS One 7(12):e52209. http://www.pubmedcentral.nih.gov/articlerender.fcgi?artid=3527415&tool=pmcentrez&rendertype=abstract. Accessed 25 Nov 2014
99. Yeh Y-C, Wang C-Z, Tang M-J (2009) Discoidin domain receptor 1 activation suppresses alpha2beta1 integrin-dependent cell spreading through inhibition of Cdc42 activity. J Cell Physiol 218(1):146–156. http://www.ncbi.nlm.nih.gov/pubmed/18780290. Accessed 25 Nov 2014
100. Staudinger LA et al (2013) Interactions between the discoidin domain receptor 1 and β1 integrin regulate attachment to collagen. Biol Open 2(11):1148–1159. http://www.pubmedcentral.nih.gov/articlerender.fcgi?artid=3828761&tool=pmcentrez&rendertype=abstract. Accessed 17 Nov 2014
101. Mihai C et al (2009) Mapping of DDR1 distribution and oligomerization on the cell surface by FRET microscopy. J Mol Biol 385(2):432–445
102. Hulmes DJ et al (1981) Electron microscopy shows periodic structure in collagen fibril cross sections. Proc Natl Acad Sci U S A 78(6):3567–3571. http://www.pubmedcentral.nih.gov/articlerender.fcgi?artid=319611&tool=pmcentrez&rendertype=abstract. Accessed 7 Dec 2013

103. Cisneros DA et al (2006) Observing growth steps of collagen self-assembly by time-lapse high-resolution atomic force microscopy. J Struct Biol 154(3):232–245. http://www.ncbi.nlm.nih.gov/pubmed/16600632. Accessed 25 Nov 2014
104. Graham HK et al (2000) Identification of collagen fibril fusion during vertebrate tendon morphogenesis. The process relies on unipolar fibrils and is regulated by collagen-proteoglycan interaction. J Mol Biol 295(4):891–902. http://www.sciencedirect.com/science/article/pii/S0022283699933843. Accessed 27 Nov 2014
105. Ricard-Blum S (2011) The collagen family. Cold Spring Harb Perspect Biol 3(1):a004978. http://www.pubmedcentral.nih.gov/articlerender.fcgi?artid=3003457&tool=pmcentrez&rendertype=abstract. Accessed 21 May 2015
106. Kadler KE, Hojima Y, Prockop DJ (1990) Collagen fibrils in vitro grow from pointed tips in the C- to N-terminal direction. Biochem J 268(2):339–43. http://www.pubmedcentral.nih.gov/articlerender.fcgi?artid=1131437&tool=pmcentrez&rendertype=abstract. Accessed 23 Nov 2014
107. Birk DE et al (1997) Collagen fibrillogenesis in situ: fibril segments become long fibrils as the developing tendon matures. Dev Dyn 208(3):291–298. http://www.ncbi.nlm.nih.gov/pubmed/9056634. Accessed 27 Nov 2014
108. Kalamajski S, Oldberg A (2010) The role of small leucine-rich proteoglycans in collagen fibrillogenesis. Matrix Biol 29(4):248–253. http://www.ncbi.nlm.nih.gov/pubmed/20080181. Accessed 16 Nov 2014
109. Borcherding MS et al (1975) Proteoglycans and collagen fibre organization in human corneoscleral tissue. Exp Eye Res 21(1):59–70. http://www.ncbi.nlm.nih.gov/pubmed/124659. Accessed 25 Nov 2014
110. Zhu J, Kaufman LJ (2014) Collagen I self-assembly: revealing the developing structures that generate turbidity. Biophys J 106(8):1822–1831. http://www.ncbi.nlm.nih.gov/pubmed/24739181. Accessed 15 May 2014
111. Perumal S, Antipova O, Orgel JPRO (2008) Collagen fibril architecture, domain organization, and triple-helical conformation govern its proteolysis. Proc Natl Acad Sci U S A 105(8):2824–2829. http://www.pubmedcentral.nih.gov/articlerender.fcgi?artid=2268544&tool=pmcentrez&rendertype=abstract. Accessed 16 Nov 2014
112. Brinckmann J et al (1999) Overhydroxylation of lysyl residues is the initial step for altered collagen cross-links and fibril architecture in fibrotic skin. J Invest Dermatol 113(4):617–621. http://www.ncbi.nlm.nih.gov/pubmed/10504450. Accessed 8 Nov 2014
113. Sricholpech M et al (2011) Lysyl hydroxylase 3 glucosylates galactosylhydroxylysine residues in type I collagen in osteoblast culture. J Biol Chem 286(11):8846–8856. http://www.pubmedcentral.nih.gov/articlerender.fcgi?artid=3058983&tool=pmcentrez&rendertype=abstract. Accessed 8 Nov 2014
114. Seidler DG et al (2006) Defective glycosylation of decorin and biglycan, altered collagen structure, and abnormal phenotype of the skin fibroblasts of an Ehlers-Danlos syndrome patient carrying the novel Arg270Cys substitution in galactosyltransferase I (beta4GalT-7). J Mol Med (Berl) 84(7):583–594. http://www.ncbi.nlm.nih.gov/pubmed/16583246. Accessed 3 June 2014
115. Amudeswari S, Liang JN, Chakrabarti B (1987) Polar-apolar characteristics and fibrillogenesis of glycosylated collagen. Coll Relat Res 7(3):215–223. http://www.ncbi.nlm.nih.gov/pubmed/3652661. Accessed 8 Nov 2014
116. Rathi AN, Padmavathi S, Chandrakasan G (1989) Influence of monosaccharides on the fibrillogenesis of type I collagen. Biochem Med Metab Biol 42(3):209–215. http://www.ncbi.nlm.nih.gov/pubmed/2597436. Accessed 8 Nov 2014
117. Sini P et al (1997) Role of decorin on in vitro fibrillogenesis of type I collagen. Glycoconj J 14(7):871–874. http://www.ncbi.nlm.nih.gov/pubmed/9511994. Accessed 8 Nov 2014
118. Bhadriraju K et al (2009) The relative roles of collagen adhesive receptor DDR2 activation and matrix stiffness on the downregulation of focal adhesion kinase in vascular smooth muscle cells. Biomaterials 30(35):6687–6694. http://www.ncbi.nlm.nih.gov/pubmed/19762078. Accessed 8 Nov 2014

119. Wall SJ et al (2005) Discoidin domain receptor 2 mediates tumor cell cycle arrest induced by fibrillar collagen. J Biol Chem 280(48):40187–40194. http://www.jbc.org/content/280/48/40187.long. Accessed 9 June 2014
120. Sricholpech M et al (2012) Lysyl hydroxylase 3-mediated glucosylation in type I collagen: molecular loci and biological significance. J Biol Chem 287(27):22998–23009. http://www.pubmedcentral.nih.gov/articlerender.fcgi?artid=3391079&tool=pmcentrez&rendertype=abstract. Accessed 2 June 2014
121. Curat CA et al (2001) Mapping of epitopes in discoidin domain receptor 1 critical for collagen binding. J Biol Chem 276(49):45952–45958. http://www.ncbi.nlm.nih.gov/pubmed/11598108. Accessed 8 Nov 2014
122. Phan TN et al (2013) Low stability and a conserved N-glycosylation site are associated with regulation of the discoidin domain receptor family by glucose via post-translational N-glycosylation. Biosci Biotechnol Biochem 77(9):1907–1916. http://www.ncbi.nlm.nih.gov/pubmed/24018687. Accessed 8 Nov 2014
123. Fu H-L et al (2014) Glycosylation at Asn211 regulates the activation state of the discoidin domain receptor 1 (DDR1). J Biol Chem 289(13):9275–9287. http://www.pubmedcentral.nih.gov/articlerender.fcgi?artid=3979393&tool=pmcentrez&rendertype=abstract. Accessed 8 Nov 2014
124. Leitinger B (2003) Molecular analysis of collagen binding by the human discoidin domain receptors, DDR1 and DDR2. Identification of collagen binding sites in DDR2. J Biol Chem 278(19):16761–16769. http://www.ncbi.nlm.nih.gov/pubmed/12611880. Accessed 14 Nov 2014
125. Yamauchi M, Shiiba M (2008) Lysine hydroxylation and cross-linking of collagen. Methods Mol Biol 446:95–108. http://www.ncbi.nlm.nih.gov/pubmed/18373252. Accessed 8 Nov 2014
126. Siegel RC (1974) Biosynthesis of collagen crosslinks: increased activity of purified lysyl oxidase with reconstituted collagen fibrils. Proc Natl Acad Sci U S A 71(12):4826–4830. http://www.pubmedcentral.nih.gov/articlerender.fcgi?artid=433990&tool=pmcentrez&rendertype=abstract. Accessed 16 Nov 2014
127. Hornstra IK et al (2003) Lysyl oxidase is required for vascular and diaphragmatic development in mice. J Biol Chem 278(16):14387–14393. http://www.ncbi.nlm.nih.gov/pubmed/12473682. Accessed 29 Oct 2014
128. Mäki JM et al (2005) Lysyl oxidase is essential for normal development and function of the respiratory system and for the integrity of elastic and collagen fibers in various tissues. Am J Pathol 167(4):927–936. http://www.pubmedcentral.nih.gov/articlerender.fcgi?artid=1603668&tool=pmcentrez&rendertype=abstract. Accessed 14 Mar 2016
129. Wang L et al (2014) Lateral growth limitation of corneal fibrils and their lamellar stacking depend on covalent collagen cross-linking by transglutaminase-2 and lysyl oxidases, respectively. J Biol Chem 289(2):921–929. http://www.pubmedcentral.nih.gov/articlerender.fcgi?artid=3887215&tool=pmcentrez&rendertype=abstract. Accessed 21 Mar 2015
130. Herchenhan A et al (2015) Lysyl oxidase activity is required for ordered collagen fibrillogenesis by tendon cells. J Biol Chem 290(26):16440–16450. http://www.ncbi.nlm.nih.gov/pubmed/25979340. Accessed 14 Mar 2016
131. Kalamajski S et al (2016) Fibromodulin interacts with collagen cross-linking sites and activates lysyl oxidase. J Biol Chem 291:7951–7960. http://www.ncbi.nlm.nih.gov/pubmed/26893379. Accessed 26 Feb 2016
132. Khosravi R et al (2014) Collagen advanced glycation inhibits its Discoidin Domain Receptor 2 (DDR2)-mediated induction of lysyl oxidase in osteoblasts. Bone 58:33–41. http://www.ncbi.nlm.nih.gov/pubmed/24120383. Accessed 8 Nov 2014
133. Wang Z, Griffin M (2012) TG2, a novel extracellular protein with multiple functions. Amino Acids 42(2-3):939–949. http://www.ncbi.nlm.nih.gov/pubmed/21818567. Accessed 14 Mar 2016
134. Fortunati D et al (2014) Cross-linking of collagen I by tissue transglutaminase provides a promising biomaterial for promoting bone healing. Amino Acids 46(7):1751–1761. http://www.ncbi.nlm.nih.gov/pubmed/24710705. Accessed 14 Mar 2016

135. Slatter DA, Farndale RW (2015) Structural constraints on the evolution of the collagen fibril: convergence on a 1014-residue COL domain. Open Biol 5(5). http://www.pubmedcentral.nih.gov/articlerender.fcgi?artid=4450265&tool=pmcentrez&rendertype=abstract. Accessed 14 Jan 2016
136. Erickson B et al (2013) Nanoscale structure of type I collagen fibrils: quantitative measurement of D-spacing. Biotechnol J 8(1)117–126. http://www.pubmedcentral.nih.gov/articlerender.fcgi?artid=3535565&tool=pmcentrez&rendertype=abstract. Accessed 14 Mar 2016
137. Wallace JM et al (2011) Nanoscale morphology of Type I collagen is altered in the Brtl mouse model of Osteogenesis Imperfecta. J Struct Biol 173(1):146–152. http://www.pubmedcentral.nih.gov/articlerender.fcgi?artid=2997118&tool=pmcentrez&rendertype=abstract. Accessed 14 Mar 2016
138. Miyagawa A et al (2001) Surface ultrastructure of collagen fibrils and their association with proteoglycans in human cornea and sclera by atomic force microscopy and energy-filtering transmission electron microscopy. Cornea 20(6):651–656. http://www.ncbi.nlm.nih.gov/pubmed/11473170. Accessed 14 Mar 2016
139. Okuyama K et al (2012) Crystal structure of (Gly-Pro-Hyp)(9): implications for the collagen molecular model. Biopolymers 97(8):607–616. http://www.ncbi.nlm.nih.gov/pubmed/22605552. Accessed 1 Dec 2013
140. Orgel JP et al (2001) The in situ supermolecular structure of type I collagen. Structure 9(11):1061–1069
141. Doucet J et al (2011) Modeling the lateral organization of collagen molecules in fibrils using the paracrystal concept. J Struct Biol 173(2):197–201. http://www.ncbi.nlm.nih.gov/pubmed/21111050. Accessed 3 Dec 2013
142. Gumez L et al (2010). Molecular structure of tail tendon fibers in TIEG1 knockout mice using synchrotron diffraction technology. J Appl Physiol (1985) 108(6):1706–1710. http://www.pubmedcentral.nih.gov/articlerender.fcgi?artid=3774169&tool=pmcentrez&rendertype=abstract. Accessed 3 Dec 2013
143. An B, Brodsky B (2016) Collagen binding to OSCAR: the odd couple. Blood 127(5):521–522. http://www.ncbi.nlm.nih.gov/pubmed/26847065. Accessed 5 Feb 2016
144. Krady MM et al (2008) Thrombospondin-2 modulates extracellular matrix remodeling during physiological angiogenesis. Am J Pathol 173(3)–879–891. http://www.pubmedcentral.nih.gov/articlerender.fcgi?artid=2527070&tool=pmcentrez&rendertype=abstract. Accessed 27 May 2015
145. Jaffe R, Deykin D (1974) Evidence for a structural requirement for the aggregation of platelets by collagen. J Clin Invest 53(3):875–883. http://www.pubmedcentral.nih.gov/articlerender.fcgi?artid=333070&tool=pmcentrez&rendertype=abstract. Accessed 19 Apr 2015
146. Sylvester MF et al (1989) Collagen banded fibril structure and the collagen-platelet reaction. Thromb Res 55(1):135–148. http://www.ncbi.nlm.nih.gov/pubmed/2506660. Accessed 19 Apr 2015
147. Salsas-Escat R, Stultz CM (2010) Conformational selection and collagenolysis in type III collagen. Proteins 78(2):325–335. http://www.ncbi.nlm.nih.gov/pubmed/19731369. Accessed 16 Nov 2014
148. Sarkar SK et al (2012) Single-molecule tracking of collagenase on native type I collagen fibrils reveals degradation mechanism. Curr Biol 22(12):1047–1056. http://www.pubmedcentral.nih.gov/articlerender.fcgi?artid=3382001&tool=pmcentrez&rendertype=abstract. Accessed 20 Nov 2014
149. Nudelman F et al (2013) In vitro models of collagen biomineralization. J Struct Biol 183(2):258–269. http://www.ncbi.nlm.nih.gov/pubmed/23597833. Accessed 19 Nov 2014
150. Addison WN et al (2015) Extracellular matrix mineralization in murine MC3T3-E1 osteoblast cultures: an ultrastructural, compositional and comparative analysis with mouse bone. Bone 71:244–256. http://www.ncbi.nlm.nih.gov/pubmed/25460184. Accessed 13 Feb 2015
151. Nudelman F et al (2010) The role of collagen in bone apatite formation in the presence of hydroxyapatite nucleation inhibitors. Nat Mater 9(12):1004–1009. http://www.pubmedcentral.nih.gov/articlerender.fcgi?artid=3084378&tool=pmcentrez&rendertype=abstract. Accessed 13 Feb 2015

152. Xu Z et al (2015) Molecular mechanisms for intrafibrillar collagen mineralization in skeletal tissues. Biomaterials 39:59–66. http://www.ncbi.nlm.nih.gov/pubmed/25477172. Accessed 17 Feb 2015
153. Li Y, Aparicio C (2013) Discerning the subfibrillar structure of mineralized collagen fibrils: a model for the ultrastructure of bone. PloS one 8(9):e76782. http://www.pubmedcentral.nih.gov/articlerender.fcgi?artid=3781166&tool=pmcentrez&rendertype=abstract. Accessed 13 Feb 2015
154. Rodriguez DE et al (2014) Multifunctional role of osteopontin in directing intrafibrillar mineralization of collagen and activation of osteoclasts. Acta Biomater 10(1):494–507. http://www.pubmedcentral.nih.gov/articlerender.fcgi?artid=3864794&tool=pmcentrez&rendertype=abstract. Accessed 13 Feb 2015
155. Deshpande AS et al (2011) Primary structure and phosphorylation of dentin matrix protein 1 (DMP1) and dentin phosphophoryn (DPP) uniquely determine their role in biomineralization. Biomacromolecules 12(8):2933–2945. http://www.pubmedcentral.nih.gov/articlerender.fcgi?artid=3171794&tool=pmcentrez&rendertype=abstract. Accessed 13 Feb 2015
156. He G, George A (2004) Dentin matrix protein 1 immobilized on type I collagen fibrils facilitates apatite deposition in vitro. J Biol Chem 279(12):11649–11656. http://www.ncbi.nlm.nih.gov/pubmed/14699165. Accessed 3 Feb 2015
157. Cocking-Johnson D, Sauk JJ (1983) The interaction of bovine dentine phosphophoryn and collagen during fibrillogenesis of collagen in vitro. Biochim Biophys Acta 742(1):49–53. http://www.ncbi.nlm.nih.gov/pubmed/6824686. Accessed 29 May 2015
158. Satoh M et al (2005) Myocardial osteopontin expression is associated with collagen fibrillogenesis in human dilated cardiomyopathy. Eur J Heart Fail 7(5):755–762. http://www.ncbi.nlm.nih.gov/pubmed/16087132. Accessed 29 May 2015
159. Mochida Y et al (2009) Decorin modulates collagen matrix assembly and mineralization. Matrix Biol 28(1):44–52. http://www.pubmedcentral.nih.gov/articlerender.fcgi?artid=2650102&tool=pmcentrez&rendertype=abstract. Accessed 29 May 2015
160. Ueno A et al (2006) Constitutive expression of thrombospondin 1 in MC3T3-E1 osteoblastic cells inhibits mineralization. J Cell Physiol 209(2):322–332. http://www.ncbi.nlm.nih.gov/pubmed/16883596. Accessed 29 May 2015
161. Lin K-L et al (2010) Transcriptional upregulation of DDR2 by ATF4 facilitates osteoblastic differentiation through p38 MAPK-mediated Runx2 activation. J Bone Miner Res 25(11):2489–2503. http://www.ncbi.nlm.nih.gov/pubmed/20564243. Accessed 16 June 2014
162. Ahmad PJ et al (2009) Discoidin domain receptor-1 deficiency attenuates atherosclerotic calcification and smooth muscle cell-mediated mineralization. Am J Pathol 175(6):2686–2696. http://www.pubmedcentral.nih.gov/articlerender.fcgi?artid=2789613&tool=pmcentrez&rendertype=abstract. Accessed 18 Feb 2015
163. Bargal R et al (2009) Mutations in DDR2 gene cause SMED with short limbs and abnormal calcifications. Am J Hum Genet 84(1):80–84. http://www.pubmedcentral.nih.gov/articlerender.fcgi?artid=2668047&tool=pmcentrez&rendertype=abstract. Accessed 18 Feb 2015
164. Ali BR et al (2010) Trafficking defects and loss of ligand binding are the underlying causes of all reported DDR2 missense mutations found in SMED-SL patients. Hum Mol Genet 19(11):2239–2250. http://www.pubmedcentral.nih.gov/articlerender.fcgi?artid=2865377&tool=pmcentrez&rendertype=abstract. Accessed 25 Jan 2015
165. Al-Kindi A et al (2014) A novel mutation in DDR2 causing spondylo-meta-epiphyseal dysplasia with short limbs and abnormal calcifications (SMED-SL) results in defective intracellular trafficking. BMC Med Genet 15:42. http://www.pubmedcentral.nih.gov/articlerender.fcgi?artid=4001364&tool=pmcentrez&rendertype=abstract. Accessed 27 Jan 2015
166. Sivakumar L, Agarwal G (2010) The influence of discoidin domain receptor 2 on the persistence length of collagen type I fibers. Biomaterials 31(18):4802–4808
167. Stuart K, Panitch A (2008) Influence of chondroitin sulfate on collagen gel structure and mechanical properties at physiologically relevant levels. Biopolymers 89(10)841–851. http://www.ncbi.nlm.nih.gov/pubmed/18488988. Accessed 19 Apr 2015
168. Gautieri A et al (2009) Molecular and mesoscale mechanisms of osteogenesis imperfecta disease in collagen fibrils. Biophys J 97(3):857–865. http://www.pubmedcentral.nih.gov/articlerender.fcgi?artid=2718154&tool=pmcentrez&rendertype=abstract. Accessed 10 Feb 2015

169. Leitinger B, Kwan APL (2006) The discoidin domain receptor DDR2 is a receptor for type X collagen. Matrix Biol 25(6):355–364. http://www.ncbi.nlm.nih.gov/pubmed/16806867. Accessed 20 Feb 2015
170. Parkin J, Des et al (2011) Mapping structural landmarks, ligand binding sites, and missense mutations to the collagen IV heterotrimers predicts major functional domains, novel interactions, and variation in phenotypes in inherited diseases affecting basement membranes. Hum Mutat 32(2):127–143. http://www.ncbi.nlm.nih.gov/pubmed/21280145. Accessed 2 June 2015
171. Hohenester E et al 2008 Structural basis of sequence-specific collagen recognition by SPARC. Proc Natl Acad Sci U S A 105(47):18273–18277. http://www.pubmedcentral.nih.gov/articlerender.fcgi?artid=2587565&tool=pmcentrez&rendertype=abstract. Accessed 2 June 2015
172. Xu L et al (2005) Activation of the discoidin domain receptor 2 induces expression of matrix metalloproteinase 13 associated with osteoarthritis in mice. J Biol Chem 280(1)548–555. http://www.ncbi.nlm.nih.gov/pubmed/15509586. Accessed 22 Feb 2015
173. Xu L et al (2010) Attenuation of osteoarthritis progression by reduction of discoidin domain receptor 2 in mice. Arthritis Rheum 62(9):2736–2744. http://www.pubmedcentral.nih.gov/articlerender.fcgi?artid=2946478&tool=pmcentrez&rendertype=abstract. Accessed 22 Feb 2015
174. Gross O et al (2004) DDR1-deficient mice show localized subepithelial GBM thickening with focal loss of slit diaphragms and proteinuria. Kidney Int 66(1):102–111. http://www.ncbi.nlm.nih.gov/pubmed/15200417. Accessed 22 Feb 2015
175. Vonk LA et al (2011) Collagen-induced expression of collagenase-3 by primary chondrocytes is mediated by integrin α1 and discoidin domain receptor 2: a protein kinase C-dependent pathway. Rheumatology (Oxford) 50(3):463–472. http://www.ncbi.nlm.nih.gov/pubmed/21075784. Accessed 22 Feb 2015
176. Hou G, Vogel W, Bendeck MP (2001) The discoidin domain receptor tyrosine kinase DDR1 in arterial wound repair. J Clin Invest 107(6):727–735
177. Lee R et al (2004) Localization of discoidin domain receptors in rat kidney. Nephron Exp Nephrol 97(2):e62–e70

# Chapter 3
# DDR Structural Biology

**Noritaka Nishida and Ichio Shimada**

## 3.1 Introduction

Receptor tyrosine kinases (RTKs) regulate fundamental cellular processes, such as cell proliferation, differentiation, migration, and metabolism. The activation of RTKs is triggered by ligand binding to the extracellular domain, which causes the autophosphorylation of cytoplasmic kinase domains, thereby leading to downstream signaling [1]. Many types of cancers are caused by aberrant activation of RTKs, induced by mutagenesis or receptor overexpression [2].

While most RTK families are activated by soluble proteins, the discoidin domain receptor (DDR) family, DDR1 and DDR2, is activated by collagen fibrils, a major component of the extracellular matrix [3, 4]. Both DDR1 and DDR2 consist of an extracellular discoidin (DS) domain and DS-like domain and an intracellular tyrosine kinase, connected by a transmembrane (TM) region (Fig. 3.1). The direct interaction of collagen with the DS domain triggers the activation of its intracellular tyrosine kinase, leading to downstream intracellular signaling [5–7]. The activation of DDRs results in the regulation of cell proliferation, migration, and extracellular matrix remodeling by upregulating both the expression and activity of matrix metalloproteinases [8–10]. The collagen-induced activation of DDRs reportedly occurs on a much slower time scale (in hours), as compared to the other RTKs (in seconds) [3, 4]. Several splice variants have been reported for DDR1: DDR1a, DDR1b, and DDR1c are the full-length functional form, whereas DDR1d, DDR1e encode truncated or kinase-inactive receptors. No splice variants are known for DDR2. In this chapter, we describe the structural information of the DDRs.

N. Nishida • I. Shimada (✉)
Graduate School of Pharmaceutical Sciences, The University of Tokyo,
Hongo 7-3-1, Bunkyo-ku, Tokyo 113-0033, Japan
e-mail: shimada@iw-nmr.f.u-tokyo.ac.jp



R. Fridman, P.H. Huang (eds.), *Discoidin Domain Receptors in Health and Disease*, DOI 10.1007/978-1-4939-6383-6_3

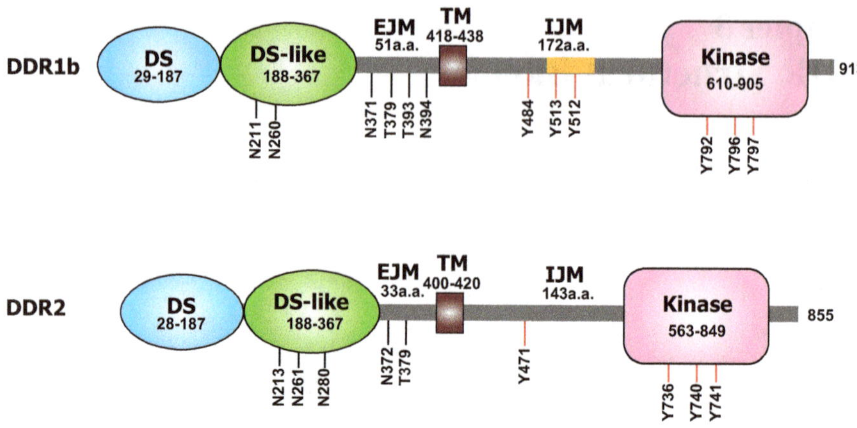


**Fig. 3.1** The Domain architectures of DDR1 and DDR2. The domain architectures of DDR1b and DDR2 are schematically shown. The positions of the predicted *N*- and *O*-glycosylation sites and phosphorylation sites are indicated by *black* and *red lines*, respectively, in the diagrams. The DDR1a splice variant lacks residues 506–542 (shown in *orange*) that include the NPXY motif

### *3.1.1 Structure of the DS Domain*

The three-dimensional structure of the DDR2-DS domain in the free state was determined, using NMR techniques [11] (Fig. 3.2a). The DDR2-DS domain adopts the β-barrel structure known as the jelly roll fold, consisting of eight major β-strands, β1–β8 (β1, 46–49; β2, 86–90; β3, 99–104; β4, 117–123; β5, 140–142; β6, 151–154; β7, 163–169; and β8, 178–185), which are arranged in two antiparallel β-sheets of five (β1–β2–β7–β4–β5) and three (β8–β3–β6) strands packed against each other (Fig. 3.2a). As predicted from its primary sequence, the overall fold is common to other DS domains (also termed the factor 5/8 type C fold) with known structures: the coagulation factors V and VIII (F5/8) C2 domains [12, 13] and the neuropilin-1 (Npn-1) b1 domain [14]. Six juxtaposed loops (L1–L6) protrude from the 'top' of the β-barrel core, forming a collagen-binding site, as described below. L2 and L6 are linked by a disulfide bridge between Cys73 and Cys177 (Fig. 3.2a). At the 'bottom' of the β-barrel core, a flat surface is formed, where the N- and C-termini are connected by a disulfide bond between Cys30and Cys185 (Fig. 3.2a). The structure of the entire DDR1 ectodomain was revealed by X-ray crystallography [15] (Fig. 3.2b). As compared to the DDR2-DS domain, the DDR1-DS domain adopts an almost identical β-barrel structure, including two intradomain disulfide bonds.

### *3.1.2 Structure of the DS-Like Domain*

Following the DS domain, another structural domain exists that is structurally similar to the DS domain, called the DS-like domain. The three-dimensional structure of the DS-like domain was revealed in the structure of the DDR1 ectodomain in

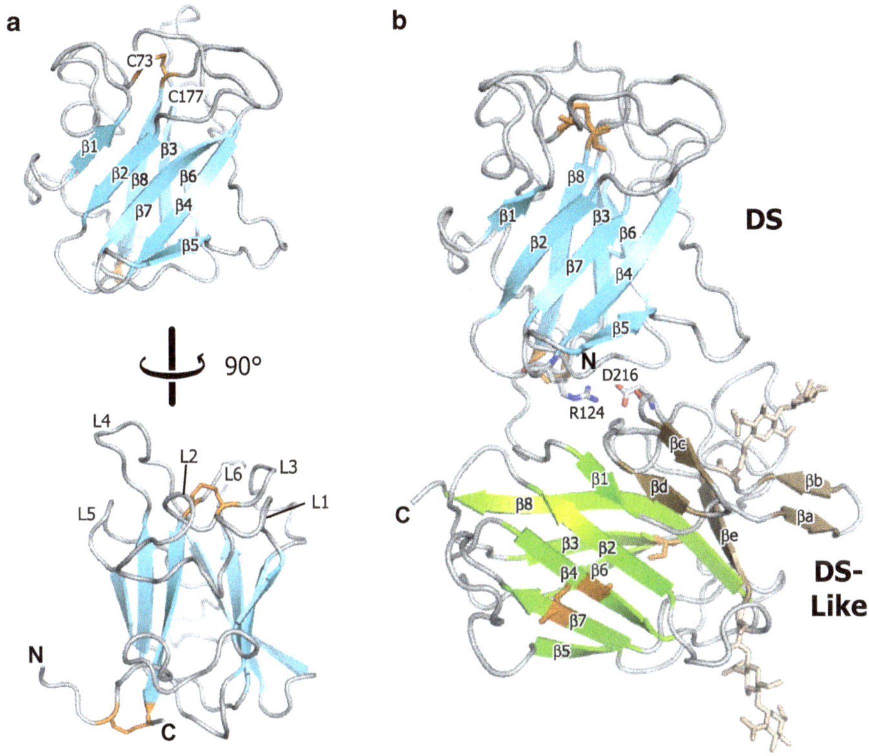


**Fig. 3.2** Structure of the DDR ectodomain. (**a**) The NMR structure of the DDR2-DS domain (PDB ID: 2Z4F). The lower panel represents a 90° rotation with respect to the upper panel. The two intradomain disulfide bonds are shown by *stick models*. The eight β-strands (β1–β8) and the six loops (L1–L6) are labeled. (**b**) The crystal structure of the entire ectodomain of DDR1 (PDB ID: 4AG4). The β-strands of the DS domain and the DS-like domain are colored *cyan* and *green*, respectively. The β-strands of the insertion loop of the DS-like domain are colored *brown*. The *N*-glycans in the DS-like domain are shown by *stick models*. The residues forming a salt bridge between the DS and DS-like domains (R124 and D216) are shown by *stick models*

complex with the Fab fragment of an inhibitory antibody [15] (Fig. 3.2b). Despite the low sequence similarity with the DS domain (only 9 %), the DS-like domain also adopts a similar eight-stranded β-barrel fold with a jellyroll topology (Fig. 3.2b). In addition, the DS-like domain contains an extra segment inserted between strands β1 and β2, containing five β-strands (denoted as βa to βe), two *N*-glycosylation sites, and a calcium-binding site (Fig. 3.2b). The DS-like domain contains an internal disulfide bond between strands β4 and β7 (Fig. 3.2b), but lacks the disulfide bond that connects the N- and C-termini of the domain. An extensive interface is formed between the bottom of the DS domain and the long insertion segment of the DS-like domain. The residues involved in the interdomain interface are mostly conserved in all vertebrate DDRs, including those forming a salt bridge (Arg124 of the DS domain and Asp216 of the DS-like domain), suggesting that the orientation of the DS and DS-like domains is stable and they behave as a single unit.

The extracellular juxtamembrane (EJM) segments (51 residues for DDR1 and 33 residues for DDR2) are inserted between the DS-like domain and the TM segment. The EJM region of DDR1 contains 12 prolines and a number of predicted *N*- and *O*-linked glycosylation sites. Assuming that the EJM segment is fully extended, the DS and DS-like domains of DDR1 would protrude by ~150 Å above the cell surface.

### 3.1.3 Collagen Recognition of the DS Domain

The primary ligand of the DDR is fibrillar collagen, a major component of extracellular matrices. Collagen is composed of either a homo- or heterotrimer of the collagen α chain which forms a long triple-helical structure (polyproline II-like helices). Under neutral pH condition, each monomeric triple-helical collagen self-assembles into large fibrillar supramolecules. The α-chain of collagen contains the Gly-X-Y triplet repeat to form a triple-helical structure, where proline at the X position and 4-hydroxyproline (Hyp) at the Y position are most thermally stable, and thus frequently seen in the native collagen sequence. Both DDR1 and DDR2 are activated by the fibrillar collagens, such as types I, II, and III [3, 4]. However, the nonfibrillar collagen types IV and X distinctively activate DDR1 and DDR2, respectively [16].

Collagen binding by DDR is exclusively mediated by the DS domain. The collagen-recognition site of the DS domain was investigated by an NMR method [11], termed the transferred cross saturation (TCS) experiment [17, 18]. This method observes the saturation transfer from the collagen fibril to the DS domain at the binding interface. Therefore, significant reductions of NMR signal intensities would be observed for the residues at the collagen-binding site (Fig. 3.3a). Such reductions were observed for Trp52, Arg105, Ile112, Glu113, Met174, Asn175, and Val176 (Red in Fig. 3.3b). Mapping of these residues on the structure of the DDR2-DS domain showed that they form a contiguous surface at the 'top' of the DDR2-DS domain in L1, L3, L4, and L6, indicating that this surface is the binding interface for collagen in the fibril form.

In another approach, the specific binding sequence of DDR1 and DDR2 was explored by using a systemic collagen peptide library, known as a "collagen Toolkit" [19]. Both DDRs recognize the triple-helical collagen peptide containing "GVMGFO," which is commonly found in the collagen types I–III [20]. In addition, DDR2 exhibits higher affinity for the modified collagen peptide, in which the methionine is replaced by the isosteric amino acid norleucine (Nle). The DDR2-DS domain was co-crystallized with the synthetic peptide; Ac-$(GPO)_3$GPRGQOGVNleGFO$(GPO)_2$G-$NH_2$, and the crystal structure was obtained at 1.6 Å resolution [21] (Fig. 3.3c, d). The overall conformation of the DS domain is mostly similar to the solution structure of the DDR2 in the free state [11], indicating no major conformational changes were induced upon ligand binding to the DS-domain.

In the crystal structure, the collagen peptide interacts with the 'top' of the DS domain, burying 530 $Å^2$ of the solvent-accessible collagen surface, with good shape complementarity. The location of the collagen-binding site at the top of the DDR2-DS domain agrees well with the results of NMR studies which used the

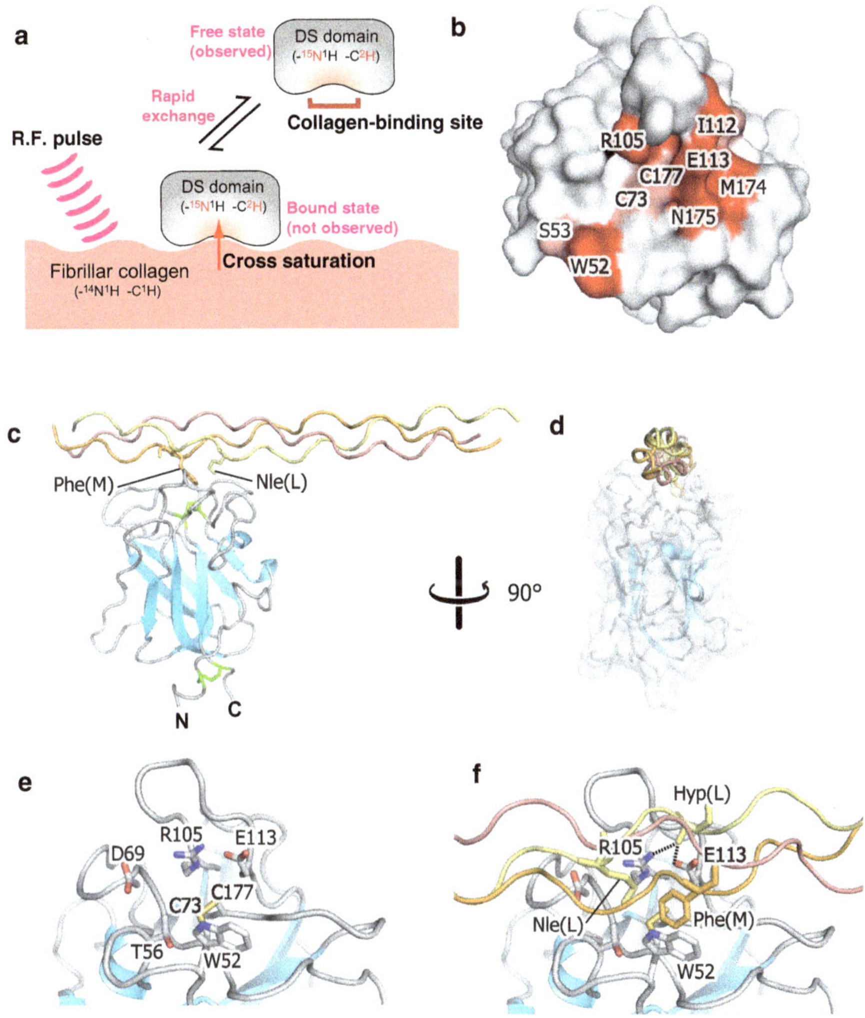


**Fig. 3.3** Collagen-binding mode of the DDR-DS domain. (**a**) A schematic representation of the transferred cross saturation experiment. The uniformly $^{2}H$, $^{15}N$-labeled DDR2-DS domain was mixed with nonlabeled type II collagen in the fibril form. The saturation of the collagen fibril by the radio frequency pulse is transmitted to the DS domain in the bound state by the cross saturation phenomena. Since the DS domain undergoes exchange between the free and collagen-bound state in a fast exchange regime, the effect of saturation can be detected as reduction of the signal intensity of the DDR2-DS domain in the free state. (**b**) The residues with significant (*red*) and moderate (*light red*) intensity reductions are shown on the surface representation of the DDR2 DS domain. (**c**) Crystal structure of the DDR2-DS domain in complex with the collagen peptide containing GPRGQOGVXGFO (X denotes norleucine (Nle)) (PDB code: 2WUH). Cartoon representation of the DS domain (*cyan*) and the collagen peptide (the leading chain: *yellow*, the middle chain: *orange*, and the trailing chain: *red*). Phe in the middle chain and Nle in the leading chain are shown by *stick models*. (**d**) A 90° rotation of (**c**). The transparent surface representation of the DS domain is also shown. (**e**, **f**) Close-up views of the collagen-binding site of the DS domain. In (**e**), the coordinate of the collagen is omitted to show the residues involved in collagen recognition (*stick model*). The hydrogen bond formed between the R104-E113 salt bridge and the hydroxyproline (Hyp) in the leading chain is shown by the *dashed line* in (**f**)

collagen fibril [11]. This result suggested that DDR recognizes the single triple helix exposed on the surface of the collagen fibril. Among the three chains of a collagen peptide (termed the leading, middle, and trailing chains), only the leading and middle chains make contact with the DS domain, whereas no contact was made by the trailing chain (Fig. 3.3c). At the interface, the Nle residue of the leading chain and the Phe residue of the middle chain are accommodated by the amphiphilic pocket, created by the L1, L2, L4, and L6 loops at the top of the DS domain (Fig. 3.3e). The surface profiles of one side of the pocket and the floor are hydrophobic (Trp52, Thr56, and Cys73-Cys177), whereas the other side of the pocket is occupied by charged residues (Asp69, Arg105, and Glu113). The Arg105 and Glu113 residues form hydrogen bond with the hydroxyl group of the hydroxyproline of the leading chain (Fig. 3.3f). These residues are strictly conserved between DDR1 and DDR2, indicating that the same recognition site is formed for DDR1. The importance of the interactions seen in the crystal structure is supported by a previous biochemical study, showing that M, F, and O of the GVMGFO motif are critical for DDR2 binding [20]. It should be noted that the Arg residue, which is four residues upstream of the GVMGFO motif, does not form any interactions with the DS domain in the crystal structure, although it was required for DDR2 signaling in biochemical studies [20]. Therefore, an alternative binding mode may be possible, in which the Arg residue in the recognition sequence directly contacts the DS domain.

### *3.1.4 Convergent Evolution of GVMGFO-Binding Sites*

Recently, it was found that the GVMGFO motif is utilized by not only the DDR2-DS domain, but also other collagen-binding proteins, such as the A3 domain of von Willebrand factor (vWF) [22] and the matricellular protein SPARC [23], although no structural similarity exists between those proteins. As described in the previous section, the DDR2 collagen-binding sites are formed by the multiple loops protruded from the β-barrel structure [21] (Fig. 3.4a). The GVMGFO-binding site of SPARC is created by a long α helix and an adjacent helical hairpin [23] (Fig. 3.4b). The GVMGFO-binding site of the A3 domain is located at the front face of the domain, which forms a concave surface to accommodate collagen the triple helix [22] (Fig. 3.4c). Despite their structural dissimilarity, all of these proteins possess a similar amphiphilic pockets that can accommodate the critical Phe side chain (the middle chain in DDR2, the trailing chain in SPARC, and the leading chain in the vWF-A3 domain) (Fig. 3.4d–f). The collagen recognition sequence of the α2 integrin I domain also includes a Phe residue (GFOGER motif) [24].

Although Phe residues are less frequently seen in the collagen primary sequence (8 in 999 residues of the α1(III) triple-helical region), it is inferred that many collagen-binding proteins utilize the Phe residue to acquire the sequence specificity, as a result of converged evolution.

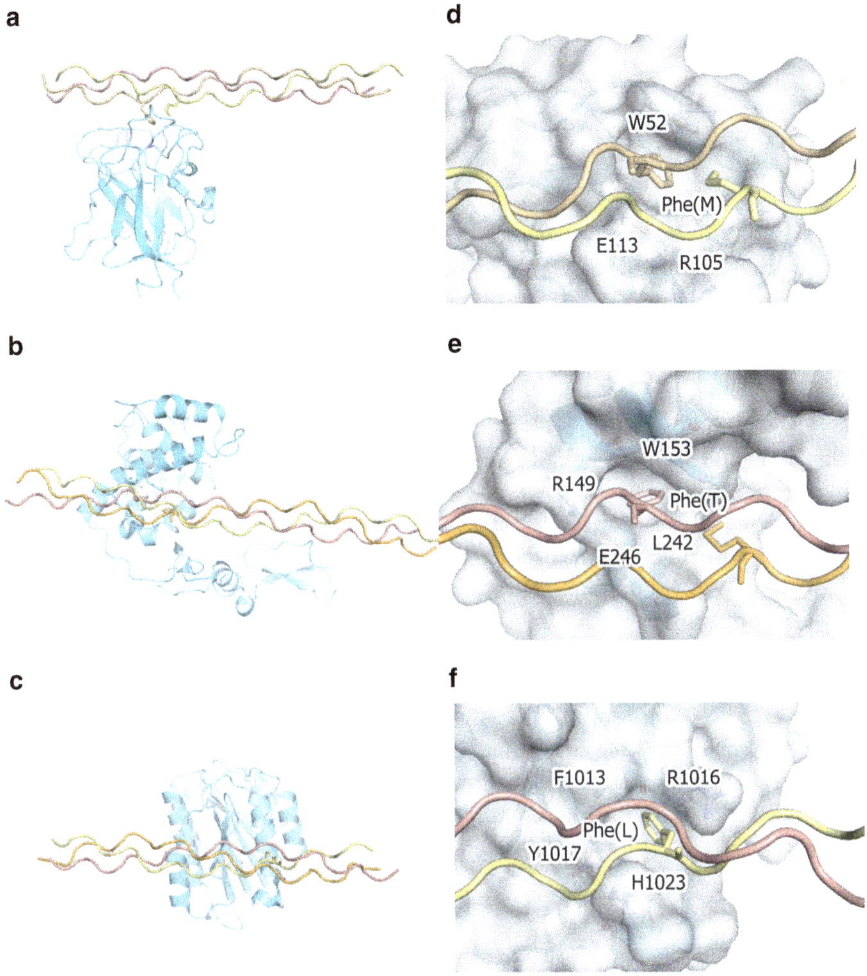


**Fig. 3.4** Recognition of the GVMGFO motif by the structurally unrelated proteins. (**a**–**c**) Crystal structures of (**a**) the DDR2-DS domain (PDB ID: 2WUH), (**b**) SPARC (PDB ID: 2 V53), and (**c**) the vWF-A3 domain (PDB ID: 4DMU) in complex with a synthetic collagen peptide containing the GVMGFO sequence. The DS domain, SPARC, and the A3 domain are shown by *ribbon diagrams*. The collagen peptide is shown by a *tube model* (the leading chain: *yellow*, the middle chain: *orange*, and the trailing chain: *red*) (**d**–**f**). Close-up views of the phenylalanine pocket of (**d**) the DDR2-DS domain, (**e**) SPARC, and (**f**) the vWF-A3 domain. The collagen chains that are not involved in interactions are omitted for clarity

## 3.2 Transmembrane and Intracellular Domains

The transmembrane segment of the DDRs is composed of a single-pass α helix. The DDRs reportedly form a pre-formed dimer on the cell surface in the absence of ligand [25, 26], and the TM helix has a strong propensity to form a dimer [27]. In addition, it was also shown that the TM helix association, through its leucine zipper motif, is necessary for DDR activation [25]. However, the structure detail of the preformed dimer, and its functional significance, remain essentially unknown.

DDRs contain an usually long intracellular juxtamembrane (IJM) region between the TM and kinase domains. There are several tyrosine phosphorylation sites in the IJM region, which serves as a docking site for downstream adaptor molecules. The splicing variants of DDR1 (DDR1b and DDR1c) contain an NPXY motif that recruits the adaptor proteins with the phosphotyrosine-binding domains, such as ShcA [4].

The kinase domain of DDR is pivotal to activate downstream signaling. As shown in Fig. 3.5a, the DDR1 kinase domain adopts a typical protein kinase fold, composed of the N-terminal lobe (N-lobe) and the C-terminal lobe (C-lobe) [28, 29]. The N-lobe is formed by a five stranded antiparallel β-sheet (β1–β5) and an α helix (αC helix), whereas the C-lobe is mainly composed of α helices. The ATP-binding site is located at the cleft between the N- and C-lobes. The activation loop that stems from the C-lobe plays an essential role in kinase activation. The activation loop contains a conserved "DFG" motif (Asp784-Phe785-Gly786) at its N-terminus, and the tyrosine residues (Tyr792, Tyr796, and Tyr797) at the middle portion are phosphorylated upon kinase activation. It is known that the activation loop adopts either an "open" or "closed" conformation, corresponding to active or the inactive state of the kinase, respectively. In the active state, the aspartate residue in the DFG motif points "in" toward the ATP-binding site ("DFG-in" conformation), whereas it flips out in the inactive state ("DFG-out" conformation). The crystal structures of the kinase domain of DDR1 were solved in complex with inhibitor compounds; imatinib and ponatinib (Fig. 3.5b) [28, 29], which are also known as an inhibitor for the ABL (Abelson) kinase for the treatment of chronic myeloid leukemia (CML). As similar to the ABL kinase, the DDR1 kinase domain adopts the inactive, DFG-out conformation in the imatinib-bound state (Fig. 3.5a). The gatekeeper residue Thr701 in the N-lobe, which confers resistance to the imatinib inhibition, forms a hydrogen bond with the aminopyrimidine in the head group of imatinib. In contrast, Thr701 did not form a hydrogen bond in the ponatinib-bound state, providing the structural explanation why the mutation in the gatekeeper residue does not affect the ponatinib inhibition. Whereas imatinib and ponatinib potently bind to both the DDR1 and ABL kinase, the DDR-IN-1 is selective for DDR1 ($IC_{50}$ = 105 nM) relative to ABL ($IC_{50}$ = 1.8 μM). Although overall conformation is similar between the ABL and DDR1 kinase, a major difference was seen in the conformation of the P-loop in the N-lobe, which makes a van der Waals contact with the pyridine group of the imatinib only in the ABL kinase. The loss of this interaction for the ABL kinase would explain the selectivity of DDR1-IN-1 for DDR1 kinase.

## 3.3 Perspectives

Thus, the isolated individual domains of DDRs have been characterized at atomic resolution. In addition, the structural basis by which each DS domain recognizes the specific sequence has also been clarified by X-ray and NMR studies. On the other

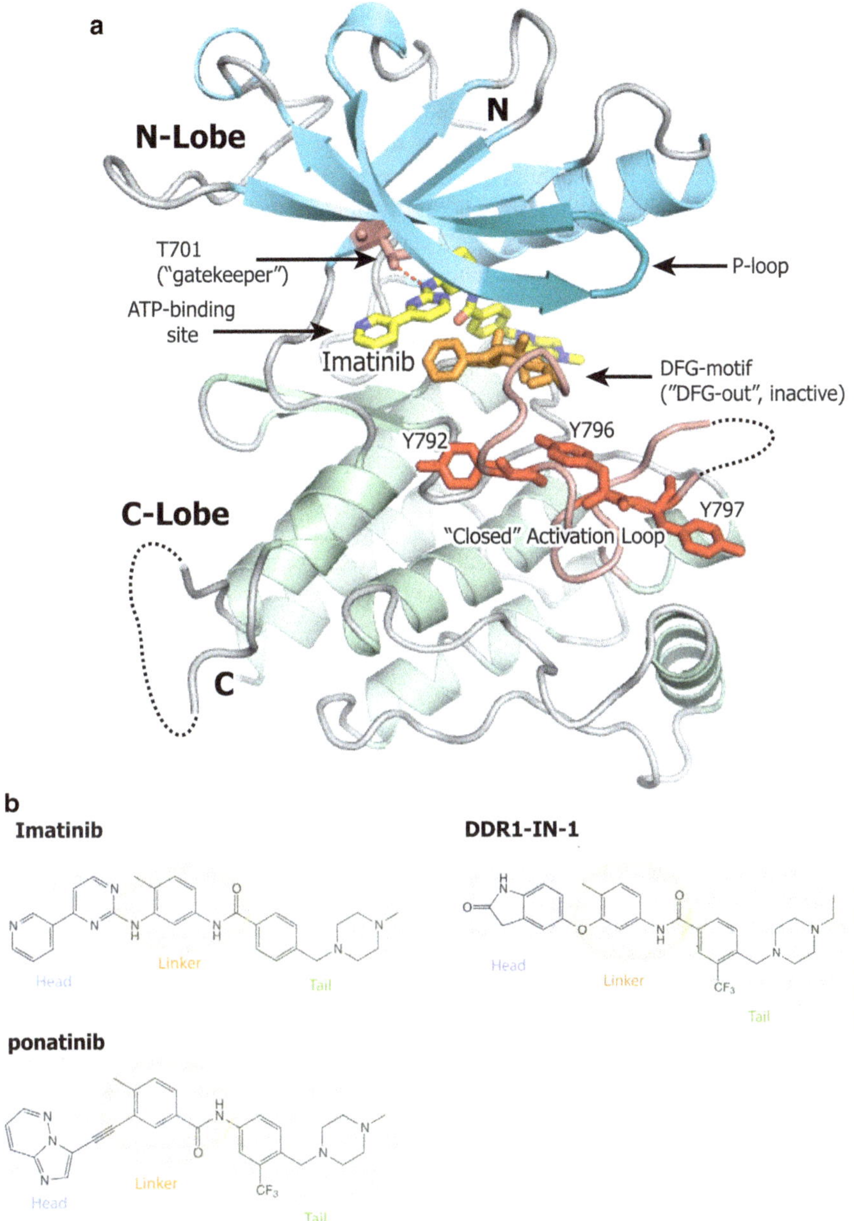


**Fig. 3.5** Structure of the DDR1 kinase domain. (**a**) Crystal structure of the DDR1 kinase domain, in complex with the type II inhibitor imatinib (PDB ID: 4BKJ). The N-lobe and the C-lobe are colored *cyan* and *green*, respectively. The inhibitor is shown by a *yellow stick*. The phosphorylation sites of the activation loop (*light red*) are shown by *red sticks*. The DFG motif and gatekeeper residues are shown by *stick models*. (**b**) Chemical structures of imatinib, ponatinib, and DDR1-IN-1. "Head," "Linker," and "Tail" segments are highlighted with different colors [29]

hand, the structural mechanisms underlying DDR activation are still mostly elusive. In the cases of other RTKs that recognize the soluble ligands, the activation mechanism has been well established: the ligand binding to the extracellular domain causes conformational changes that induce the receptor dimerization, thereby enabling the intracellular kinase domain to adopt an active conformation [1]. However, this mechanism cannot be applied to DDRs, which recognize the insoluble collagen molecule. In addition, no conformational change was observed between the DDR2-DS domain in the free state and that in the complex with a collagen peptide [21], suggesting that ligand binding does not induce an allosteric conformational change that would transmit a signal toward the TM and intracellular domains. In the crystal structure of the DDR1 ectodomain in complex with its inhibitory antibody, the long insertion loop of the DS-like domain of one monomer contacted the side face of the DS domain of the neighboring monomer in the crystal lattice [15]. Interestingly, mutations of the DS domain residues (R32E/L152E), but not those of the DS-like domain residues (L247E/R248E), abolished the collagen-induced activation of DDR1, suggesting that the lateral interaction through the patch in the DS domain is required to form the signaling complex. The importance of the lateral interaction of DDRs is also expected, because the DDR would accumulate on the collagen ligand, on which the DDR binding site is densely distributed. Further structural and biochemical experiments will be necessary to clarify the activation mechanism of the DDRs, which should be distinct from the other RTKs.

## References

1. Schlessinger J (2000) Cell signaling by receptor tyrosine kinases. Cell 103(2):211–225
2. Krause DS, Van Etten RA (2005) Tyrosine kinases as targets for cancer therapy. N Engl J Med 353(2):172–187
3. Shrivastava A, Radziejewski C, Campbell E, Kovac L, McGlynn M, Ryan TE, Davis S, Goldfarb MP, Glass DJ, Lemke G, Yancopoulos GD (1997) An orphan receptor tyrosine kinase family whose members serve as nonintegrin collagen receptors. Mol Cell 1(1):25–34
4. Vogel W, Gish GD, Alves F, Pawson T (1997) The discoidin domain receptor tyrosine kinases are activated by collagen. Mol Cell 1(1):13–23
5. Ikeda K, Wang LH, Torres R, Zhao H, Olaso E, Eng FJ, Labrador P, Klein R, Lovett D, Yancopoulos GD, Friedman SL, Lin HC (2002) Discoidin domain receptor 2 interacts with Src and Shc following its activation by type I collagen. J Biol Chem 277(21):19206–19212
6. Leitinger B (2003) Molecular analysis of collagen binding by the human discoidin domain receptors, DDR1 and DDR2. Identification of collagen binding sites in DDR2. J Biol Chem 278(19):16761–16769
7. Yang K, Kim JH, Kim HJ, Park IS, Kim IY, Yang BS (2005) Tyrosine 740 phosphorylation of discoidin domain receptor 2 by Src stimulates intramolecular autophosphorylation and Shc signaling complex formation. J Biol Chem 280(47):39058–39066
8. Ferri N, Carragher NO, Raines EW (2004) Role of discoidin domain receptors 1 and 2 in human smooth muscle cell-mediated collagen remodeling: potential implications in atherosclerosis and lymphangioleiomyomatosis. Am J Pathol 164(5):1575–1585
9. Xu L, Peng H, Wu D, Hu K, Goldring MB, Olsen BR, Li Y (2005) Activation of the discoidin domain receptor 2 induces expression of matrix metalloproteinase 13 associated with osteoarthritis in mice. J Biol Chem 280(1):548–555

10. Olaso E, Lin HC, Wang LH, Friedman SL (2011) Impaired dermal wound healing in discoidin domain receptor 2-deficient mice associated with defective extracellular matrix remodeling. Fibrogenesis Tissue Repair 4(1):5
11. Ichikawa O, Osawa M, Nishida N, Goshima N, Nomura N, Shimada I (2007) Structural basis of the collagen-binding mode of discoidin domain receptor 2. EMBO J 26(18):4168–4176
12. Macedo-Ribeiro S, Bode W, Huber R, Quinn-Allen MA, Kim SW, Ortel TL, Bourenkov GP, Bartunik HD, Stubbs MT, Kane WH, Fuentes-Prior P (1999) Crystal structures of the membrane-binding C2 domain of human coagulation factor V. Nature 402(6760):434–439
13. Pratt KP, Shen BW, Takeshima K, Davie EW, Fujikawa K, Stoddard BL (1999) Structure of the C2 domain of human factor VIII at 1.5 A resolution. Nature 402(6760):439–442
14. Lee CC, Kreusch A, McMullan D, Ng K, Spraggon G (2003) Crystal structure of the human neuropilin-1 b1 domain. Structure 11(1):99–108
15. Carafoli F, Mayer MC, Shiraishi K, Pecheva MA, Chan LY, Nan R, Leitinger B, Hohenester E (2012) Structure of the discoidin domain receptor 1 extracellular region bound to an inhibitory Fab fragment reveals features important for signaling. Structure 20(4):688–697
16. Leitinger B, Kwan AP (2006) The discoidin domain receptor DDR2 is a receptor for type X collagen. Matrix Biol 25(6):355–364
17. Takahashi H, Nakanishi T, Kami K, Arata Y, Shimada I (2000) A novel NMR method for determining the interfaces of large protein-protein complexes. Nat Struct Biol 7(3):220–223
18. Nishida N, Sumikawa H, Sakakura M, Shimba N, Takahashi H, Terasawa H, Suzuki EI, Shimada I (2003) Collagen-binding mode of vWF-A3 domain determined by a transferred cross-saturation experiment. Nat Struct Biol 10(1):53–58
19. Farndale RW, Lisman T, Bihan D, Hamaia S, Smerling CS, Pugh N, Konitsiotis A, Leitinger B, de Groot PG, Jarvis GE, Raynal N (2008) Cell-collagen interactions: the use of peptide Toolkits to investigate collagen-receptor interactions. Biochem Soc Trans 36(Pt 2):241–250
20. Konitsiotis AD, Raynal N, Bihan D, Hohenester E, Farndale RW, Leitinger B (2008) Characterization of high affinity binding motifs for the discoidin domain receptor DDR2 in collagen. J Biol Chem 283(11):6861–6868
21. Carafoli F, Bihan D, Stathopoulos S, Konitsiotis AD, Kvansakul M, Farndale RW, Leitinger B, Hohenester E (2009) Crystallographic insight into collagen recognition by discoidin domain receptor 2. Structure 17(12):1573–1581
22. Brondijk TH, Bihan D, Farndale RW, Huizinga EG (2012) Implications for collagen I chain registry from the structure of the collagen von Willebrand factor A3 domain complex. Proc Natl Acad Sci U S A 109(14):5253–5258
23. Giudici C, Raynal N, Wiedemann H, Cabral WA, Marini JC, Timpl R, Bachinger HP, Farndale RW, Sasaki T, Tenni R (2008) Mapping of SPARC/BM-40/osteonectin-binding sites on fibrillar collagens. J Biol Chem 283(28):19551–19560
24. Emsley J, Knight CG, Farndale RW, Barnes MJ, Liddington RC (2000) Structural basis of collagen recognition by integrin alpha2beta1. Cell 101(1):47–56
25. Noordeen NA, Carafoli F, Hohenester E, Horton MA, Leitinger B (2006) A transmembrane leucine zipper is required for activation of the dimeric receptor tyrosine kinase DDR1. J Biol Chem 281(32):22744–22751
26. Abdulhussein R, Koo DH, Vogel WF (2008) Identification of disulfide-linked dimers of the receptor tyrosine kinase DDR1. J Biol Chem 283(18):12026–12033
27. Finger C, Escher C, Schneider D (2009) The single transmembrane domains of human receptor tyrosine kinases encode self-interactions. Sci Signal 2(89):ra56
28. Kim HG, Tan L, Weisberg EL, Liu F, Canning P, Choi HG, Ezell SA, Wu H, Zhao Z, Wang J, Mandinova A, Griffin JD, Bullock AN, Liu Q, Lee SW, Gray NS (2013) Discovery of a potent and selective DDR1 receptor tyrosine kinase inhibitor. ACS Chem Biol 8(10):2145–2150
29. Canning P, Tan L, Chu K, Lee SW, Gray NS, Bullock AN (2014) Structural mechanisms determining inhibition of the collagen receptor DDR1 by selective and multi-targeted type II kinase inhibitors. J Mol Biol 426:2457–2470

# Chapter 4
# DDR Mouse Models

**Kiyoshi Kano, Ken Takeshi Kusakabe, and Yasuo Kiso**

## 4.1 Introduction

DDRs have been identified as collagen-binding receptors and subsequently confirmed as endogenous receptors, which regulate cell proliferation, cell adhesion, migration, as well as extracellular matrix remodeling, mainly in in vitro studies. DDR mouse models have been a powerful tool to elucidate the additional biological and physiological functions of DDRs associated with development and diseases by using phenotypic analyses.

## 4.2 Body Size

DDR1- and DDR2-knockout mice generated by gene targeting are viable; interestingly, their body size is smaller, which is one of distinguishing phenotypes of DDR loss. In addition, genetically modified DDR2 mice exhibit alterations in body size, including bone size, body weight, and fat volume. These mouse models suggest that DDRs play an important role in body growth and local and systemic development.

K. Kano (✉)
Laboratory of Developmental Biology, Joint Faculty of Veterinary Medicine, Yamaguchi University, Yamaguchi 753-8515, Japan

Laboratory of Veterinary Anatomy, Joint Faculty of Veterinary Medicine, Yamaguchi University, Yamaguchi 753-8515, Japan
e-mail: kanokiyo@yamaguchi-u.ac.jp

K.T. Kusakabe • Y. Kiso
Laboratory of Veterinary Anatomy, Joint Faculty of Veterinary Medicine, Yamaguchi University, Yamaguchi 753-8515, Japan



R. Fridman, P.H. Huang (eds.), *Discoidin Domain Receptors in Health and Disease*, DOI 10.1007/978-1-4939-6383-6_4

### 4.2.1 DDR1-Deficient Mice

DDR1 gene knockout mice have been produced using common gene targeting techniques. Heterozygous offspring have been crossed with mice on a 129/Sv, ICR, and C57BL/6 backgrounds, to generate inbred and outbred DDR1-deficient mice. DDR1-deficient homozygote mice are viable but smaller than their heterozygous littermates. DDR1-deficient female mice have a 35 % lower body weight, but DDR1-deficient male mice are only about 10 % smaller than wild-type mice, owing to an additional weight gain after puberty [1]. In both sexes, all organs are proportionately smaller. The skeletons of DDR1-deficient mice appear normal at 10 weeks of age, but a detailed analysis of DDR1-deficient mice revealed that the calcification of the fibula bone is reduced in the majority of cases. No difference in the extent of the zone of hypertrophic cartilage, chondrocyte proliferation rate, or chondrocyte apoptosis in the tibia has been observed between DDR1-deficient mice and wild-type mice. These data imply that hormonal changes affect bone and tissue growth in DDR1-deficient mice.

### 4.2.2 DDR2-Deficient Mice

DDR2-deficient mice have been produced by homologous recombination of a targeting vector into embryonic stem cells, to disrupt the DDR2 gene by replacing exon K1, encoding the ATP-binding region of the kinase domain, to a neomycin resistance cassette. Newborn DDR2-deficient mice do not have an abnormal appearance, but they gradually fail to develop, resulting in a proportionally smaller body size and reduced weight starting 10 days after birth. Adult mutant mice show a 30–40 % reduction in weight compared to heterozygous littermates in both sexes [2]. The length of the long bones of adult DDR2-deficient mice is reduced, and the axis skeleton and skull bones are also shorter. To elucidate the cellular mechanism underlying the reduced long bone growth in DDR2-deficient mice, chondrocyte proliferation, osteoblast differentiation, and the height of the growth plate have been analyzed. No difference was found in osteoblast differentiation, but both the chondrocyte proliferation and growth-plate height were found to be reduced in the metatarsals of DDR2-deficient mice at 2 weeks of age. The reduction of chondrocyte proliferation, and not cellular differentiation within the growth plate, is the major reason for the decreased bone size in DDR2-deficient mice. The evidence suggests that DDR2 contributes mainly to the regulation of chondrocyte proliferation as an extracellular matrix receptor in the skeletal structure.

### 4.2.3 Smallie: DDR2-Deficient Mutant Mice

*Smallie* (*slie*), a spontaneous, autosomal recessive mutation, which results in dwarfing and sterility, was discovered in a Jackson Laboratory colony [3]. *Slie* mutant mice arose spontaneously in a BKSChpLt(HRS)$^{\mathrm{Tg(Ins2\text{-}Cpe)\ 1LtCpe}\mathit{fat}}$/LtJng line, and an unaffected heterozygous sibling was crossed with a C57BLKS/J mouse to produce *slie* carriers, because the homozygous mutants were infertile. To briefly review this colony, the fat mutation arose spontaneously in *Cpe* in the HRS inbred mouse strain, and this mutation was transferred to the C57BLKS/J strain through congenic backcrossing experiments in the Jackson Laboratory of Drs. Dorothy Chapman (Chp) and Ed Leiter (Lt). In addition, these mice carry the $Cpe^{fat}$ transgene, under the control of the Ins2 promoter, which was generated by Dr. Leiter. The *slie* mutation spontaneously arose in a colony in the laboratory of Dr. Jurgen K. Naggert (Jng), in mice provided by Dr. Leiter. Selective breeding experiments have demonstrated that the dwarfing phenotype is independent of the presence of the $Cpe^{\mathrm{fat}}$ mutation or the transgene Ins2-Cpe. Both $Cpe^{fat}$ and Ins2-Cpe have been removed from the *slie* colony through selective breeding. To determine the chromosomal location of *slie*, F1 progeny mice were generated by in vitro fertilization of BKS(HRS)-*slie*/Jng homozygotes to NOD.NON-$H2^{nb1}$/LtJ mice. The heterozygous F1 mice, which did not exhibit dwarfism, were intercrossed for the initial linkage mapping study. The resultant F2 progeny mice were weighed at 10 weeks of age to identify the affected animals. The fine structure genetic map encompasses an estimated 1.94 Mb of physical region on Chromosome 1, which contains the *slie* mutation. According to annotations in the *Mus musculus* genome assembly, the *slie* critical region contains 23 gene loci. Among these candidates, the relative levels of *Ddr2* gene expression were quite low in *slie* mutants. Whereas *Ddr2* expression was readily detected in wild-type mice, exons 1–17 of *Ddr2* did not amplify from the *slie* mutant genomic DNA by PCR. A genomic sequencing analysis detected a 150 kb deletion that extends into the *Ddr2* gene transcript. *Slie* homozygous mutant mice have skeletal abnormalities that include craniofacial deformities, such as protuberant eyes and snub noses (Fig. 4.1). After weaning, their weight gain is slower and they lack the juvenile growth spurt observed in control littermates. Whereas the total body mass is reduced in *slie* homozygotes, as assessed by dual X-ray absorptiometry, the percentage of body fat is significantly reduced with an accompanying increase in lean muscle mass, these mice compared to wild-type mice at 5–6 months of age. Similarly, the bone mineral content, but not the density, is reduced in *slie* homozygotes compared to wild-type mice. The results suggest that the absence of DDR2 leads to a growth retardation phenotype in *slie* mice, as seen in DDR2-deficient mice.

### 4.2.4 Dominant-Negative DDR2 Transgenic Mice

To investigate the molecular role of DDR2 in endochondral cellular proliferation in vivo, a transgenic mouse was created, in which the expression of the dominant-negative DDR2 protein is induced, to evaluate the role of DDR2 in cellular proliferation [4]. The dominant-negative DDR2 protein was made from a kinase-dead DDR2 mutant (KD-DDR2), which is a truncated form lacking the kinase domain but retaining the extracellular and transmembrane domains. Transgenic mice were produced by microinjecting several hundred molecules of the DNA fragment into the pronuclei of fertilized eggs from F1 hybrid mice (C57BL/6 × DBA).

**Fig. 4.1** Four-month-old wild-type and *slie* mutant mice (*left*)

Unexpectedly, the body size and the skeleton length of KD-DDR2 overexpressed mice were not significantly different compared to the littermates. On the other hand, whereas the layer of hypertrophic chondrocytes in KD-DDR2 transgenic mice was not significantly thicker than that of normal littermates, the layer of proliferative chondrocytes in KD-DDR2 transgenic mice was significantly thicker than that of normal littermates. These data indicate that DDR2 plays an important role in endochondral ossification, but neither a critical nor an essential role in total body and skeleton size determination. The greater thickness of the proliferative chondrocyte layer in KD-DDR2 transgenic mice might be an anomaly in the reduction of chondrocyte proliferation in DDR2-deficient mice [2]. This difference might stem from the regional and functional differences in DDR2 shortage; these differences suggest that DDR2 plays various roles under different conditions, either systemically or locally.

### *4.2.5 DDR2-Overexpressing Transgenic Mice*

What happens when DDRs are overexpressed in mice? To investigate the systemic role of DDR2 in body size regulation, a transgenic mouse in which the DDR2 protein is overexpressed was produced, then screened for abnormalities using a systematic mouse abnormality screening system [5]. Transgenic mice were produced by microinjecting several 100 molecules of the DNA fragment into the pronuclei of fertilized eggs from F1 hybrid mice (C57BL/6×DBA), as described above. The transgenic mice were screened for abnormalities using the "Japan Mouse Clinic," a systematic mouse abnormality screening established and performed at the RIKEN BioResource Center (http://www.brc.riken.jp/lab/jmc/mouse_clinic/en/index.html). The modified-SHIPRA screen revealed that only the parameter of body size was significantly different among the genotypes. The body length was significantly increased, whereas the body weight was significantly decreased, in the DDR2-overexpressing transgenic mice compared to wild-type controls. In the transgenic mice, the epididymal fat pads were also significantly smaller than the normal littermate mice. This finding suggests that the decrease in body weight is correlated with the decrease in fat tissues and in cholesterol in transgenic mice. Clinical biochemical tests have revealed that the level of leptin, which is secreted from fat and known as a regulator of appetite and body weight [6–9], is higher in the transgenic mice compared to their wild-type littermates. These data suggest that an excess of DDR2 results in an increase of leptin in adipocyte cells, and ultimately affects the whole body weight in DDR2 transgenic mice. These data suggest that DDR2 plays a systemic role in the regulation of body size by affecting skeletal formation and fat metabolism.

## 4.3 Reproduction

Both DDR1- and DDR2-deficient mice exhibit infertility with various abnormalities, including anomalies in reproductive functions to a varying degree, which imply that DDRs might have unique functions to sustain the normal reproductive system.

### *4.3.1 Implantation in DDR1-Deficient Mice*

DDR1-deficient female mice generally fail to mate to either mutant or wild-type control male mice [1]. Although virginal plugs have appeared in almost all female mice after mating, only 20 % of DDR1-deficient female mice are able to give birth. A normal number of blastocysts have been found in the uterus of DDR-deficient female mice before implantation at 3.5 days postcoitum (d. p. c.), but a swollen decidua was not found at 4.5 d. p. c. Considering that DDR1-deficient blastocysts transferred into the uterus of pseudopregnant wild-type mice progress to normal litters, DDR1-deficient female mice might have a maternal deficiency that affects the implantation process. Proteases such as those of the MMP family are essential for implantation to remodel the extracellular matrix of the maternal uterine stroma. DDR1 was originally localized to the decidua in the pre- and postimplantation uterus, where the DDR1 ligands, including type VI collagen, are specifically expressed in the uterine epithelium around the implantation region. DDR1 might play an important role in reconstruction throughout MMP production in the peri-implantation adhesion between the uterine wall and the blastocyst.

### 4.3.2 Mammary Glands in DDR1-Deficient Mice

All of the pups from DDR1-deficient mice have a nutritional deficiency 1 day after birth and die within a few days, even if they have been nursed by their mother with no milk spot in their stomach [1]. When pups delivered from DDR1-deficient mice were transferred to wild-type normal foster mothers shortly after birth, they were able to grow normally. In the mammary glands of DDR1-deficient mice, the alveolar structure is much more condensed, and fewer lipid vesicles are observed during the late gestation period, compared to wild-type female mice. After delivery, the adipose tissues generally disappear and are replaced by alveolar structures filled with milk in control female mice. However, the mammary glands of DDR1-deficient mice largely consist of adipocytes, and the alveoli are predominantly condensed. The alveoli and epithelium of DDR1-deficient mice start to collapse and regress at day 2 after birth. In the developing mammary glands of DDR1-deficient virgin mice, the outgrowth of the mammary ducts is delayed, but the primary ducts and the terminal end buds continue to proliferate, resulting in an increase in the number of alveolar ducts at 3 weeks of age. Although the fat pads are filled with epithelial ducts, the number and diameter of the ducts are largely increased, and the extracellular matrix deposition is substantially increased in the developing mammary glands of mature DDR1-deficient mice. The abnormal development of the mammary glands may be the reason why DDR1-deficient mice secrete very little milk, although the milk protein transcripts were detected at normal levels. These results suggest that DDR1 might be a crucial factor in the regulation of ductal development in the mammary glands.

### 4.3.3 Ovaries in Smallie Mice

It is not known whether the reproductive phenotypes observed in *slie* mutant mice are present in DDR2-deficient mice, because similar studies of the reproductive system have not been reported in DDR2-deficient mice. This is because matings between male and female *slie* mutants have failed to produce offspring. Homozygous *slie* mutant females mated to either homozygous *slie* or wild-type males have resulted in no pregnancies or live pups [3]. DDR2 is expressed in the ovaries of adult wild-type female mice in the interstitial and thecal cells, but not in the cumulus cells, the inner layer of granulosa cells, or the oocytes. In ovaries from *slie* mutant mice, the foremost difference was the striking absence of corpora lutea, in contrast to wild-type mice (Fig. 4.2). Conversely, no significant deviation was found in the amount of preantral, antral, and Graffian follicles at these ages. The basal levels and the gene expression of the pituitary and hypothalamic hormones, as well as the gene expression of the hypothalamic releasing hormones, are not significantly different between *slie* and wild-type mice. The administration of exogenous gonadotropins induced follicular growth at the Graffian stage, but it

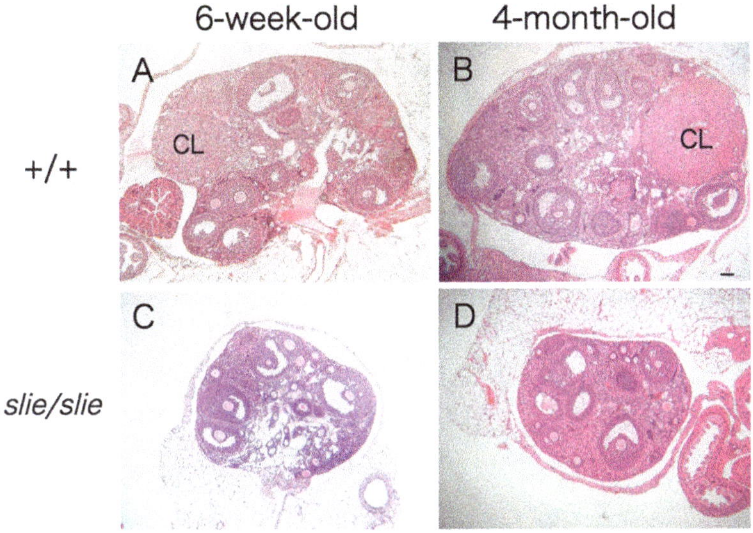


**Fig. 4.2** Anovulation in female *slie* mutants. Six-week-old and 4-month-old wild-type (**a**, **c**) and mutant (**b**, **d**) ovary sections, stained with hematoxylin and eosin, depicting follicular development, CL, corpus lutea, scale bar, 50 μm

failed to alter the anovulatory phenotype of *slie* homozygotes, as the full luteinization of the thecal cells did not occur and no corpora lutea was formed. These results suggest that the absence of corpora lutea is caused by an inability to respond to gonadotrophic signals in *slie* mutant ovaries. Following exogenous gonadotropin administration, both female and male *slie* and wild-type mice secreted estradiol, progesterone, and testosterone, respectively, in response to the exogenous gonadotropins, but the circulating steroid levels in adult *slie* mutants were lower than in the adult wild-type mice.

A systems biology approach was performed to identify the biological networks affected by the *slie* mutation in ovaries, using a microarray analysis [10]. A transcriptome analysis indicated several altered gene categories in *slie* mutants, including gonadal development, ovulation, anti-apoptosis, and steroid hormones. DDR2 signaling pathways did not activate *Mmp* genes, as no difference in their expression was found in *slie* mutants compared to wild-type mice. In hormonal signaling pathways involved in ovulation, the activity of the luteinizing hormone/choriogonadotropin (LH) receptor was decreased in the somatic cells, but not in the oocytes of *slie* mutants, resulting in an intrinsic defect in germ cells. The overlapping spatial and temporal expression of DDR2 and the LH receptor may permit their mutual coregulation during specific periods of ovulation. A lack of DDR2 signaling in *slie* mutants most likely triggers anovulation, by altering the expression of the LH receptor. Considering that *slie* mutant oocytes are competent to complete meiosis and fertilization in vitro, in spite of the ovulation of significantly

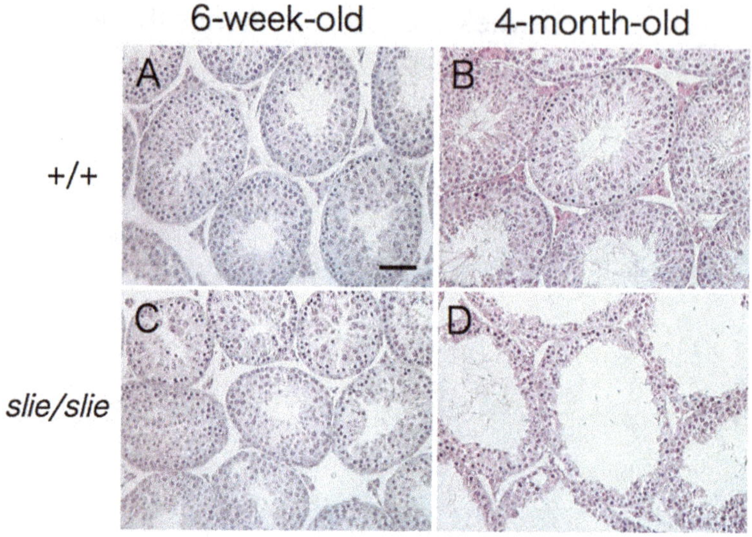


**Fig. 4.3** Gonadal abnormalities in male *slie* mutant mice. Cross sections of testes from 6-week-old and 4-month old wild-type (**a**, **b**) and *slie* mutant (**c**, **d**) mice, stained with hematoxylin and eosin, scale bar, 50 μm

fewer oocytes, these data identify DDR2 as a novel critical player in ovarian function, which acts upon classical endocrine pathways and LH signaling in somatic rather than in germline cells.

### *4.3.4 Testes in Smallie Mice*

In homozygous *slie* males, fewer spermatids, along with atrophy of the spermatogonia, Sertoli and Leydig cells are observed (Fig. 4.3) [11]. The loss of DDR2 results in a progressive increase in apoptosis in spermatogenic cells, through the process of spermatogenesis. As described above, the absence of DDR2 induces a reduction in LH receptor expression in *slie* mutant females. This result suggests a correlation between DDR2 and the LH receptor in female gonads. In males, the localization of DDR2 and the LH receptor might reflect their interaction in interstitial Leydig cells during spermatogenesis, as in female gonads. Older Leydig cells in *slie* mutant mice fail to express the LH receptor, thereby triggering reductions in the expression levels of specific steroidogenic enzyme genes, and ultimately the inability to transduce signals from LH in the testes. DDR2 might be a survival and maintenance factor required for the viability of spermatogenic cells. These results suggest that DDR2 signaling also plays a critical role in the maintenance of male spermatogenesis throughout the LH signaling pathway.

Taken together, these results obtained from the study of the *slie* mutant reproductive system suggest that the absence of DDR2 leads to gonadal dysfunction, due to peripheral defects in the hormone response pathways.

## 4.4 Skin Wound Healing

The wound-healing response depends on the regulated linkage between recruited fibroblasts and the collagenous extracellular matrix. DDR2 expression is upregulated in wound healing following cutaneous burns [12]. In a skin wound-healing model, DDR2-deficient mice exhibit a reduced proliferative response in epidermal fibroblasts and keratinocytes compared to wild-type littermates [2]. Closure of the skin burn wounds is also significantly delayed in DDR2-deficient mice compared to wild-type mice [13]. All of the skin wound healing indices, that is, the expression of α-smooth muscle actin, the MMP2 activity, the tensile strength, and several collagen levels were significantly decreased in skin tissue extracts from DDR2-deficient mice following a cutaneous wound compared to wild-type littermates. This suggests the defective recruitment of skin fibroblasts in the skin of DDR2-deficient mice [14].

These results indicate that the study of DDR2-deficient mice provides a suitable model to elucidate the role of DDR2 in modulating the proliferative response in situations associated with extreme matrix deposition and degradation, such as a tissue injury, where the extracellular matrix provides a signal to increase proliferation.

## 4.5 Osteoarthritis

Osteoarthritis is a weakening disease that results from the progressive loss of articular cartilage. Mutations in type IX and XI collagen have resulted in early-onset osteoarthritis with a wide spectrum of osteochondrodysplasia in human genetic studies. The chondrocytes in the articular cartilage are affected by extracellular interactions such as cell–matrix interactions, as well as interactions between integrins and DDRs, rather than by direct intercellular communication [15]. Along with single gene-targeting models, double gene-targeting mouse models with a collagen deficiency, which represent an osteoarthritis model, as well as DDR deficiency models, are suitable to study the function of DDR in the pathogenesis of osteoarthritis.

### *4.5.1 Osteoarthritis in DDR2-Deficient Mice*

DDR2 expression is increased in the chondrocytes of the articular cartilage of knee joints in mice that have developed osteoarthritis as a result of a heterozygous mutation in type XI collagen [16, 17], suggesting that DDR2 might be correlated with the disease. The expression and activity of matrix metalloproteinase 13 (MMP-13) is also increased in the knee cartilage of mutant mice. The upregulation of MMP-13 might be one of the common events in osteoarthritis progression, because the deletion of MMP-13 results in a delay in osteoarthritis progression.

To elucidate the function of DDR2 in articular cartilage, double-heterozygous (type XI collagen- and DDR2-deficient) mutant mice were generated [18]. Heterozygous mutant DDR2 mice appear normal in size, but homozygous mutant mice are inappropriate for the study of osteoarthritis because the phenotypes of DDR2-deficient mice include dwarfism and a reduction in chondrocyte proliferation in the cartilage growth plate. The double heterozygous mutation in mice has shown that the rate of osteoarthritis progression, which is caused by a type XI collagen insufficiency, is considerably delayed in the mouse knee joint. A microsurgery was able to mechanically mimic osteoarthritis. The progression toward osteoarthritis was also considerably delayed in heterozygous DDR2-deficient mice compared to their wild-type littermates, following a surgery on the knee joint between the medial meniscus and the anterior tibial plateau. These results suggest that reduction of DDR2 might play a positive role in the attenuation of the degeneration of osteoarthritis of the knee joint that has been induced by a lack of type XI collagen or by surgical destabilization of the medial meniscus.

### 4.5.2 Osteoarthritis in DDR1-Deficient Mice

Contrary to DDR2, DDR1 has been suggested to play a protective role in osteoarthritis. DDR1-deficient mice show several features typical of osteoarthritis pathogenesis, including surface fissures, loss of proteoglycans, chondrocyte cluster formation, collagen type I upregulation, and atypical collagen fibril arrangements in the temporomandibular joint [19]. The frequency of osteoarthritis degeneration is also increased at a younger age in this model, compared to other mouse models previously used to study temporomandibular disorders in DDR1-deficient mice. An analysis of a three-dimensional reconstruction of the mandibular condyles in DDR1-deficient mice indicated a scabrous surface of the subchondral bone and flattened mandibular condyles, which are typical alterations in temporomandibular disorders. An ultrastructural analysis revealed that the collagen fiber arrangement was altered in the superficial layer of DDR1-deficient mice, compared to the parallel fiber alignment observed in the wild-type littermates. A microarray analysis of cartilage tissue samples from the mandibles of DDR1-deficient mice and wild-type littermates showed that the overall changes in gene expression resemble the typical pattern of osteoarthritis. Interestingly, DDR1-null chondrocytes induced the compensatory expression of DDR2, which is linked to the upregulated expression of MMP13 and to articular degeneration.

Although DDR1-deficient mice exhibit osteoarthritis of the temporomandibular joints at a young age, no symptom of osteoarthritis has ever been observed in the knee joints of double-heterozygous (type XI collagen and DDR2-deficient) mutant mice at this stage. The differences in the structural or embryonic origin between the two joints might help in the phenotypical identification.

These data suggest that DDR1-deficient mice represent a good animal model, which contributes to the study of osteoarthritis in the temporomandibular disorders in vivo, and which can supply new treatment options for the disease.

Taken together, these findings from the study of osteoarthritis in DDR1- and DDR2-deficient mice suggest that DDR2 therapy provides the best treatment among members of the DDR family to cure the various symptoms of osteoarthritis, either with a specific tyrosine kinase inhibitor or with a molecule that prevents the binding to collagen type II.

## 4.6 Atherosclerosis

Atherosclerosis is a fibroinflammatory disease of the arterial wall. The responses of vascular smooth muscle cells during arterial wound repair are regulated by collagens as important signaling molecules. Following a mechanical vascular injury, the intimal thickening area was considerably attenuated in DDR1-deficient mice; however, there were no significant differences in the medial area, adventitial area, or external elastic lamina perimeter between DDR1-deficient and wild-type mice [20]. The density of the cells in the intima was almost twice as high in DDR1-deficient mice compared to wild-type mice, suggesting that matrix accumulation was decreased in the artery of the DDR1-deficient mice. The deposition of collagen fibrils in the injured arteries of the DDR1-deficient mice was substantially decreased compared to wild-type mice, as assessed using picrosirius red staining and polarized-light microscopy to visualize the collagen birefringence. These results suggest that DDR1 plays a critical role in the hyperplastic intima following an artery injury.

Double-targeted mouse with lipoprotein receptor-deficiency, which mimic atherosclerosis and DDR deficiency, also provide a good model to study the role of DDRs in the pathogenesis of atherosclerosis. Low-density lipoprotein receptor-deficient (*Ldlr–/–*) mice fed on an atherogenic diet displayed the development of complex atherosclerotic lesions, including the accumulation of lipids in vascular smooth muscle cells, macrophages, and extracellular matrix. Both DDR1 and DDR2 mRNA was expressed in *Ldlr–/–* mice [21]. Double-homozygous (DDR1 and low-density lipoprotein receptor-deficient, *Ddr1–/–*; *Ldlr–/–*) mutant mice fed an atherogenic diet exhibited drastically reduced atherosclerotic plaque development compared to *Ddr1+/+ Ldlr–/–* mice. Interestingly, the DDR1 expression level was absent in the plaque of *Ddr1–/–*; *Ldlr–/–*, but that of DDR2 was not different in the plaque between *Ddr1+/+*; *Ldlr–/–* and *Ddr1–/–*; *Ldlr–/–* mice.

The accelerated accumulation of fibrillar collagen and elastin in artherosclerotic plaques was observed in the *Ddr1–/–*; *Ldlr–/–* mice compared to the *Ddr1+/+*; *Ldlr–/–* mice. This was confirmed by the data on the expression of matrix synthesis molecules, procollagen and tropoelastin, which revealed that they were increased in *Ddr1–/–*; *Ldlr–/–* plaques at an early atherosclerotic stage. On the other hand, the proteolytic activity in the plaque of *Ddr1–/–*; *Ldlr–/–* mice was decreased, which is related to the reduction in the macrophage content of the plaque, because macrophages are a major source of proteolytic enzymes, such as members of the MMP family, and cysteine and serine proteases, in the artherosclerotic plaques. The elimination of DDR1 substantially changed the plaque constitution, which was similar to the content of smooth muscles, but with less macrophage accumulation in *Ddr1–/–*; *Ldlr–/–* mice, which is consistent with the intact role of DDR1 in the mediation of macrophage recruitment, invasion and persistence, especially at an early stage in the artherosclerotic plaque.

A sex-mismatched bone marrow transplant was performed using *Ddr1–/–* ; *Ldlr–/–* mice and their *Ddr1+/+*; *Ldlr–/–* littermates, to investigate the independent role of DDR1 on bone marrow-derived cells, such as macrophages, during atherosclerosis [22]. Atherosclerotic plaques from chimeric mice with DDR1-deficient bone marrow (*Ddr1–/–*; *Ldlr–/–* to *Ddr1+/+*; *Ldlr–/–*; *Ddr1–/–* => *Ddr1+/+*) had remarkably smaller descending aortas compared to *Ddr1+/+*⇒*Ddr1+/+* mice, 12 weeks following the administration of an atherogenic diet. However, the matrix composition was similar in the two genotypes. The accumulation of bone marrow-derived macrophages was reduced in the aortic sinus lesions of *Ddr1–/–*⇒*Ddr1+/+* mice, compared to *Ddr1+/+*⇒*Ddr1+/+* mice. By contrast, there was no effect on the number of cells of the host-derived resident vessel wall in bone marrow-specific DDR1-deficient mice, which suggests that the absence of DDR1 in bone marrow cells allows the attenuation of lesion growth, by regulating macrophage accumulation in the atherosclerotic plaques. The proportion of BrdU-positive bone marrow and circulating monocytes was comparable between *Ddr1–/–*⇒*Ddr1+/+* and *Ddr1+/+*⇒*Ddr1+/+* mice, but the proportion of the ascending aortic area occupied by a fatty streak from a lesion was significantly decreased in size in *Ddr1–/–*⇒*Ddr1+/+* mice compared to *Ddr1+/+*⇒*Ddr1+/+* mice, which suggests that DDR1 regulates the generation of a fatty streak, irrespective of the infiltration of monocytes and macrophages. These data provide support to the suggestion that DDR1 plays a critical role in promoting macrophage accumulation and lesion growth in early-stage atherogenesis.

As another experiment of reverse-oriented transplantation, bone marrow from *Ddr1+/+* ; *Ldlr–/–* mice was transplanted into *Ddr1–/–*; *Ldlr–/–* hosts, to generate chimeric mice with a deficiency in DDR1 specific to the cells of the resident vessel wall (*Ddr1+/+*⇒*Ddr1–/–*). The atherosclerotic plaque size was considerably increased in the cells of the resident vessel wall, but the lipid content of the lesions was decreased in *Ddr1+/+*⇒*Ddr1–/–* mice compared to *Ddr1+/+*⇒*Ddr1+/+* control mice, 12 weeks after having placed the mice on an atherogenic diet [23]. The extracellular matrix, comprising collagen, elastin, and proteoglycan, accumulated and the relative fibrous cap thickness was significantly increased in the plaques of *Ddr1+/+*⇒*Ddr1–/–* mice, which suggests that the cellular composition of the lesions shifted toward increased numbers of vessel wall-derived smooth muscle cells compared to bone marrow-derived macrophages in the vessel wall. These data support the suggestion that DDR1 acts as a negative regulator of matrix turnover by attenuating the proliferation, migration, and matrix accumulation during atherogenesis on the resident vessel walls.

Arterial calcification within the intimal layer is also an aspect of advanced atherosclerotic disease, including the risk of coronary arterial disorders. It is known that type I collagen promotes vascular smooth muscle cell-mediated calcification. *Ddr1−/−*; *Ldlr−/−* mice exhibit a significant attenuation of artherosclerotic intimal calcification in the aortic arch, which is independent of the serum calcium concentration but correlated with decreased inflammation [24]. This result suggests that DDR1 is also an important positive mediator of artherosclerotic intimal calcification by promoting mineralization.

Taken together, these results suggest that the DDR1, *Ldlr*-deficiency double-targeting model may have advantages when it comes to analyzing the molecular mechanism underlying atherosclerotic disease. DDR1 plays an important role in regulating the pathology of atherogenic diseases by modulating the number of macrophages and a decrease in extracellular matrix deposition.

## 4.7 Fibrosis

### *4.7.1 Lung Fibrosis*

DDR1 has been implicated in pulmonary fibrosis through its expression in the pulmonary epithelium and its increasing expression pattern in the bronchoalveolar lavage cells of patients with idiopathic pulmonary fibrosis disease. Following bleomycin administration, the ultrastructure of epithelial and endothelial cells of DDR1-deficient mice were highly normal, but wild-type littermates were substantially injured due to interalveolar thickening of the septa and a reduction of the alveolar space [25]. In the analysis of fibrosis following bleomycin administration, wild-type mice exhibited increased collagen-rich nodular deposition, which was not the case in the injured DDR1-deficient mice. Myofibroblast expansion was reduced in DDR1-deficient mice treated with bleomycin for 2 weeks, compared to wild-type mice. Cellular apoptosis and proliferation was significantly decreased in the pulmonary tissues of DDR1-deficient mice. Inflammation, which was assessed by a quantification of the number of lymphocytes and macrophages within the bronchoalveolar lavage, was decreased in DDR1-deficient mice compared to wild-type mice. These data suggest that DDR1 is a positive regulator in the development of lung inflammation and fibrosis.

### *4.7.2 Kidney Fibrosis*

In the kidneys, DDR1 is found in the basolateral membranes of nephron segments in the tubules connecting to the renal papilla in normal rats and is upregulated within the glomeruli in remnant rats kidney, suggesting that DDR1 plays a crucial role in cell–matrix interactions and kidney injury [26]. DDR1-deficient mice are useful to understand the unique functions of DDR1 in renal fibrosis and inflammation.

Adult DDR1-deficient mice show no obvious signal of abnormal renal function, such as edema, weight loss, or abnormal water intake, but exhibit proteinuria in high-molecular-weight proteins and urinary acanthocytes [27]. Histological and immunohistochemical analyses have revealed modest abnormalities in kidney structure, including a minor increment of intracellular aerocysts in the tubular epithelial cells of DDR1-deficient mice. An electron microscopy analysis has revealed a localized, subepithelial, mushroom-like isodense thickening of the glomerular basement membrane in DDR1-deficient mice. Within this area, a partial loss of the slit diaphragm between the podocyte foot processes can be observed. These data support the suggestion that DDR1 has an important function in the synthesis of basement membrane proteins and in maintaining the slit diaphragm anchored to the podocyte and glomerular basement membrane structure.

In chronic renal failure, hypertension is initiated by inflammation that develops to increase the synthesis and accumulation of collagen within the renal tissue. Following exposure to angiotensin II, DDR1 expression is increased in the renal cortex and within the glomeruli in normal mice. Although the blood pressure response to the continuous administration of angiotensin II is similar to that in wild-type mice, the renal cortex lesions and the infiltration of macrophages and lymphocytes are drastically decreased in the hypertensive DDR1-deficient mice, compared to wild-type mice [28]. Following angiotensin II treatment for 4 weeks, the expression of collagen I and IV was also decreased in the renal vessels and glomeruli of DDR1-deficient mice, compared to wild-type mice. These data suggest that DDR1 might take part in promoting inflammation and fibrosis through the synthesis of collagen and by recruiting cytokines in renal fibrosis diseases. The induction of unilateral ureteral obstruction (UUO) also leads to renal perivascular and interstitial inflammation and fibrosis in normal mice, but these severe symptoms of fibrosis are alleviative, and proinflammatory cytokines are drastically decreased in DDR1-deficient mice, compared to wild-type mice [29]. A deficiency in DDR1, induced by the injection of alloimmune sheep nephrotoxic serum, also protects from the development of crescentic glomerulonephritis [30].

Alport syndrome, which is caused by mutations in the COL4A3, 4 or 5 genes encoding type IV collagen α3/α4/α5-chains, is a chronic disease that is associated with proteinuria, hematuria, renal disorders, and deafness. COL4A3-deficient mice (*Col4a3–/–*) represent a model to examine renal degeneration in Alport syndrome. Double-homozygous (DDR1 and COL3A4-deficient, *Ddr1–/–*; *Col4a3–/–*) mutant mice show a prolonged lifespan, until they die from progressive renal fibrosis, compared to *Ddr1+/+*; *Col4a3–/–* mice [31]. Uremia is delayed and proteinuria is reduced in double-homozygous *Ddr1–/–*; *Col4a3–/–* mice. The kidneys of 9.5-week-old *Ddr1+/+*; *Col4a3–/–* mice are 33 % smaller than those of *Ddr1–/–*; *Col4a3–/–* mice. Severe glomerular, periglomerular, and tubulointerstitial fibrosis, observed in the kidneys of *Ddr1+/+*; *Col4a3–/–* mice are not found in those of *Ddr1–/–*; *Col4a3–/–* mice. An electron microscopy analysis has revealed that *Ddr1+/+*; *Col4a3–/–* mice exhibit a thickening and splitting of the glomerular basement membrane, which is a typical symptom of Alport syndrome, and the loss of the podocyte foot process. *Ddr1–/–*; *Col4a3–/–* mice, by contrast, show less severe phenotypes of the glomerular basement membrane and podocytes.

Extracellular matrix deposition of fibronectin and laminin was less pronounced in *Ddr1−/−; Col4a3−/−* mice compared to *Ddr1+/+; Col4a3−/−* mice. The expression levels of fibrotic, proliferative and inflammatory molecules are also decreased in *Ddr1−/−; Col4a3−/−* mice compared to *Ddr1+/+; Col4a3−/−* mice. A deficiency of DDR1 reduces the glomerular and tubulointerstitial matrix deposition and fibrosis, via the downregulation of the profibrotic cytokines TGF-β and CTGF, with a subsequent reduction in the tubulointerstitial infiltration of T-lymphocytes and macrophages. These data suggest that a deficiency in DDR1 relieves various Alport syndrome symptoms, such as renal fibrosis and inflammation, which indicates that DDR1 is an important collagen receptor in fibrosis and inflammation in the podocyte–matrix interaction.

Taken together, these results suggest that DDR1-deficient mice develop resistance to fibrosis and inflammation followed by macrophage infiltration, which in turn implies that DDR1 might play an important proinflammatory role in kidney damage.

### *4.7.3 Liver Fibrosis*

Contrary to the positive role of DDR1 in fibrosis, DDR2 has been reported as an inhibitor of fibrosis in the liver [32]. The administration of carbon tetrachloride ($CCl_4$) increases collagen deposition and the extracellular matrix remodeling response, related to the altered recruitment of activated hepatic stellate cells in the livers of DDR2-deficient mice. Immunomodulatory and fibrogenic genes are upregulated, and chemotactic migration and proliferation is induced. However, the extracellular matrix remodeling is reduced in the hepatic stellate cells following a $CCl_4$ injection. DDR2 expression and phosphorylated DDR2 levels are raised in hepatic stellate cells following a $CCl_4$ injection. These data suggest that DDR2 is plays an important role in suppressing fibrosis in chronic injury of the normal liver.

## 4.8 Auditory Sensation

DDR1-deficient mice have an alternative phenotype that is associated with auditory sensation. DDR1-deficient mice develop a fundamental hearing loss by 2 months of age [33]. In normal mice, DDR1 is mainly expressed in the basal cells of the stria vascularis, type III fibrocytes, and in the cells lining the basilar membrane of the organ of Corti in the normal cochlea. In DDR1-deficient mice, a number of distinguishable alterations on the stria vascularis are observed, for example, the morphological alterations of the basal cells and the increase in electron-dense matrix deposition within the stria vascularis. These cytological alterations may be responsible for the hearing loss in DDR1-deficient mice. These results suggest that DDR1 might play an essential role in tissue composition maintenance and in regulating collagen deposition in the inner ear.

## 4.9 Heart Structure and Function

DDR2 is expressed in the mesenchymal cells throughout the body, and also in cardiac fibroblasts in the heart. To elucidate the role of DDR2 in heart function and structure, DDR2-deficient mice were produced by the homologous recombination of a targeting vector into embryonic stem cells [34]. The heart size in DDR2-null mice was reduced due to shorter cardiomyocytes, and not simply because of a reduced body size. The DDR2-null heart was physiologically affected, with alterations such as a lower heart rate, reduced contractility, and slower relaxation. The normalized cardiac collagen mass was not different between DDR2-null and normal mice, but the proportion of the collagen area in DDR2-null mice was significantly higher, which suggests that collagen density was reduced due to an altered cardiac fibroblast function. These results suggest that DDR2 plays an important role in collagen deposition to maintain the cardiac structure and function.

## 4.10 Conclusion

DDR mouse models shed light on the functional importance of the DDRs, such as their contribution to normal and homeostatic development, and highlight their enhancement or attenuation of diseases, including fibrosis, atherosclerosis, and osteoarthritis, in various conditions.

Studies on DDR mouse models might help elucidate the molecular functions of this receptor and help to develop molecular therapies for DDR-related diseases.

## References

1. Vogel WF, Aszodi A, Alves F, Pawson T (2001) Discoidin domain receptor 1 tyrosine kinase has an essential role in mammary gland development. Mol Cell Biol 21(8):2906–2917. doi:10.1128/MCB.21.8.2906-2917.2001
2. Labrador JP, Azcoitia V, Tuckermann J, Lin C, Olaso E, Manes S, Bruckner K, Goergen JL, Lemke G, Yancopoulos G, Angel P, Martinez C, Klein R (2001) The collagen receptor DDR2 regulates proliferation and its elimination leads to dwarfism. EMBO Rep 2(5):446–452. doi:10.1093/embo-reports/kve094
3. Kano K, Marin de Evsikova C, Young J, Wnek C, Maddatu TP, Nishina PM, Naggert JK (2008) A novel dwarfism with gonadal dysfunction due to loss-of-function allele of the collagen receptor gene, *Ddr2*, in the mouse. Mol Endocrinol 22(8):1866–1880. doi:10.1210/me.2007-0310
4. Kawai I, Hisaki T, Sugiura K, Naito K, Kano K (2012) Discoidin domain receptor 2 (DDR2) regulates proliferation of endochondral cells in mice. Biochem Biophys Res Commun 427(3):611–617. doi:10.1016/j.bbrc.2012.09.106
5. Kawai I, Matsumura H, Fujii W, Naito K, Kusakabe K, Kiso Y, Kano K (2014) Discoidin domain receptor 2 (DDR2) regulates body size and fat metabolism in mice. Transgenic Res 23(1):165–175. doi:10.1007/s11248-013-9751-2

6. Hummel KP, Dickie MM, Coleman DL (1966) Diabetes, a new mutation in the mouse. Science 153(3740):1127–1128
7. Coleman DL, Hummel KP (1969) Effects of parabiosis of normal with genetically diabetic mice. Am J Physiol 217(5):1298–1304
8. Friedman JM, Leibel RL, Bahary N (1991) Molecular mapping of obesity genes. Mamm Genome 1(3):130–144
9. Zhang Y, Proenca R, Maffei M, Barone M, Leopold L, Friedman JM (1994) Positional cloning of the mouse obese gene and its human homologue. Nature 372(6505):425–432. doi:10.1038/372425a0
10. Matsumura H, Kano K, Marin de Evsikova C, Young JA, Nishina PM, Naggert JK, Naito K (2009) Transcriptome analysis reveals an unexpected role of a collagen tyrosine kinase receptor gene, *Ddr2*, as a regulator of ovarian function. Physiol Genomics 39(2):120–129. doi:10.1152/physiolgenomics.00073.2009
11. Kano K, Kitamura A, Matsuwaki T, Morimatsu M, Naito K (2010) Discoidin domain receptor 2 (DDR2) is required for maintenance of spermatogenesis in male mice. Mol Reprod Dev 77(1):29–37. doi:10.1002/mrd.21093
12. Feezor RJ, Paddock HN, Baker HV, Varela JC, Barreda J, Moldawer LL, Schultz GS, Mozingo DW (2004) Temporal patterns of gene expression in murine cutaneous burn wound healing. Physiol Genomics 16(3):341–348. doi:10.1152/physiolgenomics.00101.2003
13. Olaso E, Lin HC, Wang LH, Friedman SL (2011) Impaired dermal wound healing in discoidin domain receptor 2-deficient mice associated with defective extracellular matrix remodeling. Fibrogenesis Tissue Repair 4(1):5. doi:10.1186/1755-1536-4-5
14. Olaso E, Labrador JP, Wang L, Ikeda K, Eng FJ, Klein R, Lovett DH, Lin HC, Friedman SL (2002) Discoidin domain receptor 2 regulates fibroblast proliferation and migration through the extracellular matrix in association with transcriptional activation of matrix metalloproteinase-2. J Biol Chem 277(5):3606–3613. doi:10.1074/jbc.M107571200
15. Goldring MB, Otero M (2011) Inflammation in osteoarthritis. Curr Opin Rheumatol 23(5):471–478. doi:10.1097/BOR.0b013e328349c2b1
16. Xu L, Peng H, Wu D, Hu K, Goldring MB, Olsen BR, Li Y (2005) Activation of the discoidin domain receptor 2 induces expression of matrix metalloproteinase 13 associated with osteoarthritis in mice. J Biol Chem 280(1):548–555. doi:10.1074/jbc.M411036200
17. Xu L, Peng H, Glasson S, Lee PL, Hu K, Ijiri K, Olsen BR, Goldring MB, Li Y (2007) Increased expression of the collagen receptor discoidin domain receptor 2 in articular cartilage as a key event in the pathogenesis of osteoarthritis. Arthritis Rheum 56(8):2663–2673. doi:10.1002/art.22761
18. Xu L, Servais J, Polur I, Kim D, Lee PL, Chung K, Li Y (2010) Attenuation of osteoarthritis progression by reduction of discoidin domain receptor 2 in mice. Arthritis Rheum 62(9):2736–2744. doi:10.1002/art.27582
19. Schminke B, Muhammad H, Bode C, Sadowski B, Gerter R, Gersdorff N, Burgers R, Monsonego-Ornan E, Rosen V, Miosge N (2013) A discoidin domain receptor 1 knock-out mouse as a novel model for osteoarthritis of the temporomandibular joint. Cell Mol Life Sci. doi:10.1007/s00018-013-1436-8
20. Hou G, Vogel W, Bendeck MP (2001) The discoidin domain receptor tyrosine kinase DDR1 in arterial wound repair. J Clin Invest 107(6):727–735. doi:10.1172/JCI10720
21. Franco C, Hou G, Ahmad PJ, Fu EY, Koh L, Vogel WF, Bendeck MP (2008) Discoidin domain receptor 1 (ddr1) deletion decreases atherosclerosis by accelerating matrix accumulation and reducing inflammation in low-density lipoprotein receptor-deficient mice. Circ Res 102(10):1202–1211. doi:10.1161/CIRCRESAHA.107.170662
22. Franco C, Britto K, Wong E, Hou G, Zhu SN, Chen M, Cybulsky MI, Bendeck MP (2009) Discoidin domain receptor 1 on bone marrow-derived cells promotes macrophage accumulation during atherogenesis. Circ Res 105(11):1141–1148. doi:10.1161/CIRCRESAHA.109.207357
23. Franco C, Ahmad PJ, Hou G, Wong E, Bendeck MP (2010) Increased cell and matrix accumulation during atherogenesis in mice with vessel wall-specific deletion of discoidin domain receptor 1. Circ Res 106(11):1775–1783. doi:10.1161/CIRCRESAHA.109.213637

24. Ahmad PJ, Trcka D, Xue S, Franco C, Speer MY, Giachelli CM, Bendeck MP (2009) Discoidin domain receptor-1 deficiency attenuates atherosclerotic calcification and smooth muscle cell-mediated mineralization. Am J Pathol 175(6):2686–2696. doi:10.2353/ajpath.2009.080734
25. Avivi-Green C, Singal M, Vogel WF (2006) Discoidin domain receptor 1-deficient mice are resistant to bleomycin-induced lung fibrosis. Am J Respir Crit Care Med 174(4):420–427. doi:10.1164/rccm.200603-333OC
26. Lee R, Eidman KE, Kren SM, Hostetter TH, Segal Y (2004) Localization of discoidin domain receptors in rat kidney. Nephron Exp Nephrol 97(2):E62–E70
27. Gross O, Beirowski B, Harvey SJ, McFadden C, Chen D, Tam S, Thorner PS, Smyth N, Addicks K, Bloch W, Ninomiya Y, Sado Y, Weber M, Vogel WF (2004) DDR1-deficient mice show localized subepithelial GBM thickening with focal loss of slit diaphragms and proteinuria. Kidney Int 66(1):102–111. doi:10.1111/j.1523-1755.2004.00712.x
28. Flamant M, Placier S, Rodenas A, Curat CA, Vogel WF, Chatziantoniou C, Dussaule JC (2006) Discoidin domain receptor 1 null mice are protected against hypertension-induced renal disease. J Am Soc Nephrol 17(12):3374–3381. doi:10.1681/ASN.2006060677
29. Guerrot D, Kerroch M, Placier S, Vandermeersch S, Trivin C, Mael-Ainin M, Chatziantoniou C, Dussaule JC (2011) Discoidin domain receptor 1 is a major mediator of inflammation and fibrosis in obstructive nephropathy. Am J Pathol 179(1):83–91. doi:10.1016/j.ajpath.2011.03.023
30. Kerroch M, Guerrot D, Vandermeersch S, Placier S, Mesnard L, Jouanneau C, Rondeau E, Ronco P, Boffa JJ, Chatziantoniou C, Dussaule JC (2012) Genetic inhibition of discoidin domain receptor 1 protects mice against crescentic glomerulonephritis. FASEB J 26(10):4079–4091. doi:10.1096/fj.11-194902
31. Gross O, Girgert R, Beirowski B, Kretzler M, Kang HG, Kruegel J, Miosge N, Busse AC, Segerer S, Vogel WF, Muller GA, Weber M (2010) Loss of collagen-receptor DDR1 delays renal fibrosis in hereditary type IV collagen disease. Matrix Biol 29(5):346–356. doi:10.1016/j.matbio.2010.03.002
32. Olaso E, Arteta B, Benedicto A, Crende O, Friedman SL (2011) Loss of discoidin domain receptor 2 promotes hepatic fibrosis after chronic carbon tetrachloride through altered paracrine interactions between hepatic stellate cells and liver-associated macrophages. Am J Pathol 179(6):2894–2904. doi:10.1016/j.ajpath.2011.09.002
33. Meyer zum Gottesberge AM, Gross O, Becker-Lendzian U, Massing T, Vogel WF (2008) Inner ear defects and hearing loss in mice lacking the collagen receptor DDR1. Lab Invest 88(1):27–37. doi:10.1038/labinvest.3700692
34. Cowling RT, Yeo SJ, Kim IJ, Park JI, Gu Y, Dalton ND, Peterson KL, Greenberg BH (2014) Discoidin domain receptor 2 germline gene deletion leads to altered heart structure and function in the mouse. Am J Physiol Heart Circ Physiol 307(5):H773–H781. doi:10.1152/ajpheart.00142.2014

# Chapter 5
# Discoidin Domain Receptors in Invertebrates

**Thomas Unsoeld, Jesse Taylor, and Harald Hutter**

## 5.1 Evolutionary Origin and Distribution of Discoidin Domain Receptors in Invertebrates

Similarity searches using discoidin domain receptor (DDR) sequences from human or mouse reveal the presence of DDRs throughout metazoans from the freshwater polyp *Hydra* [1] to vertebrates (Table 5.1). DDRs are not found in plants and unicellular eukaryotes such as yeast. Basic metazoans such as placozoa and porifera seem to have kinases very similar to the kinase domain of human DDRs. However, these predicted genes lack transmembrane and discoidin domains and are therefore likely not *bona fide* DDRs. Current genome sequence data place the evolutionary origin of DDRs close to the origin of metazoa.

Vertebrates have two DDR genes, whereas the genomes of nonvertebrate chordates such as the lancelet *Branchiostoma floridae* [2], the vase tunicate *Ciona intestinalis* [3], and echinoderms such as the purple sea urchin *Strongylocentrotus purpuratus* [4] have only one DDR gene (Table 5.1). This suggests that a duplication of an ancestral DDR gene occurred in the chordate lineage leading to vertebrates. Nematodes such as *C. elegans* also have two DDR genes, but sequence comparisons suggest that an ancestral DDR gene has been duplicated independently in the nematode lineage ([5], Fig. 5.1). The genome of another nematode, *Pristionchus pacificus* [6], does not seem to contain any DDR genes, suggesting that gene duplication as well as gene loss has occurred within the phylum Nematoda. A variable, but small number of DDRs is found within arthropods, ranging from one gene in *Drosophila melanogaster* [7] to four in *Nasonia vitripennis*, a parasitoid wasp [8]. Frequently one and occasionally two DDRs are found in polychaetes, molluscs, and

T. Unsoeld • J. Taylor • H. Hutter (✉)
Department of Biological Sciences, Simon Fraser University, 8888 University Drive, Burnaby, BC, Canada V5A 1S6
e-mail: hutter@sfu.ca



R. Fridman, P.H. Huang (eds.), *Discoidin Domain Receptors in Health and Disease*, DOI 10.1007/978-1-4939-6383-6_5

**Table 5.1** Number of DDRs in selected metazoan species

| Species | Common name | # of ddrs[a] | Sequence identifier[b] |
|---|---|---|---|
| *Hydra vulgaris* | Fresh-water polyp | 1 | XP_002168413.2 |
| *Crassostrea gigas* | Pacific oyster | 1 | EKC35223.1 |
| *Lottia gigantea* | Owl limpet | 1 | ESO96071.1 |
| *Capitella teleta* | Polychaete worm | 2 | ELU07385.1[d] |
| | | | ELT94315.1[e] |
| *Caenorhabditis elegans* | Roundworm | 2 | NP_508926.2 (DDR-1) |
| | | | NP_508572.1 (DDR-2) |
| *Daphnia pulex* | Water flea | 3 | DAPPUDRAFT_332239 |
| | | | DAPPUDRAFT_303380 |
| | | | DAPPUDRAFT_115845 |
| *Metaseiulus occidentalis* | Mite | 1 | XP_003743103.1 |
| *Ixodes scapularis* | Deer tick | 1 | XP_002413316.1 |
| *Drosophila melanogaster* | Fruit fly | 1 | FBpp0078876 |
| *Aedes aegypti* | Mosquito | 1 | XP_001661958.1 |
| *Apis mellifera* | Honey bee | 2 | XP_006568430.1 |
| | | | XP_006565578.1 |
| *Bombus terrestris* | Bumble bee | 2 | XP_003396649.1 |
| | | | XP_003396508.1 |
| *Nasonia vitripennis* | Parasitoid wasp | 4 | XP_001601637.2 |
| | | | XP_001601514.2 |
| | | | XP_001603879.2 |
| | | | XP_001603657.2 |
| *Tribolium castaneum* | Red flour beetle | 2 | XP_968938.1[f] |
| | | | XP_968793.2[g] |
| *Strongylocentrotus purpuratus* | Purple sea urchin | 1 | XP_783985.3 |
| *Ciona intestinalis* | Vase tunicate | 1 | XP_002123027.2 |
| *Branchiostoma floridae* | Lancelet | 1 | XP_002596288.1[c] |
| *Mus musculus* | Mouse | 2 | Q03146.2 (DDR1) |
| | | | Q62371.2 (DDR2) |
| *Homo sapiens* | Human | 2 | Q08345.1 (DDR1) |
| | | | Q16832.2 (DDR2) |

[a]Identified by BLASTp searches with the human DDR proteins against species-specific NCBI protein databases
[b]NCBI database identifier
[c]XP_002109288.1 only contains a kinase domain, which is most similar to the kinase domain in human DDRs
[d]Sequence for "Ct 1" in Table 5.2
[e]Sequence for "Ct 2" in Table 5.2
[f]Sequence for "Tc 1" in Table 5.2
[g]Sequence for "Tc 2" in Table 5.2

cnidarians (Table 5.1). Overall, the number of DDRs throughout metazoans is remarkably constant. DDRs have neither been lost in larger phylogenetic groups nor been substantially expanded. However, duplications of an ancestral DDR gene seem to have occurred independently in several phyla (Fig. 5.1).

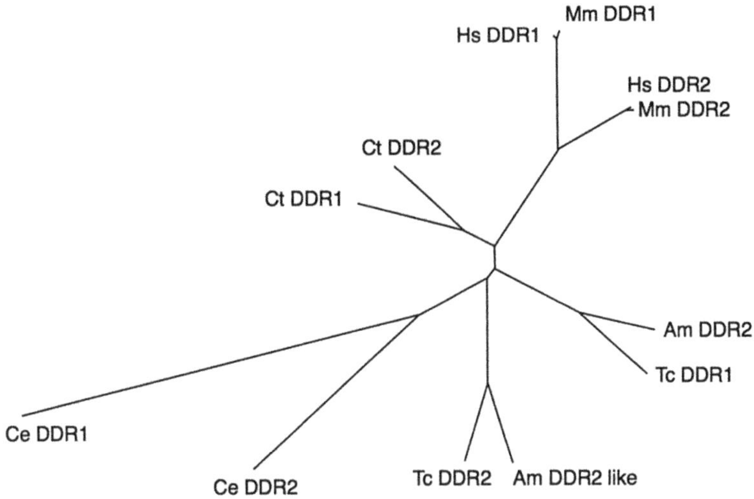


**Fig. 5.1** Pylogenetic tree of selected vertebrate and invertebrate DDRs. DDR protein sequences were aligned using ClustalW2 using default options (https://www.ebi.ac.uk/Tools/msa/clustalw2/). The phylogenetic tree was generated using ClustalW2—Phylogeny with distance correction and gaps excluded (https://www.ebi.ac.uk/Tools/services/web/toolform.ebi?tool=clustalw2_phylogeny). Phylohedron was used to draw the tree (http://iubio.bio.indiana.edu/treeapp/treeprint-sample1.html), which was redrawn using Keynote (Apple Inc) for higher resolution. See Table 5.1 for sequence accession numbers. *Abbreviations*: *Hs Homo sapiens*, *Mm Mus musculus*, *Ce Caenorhabditis elegans*, *Am Apis mellifera*, *Tc Tribolium castaneum*, *Ct Capitella teleta*

**Table 5.2** Sequence similarities in the kinase and Sequence of DDRs

| Sequence similarity[a] | | | | | | | | | | | | |
|---|---|---|---|---|---|---|---|---|---|---|---|---|
| | Hs 1 (%) | Hs 2 (%) | Ci (%) | Tc 1 (%) | Tc 2 (%) | Dm (%) | Ce 1 (%) | Ce 2 (%) | Ct 1 (%) | Ct 2 (%) | Cg (%) | Hv (%) |
| Hs[b] 1 | | 75 | 64 | 58 | 55 | 57 | 59 | 55 | 57 | 59 | 58 | 44 |
| Hs[b] 2 | 79 | | 63 | 59 | 58 | 52 | 60 | 56 | 61 | 63 | 57 | 45 |
| Ci[c] | 71 | 74 | | 53 | 58 | 59 | 58 | 56 | 58 | 56 | 60 | 45 |
| Tc[d] 1 | 67 | 65 | 66 | | 66 | 71 | 64 | 58 | 62 | 69 | 63 | 44 |
| Tc[d] 2 | 58 | 59 | 57 | 60 | | 69 | 62 | 59 | 66 | 71 | 63 | 40 |
| Dm[e] | 62 | 62 | 58 | 65 | 60 | | 66 | 59 | 64 | 65 | 67 | 44 |
| Ce[f] 1 | 44 | 43 | 41 | 45 | 50 | 49 | | 66 | 65 | 67 | 68 | 43 |
| Ce[f] 2 | 59 | 62 | 62 | 58 | 58 | 52 | 48 | | 56 | 62 | 58 | 47 |
| Ct[g] 1 | 70 | 70 | 66 | 71 | 62 | 65 | 44 | 60 | | 77 | 72 | 47 |
| Ct[g] 2 | 71 | 73 | 67 | 73 | 62 | 69 | 46 | 59 | 77 | | 75 | 41 |
| Cg[h] | 75 | 76 | 69 | 75 | 63 | 70 | 46 | 60 | 76 | 82 | | 44 |
| Hv[i] | 60 | 63 | 59 | 58 | 55 | 57 | 49 | 54 | 62 | 64 | 64 | |

[a]% similar or identical amino acids in pairwise BLAST comparisons with the discoidin domains (upper-right half) and kinase domains (lower-left half) as predicted by SMART

[b]*Homo sapiens*; note human and mouse DDR sequences are 96 % identical

[c]*Ciona intestinalis*

[d]*Tribolium castaneum*

[e]*Drosophila melanogaster*

[f]*Caenorhabditis elegans*

[g]*Capitella teleta*

[h]*Crassostrea gigas*

[i]*Hydra vulgaris*

Sequence comparisons of the DDR kinase domains (Table 5.2) reveal that DDRs within a species (e.g., *C. elegans*) are typically as divergent as DDRs from distant phylogenetic groups (e.g., humans). This suggests that gene duplications within the DDR gene family were not recent events. With the exception of *C. elegans* DDR-1, sequence similarities of invertebrate DDR kinase domains to the human sequences are in the range of 58–75 %. The kinase domain of the *C. elegans* DDR-1 protein shows by far the lowest similarity to other DDR kinase domains and is even more divergent than the *Hydra vulgaris* DDR kinase domain.

The discoidin (DS) domains of DDRs within the same species tend to be more similar to each other than to domains from other phylogenetic groups (Table 5.2). Similarities between invertebrate and human DS domains are somewhat lower than for the kinase domains and range from 44 to 64 %. The *Hydra vulgaris* DS domain is most divergent with only 44 % similarity to human DDR1, whereas the DS domain of *C. elegans* DDR-1 is not noticeably different from other invertebrate DS domains in this respect. Excluding the least (*C. intestinalis*) and most divergent (*H. vulgaris*) sequences, the invertebrate sequences show little variation in their similarity to the human DS domains (52–63 % similarity).

With the exception of the *C. elegans* DDRs (see below) invertebrate DDRs have not been functionally characterized so far. First glimpses into the potential roles of DDRs in *Drosophila* come from large-scale RNAi studies. The *Ddr* gene in *D. melanogaster* does not seem to be essential for survival since RNAi against *Ddr* does not result in lethality [9]. In a large-scale RNAi screen, *Ddr* was found to be required to maintain morphological complexity of cells in culture [10]. In an independent screen for kinases required for cell cycle progression, RNAi against *Ddr* resulted in chromosomal abnormalities in cell culture [11]. The in vivo function of *Ddr* in *Drosophila* is unclear at this point. The sea urchin DDR gene is expressed by primary mesenchyme cells in the embryo [12]. This points to a possible role during embryonic development, which remains to be explored by functional studies.

## 5.2 Discoidin Domain Receptors in *C. elegans*

The *C. elegans* genome contains two DDR genes: *ddr-1* and *ddr-2*. Sequence comparisons with vertebrate DDRs reveal that the general domain organization is quite similar with an N-terminal Discoidin domain (DS) and an intracellular C-terminal kinase domain separated by a transmembrane domain (Fig. 5.2). The extracellular

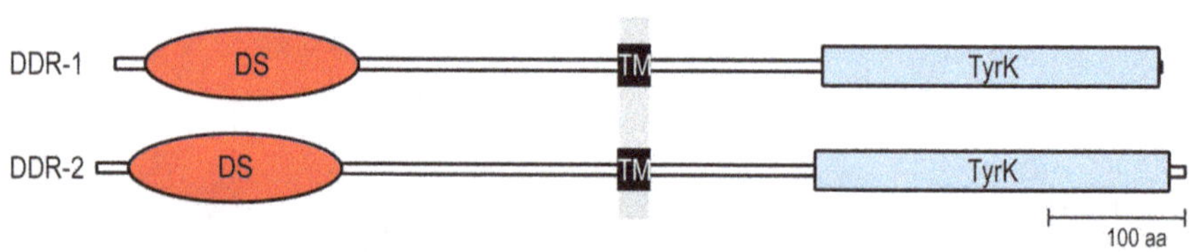


**Fig. 5.2** Domain organization of nematode DDRs. Domain organization of the *C. elegans* DDR-1 and DDR-2 proteins. *DS* discoidin domain, *TM* transmembrane domain, *TyrK* tyrosine kinase

domains of human DDRs (hDDR1 and hDDR2) have been characterized in regard to amino acids required for collagen binding and receptor activation [13]. A sequence comparison reveals that some, but not all of these amino acids, are conserved in nematode DDRs, suggesting that collagens might serve as ligands for nematode DDRs. In mammals, DDRs have been shown to bind a number of proteins through conserved binding motifs on the intracellular domain. These include PI3K through its p85 subunit, the adaptor proteins NCK1/2 and ShcA, SHP-2 phosphatase and CSK [14, Lemeer, 2011 #1696]. In *C. elegans,* homologs exist for many known interacting proteins and several of the binding motifs are conserved as well [5], suggesting that *C. elegans* DDRs could share at least some of the signaling pathways identified in vertebrates. For example in humans ShcA, an adaptor protein, has been shown to bind hDDR1 through a conserved phosphotyrosine binding (PTB) domain at tyrosine 513 [14]. Binding by ShcA at this site has been implicated in a number of cellular and extracellular processes including collagen deposition, cell differentiation, and cell migration. [15–18]. A similarly arranged motif is present in DDR-1 [5], which might be utilized by the *C. elegans* ShcA homolog (SHC-1) for downstream signaling. Proposed binding sites for NCK1/2 are found in both human DDRs [5, 19] and also in *C. elegans* DDR-2 ($^{513}$YAEP), but not in DDR-1. Loss-of-function of *C. elegans nck-1* leads to defects in axon navigation and neuronal cell migrations [20], similar to phenotypes observed in *ddr-2* mutant animals [21]. The overlapping neuronal phenotypes may hint at a potential role for NCK-1 in DDR-2 signaling in *C. elegans*. DDR-1 contains a predicted conventional binding site for Src family kinases (SFKs) containing Src-homology-2 (SH2) domains. Previous studies have suggested hDDR2 activation in mammals may occur through a Src-dependent mechanism [22, 23]. Two nonconventional binding sites in mammalian DDRs were recently discovered for SFKs including c-Src [19] indicating the binding sites may have evolved separately. The Rho/Rac guanine nucleotide exchange factor VAV1 has also been shown to interact with mammalian DDR1 [19]. A conserved binding motif for Vav exists in *C. elegans* DDR-2, but not in DDR-1. Finally, the binding motif for the PI3K p85 regulatory subunit has been conserved among all forms of DDR in both humans and *C. elegans*. Currently, no physical binding data in invertebrate models exist to verify proposed interactions. However, the conservation of several potential binding motifs in at least one of the two nematode DDRs hints at conserved signaling mechanisms in vertebrates and nematodes. While signaling pathways used by DDRs in *C. elegans* are currently unknown, initial functional studies have implicated DDRs in axonal navigation during nervous system development (see below).

### 5.2.1 *The* C. elegans *Nervous System*

The nervous system in *C. elegans* consists of only 302 neurons, which are grouped in 118 distinct classes according to their synaptic connections and morphology [24]. In contrast to neurons in vertebrates, most neurons in *C. elegans* have a simple morphology and extend only one or two processes, which are frequently unbranched.

The entire nervous system of *C. elegans* has been reconstructed from serial electron microscopic sections; as a result, all neuronal connections and the location of every cell body have been determined [24–27]. The position of neuronal cell bodies and even individual axons is remarkably invariant from animal to animal, which simplifies the detection of developmental defects such as misguided axons. Since *C. elegans* is transparent, neurons and neuronal processes can be visualized in living animals using fluorescent markers, which are expressed under the control of neuronal promoters [28]. Information flows through the nervous system from sensory neurons located mainly in the head ganglia via interneurons to motor neurons, which innervate body wall muscle cells and control movement of the animal in response to sensory input.

The majority of neurons in *C. elegans* are clustered in a series of ganglia in the head and tail (Fig. 5.3a). The head ganglia consist mainly of sensory neurons and interneurons. Some interneurons send axons into the major longitudinal nerve tract,

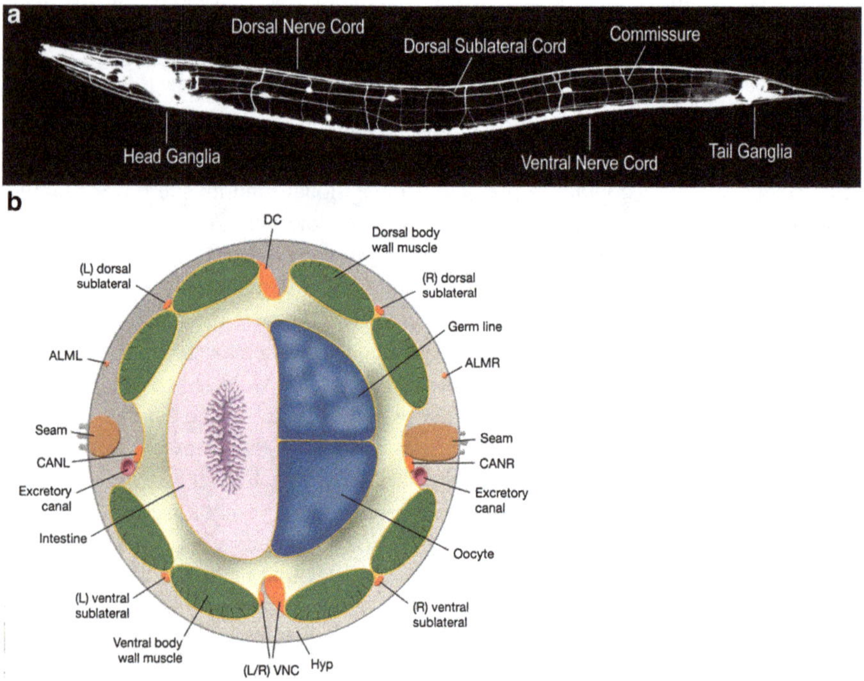


**Fig. 5.3** *C. elegans* nervous system. (**a**) Confocal image of an adult *C. elegans* animal expressing GFP in the entire nervous system. Lateral view, anterior is to the left. (**b**) Schematic cross sections through the anterior midbody region showing the location of longitudinal axon tracts. Neurons are bounded by the basement membrane of the hypodermis (Hyp). The lateral touch receptor neurons (ALMR and ALML) run on the peripheral side of the hypodermis, whereas ventral nerve cord (VNC), dorsal nerve cord (DC), and canal-associated nerves (CANR and CANL) run on the inside between the hypodermis and pseudocoelom. The sublateral (subdorsal and subventral) cords lie just inside a thin layer of hypodermis near body wall muscles. (*Orange*) basement membrane; (*yellow*) pseudocoelom; and (*green*) body wall muscle. Image reproduced with permission from Wormatlas (http://www.wormatlas.org/hermaphrodite/nervous/Neuroframeset.html)

the ventral nerve cord (VNC), where they connect to motor neurons. Outgrowing axons are in contact with the basement membrane separating ectodermal and mesodermal tissues [24], which provides a substrate to grow upon and also contains navigational cues [29]. The VNC consists of two fascicles flanking the somata of motor neurons that are aligned in a row along the ventral midline. The majority of axons (about 50) run in the right nerve tract, whereas only four axons extend in the left fascicle. On the dorsal side a single nerve tract, the dorsal nerve cord (DNC), contains axons of motor neurons that have extended from motor neuron cell bodies on the ventral midline in order to establish synapses with dorsal body wall muscle cells. In addition, several smaller tracts, each containing only a few axons, run in various lateral positions (Fig. 5.3b). The overall simplicity of the nervous system and its stereotypic development make *C. elegans* an excellent model system to study the basic molecular principles of nervous system development.

### 5.2.2 *Discoidin Domain Receptor Expression in* C. elegans

Gene expression profiling of various embryonic tissues of *C. elegans* found that *ddr-1* and *ddr-2* are strongly expressed in embryonic neurons [30]. More recently, reporter gene studies showed that both DDRs in *C. elegans* are expressed in various tissues during development and in the adult animal [21]. *ddr-1* expression is found in epidermal cells at the beginning of morphogenesis during embryonic development (Fig. 5.4a, b). At the time when neurons differentiate and send out their axons, *ddr-1* reporter gene expression is also found in a number or neurons in the head and the tail (Fig. 5.4d). Postembryonically, *ddr-1* expression is mostly limited to the nervous system. Expression can be found in many neurons in the head and tail ganglia, motor neurons of the ventral nerve cord as well as in all major longitudinal nerve tracts (Fig. 5.4e, f). Expression outside the nervous system is detectable in the stomatointestinal muscles (Fig. 5.4e) and the pharynx, the feeding organ of *C. elegans*.

*ddr-2* reporter constructs are expressed at the end of gastrulation [21] (Fig. 5.4h, i). Expression is particularly strong in lateral epidermal cells (seam cells) and at later stages also in a number of neurons in the head (Fig. 5.4j–l). Postembryonically expression in seam cells is maintained (Fig. 5.4l). In the nervous system, expression is limited to a subgroup of neurons in the head and tail and some motor neurons in the ventral nerve cord (Fig. 5.4l). Additional expression is found in some gland cells and several nonneuronal cells in the tail. GFP fusion protein of DDR-2 localizes predominantly to the major longitudinal nerve tracts and neuronal cell bodies (Fig. 5.4g, m). Likewise, DDR-1 GFP fusion protein localizes to axons [21], suggesting that DDRs could function as receptors in neurons.

Mammalian DDRs are typically more widely expressed. Epithelial tissues primarily express DDR1, and DDR2 is predominantly expressed in connective tissues [31]. DDR1 mRNA is expressed in most tissues during development and in adults. Particular high mRNA levels are found in the kidney, placenta, spleen, lung, skin

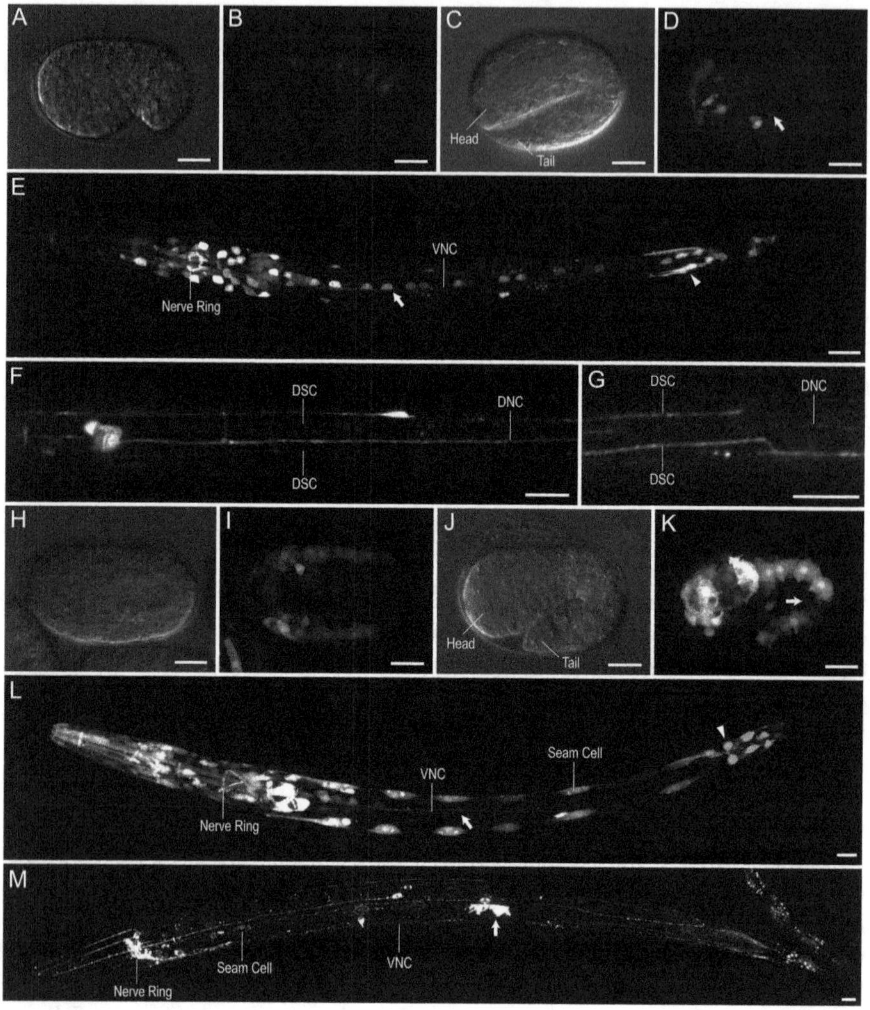


**Fig. 5.4** Expression of *ddr-1* and *ddr-2* in *C. elegans*. (**a, b**) Onset of *ddr-1* expression in the embryo during "lima-bean-stage". (**c, d**) Twofold-stage embryo showing an extending axon (*arrow*). (**e**) L1 larva, *ddr-1* is expressed in neurons in head and tail ganglia, motor neurons (*arrow*) and stomatointestinal muscle (*arrowhead*). (**f**) L1 larva, *ddr-1* is expressed in dorsal nerve cord (DNC) and dorsal sublateral cords (DSC). (**h, i**) *ddr-2* expression starts during late gastrulation. (**j, k**) 1.5-fold-stage embryo with an extending axon (*arrow*). (**l**) L1 larvae, *ddr-2* is expressed in neurons in head and tail, motor neurons (*arrow*), rectal gland cells (*arrowhead*), and seam cells. (**g, m**) L2 larvae; DDR- 2::GFP fusion protein localizes to the nerve ring, all major longitudinal axon tracts, seam cells, and most likely uterine valve cells of the vulva (*arrow*). (**a–d, h–k, m**) lateral views; (**e, l**) ventral views; (**f**) dorsal view; and (**g**) dorsolateral view. In all pictures, anterior is to the left. (**f, g**) single focal plain. Scale bar: 10 μm. Adapted from Developmental Biology, Vol. 374, Thomas Unsoeld, Ja-On Park, Harald Hutter, 'Discoidin domain receptors guide axons along longitudinal tracts in *C. elegans*', Pages 142–152, © Elsevier (2013)

epithelia, and brain [32]. DDR2 mRNA is expressed most intensely in the kidney, lung, and skeletal and heart muscles [32].

Both mammalian DDRs, like their counterparts in *C. elegans*, are expressed in the nervous system. DDR2 expression was found in the developing and mature brain of rats [33]. DDR1 is expressed in proliferating areas in the central nervous system of rats and mice [34, 35]. The similarities in neuronal expression between mammals and *C. elegans* might point to an evolutionary conserved function of discoidin domain receptors in the nervous system.

### *5.2.3* C. elegans *DDRs Function in the Guidance of Axons Along Longitudinal Nerve Tracts*

The first insight in DDR function in *C. elegans* was recently obtained from the study of loss-of-function mutants of *ddr-1* and *ddr-2* [21]. Animals lacking functional DDR-1 or DDR-2 look superficially wild type and do not show obvious morphological defects. However, *ddr-2* mutants exhibit significant defects in the main longitudinal nerve tracts. Individual processes or whole fascicles fail to navigate along their normal trajectories and instead join other nerve tracts in their vicinity. *ddr-1* single mutants show no significant axonal defects on their own, but mutants lacking both functional DDR-1 and DDR-2 display strongly enhanced guidance defects in various nerve tracts compared to *ddr-2* single mutants, indicating both DDRs play a role in axon guidance.

The most penetrant nervous system defects of discoidin domain receptor mutants are found in the left VNC [21]. The outgrowth of nerve processes in the VNC of *C. elegans* occurs successively and has been reconstructed from serial sectioned embryos of various developmental stages [36]. The left VNC tract is established by the axon of the interneuron PVPR from the posterior end (Fig. 5.5a). The PVPR pioneer axon is closely followed by the process of the interneuron PVQL [36]. Postembryonically, the axon of HSNL joins the prelaid track formed by the axons of PVPR and PVQL and follows in anterior direction toward the nerve ring [37]. In *ddr-2* mutant animals, the axon of the PVPR pioneer neuron frequently crosses the ventral midline and joins the right VNC fascicle or fails to establish the left axon tract altogether (Fig. 5.5c, e). In animals with pioneer defects, the later outgrowing axons from the PVQL and HSNL neurons follow the misguided PVPR pioneer axon (Fig. 5.5g, i), suggesting that defects in follower axons are secondary consequences of a primary defect in the pioneer.

Similar to the left tract, axon outgrowth in the right VNC fascicle follows a distinct sequential pattern: A single pioneer axon (Fig. 5.5a), in this instance coming from the anterior, initially establishes the tract and is sequentially followed by various processes that grow out later in development [36]. Mutants lacking functional DDR-2 show only mild navigation defects in the right VNC pioneer [21]. Likewise,

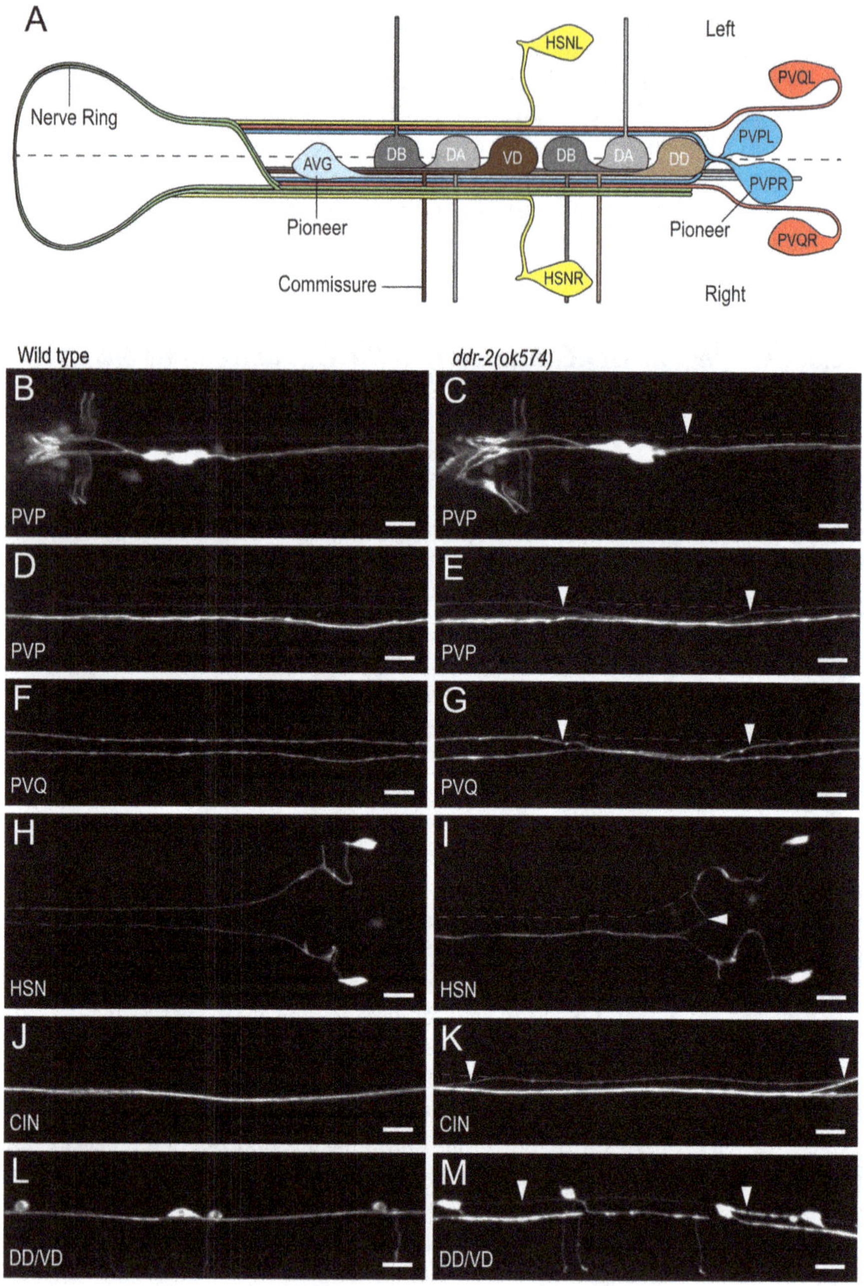


**Fig. 5.5** Axonal defects in the ventral nerve cord of *ddr-2* mutants. (**a**) Schematic of the ventral nerve cord (VNC) of *C. elegans*. The two VNC axon tracts are separated by motor neuron cell bodies (DA, DB, DD, and VD). The right VNC tract is pioneered by the axon of AVG, the left tract by the axon of PVPR. The PVP axons cross the ventral midline and extend in the respective contralateral tract. All axons with the exception of the postembryonic HSN axons grow out in the

the defects in later extending nerve processes are less penetrant compared to the left fascicle. Axons from command interneurons of the motor circuit and ventral processes of motor neurons that normally extend exclusively in the right VNC tract cross the ventral midline and join the left VNC fascicle (Fig. 5.5k, m). DDR-1 is not required for axon navigation of the pioneer and the majority of followers in the right VNC tract with the exception of a subset of motor neurons, which seem to depend on both DDR-1 and DDR-2 [21].

In addition to their importance in the ventral nerve cord, discoidin domain receptors are also required for the guidance of processes along several other longitudinal nerve tracts [21]. The dorsal nerve cord is formed on the left side of the dorsal midline (Fig. 5.6a) by motor neuron axons that extend longitudinally after growing circumferentially from the VNC [36]. Both DDR-1 and DDR-2 are involved in navigation along the dorsal nerve cord. In a quarter of *ddr-2* mutant animals, some motor neuron processes extend aberrantly on the right side rather than the left side of the dorsal midline. In *ddr-1 ddr-2* double mutants, nearly half of the animals extend processes on the wrong side of the dorsal midline (Fig. 5.6c).

Mutations in discoidin domain receptors also affect navigation along the two dorsal sublateral cords. In the anterior half of the animal, the dorsal sublateral cords run in parallel to the dorsal nerve cord. At the midbody region, the dorsal sublateral cords turn ventrally and join the more lateral running ALN axon tract (Fig. 5.6d). In a significant number of *ddr-2* mutant animals and even more frequently in *ddr-1 ddr-2* double mutant animals, processes from the right dorsal sublateral cords turn dorsally rather than ventrally and extend along the right side of the dorsal midline (Fig. 5.6e). Similarly but less frequently, processes from the left dorsal sublateral cord aberrantly grow dorsally and join the dorsal nerve cord. Other processes from both dorsal sublateral cords leave their normal trajectory or converge with the ALN tract at a more anterior or posterior position. On the ventral side, only one of the sublateral cords running in parallel of the ventral nerve cord is mildly affected in DDR-2 loss-of-function mutants. DDR-1 does not seem to be required for the navigation of processes in either ventral sublateral cord. Motor neuron commissures traveling in dorsoventral direction are unaffected in *ddr*-mutants, indicating that navigation in dorso-ventral direction is independent of DDR function.

**Fig. 5.5** (continued) embryo. (**b–m**) Axon trajectories in (**b**, **d**, **f**, **h**, **j**, **l**) wild type and (**c**, **e**, **g**, **i**, **k**, **m**) *ddr-2* loss-of-function mutants visualized by fluorescent reporter constructs. (**b**, **d**) In wild type, the PVPR axon is in the left VNC fascicle. (**c**) In *ddr-2* mutant animals, PVPR fails to pioneer the left tract (*arrowhead*) and instead grows in the right tract or (**e**) switches between the two tracts (*arrowheads*). (**g**) The PVQL axon follows the misled PVPR pioneer depicted in panel (**e**) (*arrowheads*). (**h**) In wild type, the HSNL axon extends in the left VNC tract. (**i**) In *ddr-2* mutant animals, the axon of HSNL joins the right VNC tract (*arrowhead*). (**j**) In wild type, command interneuron axons extend in the right VNC tract. (**k**) In *ddr-2* mutant animals, axons switch into the left VNC tract (*arrowheads*). (**l**) In wild type, the longitudinal processes of motor neurons (DD/VD) extend into the right VNC tract. (**m**) In *ddr-2* mutant animals, some motor neuron processes join the left tract (*arrowheads*). *Dashed lines* in pictures reflect normal axon trajectories. All pictures and the diagram show ventral views; anterior is to the left. Scale bar: 10 μm. Adapted from Developmental Biology, Vol. 374, Thomas Unsoeld, Ja-On Park, Harald Hutter, 'Discoidin domain receptors guide axons along longitudinal tracts in *C. elegans*', Pages 142–152, © Elsevier (2013)

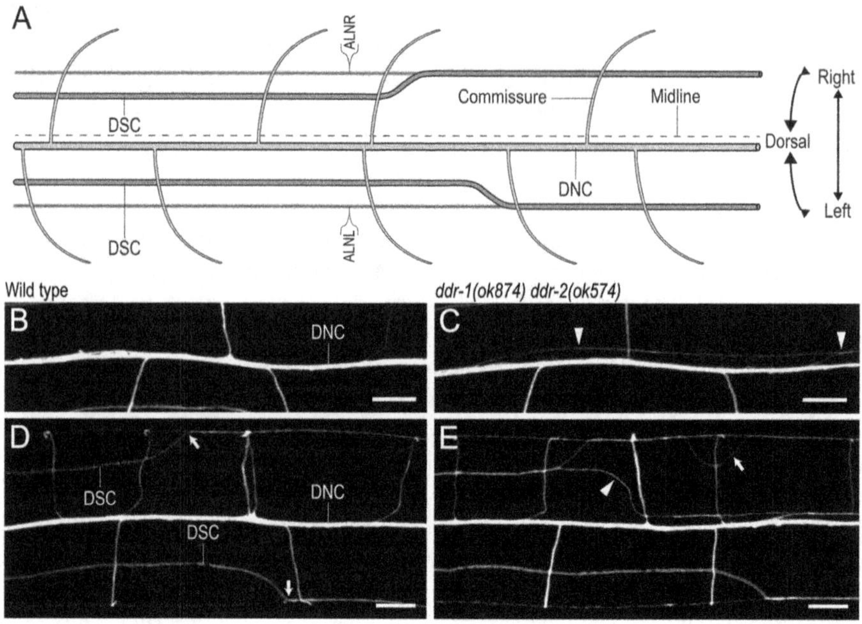


**Fig. 5.6** Axonal defects in the dorsal and dorsal sublateral nerve cords in *ddr-1 ddr-2* double mutants. (**a**) Schematic of dorsal nerve cord and dorsal sublateral cords of *C. elegans*. Commissural axons from ventral motor neurons migrate circumferentially along both sides of the body toward the dorsal midline and form the dorsal nerve cord (DNC) on the left side of the midline. The two dorsal sublateral cords (DSC) flank the dorsal nerve cord on either side. Close to the midbody region the sublateral tracts move in a more lateral position and join the ALN axon tracts. (**b–e**) Wild type and *ddr-1(ok874) ddr-2(ok574)* double mutant animals labeled with pan-neuronal GFP. (**b**) In wild-type animals, the dorsal nerve cord is tightly fasciculated, all processes extend on the left side of the dorsal midline. (**c**) In *ddr-1 ddr-2* double mutants, some processes extend on the right side of the midline (*arrowheads*). (**d**) Axons from the dorsal sublateral cords join the ALN axon trajectories (*arrows*). (**e**) In *ddr-1 ddr-2* double mutants, axons from the right dorsal sublateral cord grow dorsally and extend along the right side of the dorsal midline (*arrowhead*); other axons leave their normal trajectory (*arrow*). All pictures and the diagram show dorsal views; anterior is to the left. Scale bar: 10 µm. Adapted from Developmental Biology, Vol. 374, Thomas Unsoeld, Ja-On Park, Harald Hutter, 'Discoidin domain receptors guide axons along longitudinal tracts in *C. elegans*', Pages 142–152, © Elsevier (2013)

The observed defects in *ddr* loss-of-function mutants correspond with the expression patterns of the two DDR receptors in *C. elegans* [21]. Both DDRs are expressed in neurons extending processes in all affected nerve tracts. So far, only the importance of DDR-2 for the guidance of the PVPR pioneer axons and its follower PVQL has been studied in more detail [21]. Both DDRs are expressed in the pioneer neuron PVPR. A genetic mosaic analysis correlating the presence or absence of a rescuing *ddr-2* transgene in PVPR with the presence or absence of axonal defects demonstrated that DDR-2 is required cell-autonomously in the PVPR neuron for the guidance of the pioneer axon [21]. In addition, DDR-2 was shown to be required in PVPR for the proper guidance of the PVQL follower axon, which follows its misguided pioneer axon in most instances in the absence of DDR-2. The requirement for DDR-2 in the pioneer neuron and the strict dependence of the PVQL axon on its pioneer PVPR demonstrate that PVQL follower defects in *ddr-2* mutants are a secondary consequence of a primary pioneer defect that result from a lack of functional DDR-2 in the pioneer.

Taken together, these studies suggest that axons in different longitudinal axon tracts depend to varying extent on DDRs for their navigation. DDRs are most important for navigation in the ventral nerve cord, in particular for the pioneer neuron establishing the left VNC tract. Since defects in *ddr*-mutants are not completely penetrant, it is likely that DDRs act partially redundantly with other receptors.

## 5.3 Collagens as Putative Ligands for DDRs in *C. elegans*

Mammalian discoidin domain receptors are activated upon binding various collagens including types I–V, VIII, X, and XI [14, 38–40]. Thus far, ligands for *C. elegans* DDRs have not been identified, but collagens are the most obvious candidates. The genome of *C. elegans* contains nearly 200 genes encoding collagens [41]. There are two main classes of collagens in *C. elegans*: cuticular and basement membrane collagens. The *C. elegans* exoskeleton is a multilayered cuticle composed of a large number of collagens covering the entire outside of the body [42]. These cuticular collagens are produced in and secreted from the underlying epidermal tissue. The majority of collagens present in *C. elegans* (about 170 [42]) are of this nematode-specific class. Due to their location on the surface of the animal, cuticular collagens could only interact with DDRs expressed in the underlying epidermal cells.

As is true in other metazoans, the basement membrane of *C. elegans* plays a vital role in support of surrounding tissues and embryonic development [43]. Basic components of the basement membrane such as laminin, type IV and type XVIII collagen and nidogen are found in *C. elegans* [43]. *C. elegans* possesses two type IV collagens, EMB-9 and LET-2. Both collagens are essential structural components of basement membranes and *emb-9* and *let-2* mutant embryos arrest development at a time before axons start to grow out [44–46]. Vertebrate DDR1 binds type IV collagen [14]. Similarly, EMB-9 and LET-2 might serve as DDR ligands in *C. elegans*. The early lethality of collagen IV mutants in *C. elegans* currently prevents a straightforward evaluation of collagen IV as putative ligand for DDRs. However, discoidin domain receptors in *C. elegans* have been found to interact genetically with *nid-1/nidogen* [21]. NID-1 is expressed in most basement membranes of *C. elegans*, but accumulates on the ventral and dorsal sublateral cords and at the muscle edges flanking the ventral and dorsal nerve cord [47, 48]. Interestingly, loss-of-function mutations in *nid-1* mutants are not lethal, but *nid-1* mutants show axonal guidance defects in the major longitudinal nerve tracts, including the ventral nerve cord, that are strikingly similar to defects in DDR mutants [48]. No direct physical interaction between Nidogen and DDRs has been reported thus far, but genetic interactions do not necessarily reflect direct biochemical interactions. Instead DDRs might interact with type IV collagen, which is known to bind to Nidogens in vertebrates [49, 50]. *cle-1*, the collagen XVIII homolog in *C. elegans* is required for cell migrations and guidance of dorsally directed commissural axons [51]. However, *cle-1* mutants do not exhibit significant defects in the ventral nerve cord and also do not enhance the *ddr-2* ventral nerve cord defects in *ddr-2*; *cle-1* double mutants [21], suggesting that in contrast to *ddrs cle-1* does not function in axon navigation in the ventral nerve cord and its pioneer axon PVPR. Whether *cle-1* is required in other longitudinal nerve tracts remains to be explored.

Of the other collagen types known to bind DDRs in vertebrates [14, 38], only few homologs exist in *C. elegans*. Apart from collagen IV and XVIII, *C. elegans* has putative homologs of vertebrate collagen type III and type XIII/XXIII/XXV. Type III collagens are fibrillar proteins found in most soft and connective tissues in vertebrates that have been implicated in DDR activation through binding at a conserved GVMGFO motif in the collagen protein [52]. In *C. elegans, dpy-14* encodes a collagen similar to the vertebrate type III alpha 1 chain collagen [53]. However, DPY-14 is thought to be a cuticular collagen [53] and lacks the GVMGFO binding motif, making it an unlikely candidate for interaction with DDRs. Types XIII, XXIII, and XXV are transmembrane collagens. Type XIII collagen is strongly expressed in the developing central and peripheral nervous system of mice including the spinal cord [54]. Interestingly, type XIII collagen is able to induce neurite outgrowth [54]. Type XXV collagen is expressed in spinal motor neurons and functions in intramuscular innervation [55]. Recently, COL-99, a *C. elegans* collagen with similarity to vertebrate type XIII/XXIII/XXV collagen, has been identified in genetic screens for genes important for axon guidance in the ventral nerve cord of *C. elegans* (our own unpublished observations). The phenotype of *col-99* mutants shows striking similarities to discoidin domain receptor mutants, raising the possibility that COL-99 might serve as ligand for DDRs in *C. elegans*. Additional research will be required to address this hypothesis.

In summary, while no ligands for DDRs have been identified in *C. elegans* to date, phenotypic observations in collagen mutants and genetic interaction studies suggest collagens, including the basement membrane collagen IV as well as the transmembrane collagen COL-99, as putative ligands for DDRs.

## References

1. Reddy PC, Bidaye SS, Ghaskadbi S (2011) Genome-wide screening reveals the emergence and divergence of RTK homologues in basal Metazoan Hydra magnipapillata. J Biosci 36(2):289–296
2. Putnam NH, Butts T, Ferrier DE, Furlong RF, Hellsten U, Kawashima T, Robinson-Rechavi M, Shoguchi E, Terry A, Yu JK, Benito-Gutierrez EL, Dubchak I, Garcia-Fernandez J, Gibson-Brown JJ, Grigoriev IV, Horton AC, de Jong PJ, Jurka J, Kapitonov VV, Kohara Y, Kuroki Y, Lindquist E, Lucas S, Osoegawa K, Pennacchio LA, Salamov AA, Satou Y, Sauka-Spengler T, Schmutz J, Shin IT, Toyoda A, Bronner-Fraser M, Fujiyama A, Holland LZ, Holland PW, Satoh N, Rokhsar DS (2008) The amphioxus genome and the evolution of the chordate karyotype. Nature 453(7198):1064–1071. doi:10.1038/nature06967, nature06967 [pii]

3. Dehal P, Satou Y, Campbell RK, Chapman J, Degnan B, De Tomaso A, Davidson B, Di Gregorio A, Gelpke M, Goodstein DM, Harafuji N, Hastings KE, Ho I, Hotta K, Huang W, Kawashima T, Lemaire P, Martinez D, Meinertzhagen IA, Necula S, Nonaka M, Putnam N, Rash S, Saiga H, Satake M, Terry A, Yamada L, Wang HG, Awazu S, Azumi K, Boore J, Branno M, Chin-Bow S, DeSantis R, Doyle S, Francino P, Keys DN, Haga S, Hayashi H, Hino K, Imai KS, Inaba K, Kano S, Kobayashi K, Kobayashi M, Lee BI, Makabe KW, Manohar C, Matassi G, Medina M, Mochizuki Y, Mount S, Morishita T, Miura S, Nakayama A, Nishizaka S, Nomoto H, Ohta F, Oishi K, Rigoutsos I, Sano M, Sasaki A, Sasakura Y, Shoguchi E, Shin-i T, Spagnuolo A, Stainier D, Suzuki MM, Tassy O, Takatori N, Tokuoka M, Yagi K, Yoshizaki F, Wada S, Zhang C, Hyatt PD, Larimer F, Detter C, Doggett N, Glavina T, Hawkins T, Richardson P, Lucas S, Kohara Y, Levine M, Satoh N, Rokhsar DS (2002) The draft genome of Ciona intestinalis: insights into chordate and vertebrate origins. Science 298(5601):2157–2167. doi:10.1126/science.1080049298/5601/2157 [pii]
4. Sodergren E, Weinstock GM, Davidson EH, Cameron RA, Gibbs RA, Angerer RC, Angerer LM, Arnone MI, Burgess DR, Burke RD, Coffman JA, Dean M, Elphick MR, Ettensohn CA, Foltz KR, Hamdoun A, Hynes RO, Klein WH, Marzluff W, McClay DR, Morris RL, Mushegian A, Rast JP, Smith LC, Thorndyke MC, Vacquier VD, Wessel GM, Wray G, Zhang L, Elsik CG, Ermolaeva O, Hlavina W, Hofmann G, Kitts P, Landrum MJ, Mackey AJ, Maglott D, Panopoulou G, Poustka AJ, Pruitt K, Sapojnikov V, Song X, Souvorov A, Solovyev V, Wei Z, Whittaker CA, Worley K, Durbin KJ, Shen Y, Fedrigo O, Garfield D, Haygood R, Primus A, Satija R, Severson T, Gonzalez-Garay ML, Jackson AR, Milosavljevic A, Tong M, Killian CE, Livingston BT, Wilt FH, Adams N, Belle R, Carbonneau S, Cheung R, Cormier P, Cosson B, Croce J, Fernandez-Guerra A, Geneviere AM, Goel M, Kelkar H, Morales J, Mulner-Lorillon O, Robertson AJ, Goldstone JV, Cole B, Epel D, Gold B, Hahn ME, Howard-Ashby M, Scally M, Stegeman JJ, Allgood EL, Cool J, Judkins KM, McCafferty SS, Musante AM, Obar RA, Rawson AP, Rossetti BJ, Gibbons IR, Hoffman MP, Leone A, Istrail S, Materna SC, Samanta MP, Stolc V, Tongprasit W, Tu Q, Bergeron KF, Brandhorst BP, Whittle J, Berney K, Bottjer DJ, Calestani C, Peterson K, Chow E, Yuan QA, Elhaik E, Graur D, Reese JT, Bosdet I, Heesun S, Marra MA, Schein J, Anderson MK, Brockton V, Buckley KM, Cohen AH, Fugmann SD, Hibino T, Loza-Coll M, Majeske AJ, Messier C, Nair SV, Pancer Z, Terwilliger DP, Agca C, Arboleda E, Chen N, Churcher AM, Hallbook F, Humphrey GW, Idris MM, Kiyama T, Liang S, Mellott D, Mu X, Murray G, Olinski RP, Raible F, Rowe M, Taylor JS, Tessmar-Raible K, Wang D, Wilson KH, Yaguchi S, Gaasterland T, Galindo BE, Gunaratne HJ, Juliano C, Kinukawa M, Moy GW, Neill AT, Nomura M, Raisch M, Reade A, Roux MM, Song JL, Su YH, Townley IK, Voronina E, Wong JL, Amore G, Branno M, Brown ER, Cavalieri V, Duboc V, Duloquin L, Flytzanis C, Gache C, Lapraz F, Lepage T, Locascio A, Martinez P, Matassi G, Matranga V, Range R, Rizzo F, Rottinger E, Beane W, Bradham C, Byrum C, Glenn T, Hussain S, Manning G, Miranda E, Thomason R, Walton K, Wikramanayke A, Wu SY, Xu R, Brown CT, Chen L, Gray RF, Lee PY, Nam J, Oliveri P, Smith J, Muzny D, Bell S, Chacko J, Cree A, Curry S, Davis C, Dinh H, Dugan-Rocha S, Fowler J, Gill R, Hamilton C, Hernandez J, Hines S, Hume J, Jackson L, Jolivet A, Kovar C, Lee S, Lewis L, Miner G, Morgan M, Nazareth LV, Okwuonu G, Parker D, Pu LL, Thorn R, Wright R (2006) The genome of the sea urchin Strongylocentrotus purpuratus. Science 314(5801):941–952. doi:10.1126/science.1133609, 314/5801/941 [pii]
5. Vogel WF, Abdulhussein R, Ford CE (2006) Sensing extracellular matrix: an update on discoidin domain receptor function. Cell Signal 18(8):1108–1116. doi:10.1016/j.cellsig.2006.02.012, S0898-6568(06)00047-7 [pii]
6. Dieterich C, Clifton SW, Schuster LN, Chinwalla A, Delehaunty K, Dinkelacker I, Fulton L, Fulton R, Godfrey J, Minx P, Mitreva M, Roeseler W, Tian H, Witte H, Yang SP, Wilson RK, Sommer RJ (2008) The Pristionchus pacificus genome provides a unique perspective on nematode lifestyle and parasitism. Nat Genet 40(10):1193–1198. doi:10.1038/ng.227, ng.227 [pii]
7. Adams MD, Celniker SE, Holt RA, Evans CA, Gocayne JD, Amanatides PG, Scherer SE, Li PW, Hoskins RA, Galle RF, George RA, Lewis SE, Richards S, Ashburner M, Henderson SN, Sutton GG, Wortman JR, Yandell MD, Zhang Q, Chen LX, Brandon RC, Rogers YH,

Blazej RG, Champe M, Pfeiffer BD, Wan KH, Doyle C, Baxter EG, Helt G, Nelson CR, Gabor GL, Abril JF, Agbayani A, An HJ, Andrews-Pfannkoch C, Baldwin D, Ballew RM, Basu A, Baxendale J, Bayraktaroglu L, Beasley EM, Beeson KY, Benos PV, Berman BP, Bhandari D, Bolshakov S, Borkova D, Botchan MR, Bouck J, Brokstein P, Brottier P, Burtis KC, Busam DA, Butler H, Cadieu E, Center A, Chandra I, Cherry JM, Cawley S, Dahlke C, Davenport LB, Davies P, de Pablos B, Delcher A, Deng Z, Mays AD, Dew I, Dietz SM, Dodson K, Doup LE, Downes M, Dugan-Rocha S, Dunkov BC, Dunn P, Durbin KJ, Evangelista CC, Ferraz C, Ferriera S, Fleischmann W, Fosler C, Gabrielian AE, Garg NS, Gelbart WM, Glasser K, Glodek A, Gong F, Gorrell JH, Gu Z, Guan P, Harris M, Harris NL, Harvey D, Heiman TJ, Hernandez JR, Houck J, Hostin D, Houston KA, Howland TJ, Wei MH, Ibegwam C, Jalali M, Kalush F, Karpen GH, Ke Z, Kennison JA, Ketchum KA, Kimmel BE, Kodira CD, Kraft C, Kravitz S, Kulp D, Lai Z, Lasko P, Lei Y, Levitsky AA, Li J, Li Z, Liang Y, Lin X, Liu X, Mattei B, McIntosh TC, McLeod MP, McPherson D, Merkulov G, Milshina NV, Mobarry C, Morris J, Moshrefi A, Mount SM, Moy M, Murphy B, Murphy L, Muzny DM, Nelson DL, Nelson DR, Nelson KA, Nixon K, Nusskern DR, Pacleb JM, Palazzolo M, Pittman GS, Pan S, Pollard J, Puri V, Reese MG, Reinert K, Remington K, Saunders RD, Scheeler F, Shen H, Shue BC, Siden-Kiamos I, Simpson M, Skupski MP, Smith T, Spier E, Spradling AC, Stapleton M, Strong R, Sun E, Svirskas R, Tector C, Turner R, Venter E, Wang AH, Wang X, Wang ZY, Wassarman DA, Weinstock GM, Weissenbach J, Williams SM, Woodage T, Worley KC, Wu D, Yang S, Yao QA, Ye J, Yeh RF, Zaveri JS, Zhan M, Zhang G, Zhao Q, Zheng L, Zheng XH, Zhong FN, Zhong W, Zhou X, Zhu S, Zhu X, Smith HO, Gibbs RA, Myers EW, Rubin GM, Venter JC (2000) The genome sequence of Drosophila melanogaster. Science 287(5461):2185–2195, 8392 [pii]

8. Werren JH, Richards S, Desjardins CA, Niehuis O, Gadau J, Colbourne JK, Beukeboom LW, Desplan C, Elsik CG, Grimmelikhuijzen CJ, Kitts P, Lynch JA, Murphy T, Oliveira DC, Smith CD, van de Zande L, Worley KC, Zdobnov EM, Aerts M, Albert S, Anaya VH, Anzola JM, Barchuk AR, Behura SK, Bera AN, Berenbaum MR, Bertossa RC, Bitondi MM, Bordenstein SR, Bork P, Bornberg-Bauer E, Brunain M, Cazzamali G, Chaboub L, Chacko J, Chavez D, Childers CP, Choi JH, Clark ME, Claudianos C, Clinton RA, Cree AG, Cristino AS, Dang PM, Darby AC, de Graaf DC, Devreese B, Dinh HH, Edwards R, Elango N, Elhaik E, Ermolaeva O, Evans JD, Foret S, Fowler GR, Gerlach D, Gibson JD, Gilbert DG, Graur D, Grunder S, Hagen DE, Han Y, Hauser F, Hultmark D, Hunter HCt, Hurst GD, Jhangian SN, Jiang H, Johnson RM, Jones AK, Junier T, Kadowaki T, Kamping A, Kapustin Y, Kechavarzi B, Kim J, Kiryutin B, Koevoets T, Kovar CL, Kriventseva EV, Kucharski R, Lee H, Lee SL, Lees K, Lewis LR, Loehlin DW, Logsdon JM Jr, Lopez JA, Lozado RJ, Maglott D, Maleszka R, Mayampurath A, Mazur DJ, McClure MA, Moore AD, Morgan MB, Muller J, Munoz-Torres MC, Muzny DM, Nazareth LV, Neupert S, Nguyen NB, Nunes FM, Oakeshott JG, Okwuonu GO, Pannebakker BA, Pejaver VR, Peng Z, Pratt SC, Predel R, Pu LL, Ranson H, Raychoudhury R, Rechtsteiner A, Reese JT, Reid JG, Riddle M, Robertson HM, Romero-Severson J, Rosenberg M, Sackton TB, Sattelle DB, Schluns H, Schmitt T, Schneider M, Schuler A, Schurko AM, Shuker DM, Simoes ZL, Sinha S, Smith Z, Solovyev V, Souvorov A, Springauf A, Stafflinger E, Stage DE, Stanke M, Tanaka Y, Telschow A, Trent C, Vattathil S, Verhulst EC, Viljakainen L, Wanner KW, Waterhouse RM, Whitfield JB, Wilkes TE, Williamson M, Willis JH, Wolschin F, Wyder S, Yamada T, Yi SV, Zecher CN, Zhang L, Gibbs RA (2010) Functional and evolutionary insights from the genomes of three parasitoid Nasonia species. Science 327(5963):343–348. doi:10.1126/science.1178028, 327/5963/343 [pii]
9. Mummery-Widmer JL, Yamazaki M, Stoeger T, Novatchkova M, Bhalerao S, Chen D, Dietzl G, Dickson BJ, Knoblich JA (2009) Genome-wide analysis of Notch signalling in Drosophila by transgenic RNAi. Nature 458(7241):987–992. doi:10.1038/nature07936, nature07936 [pii]
10. Yin Z, Sadok A, Sailem H, McCarthy A, Xia X, Li F, Garcia MA, Evans L, Barr AR, Perrimon N, Marshall CJ, Wong ST, Bakal C (2013) A screen for morphological complexity identifies regulators of switch-like transitions between discrete cell shapes. Nat Cell Biol 15(7):860–871. doi:10.1038/ncb2764, ncb2764 [pii]

11. Bettencourt-Dias M, Giet R, Sinka R, Mazumdar A, Lock WG, Balloux F, Zafiropoulos PJ, Yamaguchi S, Winter S, Carthew RW, Cooper M, Jones D, Frenz L, Glover DM (2004) Genome-wide survey of protein kinases required for cell cycle progression. Nature 432(7020):980–987. doi:10.1038/nature03160, nature03160 [pii]
12. Zhu X, Mahairas G, Illies M, Cameron RA, Davidson EH, Ettensohn CA (2001) A large-scale analysis of mRNAs expressed by primary mesenchyme cells of the sea urchin embryo. Development 128(13):2615–2627
13. Carafoli F, Hohenester E (2012) Collagen recognition and transmembrane signalling by discoidin domain receptors. Biochim Biophys Acta 1834(10):2187–2194. doi:10.1016/j.bbapap.2012.10.014, S1570-9639(12)00255-5 [pii]
14. Vogel W, Gish GD, Alves F, Pawson T (1997) The discoidin domain receptor tyrosine kinases are activated by collagen. Mol Cell 1(1):13–23, S1097-2765(00)80003-9 [pii]
15. Vogel W, Brakebusch C, Fassler R, Alves F, Ruggiero F, Pawson T (2000) Discoidin domain receptor 1 is activated independently of beta(1) integrin. J Biol Chem 275(8):5779–5784
16. Matsuyama W, Kamohara H, Galligan C, Faure M, Yoshimura T (2003) Interaction of discoidin domain receptor 1 isoform b (DDR1b) with collagen activates p38 mitogen-activated protein kinase and promotes differentiation of macrophages. FASEB J 17(10):1286–1288. doi:10.1096/fj.02-0320fje02-0320fje [pii]
17. Koo DH, McFadden C, Huang Y, Abdulhussein R, Friese-Hamim M, Vogel WF (2006) Pinpointing phosphotyrosine-dependent interactions downstream of the collagen receptor DDR1. FEBS Lett 580(1):15–22. doi:10.1016/j.febslet.2005.11.035, S0014-5793(05)01398-0 [pii]
18. Slack BE, Siniaia MS, Blusztajn JK (2006) Collagen type I selectively activates ectodomain shedding of the discoidin domain receptor 1: involvement of Src tyrosine kinase. J Cell Biochem 98(3):672–684. doi:10.1002/jcb.20812
19. Lemeer S, Bluwstein A, Wu Z, Leberfinger J, Muller K, Kramer K, Kuster B (2011) Phosphotyrosine mediated protein interactions of the discoidin domain receptor 1. J Proteomics 75(12):3465–3477. doi:10.1016/j.jprot.2011.10.007, S1874-3919(11)00493-3 [pii]
20. Mohamed AM, Chin-Sang ID (2011) The C. elegans nck-1 gene encodes two isoforms and is required for neuronal guidance. Dev Biol 354(1):55–66. doi:10.1016/j.ydbio.2011.03.019, S0012-1606(11)00183-7 [pii]
21. Unsoeld T, Park JO, Hutter H (2012) Discoidin domain receptors guide axons along longitudinal tracts in C. elegans. Dev Biol 374(1):142–152. doi:10.1016/j.ydbio.2012.11.001, S0012-1606(12)00604-5 [pii]
22. Ikeda K, Wang LH, Torres R, Zhao H, Olaso E, Eng FJ, Labrador P, Klein R, Lovett D, Yancopoulos GD, Friedman SL, Lin HC (2002) Discoidin domain receptor 2 interacts with Src and Shc following its activation by type I collagen. J Biol Chem 277(21):19206–19212. doi:10.1074/jbc.M201078200M201078200 [pii]
23. Yang K, Kim JH, Kim HJ, Park IS, Kim IY, Yang BS (2005) Tyrosine 740 phosphorylation of discoidin domain receptor 2 by Src stimulates intramolecular autophosphorylation and Shc signaling complex formation. J Biol Chem 280(47):39058–39066. doi:10.1074/jbc.M506921200, M506921200 [pii]
24. White J, Southgate E, Thomson J, Brenner S (1986) The structure of the nervous system of the nematode Caenorhabditis elegans. Philos Trans R Soc Lond B 314:1–340
25. Albertson DG, Thomson JN (1976) The pharynx of Caenorhabditis elegans. Philos Trans R Soc Lond B Biol Sci 275(938):299–325
26. Hall DH, Russell RL (1991) The posterior nervous system of the nematode Caenorhabditis elegans: serial reconstruction of identified neurons and complete pattern of synaptic interactions. J Neurosci 11(1):1–22
27. White JG, Southgate E, Thomson JN, Brenner S (1976) The structure of the ventral nerve cord of Caenorhabditis elegans. Philos Trans R Soc Lond B Biol Sci 275(938):327–348
28. Chalfie M, Tu Y, Euskirchen G, Ward WW, Prasher DC (1994) Green fluorescent protein as a marker for gene expression. Science 263(5148):802–805

29. Yurchenco PD, Wadsworth WG (2004) Assembly and tissue functions of early embryonic laminins and netrins. Curr Opin Cell Biol 16(5):572–579. doi:10.1016/j.ceb.2004.07.013S0955-0674(04)00112-7 [pii]
30. McKay SJ, Johnsen R, Khattra J, Asano J, Baillie DL, Chan S, Dube N, Fang L, Goszczynski B, Ha E, Halfnight E, Hollebakken R, Huang P, Hung K, Jensen V, Jones SJ, Kai H, Li D, Mah A, Marra M, McGhee J, Newbury R, Pouzyrev A, Riddle DL, Sonnhammer E, Tian H, Tu D, Tyson JR, Vatcher G, Warner A, Wong K, Zhao Z, Moerman DG (2003) Gene expression profiling of cells, tissues, and developmental stages of the nematode C. elegans. Cold Spring Harb Symp Quant Biol 68:159–169
31. Alves F, Vogel W, Mossie K, Millauer B, Hofler H, Ullrich A (1995) Distinct structural characteristics of discoidin I subfamily receptor tyrosine kinases and complementary expression in human cancer. Oncogene 10(3):609–618
32. Leitinger B (2014) Discoidin domain receptor functions in physiological and pathological conditions. Int Rev Cell Mol Biol 310:39–87. doi:10.1016/B978-0-12-800180-6.00002-5, B978-0-12-800180-6.00002-5 [pii]
33. Lai C, Lemke G (1994) Structure and expression of the Tyro 10 receptor tyrosine kinase. Oncogene 9(3):877–883
34. Sanchez MP, Tapley P, Saini SS, He B, Pulido D, Barbacid M (1994) Multiple tyrosine protein kinases in rat hippocampal neurons: isolation of Ptk-3, a receptor expressed in proliferative zones of the developing brain. Proc Natl Acad Sci U S A 91(5):1819–1823
35. Zerlin M, Julius MA, Goldfarb M (1993) NEP: a novel receptor-like tyrosine kinase expressed in proliferating neuroepithelia. Oncogene 8(10):2731–2739
36. Durbin RM (1987) Studies on the development and organisation of the nervous system of Caenorhabditis elegans. Ph.D. Thesis, University of Cambridge
37. Garriga G, Desai C, Horvitz HR (1993) Cell interactions control the direction of outgrowth, branching and fasciculation of the HSN axons of Caenorhabditis elegans. Development 117(3):1071–1087
38. Shrivastava A, Radziejewski C, Campbell E, Kovac L, McGlynn M, Ryan TE, Davis S, Goldfarb MP, Glass DJ, Lemke G, Yancopoulos GD (1997) An orphan receptor tyrosine kinase family whose members serve as nonintegrin collagen receptors. Mol Cell 1(1):25–34, S1097-2765(00)80004-0 [pii]
39. Hou G, Vogel W, Bendeck MP (2001) The discoidin domain receptor tyrosine kinase DDR1 in arterial wound repair. J Clin Invest 107(6):727–735. doi:10.1172/JCI10720
40. Leitinger B, Kwan AP (2006) The discoidin domain receptor DDR2 is a receptor for type X collagen. Matrix Biol 25(6):355–364. doi:10.1016/j.matbio.2006.05.006, S0945-053X(06)00058-8 [pii]
41. Consortium CS (1998) Genome sequence of the nematode C. elegans: a platform for investigating biology. Science 282(5396):2012–2018
42. Page AP, Johnstone IL (2007) The cuticle. WormBook 1–15. doi:10.1895/wormbook.1.138.1
43. Kramer JM (2005) Basement membranes. WormBook 1–15. doi:10.1895/wormbook.1.16.1
44. Guo XD, Johnson JJ, Kramer JM (1991) Embryonic lethality caused by mutations in basement membrane collagen of C. elegans. Nature 349(6311):707–709. doi:10.1038/349707a0
45. Gupta MC, Graham PL, Kramer JM (1997) Characterization of alpha1(IV) collagen mutations in Caenorhabditis elegans and the effects of alpha1 and alpha2(IV) mutations on type IV collagen distribution. J Cell Biol 137(5):1185–1196
46. Sibley MH, Graham PL, von Mende N, Kramer JM (1994) Mutations in the alpha 2(IV) basement membrane collagen gene of Caenorhabditis elegans produce phenotypes of differing severities. EMBO J 13(14):3278–3285
47. Kang SH, Kramer JM (2000) Nidogen is nonessential and not required for normal type IV collagen localization in Caenorhabditis elegans. Mol Biol Cell 11(11):3911–3923
48. Kim S, Wadsworth WG (2000) Positioning of longitudinal nerves in C. elegans by nidogen. Science 288(5463):150–154
49. Aumailley M, Battaglia C, Mayer U, Reinhardt D, Nischt R, Timpl R, Fox JW (1993) Nidogen mediates the formation of ternary complexes of basement membrane components. Kidney Int 43(1):7–12

50. Kohfeldt E, Sasaki T, Gohring W, Timpl R (1998) Nidogen-2: a new basement membrane protein with diverse binding properties. J Mol Biol 282(1):99–109. doi:10.1006/jmbi.1998.2004, S0022-2836(98)92004-6 [pii]
51. Ackley BD, Crew JR, Elamaa H, Pihlajaniemi T, Kuo CJ, Kramer JM (2001) The NC1/endostatin domain of Caenorhabditis elegans type XVIII collagen affects cell migration and axon guidance. J Cell Biol 152(6):1219–1232
52. Xu H, Raynal N, Stathopoulos S, Myllyharju J, Farndale RW, Leitinger B (2011) Collagen binding specificity of the discoidin domain receptors: binding sites on collagens II and III and molecular determinants for collagen IV recognition by DDR1. Matrix Biol 30(1):16–26. doi:10.1016/j.matbio.2010.10.004, S0945-053X(10)00154-X [pii]
53. Gallo M, Mah AK, Johnsen RC, Rose AM, Baillie DL (2006) Caenorhabditis elegans dpy-14: an essential collagen gene with unique expression profile and physiological roles in early development. Mol Genet Genomics 275(6):527–539. doi:10.1007/s00438-006-0110-3
54. Sund M, Vaisanen T, Kaukinen S, Ilves M, Tu H, Autio-Harmainen H, Rauvala H, Pihlajaniemi T (2001) Distinct expression of type XIII collagen in neuronal structures and other tissues during mouse development. Matrix Biol 20(4):215–231, S0945-053X(01)00134-2 [pii]
55. Tanaka T, Wakabayashi T, Oizumi H, Nishio S, Sato T, Harada A, Fujii D, Matsuo Y, Hashimoto T, Iwatsubo T (2014) CLAC-P/collagen type XXV is required for the intramuscular innervation of motoneurons during neuromuscular development. J Neurosci 34(4):1370–1379. doi:10.1523/JNEUROSCI.2440-13.2014, 34/4/1370 [pii]

# Part II
# Discoidin Domain Receptors in Cancer

# Chapter 6
# Discoidin Domain Receptors and Disease

**Gregory D. Longmore and Whitney R. Grither**

## 6.1 DDRs and Development

The DDRs both play key roles in development. Comparison of $Ddr1^{-/-}$ and $Ddr2^{-/-}$ mice reveals overlapping and unique developmental functions. An important reason for differences could be that DDR1 is expressed primarily in epithelial cells while DDR2 is expressed primarily by mesenchymal cells.

Mice lacking DDR1 are small, infertile (females), and have abnormal mammary gland development [1]. The cellular basis for dwarfism is not clear, but the size reduction is greater in females than males, suggesting possible abnormalities in endocrine or metabolic physiology. Female infertility is due to uterine wall defects that preclude blastocyst implantation. In the uterus DDR1 is present in muscles cells of the myometrium but not glandular epithelium or stromal cells. The implantation defect could reflect an inability to remodel the ECM due to absence of MMP2 production and altered collagen production and, or the inability of blastocysts to adhere to the uterine epithelium [1]. Female DDR1–/– mice are unable to lactate due to increased epithelial proliferation and branching of ducts [1, 2]. DDR1 influences the sensitivity of mammary epithelial cells to prolactin stimulation [2] and transplant

G.D. Longmore (✉)
Departments of Medicine (Oncology), The ICCE Institute, Washington University, St. Louis, MO 63110, USA

Departments of Cell Biology and Physiology, The ICCE Institute, Washington University, St. Louis, MO 63110, USA
e-mail: glongmor@dom.wustl.edu

W.R. Grither
Departments of Medicine (Oncology), The ICCE Institute, Washington University, St. Louis, MO 63110, USA

Departments of Biochemistry, The ICCE Institute, Washington University, St. Louis, MO 63110, USA



R. Fridman, P.H. Huang (eds.), *Discoidin Domain Receptors in Health and Disease*, DOI 10.1007/978-1-4939-6383-6_6

experiments demonstrate that this is an epithelial cell intrinsic defect, not the result of changes to the breast environment [2]. In the breast DDR1 is present primarily in epithelial cells. There are mild defects in kidney glomeruli. There are discontinuous areas of excessive basement membrane deposition and where present an associated loss in slit diaphragm production by glomeruli podocyte foot processes [3]. Despite these changes only mild renal failure was observed. There are also defects in auditory function due to morphologic changes in the inner ear as a result of excessive ECM accumulation [4]. This could be due to excessive ECM production or abnormal remodeling by cells lacking DDR1.

The role of DDR2 in mouse and human development has been analyzed through multiple mutations in DDR2. First, the Ddr2 gene has been genetically deleted in mice [5]. Second, a mouse strain harboring a spontaneous autosomal-recessive deletion of a chromosomal region containing the Ddr2 gene have been described (Smallie or *slie* mice) [6]. Finally, in humans suffering from the rare spondylo-*meta*-epiphyseal dysplasia with short limbs syndrome (SMED-SL) missense and splice-site mutations in Ddr2 have been identified [7, 8]. Ubiquitous Ddr2$^{-/-}$ mice are dwarfs, sterile, and exhibit delayed skin wound healing [5, 6, 9]. Dwarfism is due to a defect in chondrocyte proliferation during postnatal bone growth that leads to shortened bones. In contrast to Ddr1 null mice, there is no difference in size between females and males. Male sterility is due to progressive diminution of spermatogenesis as a result of decreased steroid hormone (testosterone) production, possibly due to altered sensitivity of Leydig cells to luteinizing hormone [10]. Female mice are anovulatory and lack proper development of the corpus luteum in ovaries [6]. In gonads, DDR2 is expressed by interstitial cells but not germ cells [6]. Delayed wound healing is due to a decrease in proliferation and migration of Ddr2$^{-/-}$ dermal fibroblasts [5, 9].

## 6.2 DDRs and Atherosclerosis

Development and progression of atherosclerosis is associated with inflammation, intimal thickening of the artery wall and instability of formed atheroma plaques. Arterial smooth muscle cells (SMC) and plaque macrophages express DDR1 [11]. In SMC, DDR1 regulates proliferation, migration, and MMP and ECM production [12, 13]. As a result, mice lacking DDR1 are protected from intimal thickening after mechanical carotid injury [12]. DDR1 is critical for atherosclerotic plaque development, progression, and calcification and this is thought to be due to DDR1 capacity to promoting plaque inflammation and fibrosis through secretion of inflammatory modulators and ECM [14–16].

Despite its presence in macrophages and SMC of atherosclerotic plaques [11], no direct evidence implicates DDR2 as contributing to atherosclerosis, in vivo. In addition, there are conflicting reports as to whether the action of DDR2 affects SMC proliferation, migration, and ECM production [17, 18].

## 6.3 DDRs and Arthritis

Osteoarthritis (OA) involves degeneration of articular cartilage. Injury results in activation of MMP13 that degrades collagen II, the major collagen present in articular cartilage. Since collagen II is a ligand for DDR1/2 several studies have explored possible roles for DDRs in the pathogenesis of OA. DDR2 expression is increased in chondrocytes of OA patients and in mouse injury models of OA [19, 20]. Activation of DDR2 in chondrocytes leads to MMP13 upregulation [20]. But overexpression of DDR2 alone in chondrocytes does not result in OA in normal mice, likely because in the absence of injury normal articular chondrocytes are not directly exposed to collagen II [21]. When these mice are injured chondrocytes are now exposed to collagen II and develop OA [21]. Genetically reducing DDR2 levels in mouse genetic models and injury models of OA attenuates OA progression [22, 23]. As a result of these observations, there is increasing interest in DDR2 as a target for treatment of OA patients.

DDR1 has been suggested have a protective effect in OA as $Ddr1^{-/-}$ showed a higher incidence of OA [24]. However, in $Ddr1^{-/-}$ mice DDR2 levels increase in chondrocytes, so it may not be the loss of DDR1 but the increased expression of DDR2 that is responsible for increased OA. Whether DDRs contributes to rheumatoid arthritis has not been experimentally tested, but DDR2 is highly expressed in synovial fibroblasts from patients with RA [25].

## 6.4 DDRs and Fibrosis

The action of both DDRs has been shown to be important for organ fibrosis: DDR1 in lung fibrosis and DDR2 in liver fibrosis. Ubiquitous $Ddr1^{-/-}$ mice are protected from bleomycin-induced lung injury [26]. In this model DDR1, which is expressed by lung airway epithelial cells, regulates both the inflammatory and fibrotic response. Whether DDR2, which is expressed by lung fibroblasts, is also critical for pathologic lung fibrosis has not been determined. For DDR2, there are conflicting reports about its role in liver fibrosis. $Ddr2^{-/-}$ mice exhibit increased liver inflammation and fibrosis in response to carbon tetrachloride injury [27]. But the same group has also shown that DDR2 is important for hepatic stellate cell activation—a profibrotic state [28]. In rats, RNAi depletion of DDR2 in liver stellate cells protects against alcohol-induced liver injury and fibrosis [29]. Snail1, a transcriptional regulator implicated in liver fibrosis in response to injury is expressed by liver hepatocytes [30]. DDR2 signaling can increase Snail1 levels and activity [31], so possibly the DDR2-Snail1 signaling pathway is critical for pathologic fibrosis.

## 6.5 DDRs and Cancer

### 6.5.1 *DDRs Expression in Cancer*

Genomic analysis and histologic evaluation of many different human tumor samples reveal that expression of DDR1 and DDR2 are either present or increased (reviewed in [32]). In many of these tumors, increased expression correlates with unfavorable outcome. These include lung, breast, head and neck, brain, gynecologic, esophageal, liver and pancreatic, prostate, fibrosarcoma, and lymphomas and leukemia (reviewed in [32]). In invasive breast cancer, for example, 50–70 % of invasive ductal tumors, across all clinical subtypes express high levels of DDR2 (normal breast epithelium does not express DDR2) [31, 33], yet clinical association of high DDR2 expression and outcome was only observed in aggressive triple negative breast cancers [34]. In some cancers, DDR2 expression has also been detected in tumor associated stromal cells such as cancer-associated fibroblasts (CAFs) [31, 34], endothelial cells [35], and myeloid cells [36]. This suggests a possible cancer role for the action of DDR2 in both cells within the tumor microenvironment as well as tumor cells. Mutations in DDR1 or DDR2 are much less common. In 3.8 % of squamous cell lung carcinomas, mutations in DDR2 were found and these occurred throughout the entire coding region [37]. Some mutations result in gain-of-function phenotypes (proliferation) when expressed in cell lines. In the breast cancer TCGA project, 5 % of patients exhibited amplified expression of DDR2, for unclear reasons, and these patients have decreased survival [31]. One AML patient of 188 whose kinome was sequenced was found to have a mutation in DDR1. The functional relevance of this mutation has not been explored [38]. Recently, DDR1 expression was found to be induced in early stage K-Ras-driven adenocarcinomas of the lung [39]. Genetic and pharmacologic inhibition of DDR1 attenuated lung tumor initiation and progression in mouse models [39].

In multiple different ex vivo tumor cell-based studies and a few in vivo mouse models of various cancers, DDR1 and DDR2 have been shown to exhibit pro- and anti-proliferative effects, positively influence cell–cell adhesion (DDR1), cell-ECM adhesion, and positively influence tumor cell invasion and migration (reviewed in [32]).

### 6.5.2 *Tumor Cell Adhesion*

Although DDR1 and DDR2 are collagen receptors and quantifiably bind collagen [40], accumulated data do not support the concept that they are themselves independent critical "adhesion" receptors, as opposed to other collagen receptors such as Integrins. DDRs aid in cell attachment but appear to not contribute to cell spreading [41]; however, they can modulate Integrin affinity [41, 42]. The mechanism(s) whereby DDRs affect Integrin function has not been discerned. DDR1 has been observed to localize at E-cadherin mediated cell–cell junctions in epithelia and therein regulate collective cell migration by interacting with Par3/Par6 to localize RhoE at junction where it antagonizes Rho/ROCK-mediated actomyosin contractility [43]. DDR1 may also increase the stability of cell surface E-cadherin as well [44].

### 6.5.3 Tumor Cell Invasion and Migration

DDR expression has been found to correlate with both developmental and cancer epithelial mesenchymal transitions (EMT). In general, during EMT induced by many different environmental signals (e.g., TGFβ, hypoxia, and collagen I) and in different epithelial and tumor cell lines, DDR1 expression decreases and DDR2 expression increases as cells become less epithelial and more mesenchymal in nature.

During development and cancer progression epithelial cells can reversibly differentiate into more mesenchymal-like cells that then invade and migrate: Epithelial to Mesenchymal Transition (EMT). EMT can be induced by a plethora of environmental signals (chemical and physical) that ultimately converge to induce transcription of a number of transcriptional regulators that are considered the inducers of EMT. These transcription factors are typically short-lived proteins that require continual environmental signals to maintain their expression. These environmental signals regulate EMT inducing proteins transcriptionally and posttranscriptionally, affect their subcellular localization, and activity. Since EMT stimulates tumor cells to invade and migrate they are constantly exposed to changing environmental signals in vivo. It is important to appreciate that in vivo EMT inducers are not overexpressed in tumors, rather their transcription is induced and protein levels stabilized by environmental signals. Genomic analysis of multiple cancers has not identified coding mutations or gene amplification in EMT inducers. Experimentally, constitutive overexpression of EMT inducers in cell lines leads to factor-independent EMT, which likely does not occur in vivo. So to sustain an EMT phenotype in cancer cells requires constant environmental signals and these can change with location. Many of the EMT inducers also have important functions in mesenchymal cells [45]. For example, Snail1 is important for CAF function and the fibrotic responses of CAFs and stellate cells [46, 47].

A number of cell-based studies have implicated DDRs as important regulators of cancer EMT. In pancreatic cancer cells, both Integrin α2β1 and DDR1 were important for collagen I-induced cell scattering [48]. In renal tubule, epithelial cells and lung cancer cells RNAi-mediated depletion of DDR2 inhibits TGFβ-mediated EMT [49]. And hypoxia induced EMT was inhibited in breast cancer cell lines transduced with DDR2 shRNAi [50]. In many of these cells, the EMT inducing factors, such as Snail1 or Zeb1/2, were expressed before DDRs were manipulated. When epithelial cells, that do not express Snail1, are induced to undergo TGFβ-mediated EMT DDR1 levels decrease and DDR2 levels increase and DDR2 expression was not necessary for EMT induction [31]. Rather DDR2, which was expressed by virtue of the induced mesenchymal differentiation program was critical to maintain EMT by sustaining the level and activity of Snail1 in tumor cells so that they could continue to invade and migrate. Activation of DDR2 by collagen I led to the posttranscriptional stabilization of Snail1 protein and its nuclear accumulation. In breast tumor cell breast transplant experiments, RNAi-mediated reduction of DDR2 levels dramatically inhibits lung metastasis of breast cancer cells that have already undergone EMT (e.g., human MDA-MB-231 and mouse 4T1 cells).

### 6.5.4 DDRs and Tumor Development, Progression, and Metastasis In Vivo

In vivo, the role of DDR1 and DDR2 in tumor growth, invasion, and metastasis has largely been evaluated in tumor cell transplant experiments. Most of these have been xenograft experiments (human in mouse) that are limited in their significance due to genetically and functionally distinct tumor environments and tumor cells. For DDR2, this could be particularly important considering that DDR2 is expressed not only by invading tumor cells but also tumor-associated stromal cells (CAFs, endothelial cells, and myeloid cells).

Reduction of DDR1 expression in human HCT116 colon cancer cells reduced tumor growth when implanted into nude (athymic) mouse subcutaneous tissue [51]. Downregulation of DDR1 in human H460 lung cancer cells reduced tumor metastasis to bone when injected into the left ventricle of athymic mice and reduced bone tumor burden when injected directly into the tibia of athymic mice [52]. Finally, genetic deletion of DDR1 reduced K-Ras-driven lung adenocarcinoma [39]. Reduction of DDR2 in human A375 melanoma cells reduced tumor burden in the liver of athymic mouse recipients following intrasplenic injection [53]. Yet, when mouse DDR2+ MCA38 colon cancer cells were injected into the spleen of ubiquitous Ddr2$^{-/-}$ mice, there were increased liver metastases [54]. This was thought to be due to increased hepatic stellate cell activation, enhanced immune suppression, and profibrogenic response in Ddr2$^{-/-}$ livers. However, it was not clear whether the Ddr2$^{-/-}$ mice were on a pure genetic background and syngeneic with MCA38 cells. If not then the inherent immune response from this mixed genetic background could explain the result. Implantation of human MDA-MB-231 breast tumor cells RNAi-depleted of DDR2 into the breast of immunocompromised mice also resulted in decreased lung metastases. Of note in this study, the size of primary breast tumors was not affected by the absence of DDR2 [50].

Syngeneic transplants of mouse breast cancer 4T1 cells RNAi-depleted of DDR2 into the breast of Balb/c mice resulted in a dramatic decrease in lung metastasis with only minimal or no change in the growth of primary tumors [31]. A single patient with squamous cell lung cancer who responded to treatment with dasatinib and erlotinib was found to have a kinase domain mutation in DDR2. Since both dasatinib and erlotinib are not specific for DDR2, this singular example does not necessarily imply that the mutation in DDR2 per se was driving this patients cancer [37].

Spontaneous tumor development has not been reported for ubiquitous $Ddr1^{-/-}$ or $Ddr2^{-/-}$ mice. Since DDR2 expression is present in invasive tumor cells and cells within the tumor stroma, the generation of mice with a conditional allele of Ddr2 will be required to dissect the cellular basis for the action of DDR2 in the regulation of cancer development, progression, and metastasis. Whether DDR1 also plays a role in the tumor stromal cells also remains an open question; however, in contrast to DDR2, DDR1 expression is more restricted to epithelial cells, not the mesenchymal cells that make up the tumor stroma.

# References

1. Vogel WF, Aszodi A, Alves F, Pawson T (2001) Discoidin domain receptor 1 tyrosine kinase has an essential role in mammary gland development. Mol Cell Biol 21:2906–2917
2. Faraci-Orf E, McFadden C, Vogel WF (2006) DDR1 signaling is essential to sustain Stat5 function during lactogenesis. J Cell Biochem 97:109–121
3. Gross O et al (2004) DDR1-deficient mice show localized subepithelial GBM thickening with focal loss of slit diaphragms and proteinuria. Kidney Int 66:102–111
4. Meyer zum Gottesberge AM, Gross O, Becker-Lendzian U, Massing T, Vogel WF (2008) Inner ear defects and hearing loss in mice lacking the collagen receptor DDR1. Lab Invest 88:27–37
5. Labrador JP et al (2001) The collagen receptor DDR2 regulates proliferation and its elimination leads to dwarfism. EMBO Rep 2:446–452
6. Kano K et al (2008) A novel dwarfism with gonadal dysfunction due to loss-of-function allele of the collagen receptor gene, Ddr2, in the mouse. Mol Endocrinol 22:1866–1880
7. Bargal R et al (2009) Mutations in DDR2 gene cause SMED with short limbs and abnormal calcifications. Am J Hum Genet 84:80–84
8. Ali BR et al (2010) Trafficking defects and loss of ligand binding are the underlying causes of all reported DDR2 missense mutations found in SMED-SL patients. Hum Mol Genet 19:2239–2250
9. Olaso E et al (2002) Discoidin domain receptor 2 regulates fibroblast proliferation and migration through the extracellular matrix in association with transcriptional activation of matrix metalloproteinase-2. J Biol Chem 277:3606–3613
10. Kano K, Kitamura A, Matsuwaki T, Morimatsu M, Naito K (2010) Discoidin domain receptor 2 (DDR2) is required for maintenance of spermatogenesis in male mice. Mol Reprod Dev 77:29–37
11. Ferri N, Carragher NO, Raines EW (2004) Role of discoidin domain receptors 1 and 2 in human smooth muscle cell-mediated collagen remodeling: potential implications in atherosclerosis and lymphangioleiomyomatosis. Am J Pathol 164:1575–1585
12. Hou G, Vogel W, Bendeck MP (2001) The discoidin domain receptor tyrosine kinase DDR1 in arterial wound repair. J Clin Invest 107:727–735
13. Hou G, Vogel WF, Bendeck MP (2002) Tyrosine kinase activity of discoidin domain receptor 1 is necessary for smooth muscle cell migration and matrix metalloproteinase expression. Circ Res 90:1147–1149
14. Ahmad PJ et al (2009) Discoidin domain receptor-1 deficiency attenuates atherosclerotic calcification and smooth muscle cell-mediated mineralization. Am J Pathol 175: 2686–2696
15. Franco C et al (2008) Discoidin domain receptor 1 (ddr1) deletion decreases atherosclerosis by accelerating matrix accumulation and reducing inflammation in low-density lipoprotein receptor-deficient mice. Circ Res 102:1202–1211

16. Franco C, Ahmad PJ, Hou G, Wong E, Bendeck MP (2010) Increased cell and matrix accumulation during atherogenesis in mice with vessel wall-specific deletion of discoidin domain receptor 1. Circ Res 106:1775–1783
17. Shyu KG, Wang BW, Chang H (2009) Hyperbaric oxygen activates discoidin domain receptor 2 via tumour necrosis factor-alpha and the p38 MAPK pathway to increase vascular smooth muscle cell migration through matrix metalloproteinase 2. Clin Sci 116:575–583
18. Hou G, Wang D, Bendeck MP (2012) Deletion of discoidin domain receptor 2 does not affect smooth muscle cell adhesion, migration, or proliferation in response to type I collagen. Cardiovasc Pathol 21:214–218
19. Xu L et al (2005) Activation of the discoidin domain receptor 2 induces expression of matrix metalloproteinase 13 associated with osteoarthritis in mice. J Biol Chem 280:548–555
20. Xu L et al (2007) Increased expression of the collagen receptor discoidin domain receptor 2 in articular cartilage as a key event in the pathogenesis of osteoarthritis. Arthritis Rheum 56:2663–2673
21. Xu L et al (2011) Intact pericellular matrix of articular cartilage is required for unactivated discoidin domain receptor 2 in the mouse model. Am J Pathol 179:1338–1346
22. Xu L et al (2010) Attenuation of osteoarthritis progression by reduction of discoidin domain receptor 2 in mice. Arthritis Rheum 62:2736–2744
23. Salazar A, Polur I, Servais JM, Li Y, Xu L (2014) Delayed progression of condylar cartilage degeneration, by reduction of the discoidin domain receptor 2, in the temporomandibular joints of osteoarthritic mouse models. J Oral Pathol Med 43:317–321
24. Schminke B et al (2014) A discoidin domain receptor 1 knock-out mouse as a novel model for osteoarthritis of the temporomandibular joint. Cell Mol Life Sci 71:1081–1096
25. Su J et al (2009) Discoidin domain receptor 2 is associated with the increased expression of matrix metalloproteinase-13 in synovial fibroblasts of rheumatoid arthritis. Mol Cell Biochem 330:141–152
26. Avivi-Green C, Singal M, Vogel WF (2006) Discoidin domain receptor 1-deficient mice are resistant to bleomycin-induced lung fibrosis. Am J Respir Crit Care Med 174:420–427
27. Olaso E, Arteta B, Benedicto A, Crende O, Friedman SL (2011) Loss of discoidin domain receptor 2 promotes hepatic fibrosis after chronic carbon tetrachloride through altered paracrine interactions between hepatic stellate cells and liver-associated macrophages. Am J Pathol 179:2894–2904
28. Olaso E et al (2001) DDR2 receptor promotes MMP-2-mediated proliferation and invasion by hepatic stellate cells. J Clin Invest 108:1369–1378
29. Luo Z et al (2013) RNA interference against discoidin domain receptor 2 ameliorates alcoholic liver disease in rats. PLoS One 8:e55860
30. Rowe RG et al (2011) Hepatocyte-derived Snail1 propagates liver fibrosis progression. Mol Cell Biol 31:2392–2403
31. Zhang K et al (2013) The collagen receptor discoidin domain receptor 2 stabilizes SNAIL1 to facilitate breast cancer metastasis. Nat Cell Biol 15:677–687
32. Valiathan RR, Marco M, Leitinger B, Kleer CG, Fridman R (2012) Discoidin domain receptor tyrosine kinases: new players in cancer progression. Cancer Metastasis Rev 31:295–321
33. Ren T, Zhang J, Zhang J, Liu X, Yao L (2013) Increased expression of discoidin domain receptor 2 (DDR2): a novel independent prognostic marker of worse outcome in breast cancer patients. Med Oncol 30:397
34. Toy KA et al (2015) Tyrosine kinase discoidin domain receptors DDR1 and DDR2 are coordinately deregulated in triple-negative breast cancer. Breast Cancer Res Treat 150:9–18
35. Zhang S et al (2014) A host deficiency of discoidin domain receptor 2 (DDR2) inhibits both tumour angiogenesis and metastasis. J Pathol 232:436–448
36. Afonso PV, McCann CP, Kapnick SM, Parent CA (2013) Discoidin domain receptor 2 regulates neutrophil chemotaxis in 3D collagen matrices. Blood 121:1644–1650
37. Hammerman PS et al (2011) Mutations in the DDR2 kinase gene identify a novel therapeutic target in squamous cell lung cancer. Cancer Discov 1:78–89

38. Tomasson MH et al (2008) Somatic mutations and germline sequence variants in the expressed tyrosine kinase genes of patients with de novo acute myeloid leukemia. Blood 111:4797–4808
39. Ambrogio C et al (2016) Combined inhibition of DDR1 and Notch signaling is a therapeutic strategy for KRAS-driven lung adenocarcinoma. Nat Med
40. Agarwal G, Kovac L, Radziejewski C, Samuelsson SJ (2002) Binding of discoidin domain receptor 2 to collagen I: an atomic force microscopy investigation. Biochemistry 41:11091–11098
41. Xu H et al (2012) Discoidin domain receptors promote alpha1beta1- and alpha2beta1-integrin mediated cell adhesion to collagen by enhancing integrin activation. PLoS One 7:e52209
42. Staudinger LA et al (2013) Interactions between the discoidin domain receptor 1 and beta1 integrin regulate attachment to collagen. Biol Open 2:1148–1159
43. Hidalgo-Carcedo C et al (2011) Collective cell migration requires suppression of actomyosin at cell-cell contacts mediated by DDR1 and the cell polarity regulators Par3 and Par6. Nat Cell Biol 13:49–58
44. Eswaramoorthy R et al (2010) DDR1 regulates the stabilization of cell surface E-cadherin and E-cadherin-mediated cell aggregation. J Cell Physiol 224:387–397
45. Rowe RG et al (2009) Mesenchymal cells reactivate Snail1 expression to drive three-dimensional invasion programs. J Cell Biol 184:399–408
46. Batlle R et al (2013) Snail1 controls TGF-beta responsiveness and differentiation of mesenchymal stem cells. Oncogene 32:3381–3389
47. Shields MA et al (2013) Snail cooperates with KrasG12D to promote pancreatic fibrosis. Mol Cancer Res 11:1078–1087
48. Shintani Y et al (2008) Collagen I-mediated up-regulation of N-cadherin requires cooperative signals from integrins and discoidin domain receptor 1. J Cell Biol 180:1277–1289
49. Walsh LA, Nawshad A, Medici D (2011) Discoidin domain receptor 2 is a critical regulator of epithelial-mesenchymal transition. Matrix Biol 30:243–247
50. Ren T et al (2014) Discoidin domain receptor 2 (DDR2) promotes breast cancer cell metastasis and the mechanism implicates epithelial-mesenchymal transition programme under hypoxia. J Pathol 234:526–537
51. Kim HG, Hwang SY, Aaronson SA, Mandinova A, Lee SW (2011) DDR1 receptor tyrosine kinase promotes prosurvival pathway through Notch1 activation. J Biol Chem 286: 17672–17681
52. Valencia K et al (2012) Inhibition of collagen receptor discoidin domain receptor-1 (DDR1) reduces cell survival, homing, and colonization in lung cancer bone metastasis. Clin Cancer Res 18:969–980
53. Badiola I, Villace P, Basaldua I, Olaso E (2011) Downregulation of discoidin domain receptor 2 in A375 human melanoma cells reduces its experimental liver metastasis ability. Oncol Rep 26:971–978
54. Badiola I, Olaso E, Crende O, Friedman SL, Vidal-Vanaclocha F (2012) Discoidin domain receptor 2 deficiency predisposes hepatic tissue to colon carcinoma metastasis. Gut 61:1465–1472

# Chapter 7
# Discoidin Domain Receptors in Normal Mammary Development and Breast Cancer Progression

**Sandamali A. Ekanayaka, Celina G. Kleer, Aliccia Bollig-Fischer, Rodrigo Fernandez-Valdivia, and Rafael Fridman**

## 7.1 Introduction

The DDRs are the only receptor tyrosine kinases (RTKs) that signal in response to collagen. This characteristic of DDRs places these receptors at the center of the signaling networks that drive cell–matrix interactions in physiological and pathological conditions. DDRs bind and are activated by both basement membrane and interstitial collagens, and therefore they can transduce collagen-initiated signals in a variety of epithelial and mesenchymal cells. The DDR family comprises two receptors, DDR1 and DDR2 [1]. DDR1 undergoes phosphorylation in response to both fibrillar (connective tissue) and nonfibrillar (basement membrane) collagens, whereas DDR2 is activated only by fibrillar collagens. DDR1 is mostly expressed by epithelial cells and DDR2 is found in cells of mesenchymal origin. Alternative splicing generates five DDR1 isoforms: DDR1a, DDR1b, and DDR1c are full-length functional receptors, whereas DDR1d and DDR1e are truncated or kinase-inactive receptors. DDR1b and DDR1c contain an additional 37 residues (including an extra tyrosine residue) within the intracellular juxtamembrane region, suggesting

Rodrigo Fernandez-Valdivia and Rafael Fridman are cosenior authors

S.A. Ekanayaka
Department of Pathology, Wayne State University, Detroit, MI 48201, USA

C.G. Kleer
Department of Pathology and Comprehensive Cancer Center, University of Michigan, Ann Arbor, MI 48109, USA

A. Bollig-Fischer
Wayne State University School of Medicine and Karmanos Cancer Institute, Detroit, MI 48201, USA

R. Fernandez-Valdivia • R. Fridman (✉)
Department of Pathology, Wayne State University School of Medicine and Karmanos Cancer Institute, Detroit, MI 48201, USA
e-mail: rfridman@med.wayne.edu



R. Fridman, P.H. Huang (eds.), *Discoidin Domain Receptors in Health and Disease*, DOI 10.1007/978-1-4939-6383-6_7

that these receptors activate distinct signaling pathways in response to collagen and may be differentially expressed in malignant cells. In contrast, a single protein has been identified for DDR2. Upon collagen-binding DDRs undergo tyrosine autophosphorylation. DDRs mediate cell–collagen interactions and regulate diverse biological function including cell adhesion, migration, and invasion in a DDR type- and cell type-dependent manners (details on structure–function relationship are discussed in various chapters). Accumulating evidence from *DDR*-deficient mice point to a critical role for these receptors in developmental and pathological processes including cancer, fibrosis, and inflammation, which are the topics of several chapters in this book. Because cell–collagen interactions are central to normal mammary gland development and function, and breast cancer progression [2–5], a significant effort was invested in defining the relative contribution of DDRs in these processes. DDRs are kinases, and as such they can be targeted by specific kinase inhibitors. They are therefore potential therapeutic targets in disease conditions with pathological DDR-mediated signaling. Emerging evidence support the use of inhibitors to target DDRs in breast cancer but these data although promising awaits a better understanding of how DDRs contribute to disease progression. This chapter attempts to provide a comprehensive overview on the expression and function of DDRs in breast tissues, from the early stages of mammary gland development to the ultimate dissemination of metastatic breast cancer cells.

## 7.2 Role of DDRs in Mammary Gland Development

The highly regenerative mammary gland tissue begins its developmental journey during embryogenesis, goes through allometric pubertal ductal morphogenesis and alveologenesis, and undergoes, potentially recurrent, pregnancy-regulated alveolar functional differentiation, lactation, and regression/involution. This morphogenetic odyssey is tightly and precisely controlled by the orchestrated action of intracellular factors, endocrine cues, intercellular signals, and microenvironmental and stromal entities [2, 3, 6–9]. Although most of the attention in mammary gland biology has been focused on mammary epithelial cell–cell interactions and hormonal regulation, likely because of their evident involvement in breast development and cancer, multiple evidence highlighted the importance of the crosstalk between mammary epithelium and mammary stroma for mammary development, homeostasis, and tumorigenesis [10]. The function of the stroma in breast biology regulation relies on the dynamic interaction between mammary epithelial cells, mammary stroma cells, and the extracellular matrix (ECM) and is, in great extent, determined by factors regulating the ECM. To date, these factors include known regulators of cell–matrix interactions such as members of the α and β integrins family [11], laminins [5], collagens [12–14], matrix metalloproteinases [15], and DDRs [16, 17].

*DDR1*: Insight into the function of DDR1 in breast development has come from mouse genetic studies. DDR1-deficient mice were generated by Vogel and colleagues by deleting the first 12 exons of the *DDR1* gene [17]. Pups from homozygous females appeared malnourished 1 day after birth, had small amounts of milk in their stomachs, and, if kept with the DDR1-deficient females, they eventually died. However, transfer of the pups to wild-type foster mothers shortly after birth prevented their death, suggesting a lactating defect in the mammary gland of DDR1-deficient females, consistent with a role for DDR1 in mammary gland function [17]. Histological analyses of mammary gland tissue derived from the DDR1 knockout mice at various stages confirmed these findings and revealed that lack of DDR1 was associated with disruption of glandular organization, and cell proliferation and differentiation [17]. Interestingly, however, whereas the 3-week-old DDR1-deficient mammary gland appeared delayed in its development and displayed a significant impairment in ductal growth compared to wild-type mammary glands, a marked increase in the number and diameter of mammary ducts was observed in the mammary gland of DDR1-deficient adult virgin female mice compared to their wild-type counterparts [17]. This marked increase in ductal growth in DDR1-deficient mammary glands was caused by an augmented epithelial cell proliferation rate (between fourfold and sixfold increase) and accompanied by a significant increase in collagen deposition within the stroma, which was observed around the mammary epithelial tissue and in the adjacent adipose tissue [17]. Interestingly, the terminal end buds (TEB) in the prepubertal DDR1-deficient mammary gland were enlarged compared to the ones present in a wild-type mammary gland. Together, these observations supported the notion that DDR1 exerts a cell proliferation suppression function in the developing (pubertal) and adult nulliparous mammary epithelium. To further strengthen this idea, it has been shown that compound deficiency for c-Jun kinases (JNK) JNK1 and JNK2 in the mammary epithelium causes an increase in mammary branching morphogenesis and a significant downregulation in the expression of DDR1 and also of integrins α1, α5, α6, and β1 mRNA in mammary epithelial cells [18]. Furthermore, inhibition of transforming growth factor (TGF)-β (TGF-β) signaling, which results in accelerated ductal elongation and side branching development [19, 20] and decreased levels of Wnt5a mRNA and protein levels in the mammary epithelium, causes a significant decrease in DDR1 phosphorylation that is mediated by Wnt5a downregulation [21].

Transplantation studies have also shed light on DDR1 function and have uncovered a dual role of DDR1 in the mammary epithelium. Faraci-Orf et al. [16] conducted a study to determine whether the mammary phenotypes displayed by DDR1-deficient mice were intrinsic to the mammary epithelium or a result of altered endocrine and/or paracrine/juxtacrine function. They found that even though the transplanted DDR1-null and wild-type nulliparous mammary epithelium comparably populated and filled the fat pad of the recipient mice, DDR1-deficient ducts showed a significant reduced number of branch points and lessen branching growth, and the TBEs remained larger when compared to the ones present in wild-type transplants [16]. These observations suggested a dual role of DDR1 in the mammary epithelium, acting as a suppressor of cell proliferation in the developing TEBs and, contrastingly, supporting cell proliferation in the mammary ducts during branching morphogenesis. Interestingly, DDR1 expression was found to be upregulated by progesterone treatment in the nulliparous murine mammary gland, and this upregulation occurred at a time where strong progesterone-induced cell proliferation was observed and when the expression of known key mediators of progesterone's action, including RANKL, Wnt4, and inhibitor of differentiation 4 (ID4), was also upregulated [22]. Whether DDR1 is a direct progesterone receptor (PR) target and whether it is a mediator in the robust proliferative response of the mammary epithelium to progesterone signal remains to be determined.

Attributed DDR1 function in the late-pregnant and lactating mammary epithelium seems to be less compounded and more in line with a cell differentiation and a milk protein synthesis promoting role. Histological analysis of DDR1-null late pregnant mammary glands revealed that a more condensed alveolar structure, and with small number of lipid vesicles, was present in DDR1-null mice compared to wild-type animals [17]. Moreover, the histological examination and molecular analysis of postpartum mammary glands revealed that—albeit expressing normal levels of mRNA transcripts for milk proteins—DDR1-deficient mammary glands failed to produce various milk proteins [17]. Interestingly, the mammary glands from postpartum DDR1-deficient animals were largely composed of adipocytes, and the alveoli were condensed, had very little milk on them, and, eventually, collapsed [17]. Remarkably, the cell differentiation defects in late-pregnant DDR1-null mammary glands are not accompanied by premature cell death [17]. Importantly, the lactating defects conferred by DDR1 deficiency [16, 17] are somewhat similar to those observed upon loss of prolactin receptor (PrlR) or its downstream effector Signal transducer and activator of transcription 5(Stat5) [23–25]. Moreover, it has been shown that DDR1-expressing mammary epithelial cells participate, upon contact with a collagen-rich matrix, in PrlR signaling by sustaining Stat5 phosphorylation and transcriptional activity [16]. However, if DDR1 function in the late-pregnant and lactating mammary epithelium is positioned upstream of Stat5, DDR1 must also be controlling other signaling networks, as it is shown that the lactation defects in targeted WAP-Cre-mediated genetic inactivation of Stat5 is accompanied by increased cell apoptosis [24]. Moreover, the lactation defect observed in mice deficient for RANKL, which has been shown to be a downstream target of Stat5 and PR [6, 22, 26, 27], is, in contrast to the one caused by absence of DDR1, accompanied by enhanced apoptosis [28]. It should be noted that DDR1-null epithelial transplants fail to undergo lobuloalveolar outgrowth upon induction of pregnancy [16], suggesting that DDR1 also acts as a mediator of pregnancy-induced cell proliferation, which could also be in concordance with the fact that DDR1 is a downstream target of progesterone signal [22]. Interestingly, it has been shown that the WW domain-containing protein 1 (WWC1) KIBRA, a transactivator of estrogen receptor (ER) [29] that is positively regulated by progesterone and is downregulated in PrlR-deficient mammary tissue [30, 31], physically interacts with DDR1 and protein kinase Cζ (pKCζ), and that this interaction is negatively regulated by collagen-triggered DDR1 phosphorylation [31].

It is also important to highlight that among the known collagen receptors, DDR1 is the one that appears to have a more prominent role in mammary gland function. For instance, mice lacking integrin α2 show modest alterations in mammary epithelial duct branching and no lactation defect [32], and mice deficient for integrin α1 display a normal lactation phenotype [33]. Finally, another important aspect of

DDR1 involvement in breast development that has been much less explored is its potential role in mammary gland stem cell function. In this regard, DDR1 was recently found, along with collagen 1α1 (Col1α1), Col9α1, several Notch signaling downstream targets, and laminins α1, α2, and α4, to be significantly upregulated in fetal mammary gland stem cells [34, 35]. Given that mammary gland stem cells are preferentially located in the TEBs [36, 37] of the developing gland, the above observations could, therefore, be in line with the enlarged TEBs phenotype displayed in pubertal DDR1-deficient mice [17]. Taken together, these data are consistent with (1) a role of DDR1 in suppression of cell proliferation within the mammary epithelium, (2) a cell differentiation promoting function during pregnancy, and (3) a role in lactation-associated lobuloalveolar functional differentiation. Thus, DDR1 may function in a convergent point for several signaling pathways governing breast development, function, homeostasis, and neoplastic conversion (Fig 7.1).

*DDR2*: The role of DDR2 in normal mammary gland development and homeostasis remains largely unexplored. Three types of mice carrying targeted and spontaneous inactivating mutations in the *DDR2* gene have shown DDR2's central role in cellular growth and endocrine and gonadal function [38–40]. Thus, whereas female mice homozygous for the *slie* mutation, which encompasses a large ~150 kb spontaneous deletion that removes most of *DDR2* gene, display a failure in the formation of the corpus luteum [38], females homozygous for a DDR2-null allele progressively become infertile with age [40]. However, a recent study from Corsa et al. found that genetic deletion of DDR2 in mice had no evident effect on mammary gland development [118]. Interestingly, previous work from the Longmore laboratory in Snail1 regulation of epithelial to mesenchymal transition (EMT) shed light into DDR2 function in the human breast. Specifically, Zhang et al. [41] showed that DDR2 expression was induced along with Snail1 in normal human mammary epithelial cells MCF-10A during TGF-β-triggered EMT. Intriguingly, it was observed that although Snail1 levels were reduced in DDR2-depleted, nonmalignant human, breast epithelial MCF10A cells, which is consistent with the Snail1-stabilizing properties of DDR2, the TGF-β-triggered EMT induction was not affected [41]. These observations, however, are in manifest contrast with the fact that Snail1 alone was able to induce EMT in MCF-10A cells [41] and indicate that DDR2 could be functionally relevant in Snail1-triggered, TGF-β-independent EMT. Remarkably, DDR2 has been found differentially expressed in human mammary stem/progenitor cells compared to differentiated mammary epithelial cells [42], and it is upregulated in hyperplastic mammary glands of MMTV-Wnt1 mice [43], which have been shown to harbor an aberrantly increased mammary stem cell pool [37], and preferentially develop mammary tumors from progenitor cells [44]. This evidence is consistent with DDR2 being a critical player in Snail1-triggered EMT in mammary epithelial cells, and plausibly, a relevant player in mammary stem cell function, and in Wnt1-triggered mammary tumorigenesis. Further studies using front-end mouse genetics and patient-derived xenografts along with cell culture and biochemical and structural assays will be required to delineate DDRs' function in the mammary epithelium and stroma as well as in breast cancer initiation, progression, and metastasis. Figure 7.1 summarizes the current known roles of and molecular pathways associated with DDR1 during mammary gland development.

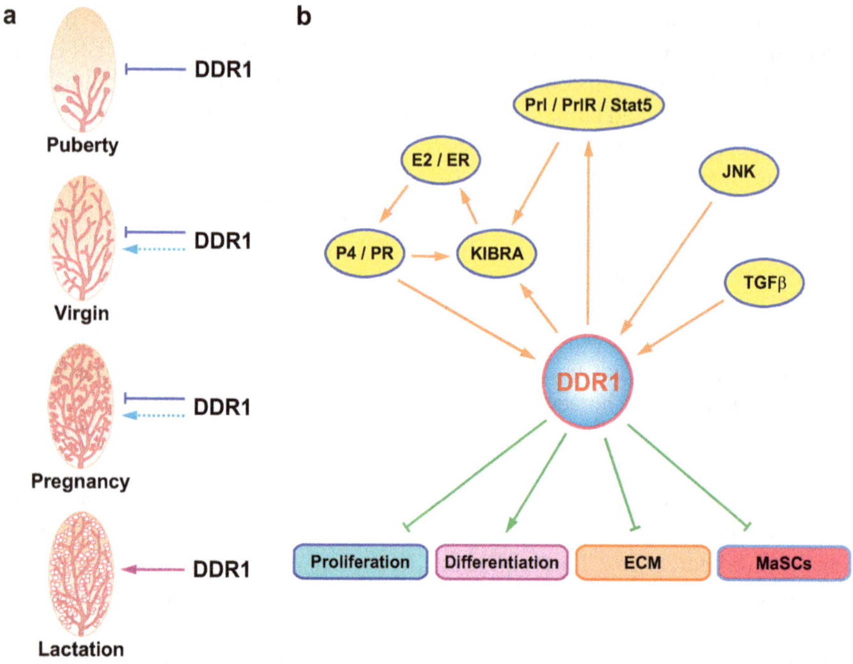


**Fig. 7.1** Proposed roles of DDR1 in mammary gland development and function. (**a**) DDR1 exerts an inhibitory function during pubertal development of the mammary gland as revealed by the increased cell proliferation observed in DDR1-deficient mice (*arrow in blue*). Similarly, the marked increased in Ki-67 positive cells in the adult virgin mammary glands from DDR1-deficient mice indicates that DDR1 has cell proliferation suppression function (*arrow in blue*). During pregnancy, DDR1 is required for proper cell proliferation of the mammary epithelium as evidenced by the increased cell proliferation, excessive filling of the fat pad, and the over condensed alveolar structures exhibited by DDR1-null pregnant female mice (*arrow in blue*). Interestingly, transplantation studies in which DDR1-deficient mammary tissue was transplanted into cleared fat pads of wild-type female mice have shown that DDR1 may have an inductive role in mammary branching morphogenesis and pregnancy-induced alveologenesis (*dashed arrows*). Remarkably, the terminal end buds generated by the transplanted DDR1-null epithelium displays an increased cell proliferation phenotype, corroborating the cell proliferation suppression function of DDR1. During lactation, DDR1 promotes alveolar differentiation and cell survival (*arrow in magenta*) as demonstrated by the failure to produce milk and the premature regression and collapse of the mammary alveoli seen in female mice lacking DDR1. (**b**) Schematics depicting DDR1 at the convergence of several signaling pathways governing important molecular and cellular processes in breast biology. Upon contact with collagen, DDR1 participates in prolactin (Prl)/prolactin receptor (PrlR) signaling by sustaining Stat5 phosphorylation and transcriptional activity. DDR1 expression has been found to be upregulated by progesterone (P4) progesterone receptor (PR) signal in the mammary gland. Importantly, KIBRA, a downstream PrlR target that acts as a transactivator of estrogen (E2)/estrogen receptor α (ER) complex and that is positively regulated by progesterone, has been shown to physically interact with DDR1 and form a complex that dissociates upon collagen-triggered DDR1 phosphorylation. DDR1 has been found to be a downstream target of JNK and TGF-β in the mammary epithelium. DDR1 has been demonstrated to have a cell proliferation suppression function in the mammary epithelium during pubertal, adult virgin and pregnancy-induced mammary development, and a cell differentiation promoting role in the lactating mammary epithelium. Extracellular matrix deposition has been shown to be negatively regulated by DDR1. Similarly, DDR1 has been found to negatively regulate cell proliferation in the terminal end buds, a mammary structure rich in epithelial stem cells

## 7.3 Expression and Role of DDRs in Breast Cancer Progression

Invasive breast carcinomas are a heterogeneous group of malignant epithelial tumors that arise in the breast parenchyma and are characterized by their invasion of adjacent tissue and metastatic ability. Invasive breast carcinomas comprise various histological morphologies and biological features, and exhibit different clinical behaviors and treatment responses. The morphological classification of invasive breast carcinomas depends on their degree of differentiation, which reflects how closely a tumor resembles normal breast glandular epithelium in structural organization, cytological features, and growth pattern [45]. Most invasive tumors in the breast are invasive ductal carcinomas, accounting for approximately 80 % of invasive breast cancer. Invasive lobular carcinomas comprise 10–15 % of all breast cancers, and the remainder constitutes special histological types including mucinous, tubular, micropapillary, and others. The morphologic features and cell proliferation status are robust surrogates for biological variables that determine and relate to their natural behaviors. Importantly, hormone receptor and human epidermal growth factor receptor 2 (HER-2/neu) status was shown to have prognostic value and predict clinical responses [46–48]. However, it has become increasingly evident that breast cancer heterogeneity extends beyond the classic immunohistochemistry-based divisions of ER, PR, and HER-2/neu. The development of the molecular classification of breast cancers [49] was recently supported by the Cancer Genome Atlas (TCGA) Program through mRNA, miRNA, DNA, and epigenetic analyses [50]. Invasive carcinomas may belong to the luminal A, luminal B, HER2, and triple negative subtypes. Of note, these subtypes are also heterogeneous and maybe further refined. For example, recent studies defined four subgroups of triple negative breast cancers (TNBCs): luminal androgen receptor, mesenchymal, basal-like immunosuppressed, and basal-like immune-activated groups [51].

In addition to the intrinsic characteristics of breast cancer cells, the interactions with the cellular and structural components of the microenvironment are crucial for breast cancer development and progression. Studies have shown that high mammographic density is associated with a twofold risk of breast cancer development [52–54]. While in the normal adult breast there are several types of collagen, breast density is mainly due to increased deposition of type I collagen [54–56]. Pathologists have noticed the presence of a desmoplastic stroma associated with invasive carcinomas decades ago, which results from increased deposition of extracellular matrix proteins [45]. It is now recognized that stromal desmoplasia alters the chemical composition and the mechanical properties of the ECM [57, 58]. Breast cancer cells respond to these changes in the ECM through deregulated signaling pathways, which promote neoplastic functions and result in loss of normal architecture, invasion, and increased proliferation [57, 58]. The tumor-associated desmoplastic stroma is particularly enriched in fibrillar collagens, which have been shown to provide a path for breast cancer cell invasion [59]. Because DDRs are part of the arsenal of cell surface receptors that mediate tumor cell–collagen interactions [60, 61], they may play an important role in breast cancer progression [41, 60, 62, 63].

### 7.3.1 Expression of DDRs in Normal and Cancerous Breast Tissues

*DDR1*: Immunohistochemical studies showed that in normal human mammary gland DDR1 protein is highly expressed in the epithelium throughout the gland, but it is not detected in the stroma [62]. Antibodies directed to the extracellular domain revealed clear membranous DDR1 localization in normal epithelial cells (our data). We also found DDR1 to be highly expressed in the acinar-like structures of MCF10A cells cultured within Matrigel, where it preferentially localizes at cell–cell contacts with a pattern similar to that displayed by E-cadherin. DDR1 was also highly expressed in tumors produced by MCF10A.DCIS.COM cells in mice (our unpublished data). In specimens of human ductal in situ carcinomas (DCIS), DDR1 protein was also expressed in the epithelial cells [62]. Thus, DDR1 expression in breast epithelial cells does not appear to be significantly altered in the early stages of breast cancer suggesting that neoplastic transformation does not involve changes in DDRs. In agreement with these studies, mRNA analyses showed that DDR1 was readily detected in normal breast tissues and in benign tumors (cystic hyperplasia and fibroadenoma). However, the benign tumors showed a trend to express higher levels of DDR1 mRNA [64]. In breast tumors, the profile of DDR1 expression appears to be complex. For instance, the studies of Ren et al. [64] showed significantly downregulation of DDR1 mRNA expression in normal vs. cancerous breast tissues. Analyses of protein expression by immunohistochemistry, however, showed a complex pattern of expression. Dejmek et al. found DDR1 to be heterogeneously expressed in invasive breast carcinomas [65]. Turashvili et al. reported that DDR1 is expressed by most ductal but not lobular carcinomas, and thus these investigators proposed that DDR1 could be an additional marker to distinguish between these two cancer subtypes [66]. In this regard, the reported association of DDR1 with ductal carcinomas appears to resemble that of E-cadherin, which is highly expressed in ductal but not in lobular carcinomas. Interestingly, DDR1 has been found to interact with E-cadherin at cell–cell contacts [67–69], and in some cancer types, absence of E-cadherin correlates with lack of DDR1 expression [69]. Thus, loss of E-cadherin may occur concomitantly with loss of DDR1. The study of Ameli et al., however, found positive DDR1 expression in both ductal and lobular carcinomas, and a lack of correlation between DDR1 levels with tumor type, grade, or receptor status [70]. Another study, however, reported that the majority of ductal carcinomas exhibited a reduced or a lack of DDR1 mRNA expression [71]. However, a strong correlation between reduced DDR1 mRNA expression and malignancy could not be established [64]. In the study of Toy et al. [62], invasive breast carcinomas displayed a heterogeneous expression of DDR1 protein, with half of the tumors expressing high levels, whereas the other half showing reduced levels of DDR1 protein [62]. Moreover, DDR1 expression was not associated with clinicopathological features in this set of tumors, and both luminal and TNBC subtypes exhibited an equal distribution of high and low DDR1-expressing tumors. Another study focusing on TNBCs found low DDR1 expression, which correlated with poor disease-free survival [72]. Yet, a significant number of invasive breast cancer tumors, including TNBC, express DDR1 [62, 66, 73], and at levels similar to those found in normal breast tissue and DCIS [62]. Since TNBCs show significant heterogeneity [74–76], differences in DDR1 expression profile may reflect this fact, and the complex role of DDR1 in various TNBC contexts. For instance, based on experimental findings, high DDR1 expression in certain TNBCs may be associated with activation of pro-survival signals, under stress conditions [77] (discussed below). On the other hand, reduced DDR1 expression in a subset of TNBCs may reflect the reported antimigratory and anti-invasive effects of DDR1 in breast cancer cell lines [72]. Interestingly, clinical association studies showed that among the various combinations of DDR expression profiles, TNBC patients displaying a $DDR1^{Low}/DDR2^{High}$ profile exhibited a significantly worse survival when compared to patients that were negative for this DDR profile [62]. This observation suggests that in certain aggressive TNBCs, concomitant expression of both DDRs may be counterproductive for disease progression. Both DDR1 and DDR2 recognize similar fibrillar collagen ligands, yet they exhibit significant structural and regulatory differences, and consequently activate distinct signaling pathways [1]. It is possible that in the course of tumor evolution there is preferential expansion of tumor cell subpopulations with a DDR profile that is compatible with cancer survival (DDR1 high?) and/or dissemination (DDR2 high?), within a defined collagen microenvironment. Understanding the signaling dynamics (cooperating or antagonizing) of DDRs within a cellular background will shed light on how breast cancer cells sort the collagen-initiated signals in conditions of single or dual DDR receptor expression, and how the activated networks may impact outcome in TNBC subtypes. Another important aspect that needs considerations, when analyzing the association of DDR1 expression with disease progression by immunohistochemical methods, is the expression and role of the DDR1 different isoforms. However, at present, there are no antibodies that can specifically distinguish between the various DDR1 isoforms in breast cancer tissues, and therefore the association between clinicopathological features and DDR1 isoforms remains unknown. This current limitation may be important because there are five structurally different DDR1 isoforms, from which DDR1a and DDR1b are the most common. DDR1a and DDR1b display structural differences within their intracellular juxtamembrane region including the presence of two additional tyrosine residues in DDR1b. Because tyrosine phosphorylation is critical in RTK-mediated signal transduction, these DDR1 isoforms may activate distinct signaling networks in response to collagen binding, and consequently may also elicit isoform-specific effects (promoter, suppressive) on cancer progression. So far the data show that a significant proportion of invasive breast tumors express DDR1 at levels similar to those found in normal breast epithelium [62, 66, 73], raising the question about its role(s) in malignant tissues. Evidence suggests that de novo expression DDR1 is part of a kinome reprograming process that takes place in TNBC cells when exposed to MEK inhibitors [77]. Thus, expression of DDR1 may be a part of the genetic program that confers survival in TNBC cells exposed to stress. However, there is no reported evidence that high expression of DDR1 in invasive cancers predicts outcome. In contrast, TNBC patients with tumors displaying low DDR1 expression were reported to have a poorer survival [62, 72].

In recent years, significant genomic data have been obtained from sequencing breast tumors from multiple patients. Analyses of these data sets revealed little evidence for the occurrence of DDR1 somatic gene aberrations in breast cancer. Moreover, the lack of evidence for DDR1 mutations in the literature is reinforced by information extracted from functional genomics databases. According to data published by The Cancer Genome Atlas (TCGA) and accessed via cBioPortal a single breast cancer sample out of 825 cases showed a somatic DDR1 mutation—a missense mutation at codon E618Q that is, however, predicted to have no functional consequence [50, 78]. Also according to TCGA data, DDR1 gene amplification was identified in one other sample; and there were no instances of DDR1 gene copy number loss [50, 78]. Furthermore, by query of a variety of breast cancer gene expression databases, results were either mixed or demonstrated no significant association for survival or recurrence outcomes and DDR1 mRNA expression levels (tested on complete datasets and subtyped datasets) [79–81]. Thus, the findings so far suggest that DDR1 plays complex roles in breast cancer, with possibly pro- and antimalignant effects depending on the tumor subtype and genetic background. Whether these potential opposite roles of DDR1 in breast cancer are isoform-specific warrant further studies.

*DDR2*: In nonmalignant (carcinoma-associated) breast tissue, DDR2 protein and mRNA are undetectable both in the epithelial and the stromal compartments [41, 62, 64]. Consistently, DDR2 is not expressed in acinar-like structures generated by cultured MCF10A cells (our unpublished data). The lack of DDR2 expression in normal breast epithelial cells is consistent with the fact that in normal breast tissues the epithelial cells are separated from the interstitial collagen matrix by an underlying basement membrane, which contains collagen IV, a ligand of DDR1 but not of DDR2. However, in the course of disease progression, breast cancer cells begin to interact with fibrillar collagens, which areDDR2 ligands, particularly during the process of degradation and invasion through the basement membrane. Consistent with this scenario, Toy et al. [62] found that whereas the majority of the DCIS lesions were negative for DDR2 protein, small groups of DCIS cells facing the tumor–stromal interface displayed specific DDR2 immunoreactivity. Possibly, de novo expression of DDR2 may aid the proinvasive phenotype by allowing invading breast cancer cells to confront the interstitial matrix. Indeed, invasive breast cancers express high levels of DDR2 mRNA [64] and protein [41, 62]. Ren et al. found elevated expression of DDR2 mRNA in 122 samples of invasive breast cancer (ductal and lobular) when compared to the adjacent nonneoplastic breast tissue or nonmalignant (cystic hyperplasia and fibroadenoma) [64]. In the invasive breast cancers, the majority of the ductal carcinomas displayed high DDR2 mRNA levels. Moreover, DDR2 expression was associated with presence of lymph node and distant metastasis, and advanced stage [64]. Importantly, breast cancer patients with high DDR2 mRNA showed worse overall survival and high risk to relapse [64]. Ren et al. also found a strong association between DDR2 levels and expression of hypoxia-inducible factor-1α (HIF-1α), a marker of hypoxia, and its target vascular endothelial growth factor (VEGF), in samples of invasive breast cancer tissues [71]. In these samples, high expression of DDR2 and low levels of E-cadherin correlated with lymph node metastases, whereas tumors with no detectable DDR2 expression, regardless of the levels of E-cadherin, displayed low incidence of lymph node metastases [71]. Based on these studies, Ren et al. suggested that the combined analyses of DDR2 and E-cadherin could distinguish metastatic from nonmetastatic breast cancer [71]. Immunohistochemical studies by Zhang et al. in invasive breast cancer found that 71 % of invasive ductal carcinomas and 21 % of invasive lobular carcinomas expressed DDR2 [41], in agreement with the studies of Ren et al. [64]. However, contrary to their studies, Zhang et al. found no association between DDR2 expression and lymph node involvement [41]. DDR2 expression was significantly associated with Snail1 positivity in the tumor cells, consistent with the experimental findings showing stabilization of Snail1 by DDR2 [41], and previous studies showing that induction of EMT in human breast epithelial cells leads to de novo expression of DDR2 and downregulation of DDR1 [82]. Indeed, Zeb1, an EMT transcription factor, was shown to be a negative transcriptional regulator of DDR1 expression [72]. In a set of 198 tumor specimens, Toy et al. found that high DDR2 protein expression was significantly associated with higher histological grade, negative ER and PR status, negative HER2/neu overexpression, and with the TNBC subtype [62]. Moreover, patients with tumors expressing high DDR2 had a significantly worse overall survival than those expressing low DDR2 after initial surgical treatment. When examined in combination with DDR1, patients with tumors displaying a specific profile of DDR1$^{Low}$/DDR2$^{High}$ protein expression had a shorter overall survival compared to other DDR expression profiles. Importantly, this specific DDR profile predicted survival independently of tumor size, TNBC phenotype, and lymphovascular invasion [62].

The evidence for somatic DDR2 aberrations in breast cancer is somewhat stronger than the data for DDR1, where according to TCGA data 4 % of breast cancer samples showed copy number amplification [50, 78], but for these same samples, gene expression did not appear to concomitantly increase with copy number amplification. Of the 825 patient samples in the study, single nucleotide variants were identified in six (at codons S123I, E361D, A407P, K616N, S674Y, and R752H) [50, 78]. Among these, codon DDR2 S123I introduces a missense mutation in the Discoidin domain and is strongly predicted to have an impact on function [83]. Queries of breast cancer gene expression databases indicated that for luminal and ERα-positive breast cancers there were associations for high DDR2 expression and increased overall patient survival [81]. However, other datasets demonstrated no significant association for survival or recurrence outcomes and DDR2 mRNA expression levels [79, 80]. These findings in current databases are in conflict with the expression data showing a strong association between high levels of DDR2 and poor outcome in women with invasive carcinomas, particularly of the TNBC subtype [41, 62, 64]. These differences highlight the need to continue our effort to unveil whether a specific profile of DDR expression (and activation) is associated with a particular breast cancer subtype, histopathological features, and/or patient outcome. Figure 7.2 illustrates the profile of DDRs in normal and cancerous breast epithelial tissues, and its association with disease progression.

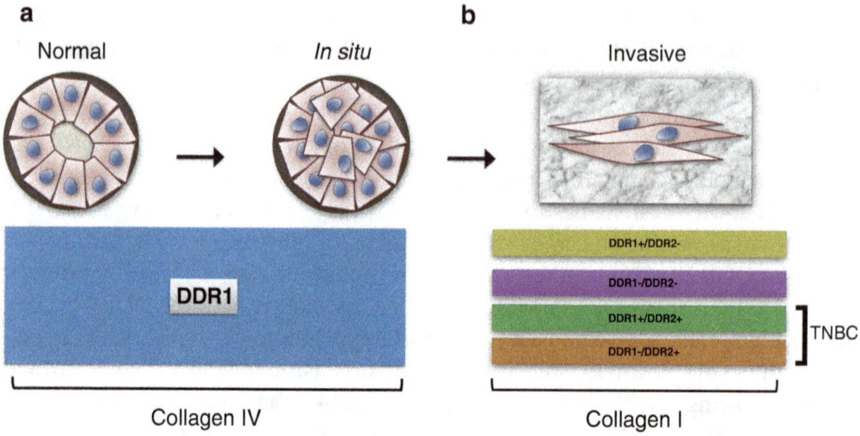


**Fig. 7.2** Expression of DDRs in normal and cancerous breast tissues. (**a**) Normal human breast tissues express DDR1 in the epithelial compartment, whereas DDR2 is not detected [41, 62]. In situ carcinomas are rich in DDR1 but lack substantial DDR2 expression [62]. In normal and in situ carcinoma tissues, the epithelial cells are within the confines of a collagen-IV-containing basement membrane. Thus, in those sites, DDR1, which is activated in response to collagen IV, may be the operative kinase. (**b**) To metastasize, invasive breast cancer cells penetrate basement membranes and subsequently transmigrate within interstitial matrices, which are enriched in fibrillar collagens. Both DDRs are therefore likely to mediate collagen signaling in disseminated breast cancer cells. Evidence suggests a complex prolife of DDR expression in invasive carcinomas [62]. These tumors appear to display a heterogeneous DDR1 profile with some tumors being highly positive and others negative for DDR1 expression. However, the biological and pathological reasons and the functional consequences for this heterogeneity remain unknown. In contrast, most invasive tumors, particularly TNBC, appear to be consistently associated with higher levels of DDR2 [41, 62, 64], which may become a worthwhile target in this tumor subtype

## *7.3.2 Role of DDRs in Experimental Models of Breast Cancer*

As discussed above, both IHC and gene expression studies have implicated DDRs in breast cancer progression. In addition, large cancer genome studies from breast cancer tissues have shown alterations in DDR genes, involving mostly copy number aberrations than point mutations or rearrangements. However, by their nature, these types of studies cannot reveal how DDRs contribute to breast cancer development and progression. In the last decade, a number of studies began to dissect the functions of DDRs in breast epithelium and in breast cancer cells in cell culture and animal models. The emerging picture however is rather complex and far from being complete. For instance, in the case of DDR1 there is a significant lack of consensus on whether or not this receptor supports malignant activities, which is also compounded by inconsistencies on the profile of basal DDR1 expression in various breast cancer cell lines, among different laboratories [84–87]. The following section summarizes current information on DDR function in experimental models of breast cancer. A special effort was taken to provide for the specific conditions (cell line, assay type, and in vitro vs. in vivo) of the findings, which we hope aid in data interpretation and a more rigorous assessment of the status of the field.

#### 7.3.2.1 Roles in Cell Proliferation, Survival, and Apoptosis

*DDR1*: DDR1 is expressed in cell lines of human [69] and mouse [88] mammary epithelial cells, where it colocalizes with E-cadherin at cell–cell junctions. DDR1 was reported to stabilize E-cadherin in cell–cell contacts, and consequently DDR1 may play a role in maintenance of epithelial cell organization in cultured cells [88]. There are conflict data regarding the role of DDR1 in proliferation of breast cancer cells. Gao et al. reported significant inhibition of cell proliferation in various breast cancer cell lines (MCF7, T47D, and MDA-MB-435S) when treated with a specific DDR1 type I kinase inhibitor [89]. However, a selective DDR1 type II kinase inhibitor (DDR1-IN-1) that inhibited collagen-induced activation showed no antiproliferative effects in T47D and SKBR3 breast cancer cells at concentrations below 10 μM, which effectively and selectively blocked receptor activation over other kinases including DDR2 [90]. Therefore, any effect of DDR1-IN-1 on cell proliferation at concentrations above 10 μM was attributed to inhibition of other kinases, possibly in working in conjunction with DDR1 [90]. Consistent with the lack of effect of DDR1 inhibition on cell proliferation, collagen stimulation of adherent T47D cells had modest effect on phosphorylation of ERK1/2, a known effector of proliferation stimuli mediated by the MAPK pathway [91]. Moreover, robust DDR1 collagen-induced activation in suspended T47D cells did not result in ERK1/2 phosphorylation [91]. Thus, a clear relationship between the extent and kinetics of DDR1 activation and the MAPK pathway could not be established in T47D cells. Surprisingly, using the same DDR1-IN-1 inhibitor in MCF7 cells, Malaguarnera et al. found partial inhibition (~20 %) of cell proliferation at doses of 0.4 and 1 μM [87]. The reasons for these conflicting results are unclear but may include issues of inhibitor selectivity and affinity, differences among cell lines, and experimental conditions, just to mention a few. It should also be noted that these studies were conducted without collagen stimulation. Therefore, in addition to the above-mentioned issues, the effect of DDR1 kinase inhibitors on collagen-stimulated cell proliferation, including in 2D and 3D collagen microenvironments, remains unclear.

Other studies utilized RNA interference (RNAi) to address the function of DDR1 in cell proliferation. Experiments aimed at identifying genes essential for cancer cell survival, using a library of shRNAs to screen 72 cancer cell lines, 29 of which were breast cancer cell lines, identified DDR1 as one the genes to be critical for cell proliferation [92]. In these studies, RNA silencing of DDR1 in Cal51, MCF7, Sk-Br-3, BT-20, HCC1954, and HCC38 breast cancer cells inhibited cell proliferation, suggesting that DDR1 is critical for cell growth in breast cancer lines belonging to different breast cancer subtypes [92]. Similar results were reported in the studies of Malaguarnera et al. [87], which showed that transient silencing of DDR1 expression reduced cell proliferation and colony formation in MCF-7, BT-474, and MDA-MB-231 breast cancer cells. In contrast, overexpression of wild-type or kinase dead (K618A) DDR1 enhanced these activities, suggesting a role for DDR1 in in vitro cell growth and survival, independently of receptor phosphorylation. Although the effects of collagen under those conditions were not tested, the investigators found that DDR1 was also required for insulin growth factor (IGF)-1-

stimulated cell proliferation [87]. Interestingly, IGF-1 stimulation of MCF7 cells resulted in collagen-independent DDR1 phosphorylation, which was ascribed to the formation of a complex between DDR1 and IGF-1R, leading to rapid DDR1 phosphorylation and internalization, in the absence of collagen. Therefore, the enhancement of proliferation and survival in these breast cancer cells by IGF-1 was mediated in part by activation of DDR1 independently of collagen but required IGF-1R.

Large-scale functional gene screen approaches identified DDR1 as one of the genes required for cell viability in several breast epithelial cell lines. For instance, shRNA screens found that DDR1 downregulation inhibits the viability of human mammary epithelial cells (HMECs) isolated from a reduction mammoplasty and immortalized with human telomerase [93]. Another study aimed at identifying essential genes in human breast epithelial cells by RNAi screening found DDR1 to be one of the genes required for the survival of MCF10A cells and breast cancer MDA-MB-435 cells [94]. Thus, according to these studies, DDR1 plays a role in support of cell survival.

DDR1 and DDR2 are among the RTKs that have been implicated in the resistance of TNBC cell lines (SUM159 and MDA-MB-231) and genetically engineered C3tag mice harboring mammary tumors to MEK inhibitors [77]. It was found that transcriptional induction of DDR1 and DDR2 expression were part of a kinome reprograming mechanism in TNBC cells, which allowed the tumor cells to escape the growth arrest induced in response to MEK inhibition. Under these conditions, kinome reprograming was mediated by the proteolytic degradation of Myc as a result of MEK–ERK inhibition. The role of DDRs in this resistance mechanism was demonstrated by siRNA knockdown of DDR1 or DDR2 (among other kinases), which restored growth inhibition in SUM159 and MDA-MB-231 cells treated with two MEK inhibitors. Thus, DDRs appear to play a key role in supporting the proliferation of TNBC cells in the presence of MEK inhibitors. Interestingly, siRNAs to DDR1 in untreated SUM159 cells had inconsistent effects on cell proliferation, in spite of knockdown of DDR1 expression of 70–90 %. Likewise, knockdown of DDR2 in untreated SUM159 and MDA-MB-231 cells had modest effect on cell growth [77]. These studies suggest that in these TNBC cell lines, DDRs are not critical for cell proliferation under basal conditions. However, DDRs appear to be critical in supporting proliferation of TNBC cells that become resistant to therapies targeting the RAF–MEK–ERK pathway. These results suggest that DDRs may constitute promising therapeutic targets in TNBC patients who fail to respond to MEK-targeted therapies. However, more studies are required to determine whether DDRs are also involved in resistance to other therapies in breast cancer patients.

Contrary to the findings suggesting a role for DDR1 in cell proliferation and survival, studies with MCF7 and ZR-75-1 breast cancer cells cultured within a 3D collagen matrix suggested a role for DDR1 in 3D collagen-induced apoptosis, in a process that was proposed to involve the induction of Bcl-2-interacting killer (BIK) protein, a proapoptotic member of the Bcl-2 family [95]. Consistently, inhibition of DDR1 phosphorylation with inhibitor DDR1-IN-1 [90] reduced apoptosis and induced BIK in cells cultured within 3D collagen [95]. Interestingly, MT1-MMP, a potent membrane-anchored, was shown to prevent apoptosis of breast cancer cells in 3D collagen [96] and to negatively regulate DDR1 activation by initiating receptor cleavage [95, 97]. Therefore, it was proposed that an MT1–MMP/DDR1 axis regulates apoptosis of breast cancer cells embedded in 3D collagen in which collagenolytic activity of the protease and DDR1 cleavage support survival of the tumor cells [95]. Although these studies were limited to two ERα-positive luminal breast cancer lines, the proposed model of DDR1 action in cell survival takes into consideration that invasive breast cancer cells thrive within a stroma enriched in collagen fibers, which elicits profound effects on cell behavior, likely to be mediated in part by DDR1 signaling. More studies are needed to address the pro- and antiapoptotic effects of DDR1 in a broad spectrum of breast cancer cell lines in various matrix conditions. In summary, the accumulating data points to a role for DDR1 in supporting cell proliferation and survival in breast cancer cells. However, there are still important inconsistences, which appear to be related to the assay conditions (plastic vs. 3D collagen) and approaches to stimulate or inhibit DDR1 expression/activity.

*DDR2*: Silencing of DDR2 expression in MDA-MB-231 and 4T1 breast cancer cells had no effect on cell proliferation in vitro [41]. However, studies on kinome reprograming showed that resistance of MDA-MB-231 and SUM159 breast cancer cells to a MEK1/2 inhibitor resulted in de novo expression of DDR2 [77], suggesting a prosurvival for DDR2. Consistently, downregulation of DDR2 resulted in strong synthetic lethality in the presence of the MEK1/2 inhibitor. These studies suggest that DDR2 may be required for cell proliferation and survival under stress conditions.

#### 7.3.2.2 Roles in Cell Migration, Invasion, and Metastasis

A key aspect of malignancy is the ability of cancer cells to migrate and invade through extracellular matrices, including basement membranes and interstitial stroma, both of which are rich in collagen proteins. Because DDRs are collagen receptors, many studies addressed the contribution of DDRs to breast cancer migration and invasion in various established assays utilizing Matrigel, a reconstituted basement membrane enriched in collagen IV and laminin, and collagen I, a component of the breast stromal matrix, as protein barriers. Matrigel is used to mimic the step of basement membrane invasion and is expected to induce the activation of DDR1 but not of DDR2 [1]. In contrast, collagen I is an activator of both DDR1 and DDR2 [1]. Therefore, if ligand-dependent DDR activation were required for tumor cell dissemination (migration, invasion, and metastasis), DDR1 would be expected to play a role during penetration of basement membranes, whereas DDR1 and DDR2 would play a role during invasion of the interstitial matrix. Thus, a breast cancer cell with a defined set of DDRs may or may not activate these receptors in migration/invasion depending on the nature of the matrix environment utilized in the assay. On the other hand, effects of DDRs may not necessary be mediated via their kinase activity, which would indicate ligand-independent effects. However, a clear understanding how DDRs elicit biological actions without activating their kinase function is still missing.

*DDR1*: The role of DDR1 in experimental models of breast cancer metastasis in mice remains to be determined. However, there are multiple reports on its role migration and invasion of breast cancer cells in in vitro assays. Castro-Sanchez et al. reported that collagen-IV-induced migration [98] and Matrigel invasion [84] of MDA-MB-231 breast cancer cells required expression of DDR1. This effect of collagen IV involved induction of the tetraspanin CD9 through a DDR1-dependent pathway, which in turn induced the expression of MMP-2 and MMP-9, two MMPs associated with tumor cell invasion [84, 98]. In agreement with a promigratory role for DDR1, Neuhaus et al. found that silencing of DDR1 mRNA in human nonmalignant HB2 breast epithelial cells and T47D and MDA-MB-468 breast cancer cells reduced cell migration, as determined using Boyden chambers assembled with uncoated filters, and fibroblast conditioned media as chemoattractant [99]. These investigators also found that DDR1 restored the antimigratory activity induced by Syk (spleen tyrosine kinase), a nonreceptor tyrosine kinase shown to elicit antimalignant effects in breast carcinoma cells [100], and shown to form a complex with DDR1 in nonmalignant HC11 mammary epithelial cells [99, 101]. Malaguarnera et al. [87] also reported promigratory effects of DDR1 in MCF-7, BT-474, and MDA-MB-231 breast cancer cells in response to IGF-1 using fibronectin- or collagen-IV-coated filters. Indeed, knockdown of DDR1 expression by siRNA reduced cell migration, whereas wild type but to a lesser extent kinase dead DDR1 overexpression enhanced migration. As with the findings with cell proliferation, DDR1 also regulated IGF-1-stimulated cell migration. Interestingly, because the ability of DDR1 to support cell migration was also observed on fibronectin substrates, the investigators concluded that DDR1 regulation of cell migration is independent of its function as a collagen receptor [87]. Studies in MDA-MB-231 cells with depleted DDR1 expression by siRNA showed that DDR1 was required for in vitro invasion of a 3D collagen I matrix [85]. This process appeared to involve DDR1-mediated formation of invadosomes, an F-actin-rich structure that was shown to focus proteolytic activity in invasive cancer cells and thus facilitates matrix invasion [102]. Moreover, the proinvasive effects of DDR1 required metalloproteinase activity [85]. Interestingly, Juin et al. showed that the involvement of DDR1 in invadosome formation in response to collagen I was independent of its kinase activity [85]. The proposed mechanism for collagen I-induced invadosome formation involved the activation of the Rho GTPase cell division control protein 42 homolog (Cdc42), which required DDR1 expression. DDR1's role in Cdc42 activation involved the action of the guanine nucleotide-exchange factor, Tuba, which colocalized with DDR1 in invadosomes [85]. However, how DDR1 mediates these effects, independently of its kinase activity, when invasive cells confront the fibrillar collagen matrix, which is known to induce receptor phosphorylation, remains unclear.

Contrary to the reports ascribing a promigratory effect of DDR1 in breast cancer cells, other studies found DDR1 to be a negative regulator of cell migration in breast epithelial cells regardless of malignancy status. Using the yeast two-hybrid system to identify DDR1 interacting proteins, Hansen et al. identified DARPP-32 (dopamine and cAMP-regulated phosphoprotein of 32 kDa), a protein initially known to play a role in regulation of neurotransmission, as a DDR1-binding protein.

Functional assays revealed that coexpression of DARPP-32 and DDR1 in breast cancer cells inhibited in vitro cell migration [86]. This effect required a Wnt5a-cAMP-PKA-mediated phosphorylation of DARPP-32 at threonine 34 [103]. Interestingly, Wnt5a, a noncanonical Wnt signaling protein, was shown to enhance cell adhesion and consequently to inhibit cell migration of breast cancer cells [104, 105]. Importantly, downregulation of Wnt5a impaired collagen-dependent DDR1 activation suggesting that the antimigratory effects of Wnt5a, and possibly its tumor suppressive roles in breast cancer, are mediated in part by DDR1 activation [65, 104–106]. In breast tumor specimens, loss of Wnt5a correlated with disease recurrence and was an independent predictor of poor prognosis [101, 107], consistent with a tumor suppressive role for Wnt5a in breast cancer progression [106–109]. If so, it will be interesting to examine whether loss of Wnt5a in aggressive breast tumors is also associated with reduced levels of phosphorylated DDR1. Nevertheless, the potential relationship between Wnt5a and DDR1 activation suggest a tumor suppressive role for DDR1 in certain breast cancers, possibly as a negative regulator of cell migration. The studies of Koh et al. also showed an antimigratory and anti-invasive effect of DDR1 in both nonmalignant and malignant breast epithelial cells [72]. In MCF10A cells, silencing of DDR1 expression enhanced in vitro cell migration and invasion, whereas overexpression of DDR1 in triple negative Hs587T and MDA-MB-231 breast cancer cells inhibited in vitro and in vivo cell invasion [72], consistent with DDR1 being a suppressor of promalignant cellular activities. Consistently, expression of mutated H-Ras in MCF10A cells, which induced EMT via Zeb1 upregulation, resulted in acquisition of a migratory and invasive phenotype with a concomitant transcriptional downregulation of DDR1 [72]. Likewise, ectopic expression of Zeb1 in MCF10A cells inhibited DDR1 expression [72]. DDR1 downregulation was also observed after treatment of MCF10A cells with TGF-β, which upregulated Snail1 and Twist1 [41]. Interestingly, earlier studies showed that TGF-β treatment induces the in vitro migration and invasion of MCF10A cells [110], raising the possibility that the promigratory effects of TGF-β were mediated in part by DDR1 downregulation. Although this possibility remains to be proven, the emerging evidence suggests that inhibition of DDR1 expression is part of the EMT program that promotes malignancy [82]. However, loss of DDR1 expression alone was not sufficient to induce EMT in parental MCF10A cells but, interestingly, it was sufficient to enhance migratory and invasive activities [72]. These results suggest that although DDR1 is not a direct regulator of EMT in breast epithelial cells, it can directly influence cell migratory activities. The molecular mechanism(s) by which DDR1 regulates cell migration in breast epithelial cells remains to be defined.

An association between DDR1 and EMT is also suggested by the findings with KIBRA, a protein that was shown to bind DDR1 in MCF10A and T47-D cells [31] and to regulate EMT via the Hippo pathway in MCF10A cells [111]. Early studies showed that DDR1 forms a stable complex with KIBRA that dissociates upon stimulation with collagen [31]. It was also shown that knockdown of KIBRA in MCF10A cells reduced collagen IV-induced ERK phosphorylation, suggesting that KIBRA, upon dissociation from the DDR1/KIBRA complex, plays a permissive role in collagen IV-induced MAPK activation [31]. However, whether the role of KIBRA in regulation of collagen IV-induced ERK phosphorylation is mediated by DDR1 activation remains to be determined. This missing information is critical because collagen may activate the MAPK pathway via other receptors (i.e., integrins) [112, 113]. Interestingly, Moleirinho et al. reported that silencing of KIBRA expression in MCF10A cells induced expression of EMT markers and enhanced cell migration, a process that was related to decreased LATS and YAP phosphorylation [111], two kinases involved in the Hippo pathway [114]. However, the role of DDR1 in EMT induced by downregulation of KIBRA remains to be determined. Regardless, these data suggest the existence of a crosstalk between DDR1, KIBRA, and the Hippo pathway in regulation of EMT in breast epithelial cells. Notably, accumulating evidence shows a key role for the Hippo pathway in dictating the cellular phenotype in response to cell–matrix interactions, in particular in response to matrix stiffness [115, 116]. Because DDRs are collagen receptors, which may play a role in mechanosensing [117], it will be interesting to investigate the role of DDRs in regulation of Hippo pathway in breast development and breast cancer progression.

Taken together, the emerging evidence points to a complex role for DDR1 in breast cancer cell behavior, and therefore at this junction there is no clear consensus as to whether DDR1 plays suppressive or promoting effects or both on disease progression. This complexity is further highlighted by the finding of a heterogeneous expression of DDR1 in breast tissues showing that approximately 50 % of invasive breast carcinomas display high levels of DDR1, whereas the other half show reduced or no expression of receptor [58]. Thus, it is reasonable to speculate that the function of DDR1 in disease progression may be dictated by the unique genomic and proteomic makeup of each particular breast tumor. Identifying the molecular and environmental factors that influence DDR1 action in breast cancer will help selecting those patients who will potentially benefit from anti-DDR1 therapies.

*DDR2*: The accumulating data point to a role for DDR2 in supporting the invasive and metastatic activities of breast cancer cells both in in vitro and in vivo assays. Using human MDA-MB 231 and mouse 4T1 breast cancer cells, and RNA silencing, Zhang et al. [41] demonstrated a key role for DDR2 in supporting in vitro tumor cell migration and invasion through a collagen matrix. Importantly, DDR2-silenced 4T1 cells inoculated into the mammary fat pad developed fewer lung metastases [41]. In agreement with these studies, Ren et al. [71] found that DDR2 supported the metastatic ability of MDA-MB-231 cells inoculated into the mammary fat pad. The effects of DDR2 on metastases were specific to the process of tumor cell dissemination because DDR2 had no effect on the growth of the primary tumors [41, 71]. The prometastatic role of DDR2 was associated with the development of EMT and the ability of DDR2 to support Snail1 protein stabilization via an ERK-mediated pathway [41]. Ren et al. also showed that hypoxic conditions enhanced DDR2 expression in SK-BR3, MDA-MB-231 MDA-MD-468, and MCF-7 cells, concomitant with a promigratory and proinvasive in vitro activity, which dependent on DDR2 expression [71]. The association between EMT and DDR2 was also shown in nonmalignant MCF10A cells, which upon treatment with TFG-β display induction of the EMT transcription factors Snail1 and Twist1, with the concomitant increase in DDR2 expression [41]. Thus, DDR2 is an EMT upregulated gene in both nonmalignant and malignant breast epithelial cells, which supports cell migration. A recent study in genetically modified mice harboring a specific deletion of *DDR2* and mammary tumor virus-polyoma middle T antigen (MMTV-PyMT) tumors demonstrated the importance of this receptor in the development of breast cancer metastases, but not in the formation of primary tumors [118]. Importantly, these studies highlighted the contribution of stromal *DDR2*, rather than tumor-derived *DDR2*, in formation of lung metastases. Specifically, transplantation of MMTV-PyMT breast cancer cells into syngeneic mice with global *DDR2* deletion significantly inhibited lung metastasis formation, regardless of the expression of *DDR2* in the tumor cells. Further analyses showed that the primary tumors growing in *DDR2*-deficient mice displayed a significant reduction in fibrillar collagen and reduced vascularization. Disruption in collagen deposition and organization was ascribed to ineffective function of the cancer-associated fibroblasts, a major collagen-producing cells, present in the *DDR2*-null mice. These cells also exhibited a reduced fibrogenic gene expression profile suggesting that DDR2 is a pro-fibrotic gene in tumor stroma. Collectively, these studies demonstrated a key role for tumor and stromal DDR2 in breast cancer metastases, and provided strong support for developing DDR2 inhibitors for the treatment of metastatic breast cancer [118]. Figure 7.3 provides a schematic representation of the reported functions of DDRs in experimental models of breast cancer.

## 7.4 Conclusions: DDRs in Breast Tissues—an Evolving Field

The importance of the ECM in mammary gland development and breast cancer progression, the pivotal role that RTKs play in cancer, and the fact that DDRs are the only kinases that recognize collagen as their ligands provide a strong impetus to decipher the action of DDRs in normal and malignant breast tissues. Overall, the accumulated evidence strongly suggests a central role for DDRs in both normal and malignant breast tissues. However, in spite of the encouraging findings, so far the overall picture remains incomplete and in many cases the evidence is limited and contradictory. Certain DDR-mediated functions have yet to be confirmed or reproduced. Issues of kinase or ligand dependency for the biological effects observed remain unclear. Significant inconsistencies exist in the experimental conditions; for instance, the contribution of collagen (2D, 3D) when analyzing effects on cell proliferation/survival are not clearly delineated. Likewise, the nature and status of the substrate and/or presence of collagen ligands in assays of cell migration and invasion are usually not addressed. Analyses of receptor activation status and kinetics in functional assays are not always examined. These uncertainties are further compounded by the fact that cell lines may express both DDRs, which may complicate data interpretation in conditions that use common ligands. Moreover, discrepancies exist in the basal levels of DDRs expressed by breast cancer cell lines. We also lack a thorough understanding of the signaling networks and downstream effectors regulated by DDRs in breast epithelial cells in response to collagen stimulation. Importantly, how these DDR signaling outcomes are affected, by the biophysical

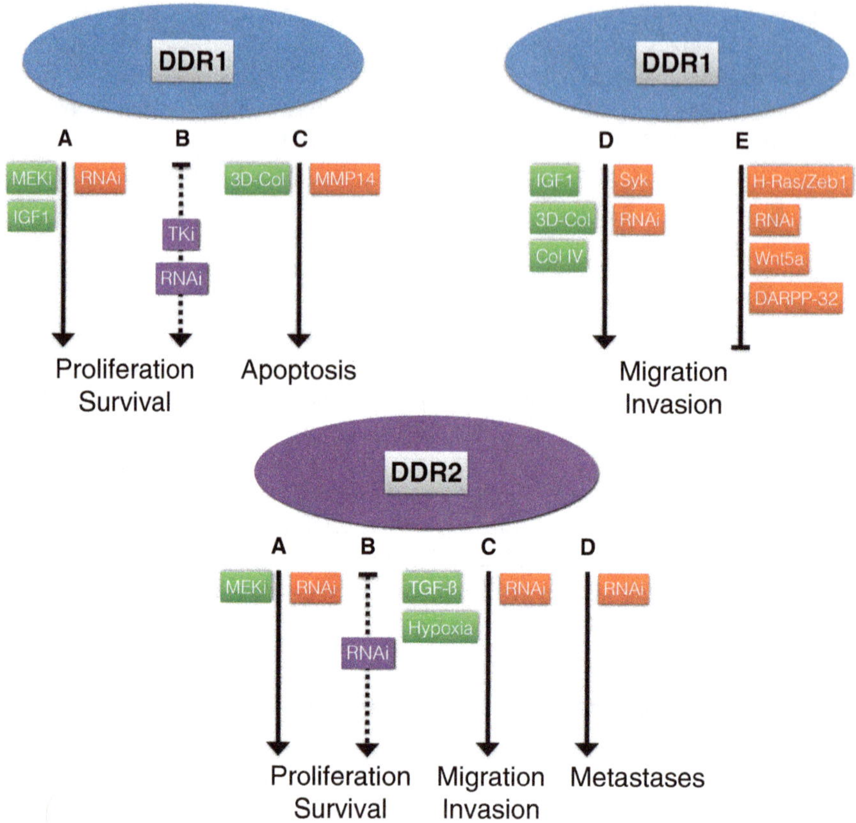


**Fig. 7.3** Potential roles of DDRs in breast cancer cells. DDR1: (**a**) DDR1 promotes or supports cell proliferation and survival in response IGF-1 or MEK1/2 inhibitors. Knockdown of DDR1 expression inhibits cell proliferation in MEK-resistant cells [77], unstimulated or IGF1-stimulated cells [87], or in genetic RNAi screen assays [92–94]. (**b**) Tyrosine kinase inhibitors (TKi) of DDR1 activity provided conflicting results (*magenta box*), with some inhibiting [87, 89] and others having no effect [90]. Likewise, downregulation of DDR1 by RNAi, under basal conditions, had no effects on cell proliferation [77]. (**c**) DDR1 mediates the induction of apoptosis in cells grown within 3D-collagen I [95]. This effect is blocked upon cleavage of DDR1 by MMP14 [96]. (**d**) DDR1 supports cell migration in response to IGF-1 [87], 3D-collagen I [85], and collagen IV [87, 98], and supports invasion through Matrigel [84]. Inhibition of DDR1 expression by RNAi blocks migration [85, 87, 99]. Syk kinase elicits its an-migratory activity by blocking DDR1 [99]. (**e**) DDR1 is an inhibitor of cell migration and invasion: H-Ras/Zeb1 induces migration and invasion by downregulating DDR1 expression [72], and Wnt5a blocks migration by impairing collagen-dependent DDR1 activation [65, 105]. Inhibition of DDR1 expression by RNAi enhances migration and invasion [72]. Coexpression of DARPP-32 and DDR1 inhibits cell migration [103]. DDR2: (**a**) DDR2 is induced in response to resistance to MEK1/2 inhibition in TNBC cells, as a part of a kinome reprograming process designed to promote cell survival. DDR2 RNAi elicits synthetic lethality when administered with MEK inhibitors [77]. (**b**) DDR2 is not required for in vitro and in vivo proliferation of breast cancer cell lines [41, 71], as determined by RNAi. (**c**) TGF-β and hypoxia induce DDR2 expression and consequently enhance in vitro migration and invasion [41, 71]. (**d**) DDR2 is required for metastatic dissemination of human breast cancer cells when inoculated into mammary fat pads [41, 71] and in the mouse MMTV-PyMT model of mammary cancer [118]. *Solid arrows* show a positive or negative effect, *dotted arrows* indicate conflicting results; *green boxes* indicate a positive effect, and *red boxes* indicate an inhibitory effect

status of the collagen matrix, which may be key in conditions in which disease progression is associated with tissue stiffness. In this regard, the role of DDRs in mechanosensing remains practically unknown. In mice studies, cell-specific deletion of DDRs in the mammary gland need to be performed to clearly define the role of DDRs in development and differentiation. In human tissues, more studies are required to better define the relative levels of expression and receptor activation in various breast cancer subtypes and in primary vs. metastatic sites, utilizing highly specific antibodies and probes. Finally, the assessment of DDRs as therapeutic targets in breast cancer awaits the development of specific kinase inhibitors, a major challenge, but one that will help to elucidate their contribution to disease progression in more rigorous ways. These are challenges that are constantly met and thus we are hopeful that new and exciting findings will illuminate the DDR field and elucidate their role and therapeutic value in normal and diseased breast tissues.

## References

1. Fu HL et al (2013) Discoidin domain receptors: unique receptor tyrosine kinases in collagen-mediated signaling. J Biol Chem 288(11):7430–7437
2. Inman JL et al (2015) Mammary gland development: cell fate specification, stem cells and the microenvironment. Development 142(6):1028–1042
3. Sternlicht MD (2006) Key stages in mammary gland development: the cues that regulate ductal branching morphogenesis. Breast Cancer Res 8(1):201
4. Oskarsson T (2013) Extracellular matrix components in breast cancer progression and metastasis. Breast 22(Suppl 2):S66–S72
5. Muschler J, Streuli CH (2010) Cell-matrix interactions in mammary gland development and breast cancer. Cold Spring Harb Perspect Biol 2(10):a003202
6. Obr AE et al (2013) Progesterone receptor and Stat5 signaling cross talk through RANKL in mammary epithelial cells. Mol Endocrinol 27(11):1808–1824
7. Fernandez-Valdivia R, Lydon JP (2012) From the ranks of mammary progesterone mediators, RANKL takes the spotlight. Mol Cell Endocrinol 357(1-2):91–100
8. Fernandez-Valdivia R et al (2005) Revealing progesterone's role in uterine and mammary gland biology: insights from the mouse. Semin Reprod Med 23(1):22–37
9. Wiseman BS, Werb Z (2002) Stromal effects on mammary gland development and breast cancer. Science 296(5570):1046–1049
10. Sakakura T, Suzuki Y, Shiurba R (2013) Mammary stroma in development and carcinogenesis. J Mammary Gland Biol Neoplasia 18(2):189–197
11. Lambert AW, Ozturk S, Thiagalingam S (2012) Integrin signaling in mammary epithelial cells and breast cancer. ISRN Oncol 2012:493283
12. Schedin P, Keely PJ (2011) Mammary gland ECM remodeling, stiffness, and mechanosignaling in normal development and tumor progression. Cold Spring Harb Perspect Biol 3(1):a003228
13. Brownfield DG et al (2013) Patterned collagen fibers orient branching mammary epithelium through distinct signaling modules. Curr Biol 23(8):703–709
14. Ingman WV et al (2006) Macrophages promote collagen fibrillogenesis around terminal end buds of the developing mammary gland. Dev Dyn 235(12):3222–3229
15. Rodriguez D, Morrison CJ, Overall CM (2010) Matrix metalloproteinases: what do they not do? New substrates and biological roles identified by murine models and proteomics. Biochim Biophys Acta 1803(1):39–54

16. Faraci-Orf E, McFadden C, Vogel WF (2006) DDR1 signaling is essential to sustain Stat5 function during lactogenesis. J Cell Biochem 97(1):109–121
17. Vogel WF et al (2001) Discoidin domain receptor 1 tyrosine kinase has an essential role in mammary gland development. Mol Cell Biol 21(8):2906–2917
18. Cellurale C et al (2012) Role of JNK in mammary gland development and breast cancer. Cancer Res 72(2):472–481
19. Crowley MR, Bowtell D, Serra R (2005) TGF-beta, c-Cbl, and PDGFR-alpha the in mammary stroma. Dev Biol 279(1):58–72
20. Joseph H et al (1999) Overexpression of a kinase-deficient transforming growth factor-beta type II receptor in mouse mammary stroma results in increased epithelial branching. Mol Biol Cell 10(4):1221–1234
21. Roarty K, Serra R (2007) Wnt5a is required for proper mammary gland development and TGF-beta-mediated inhibition of ductal growth. Development 134(21):3929–3939
22. Fernandez-Valdivia R et al (2008) Transcriptional response of the murine mammary gland to acute progesterone exposure. Endocrinology 149(12):6236–6250
23. Brisken C et al (1999) Prolactin controls mammary gland development via direct and indirect mechanisms. Dev Biol 210(1):96–106
24. Cui Y et al (2004) Inactivation of Stat5 in mouse mammary epithelium during pregnancy reveals distinct functions in cell proliferation, survival, and differentiation. Mol Cell Biol 24(18):8037–8047
25. Miyoshi K et al (2001) Signal transducer and activator of transcription (Stat) 5 controls the proliferation and differentiation of mammary alveolar epithelium. J Cell Biol 155(4):531–542
26. Mukherjee A et al (2010) Targeting RANKL to a specific subset of murine mammary epithelial cells induces ordered branching morphogenesis and alveologenesis in the absence of progesterone receptor expression. FASEB J 24(11):4408–4419
27. Santos SJ, Haslam SZ, Conrad SE (2010) Signal transducer and activator of transcription 5a mediates mammary ductal branching and proliferation in the nulliparous mouse. Endocrinology 151(6):2876–2885
28. Fata JE et al (2000) The osteoclast differentiation factor osteoprotegerin-ligand is essential for mammary gland development. Cell 103(1):41–50
29. Rayala SK et al (2006) Essential role of KIBRA in co-activator function of dynein light chain 1 in mammalian cells. J Biol Chem 281(28):19092–19099
30. Harris J et al (2006) Socs2 and elf5 mediate prolactin-induced mammary gland development. Mol Endocrinol 20(5):1177–1187
31. Hilton HN et al (2008) KIBRA interacts with discoidin domain receptor 1 to modulate collagen-induced signalling. Biochim Biophys Acta 1783(3):383–393
32. Chen J et al (2002) The alpha(2) integrin subunit-deficient mouse: a multifaceted phenotype including defects of branching morphogenesis and hemostasis. Am J Pathol 161(1):337–344
33. Gardner H et al (1996) Deletion of integrin alpha 1 by homologous recombination permits normal murine development but gives rise to a specific deficit in cell adhesion. Dev Biol 175(2):301–313
34. Spike BT et al (2012) A mammary stem cell population identified and characterized in late embryogenesis reveals similarities to human breast cancer. Cell Stem Cell 10(2):183–197
35. Wansbury O et al (2011) Transcriptome analysis of embryonic mammary cells reveals insights into mammary lineage establishment. Breast Cancer Res 13(4):R79
36. Bai L, Rohrschneider LR (2010) s-SHIP promoter expression marks activated stem cells in developing mouse mammary tissue. Genes Dev 24(17):1882–1892
37. Shackleton M et al (2006) Generation of a functional mammary gland from a single stem cell. Nature 439(7072):84–88
38. Kano K et al (2008) A novel dwarfism with gonadal dysfunction due to loss-of-function allele of the collagen receptor gene, Ddr2, in the mouse. Mol Endocrinol 22(8):1866–1880
39. Labrador JP et al (2001) The collagen receptor DDR2 regulates proliferation and its elimination leads to dwarfism. EMBO Rep 2(5):446–452

40. Cowling RT et al (2014) Discoidin domain receptor 2 germline gene deletion leads to altered heart structure and function in the mouse. Am J Physiol Heart Circ Physiol 307(5): H773–H781
41. Zhang K et al (2013) The collagen receptor discoidin domain receptor 2 stabilizes SNAIL1 to facilitate breast cancer metastasis. Nat Cell Biol 15(6):677–687
42. Dontu G et al (2003) In vitro propagation and transcriptional profiling of human mammary stem/progenitor cells. Genes Dev 17(10):1253–1270
43. Huang S et al (2005) Changes in gene expression during the development of mammary tumors in MMTV-Wnt-1 transgenic mice. Genome Biol 6(10):R84
44. Li Y et al (2003) Evidence that transgenes encoding components of the Wnt signaling pathway preferentially induce mammary cancers from progenitor cells. Proc Natl Acad Sci U S A 100(26):15853–15858
45. Rosen PP (2009) Rosen's breast pathology, 3rd edn. Wolters Kluwer/Lippincott Williams & Wilkins, Philadelphia, p 1116
46. Elledge RM et al (1998) HER-2 expression and response to tamoxifen in estrogen receptor-positive breast cancer: a Southwest Oncology Group Study. Clin Cancer Res 4(1):7–12
47. Osborne CK (1998) Tamoxifen in the treatment of breast cancer. N Engl J Med 339(22):1609–1618
48. Slamon D, Pegram M (2001) Rationale for trastuzumab (Herceptin) in adjuvant breast cancer trials. Semin Oncol 28(1 Suppl 3):13–19
49. Perou CM et al (2000) Molecular portraits of human breast tumours. Nature 406(6797):747–752
50. Cancer Genome Atlas Network (2012) *Comprehensive molecular portraits of human breast tumours*. Nature 490(7418):61–70
51. Burstein MD et al (2015) Comprehensive genomic analysis identifies novel subtypes and targets of triple-negative breast cancer. Clin Cancer Res 21(7):1688–1698
52. Maskarinec G et al (2013) Mammographic density as a predictor of breast cancer survival: the Multiethnic Cohort. Breast Cancer Res 15(1):R7
53. Tice JA et al (2013) Benign breast disease, mammographic breast density, and the risk of breast cancer. J Natl Cancer Inst 105(14):1043–1049
54. Ursin G et al (2005) Greatly increased occurrence of breast cancers in areas of mammographically dense tissue. Breast Cancer Res 7(5):R605–R608
55. Boyd NF et al (2001) Mammographic density as a marker of susceptibility to breast cancer: a hypothesis. IARC Sci Publ 154:163–169
56. Martin LJ, Boyd NF (2008) *Mammographic density. Potential mechanisms of breast cancer risk associated with mammographic density: hypotheses based on epidemiological evidence*. Breast Cancer Res 10(1):201
57. Paszek MJ et al (2005) Tensional homeostasis and the malignant phenotype. Cancer Cell 8(3):241–254
58. Provenzano PP, Eliceiri KW, Keely PJ (2009) Shining new light on 3D cell motility and the metastatic process. Trends Cell Biol 19(11):638–648
59. Provenzano PP et al (2006) Collagen reorganization at the tumor-stromal interface facilitates local invasion. BMC Med 4(1):38
60. Valiathan RR et al (2012) Discoidin domain receptor tyrosine kinases: new players in cancer progression. Cancer Metastasis Rev 31(1-2):295–321
61. Vogel W et al (1997) The discoidin domain receptor tyrosine kinases are activated by collagen. Mol Cell 1(1):13–23
62. Toy KA et al (2015) Tyrosine kinase discoidin domain receptors DDR1 and DDR2 are coordinately deregulated in triple-negative breast cancer. Breast Cancer Res Treat 150(1):9–18
63. Yeh YC, Wang CZ, Tang MJ (2009) Discoidin domain receptor 1 activation suppresses alpha-2beta1 integrin-dependent cell spreading through inhibition of Cdc42 activity. J Cell Physiol 218(1):146–156
64. Ren T et al (2013) Increased expression of discoidin domain receptor 2 (DDR2): a novel independent prognostic marker of worse outcome in breast cancer patients. Med Oncol 30(1):397

65. Dejmek J et al (2003) Wnt-5a and G-protein signaling are required for collagen-induced DDR1 receptor activation and normal mammary cell adhesion. Int J Cancer 103(3):344–351
66. Turashvili G et al (2007) Novel markers for differentiation of lobular and ductal invasive breast carcinomas by laser microdissection and microarray analysis. BMC Cancer 7:55
67. Wang CZ, Yeh YC, Tang MJ (2009) DDR1/E-cadherin complex regulates the activation of DDR1 and cell spreading. Am J Physiol Cell Physiol 297(2):C419–C429
68. Eswaramoorthy R et al (2010) DDR1 regulates the stabilization of cell surface E-cadherin and E-cadherin-mediated cell aggregation. J Cell Physiol 224(2):387–397
69. Hidalgo-Carcedo C et al (2011) Collective cell migration requires suppression of actomyosin at cell-cell contacts mediated by DDR1 and the cell polarity regulators Par3 and Par6. Nat Cell Biol 13(1):49–58
70. Ameli F, Rose IM, Masir N (2015) Expression of DDR1 and DVL1 in invasive ductal and lobular breast carcinoma does not correlate with histological type, grade and hormone receptor status. Asian Pac J Cancer Prev 16(6):2385–2390
71. Ren T et al (2014) Discoidin domain receptor 2 (DDR2) promotes breast cancer cell metastasis and the mechanism implicates epithelial-mesenchymal transition programme under hypoxia. J Pathol 234(4):526–537
72. Koh M et al (2015) Discoidin domain receptor 1 is a novel transcriptional target of ZEB1 in breast epithelial cells undergoing H-Ras-induced epithelial to mesenchymal transition. Int J Cancer 136(6):E508–E520
73. Morikawa A et al (2015) Expression of beclin-1 in the microenvironment of invasive ductal carcinoma of the breast: correlation with prognosis and the cancer-stromal interaction. PLoS One 10(5):e0125762
74. Lehmann BD et al (2011) Identification of human triple-negative breast cancer subtypes and preclinical models for selection of targeted therapies. J Clin Invest 121(7):2750–2767
75. Al-Ejeh F et al (2014) Kinome profiling reveals breast cancer heterogeneity and identifies targeted therapeutic opportunities for triple negative breast cancer. Oncotarget 5(10):3145–3158
76. Mayer IA et al (2014) New strategies for triple-negative breast cancer—deciphering the heterogeneity. Clin Cancer Res 20(4):782–790
77. Duncan JS et al (2012) Dynamic reprogramming of the kinome in response to targeted MEK inhibition in triple-negative breast cancer. Cell 149(2):307–321
78. Cerami E et al (2012) The cBio cancer genomics portal: an open platform for exploring multidimensional cancer genomics data. Cancer Discov 2(5):401–404
79. Gentles AJ et al (2015) The prognostic landscape of genes and infiltrating immune cells across human cancers. Nat Med 21(8):938–945
80. Gyorffy B et al (2013) Online survival analysis software to assess the prognostic value of biomarkers using transcriptomic data in non-small-cell lung cancer. PLoS One 8(12):e82241
81. Jezequel P et al (2013) *bc-GenExMiner 3.0: new mining module computes breast cancer gene expression correlation analyses*. Database (Oxford) 2013:bas060
82. Taube JH et al (2010) Core epithelial-to-mesenchymal transition interactome gene-expression signature is associated with claudin-low and metaplastic breast cancer subtypes. Proc Natl Acad Sci U S A 107(35):15449–15454
83. Reva B, Antipin Y, Sander C (2011) Predicting the functional impact of protein mutations: application to cancer genomics. Nucleic Acids Res 39(17):e118
84. Castro-Sanchez L et al (2011) Role of DDR1 in the gelatinases secretion induced by native type IV collagen in MDA-MB-231 breast cancer cells. Clin Exp Metastasis 28(5):463–477
85. Juin A et al (2014) Discoidin domain receptor 1 controls linear invadosome formation via a Cdc42-Tuba pathway. J Cell Biol 207(4):517–533
86. Hansen C et al (2006) Phosphorylation of DARPP-32 regulates breast cancer cell migration downstream of the receptor tyrosine kinase DDR1. Exp Cell Res 312(20):4011–4018
87. Malaguarnera R et al (2015) Novel cross talk between IGF-IR and DDR1 regulates IGF-IR trafficking, signaling and biological responses. Oncotarget 6(18):16084–16105
88. Yeh YC et al (2011) DDR1 triggers epithelial cell differentiation by promoting cell adhesion through stabilization of E-cadherin. Mol Biol Cell 22(7):940–953

89. Gao M et al (2013) Discovery and optimization of 3-(2-(Pyrazolo[1,5-a]pyrimidin-6-yl)ethynyl)benzamides as novel selective and orally bioavailable discoidin domain receptor 1 (DDR1) inhibitors. J Med Chem 56(8):3281–3295
90. Kim HG et al (2013) Discovery of a potent and selective DDR1 receptor tyrosine kinase inhibitor. ACS Chem Biol 8(10):2145–2150
91. L'Hote CG, Thomas PH, Ganesan TS (2002) Functional analysis of discoidin domain receptor 1: effect of adhesion on DDR1 phosphorylation. FASEB J 16(2):234–236
92. Marcotte R et al (2012) Essential gene profiles in breast, pancreatic, and ovarian cancer cells. Cancer Discov 2(2):172–189
93. Schlabach MR et al (2008) Cancer proliferation gene discovery through functional genomics. Science 319(5863):620–624
94. Silva JM et al (2008) Profiling essential genes in human mammary cells by multiplex RNAi screening. Science 319(5863):617–620
95. Assent D et al (2015) A membrane-type-1 matrix metalloproteinase (MT1-MMP)-discoidin domain receptor 1 axis regulates collagen-induced apoptosis in breast cancer cells. PLoS One 10(3):e0116006
96. Maquoi E et al (2012) MT1-MMP protects breast carcinoma cells against type I collagen-induced apoptosis. Oncogene 31(4):480–493
97. Fu HL et al (2013) Shedding of discoidin domain receptor 1 by membrane-type matrix metalloproteinases. J Biol Chem 288(17):12114–12129
98. Castro-Sanchez L et al (2010) Native type IV collagen induces cell migration through a CD9 and DDR1-dependent pathway in MDA-MB-231 breast cancer cells. Eur J Cell Biol 89(11):843–852
99. Neuhaus B et al (2011) Migration inhibition of mammary epithelial cells by Syk is blocked in the presence of DDR1 receptors. Cell Mol Life Sci 68(22):3757–3770
100. Krisenko MO, Geahlen RL (2015) Calling in SYK: SYK's dual role as a tumor promoter and tumor suppressor in cancer. Biochim Biophys Acta 1853(1):254–263
101. Dejmek J et al (2005) Expression and signaling activity of Wnt-5a/discoidin domain receptor-1 and Syk plays distinct but decisive roles in breast cancer patient survival. Clin Cancer Res 11(2 Pt 1):520–528
102. Paz H, Pathak N, Yang J (2014) Invading one step at a time: the role of invadopodia in tumor metastasis. Oncogene 33(33):4193–4202
103. Hansen C et al (2009) Wnt-5a-induced phosphorylation of DARPP-32 inhibits breast cancer cell migration in a CREB-dependent manner. J Biol Chem 284(40):27533–27543
104. Safholm A et al (2006) A formylated hexapeptide ligand mimics the ability of Wnt-5a to impair migration of human breast epithelial cells. J Biol Chem 281(5):2740–2749
105. Jonsson M, Andersson T (2001) Repression of Wnt-5a impairs DDR1 phosphorylation and modifies adhesion and migration of mammary cells. J Cell Sci 114(Pt 11):2043–2053
106. Serra R et al (2011) Wnt5a as an effector of TGFbeta in mammary development and cancer. J Mammary Gland Biol Neoplasia 16(2):157–167
107. Jonsson M et al (2002) Loss of Wnt-5a protein is associated with early relapse in invasive ductal breast carcinomas. Cancer Res 62(2):409–416
108. McDonald SL, Silver A (2009) The opposing roles of Wnt-5a in cancer. Br J Cancer 101(2):209–214
109. Zhu N et al (2014) Challenging role of Wnt5a and its signaling pathway in cancer metastasis (Review). Exp Ther Med 8(1):3–8
110. Kim ES, Kim MS, Moon A (2004) TGF-beta-induced upregulation of MMP-2 and MMP-9 depends on p38 MAPK, but not ERK signaling in MCF10A human breast epithelial cells. Int J Oncol 25(5):1375–1382
111. Moleirinho S et al (2013) KIBRA exhibits MST-independent functional regulation of the Hippo signaling pathway in mammals. Oncogene 32(14):1821–1830
112. Barczyk M, Carracedo S, Gullberg D (2010) Integrins. Cell Tissue Res 339(1):269–280
113. Leitinger B (2011) Transmembrane collagen receptors. Annu Rev Cell Dev Biol 27:265–290

114. Yu FX, Zhao B, Guan KL (2015) Hippo pathway in organ size control, tissue homeostasis, and cancer. Cell 163(4):811–828
115. Dupont S (2015) Role of YAP/TAZ in cell-matrix adhesion-mediated signalling and mechanotransduction. Exp Cell Res 343(1):42–53
116. Dupont S et al (2011) Role of YAP/TAZ in mechanotransduction. Nature 474(7350):179–183
117. Ghosh S et al (2013) Regulation of adipose oestrogen output by mechanical stress. Nat Commun 4:1821
118. Corsa CAS et al (2016) The action of Discoidin Domain Receptor 2 in basal tumor cells and stromal cancer-associated fibroblasts is critical for breast cancer metastases. Cell Rep 15(11):2510–2523

# Chapter 8
# Discoidin Domain Receptors in Lung Cancer

**Yingtao Zhang, Agnes Malysa, and Gerold Bepler**

Lung cancer is the leading cause of cancer-related deaths in the United States accounting for 27 % of the estimated 589,430 deaths in 2015 (Cancer Facts & Figures 2015; www.ACS.org). Because of differences in clinical presentation and therapeutic approaches, lung cancer has been divided into two major types, non-small cell lung cancer (NSCLC) and small cell lung cancer (SCLC). While SCLC is a single histopathological entity, NSCLC comprises adenocarcinoma (50 % of lung cancers), squamous cell carcinoma (30 %), large cell carcinoma (10 %), and a variety of other and less common subtypes and accounts for 85 % of lung cancers. Advances in molecular phenotyping have revealed striking differences in the frequencies of specific gene mutations among these subtypes. Most notable are mutations in receptor tyrosine kinase (RTK) genes, such as the epidermal growth factor receptor (EGFR) and the anaplastic lymphoma kinase (ALK), which have been exploited as targets for remarkably effective novel therapeutics [1, 2]. There is evolving evidence suggesting that discoidin domain receptors (DDRs) play an important role in the progression and metastatic spread of lung cancer, in particular in NSCLC [3], which may provide an opportunity for the development of therapies that target DDRs.

## 8.1 DDR Expression in Lung Cancer

Several investigations have detected overexpression of DDR1 in NSCLC, and some have described an association between overexpression and patients' survival (Table 8.1). Ford et al. [4] found higher levels of DDR1 mRNA and lower levels of

---

Y. Zhang • A. Malysa • G. Bepler, M.D., Ph.D. (✉)
Karmanos Cancer Institute, Wayne State University,
4100 John R Street, Detroit, MI 48301, USA
e-mail: beplerg@karmanos.org



R. Fridman, P.H. Huang (eds.), *Discoidin Domain Receptors in Health and Disease*, DOI 10.1007/978-1-4939-6383-6_8

**Table 8.1** Tabulary summary of data reviewed

| DDR expression and mutations | Assay type or mutations detected | Histology | Specimen type | Specimen fixation | Treatment received | Number of patients | Tumor compared to normal | Cut point for survival analysis | Survival impact | Author |
|---|---|---|---|---|---|---|---|---|---|---|
| DDR1 expression | RNA by qPCR | Non-small cell | Resection | Frozen (?) | Surgery only | 146 | Tumor > normal | 50 % high vs. 50 % low | High levels = long survival | Ford et al. [4] |
| DDR1 expression | Protein by IHC | Adeno-squamous and squamous | Resection | Formalin (?) | Surgery (other treatments not specified) | 155 | Tumor > normal | 65 % presence vs. 35 % absence | Presence = poor survival | Yang et al. [5] |
| DDR1 expression | RNA by qPCR | Adeno-squamous and squamous | Resection | Frozen | Surgery only | 82 | Tumor > normal | 61 % high vs. 39 % low | High levels = lymph node spread and poor survival | Miao et al. [6] |
| DDR1 expression | Protein by IHC | Non-small cell | Resection | Formalin | Surgery (adjuvant therapy in 36) | 83 | Not specified | High (upper quartile) vs. low (lower quartile) | High levels = poor survival in stage I only | Valencia et al. [7] |
| DDR1 mutations | DNA sequencing: W385C, F866Y | Adeno-squamous | Resection | Formalin | Not reported | 188 | Normal not analyzed | Not applicable | Mutation frequency: 2/188 (1.1 %) | Ding et al. [20] |
| DDR1 mutations | DNA sequencing: A496S, R824W | Adeno-squamous, squamous, large cell, carcinoid | Not reported | Not reported | Not reported | 26 | Normal analyzed | Not applicable | Mutation frequency: 2/26 (7.7 %) | Davies et al. [21] |
| DDR1 mutations | RNA sequencing of selected regions: no mutations detected | Non-small cell | Resection | Frozen (?) | Surgery only | 23 | Normal analyzed | Not applicable | Mutation frequency: 0/23 (0 %) | Ford et al. [4] |

| | | | | | | | | | | |
|---|---|---|---|---|---|---|---|---|---|---|
| DDR2 expression | RNA by qPCR | Non-small cell | Resection | Frozen (?) | Surgery only | 146 | Normal > tumor | 50 % high vs. 50 % low | No impact on survival or other clinical parameters | Ford et al. [4] |
| DDR2 expression | RNA by qPCR | Squamous cell | Resection | Frozen | Surgery (other treatments not specified) | 92 | Normal > tumor | Not specified | High levels = higher stage (not statistically significant) | Sasaki et al. [14] |
| DDR2 expression | RNA by qPCR | Squamous cell | Resection | Frozen | Surgery (other treatments not specified) | 54 | Normal > tumor | 50 % high vs. 50 % low | Low levels = higher stage (discrepancy between text and Fig. 1); no impact on survival | Miao et al. [15] |
| DDR2 expression | Protein by IHC | Not reported | Not reported | Formalin | Not reported | Not reported | Tumor > normal | Not applicable | Not reported | Kim et al. [16] |
| DDR2 mutations | DNA sequencing: N456S | Adeno-squamous | Resection | Formalin | Not reported | 188 | Normal not analyzed | Not applicable | Mutation frequency: 1/188 (0.5 %) | Ding et al. [20] |
| DDR2 mutations | DNA sequencing: R105S | Adeno-squamous, squamous, large cell, carcinoid | Not reported | Not reported | Not reported | 26 | Normal analyzed | Not applicable | Mutation frequency: 1/26 (3.8 %) | Davies et al. [21] |
| DDR2 mutations | DNA sequencing: L63V, I120M, D125Y, L239R, G253C, C580Y, I638F, T765P, S768R, G774E/V, | Squamous cell | Resection | Not reported | Surgery (other treatments not specified) | 277 (plus one with s768r) | Normal analyzed in 20 patients | Not applicable | Mutation frequency: 9/277 (3.2 %) s768r associated with response to dasatinib | Hammerman et al. [22] |

(continued)

**Table 8.1** (continued)

| DDR expression and mutations | Assay type or mutations detected | Histology | Specimen type | Specimen fixation | Treatment received | Number of patients | Tumor compared to normal | Cut point for survival analysis | Survival impact | Author |
|---|---|---|---|---|---|---|---|---|---|---|
| DDR2 mutations | DNA sequencing: S768R | Squamous cell | Biopsy | Not reported | Not resection | 1 | Normal not analyzed | Not applicable | S768r associated with response to dasatinib | Pitini et al. [23] |
| DDR2 mutations | RNA sequencing of selected regions: no mutations detected | Non-small cell | Resection | Frozen (?) | Surgery only | 23 | Normal analyzed | Not applicable | Mutation frequency: 0/23 (0 %) | Ford et al. [4] |
| DDR2 mutations | RNA sequencing of selected regions: no mutations detected | Squamous cell | Resection | Frozen | Surgery (other treatments not specified) | 166 | Normal not analyzed | Not applicable | Mutation frequency: 0/166 (0 %) | Sasaki et al. [14] |
| DDR2 mutations | DNA sequencing: S131C, G531V, T681I | Squamous cell | Resection | Frozen | Surgery (other treatments not specified) | 86 | Normal analyzed (?) | Not applicable | Mutation frequency: 4/86 (4.7 %) no association with clinical parameters | Miao et al. [15] |
| DDR2 mutations | DNA sequence analysis by mass spectrometry: L63V, G505S, G774V | Adeno-squamous, squamous, large cell | Resection | Frozen and formalin | Surgery only | 472 | Normal analyzed | Not applicable | Mutation frequency: 3/472 (0.6 %) l63v and g505s are germ line polymorphisms | Bollig-Fischer et al. [28] |

DDR2 mRNA in 23 tumors compared to matched normal lung specimens. They also found that high levels of DDR1 expression were associated with good overall and disease-free survival in a multivariable analysis in a cohort of 146 patients that had been treated with surgical resection without additional systemic therapy, which suggests that DDR1 expression is an independent prognostic biomarker of favorable outcome. A similar association between DDR2 expression and patients' survival was not detected.

In contrast, Yang et al. [5] described that DDR1 expression, as assessed by immunohistochemistry with a commercial antiserum (sc-532, Santa Cruz Biotechnology) in 155 patients treated with surgical resection, was found in 65 % of cases and was associated with metastatic spread to lymph nodes and poor overall survival. These authors also transiently transfected two lung cancer cell lines with DDR1 isoforms and found increased migration particularly in the presence of collagen. They concluded that DDR1 expression contributes to poor prognosis in patients with NSCLC as a result of increased invasiveness. Similar results were described by Miao et al. [6] where in vitro studies of DDR1 promoted cell invasion via epithelial-to-mesenchymal transition. These authors found an association between lymph node metastatic spread and elevated DDR1 mRNA levels in the 82 patients studied and survival appeared to be worse for patients with high DDR1 levels. Finally, Valencia et al. [7] conducted an immunohistochemical analysis in a cohort of 83 lung cancer patients and found that high DDR1 levels were associated with poor survival in a subset of patients with stage I disease. The authors also described increased cell survival, homing, and colonization in a mouse model of human lung cancer metastasis to bone [7].

Investigations seeking to establish an association between DDR2 expression and NSCLC patients' outcome are similarly confusing (reviewed in [8]). Studies conducted in patients with breast cancer [9–11], prostate cancer [12], and head and neck squamous cell carcinoma [13] found that DDR2 is overexpressed, promotes metastasis formation, and is associated with poor survival.

In contrast, several studies [4, 14, 15] showed that DDR2 mRNA levels were decreased in tumor compared to matched normal lung samples in NSCLC. Sasaki et al. [14] analyzed DDR2 mRNA levels in tumor and adjacent normal lung samples from 92 patients treated with surgical resection for squamous cell carcinoma of the lung (including four adeno-squamous carcinomas). Surprisingly, the authors showed that DDR2 mRNA levels were decreased in tumor samples as compared to the matched normal lung samples. In addition, there was no significant association between tumor stage and DDR2 mRNA levels, although there was a trend toward an increase in levels with higher tumor stages (median levels 1.0 for stage I and stage II and 1.5 for stage III). Although Miao et al. [15] reported similar results, i.e., lower DDR2 mRNA levels in squamous cell carcinoma compared to matched normal lung specimens in 54 patients, these authors found lower levels in patients with higher tumor stages and statistically significant lower DDR2 mRNA levels in patients with as compared to without lymph node metastases (note that there is a discrepancy between the text and the results shown in Fig. 8.1 of this reference). Patients' survival for those with high as compared to low DDR2 levels was indistinguishable with the median expression level used to categorize patients into the high and low groups.

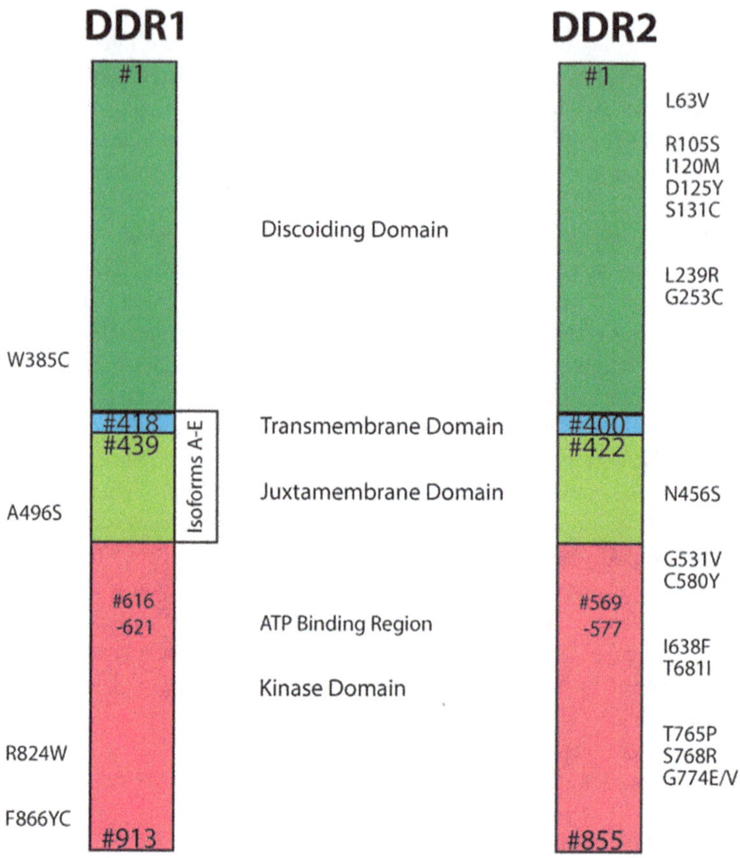


**Fig. 8.1** Schematic depiction of DDR1 and DDR2 with described mutations

However, a recent report [16] stated that DDR2 expression was upregulated in lung cancer compared to normal tissue specimens by immunohistochemical analysis of 30 normal and 29 cancer specimens; however, the number of lung cancers analyzed was not specified.

In summary, currently available data do not allow for any conclusion as to the prognostic utility of DDR1 or DDR2 levels in patients with NSCLC (Table 8.1). This is most likely a result of the use of different technologies for assessment of DDR levels, the use of specimens collected under a variety of conditions, and the use of different reagents for DDR quantitation.

## 8.2 DDR Activation and Signaling Pathways in Lung Cancer

Using a phospho-proteomic approach on 41 NSCLC cell lines and 150 pathological specimens, Rikova et al. [17] found that DDR1 and DDR2 were among the most highly phosphorylated RTKs, such as EGFR, ALK, and MET. Although the mechanistic role of DDRs in lung cancer pathogenesis remains uncertain, the study suggests that DDRs are important in the maintenance of the malignant phenotype and focuses attention on their potential role in lung cancer cell survival, adaptability, and progression and perhaps as targets for effective novel therapies. While investigating regulatory mechanisms of DDR expression, Ruiz and Jarai [18] demonstrated that DDR1 expression in human lung fibroblasts is induced by type I collagen and dependent on DDR2, JAK2, and MAPK pathway activation. The authors also demonstrated that ERK1/ERK2 inhibition suppressed DDR1 expression, which underscores the complex interrelationship of RTK signaling pathways and the importance of context when investigating the role of DDRs in lung cancer progression, prognosis, and therapeutic interventions. This is further highlighted by a recent study [19] that analyzed global tyrosine phosphorylation changes after collagen stimulation and identified 45 signaling effectors downstream of this receptor, including substrates such as SHP-2, SHIP-2, PIK3C2A, and PLCL2. The investigators demonstrated that SHP-2 is a key signaling node downstream of the DDR2 receptor, which may be therapeutically relevant.

## 8.3 Somatic DDR Mutations and Targeted Therapy in Lung Cancer

Several somatic mutations in the DDR1 and DDR2 genes have been reported in tumor specimens from patients with NSCLC (Fig. 8.1, Table 8.1). Ding et al. [20] screened 623 cancer-relevant genes for somatic mutations in coding exons and splice sites in 188 primary lung adenocarcinomas and found 2 novel somatic missense mutations in the DDR1 gene, W385C and F866Y, and 1 in the DDR2 gene, N456S. Davies et al. [21] screened the coding regions of 518 protein kinase genes for mutations in 26 primary lung cancers (7 adenocarcinomas, 7 squamous cell carcinomas, 6 large cell carcinomas, and 6 carcinoids) and 7 lung cancer cell lines (6 adenocarcinomas, 1 neuroendocrine cancer) and reported two somatic DDR1 missense mutations, A496S in a patient with squamous cell carcinoma and R824W in the neuroendocrine cancer cell line NCI-H1770, as well as one somatic DDR2 missense mutation R105S in a patient with large cell carcinoma. These mutations map to the cytosolic juxtamembrane region in DDR1 and the discoidin 1 domain in DDR2. A functional characterization of these mutations was not reported.

Focusing on squamous cell carcinoma, Hammerman et al. [22] assessed the mutation frequency in 20 primary tumors for 201 tyrosine kinase genes. Among the 25 genes with somatic missense mutations was DDR2. Further analysis of this gene in 257 primary tumors and 13 cell lines revealed a total of 11 point mutations. Nine of the mutations (L63V, I120M, D125Y, L239R, G253C, C580Y, I638F, T765P, and G774E/V) were identified among the 277 primary cancers resulting in a mutation frequency of 3.2 % (9/277) and two (L239R in HCC-366 and I638F in NCI-H2286) among the 13 cell lines. These mutations were distributed across the entire gene including the globular domains and the juxtamembrane region. Functional in vitro studies revealed that the L239R and I638F mutations were consistent with a gain of function phenotype and sensitization to the multikinase inhibitor dasatinib. In addition, the authors observed a tumor response to a combination of dasatinib and erlotinib in a female patient with squamous cell carcinoma, no evidence for an EGFR mutation, and a novel DDR2 S768R mutation. DNA from a normal tissue specimen for validation of DDR2 S768R as a somatic mutation was not available. Pitini et al. [23] reported a second female patient with squamous cell carcinoma of the lung harboring a DDR2 S768R mutation with an excellent response to dasatinib. This patient had synchronous chronic myelogenous leukemia, which responded to dasatinib. Unfortunately, constitutional DNA was not evaluated for DDR2 mutations. Phase II clinical trials of dasatinib in lung cancer (NCT00787267, NCT01491633, NCT01514864) have been terminated for various reasons including lack of efficacy (reviewed in [24]). These results suggest that DDR2 mutations may be clinically actionable through treatment with selective small-molecule tyrosine kinase inhibitors.

Beauchamp et al. [25] investigated acquired resistance to dasatinib therapy and identified acquisition of a T654I mutation in DDR2 and loss of NF1 as potential mechanisms of resistance. The T654I mutation can increase ATP affinity, thus reducing effective drug interaction [26], and NF1 loss activates an alternative pathway resulting in maintenance of ERK1/ERK2 survival signaling. A characterization of dasatinib-regulated global tyrosine phosphorylation revealed additional potential resistance mechanisms in DDR2-mutant cancers [27]. These results provide a rationale for combination regimens in patients with DDR2 mutations that may include among others ERK1/ERK2 pathway inhibitors to overcome dasatinib resistance.

Additional DDR2 mutations have been identified in primary squamous cell carcinoma specimens of Chinese patients. The investigators sequenced DDR2 in 86 tumor samples and identified 3 novel DDR2 somatic mutations (S131C, G531V, and T681I) in 4 patients [15]. The reported frequency of DDR2 mutations in this investigation is thus 4.7 % (4/86). It is unclear if constitutional normal DNA was sequenced in these patients to confirm that the mutations were indeed somatic. In vitro investigations suggested that the S131C mutation can promote cell proliferation, migration, and invasion through induction of MMP-2 and suppression of E-cadherin expression.

However, in studies conducted in 23 NSCLCs in Canada [4] and 166 squamous cell carcinomas in Japan [14], no DDR1 (discoidin domain [4]) or DDR2 (juxtamembrane and kinase domain [4]; discoidin domain [14]) missense mutations were detected. In contrast to the earlier mentioned studies, which utilized tumor-derived DNA for sequencing of the entire coding region, these authors sequenced selected regions in cDNA generated from tumor-derived RNA with oligo-dT priming for reverse transcription. Since oligo-dT priming biases the abundance of available cDNA for subsequent analysis toward the 3'-end of mRNA, it is possible that some mutations, in particular those in the juxtamembrane and discoidin domains, went undetected.

A recently reported study found 3 cases with DDR2 mutations in a total of 472 NSCLC patients [28]. These authors used DNA and specifically queried for L63V, I120M, D125Y, L239R, G253C, G505S, C580Y, I638F, T765P, and G774E/V mutations. L63V and G505S, each detected in two distinct patients with squamous cell carcinoma, were identified as single nucleotide polymorphism since the same sequence alteration was present in DNA from corresponding normal tissue specimens. G774V, detected in a patient with squamous cell carcinoma, was identified as a somatic mutation since it was not present in normal tissue-derived DNA.

## 8.4 Conclusion

Taken together, DDR1 and DDR2 are biologically important RTKs that impact cellular survival, migration, and invasion in response to extracellular signals. The cellular signaling pathways used to effectuate these responses share commonalities with other RTKs, have alternative pathways, and depend on the overall context of microenvironmental and cellular factors. The clinical value of DDR1 and DDR2 expression levels in patients with lung cancer is uncertain despite multiple investigations, which is likely a result of none standardized biospecimen processing, bioassay technology usage, and data analysis. The frequency and clinical utility of the described somatic mutations requires further investigation, which should include germ line DNA analysis and detailed racial and ethnic considerations. Such studies are necessary to evaluate the future promise of DDRs as novel therapeutic targets for patients with lung cancer.

## References

1. Mok TS, Wu YL, Thongprasert S et al (2009) Gefitinib or carboplatin-paclitaxel in pulmonary adenocarcinoma. N Engl J Med 361:947–957
2. Shaw AT, Kim D-W, Mehra R et al (2014) Ceritinib in ALK-rearranged non-small-cell lung cancer. N Engl J Med 370:1189–1197
3. Valiathan RR, Marco M, Leitinger B, Kleer CG, Fridman R (2012) Discoidin domain receptor tyrosine kinases: new players in cancer progression. Cancer Metastasis Rev 31:295–321
4. Ford CE, Lau SK, Zhu CQ, Andersson T, Tsao MS, Vogel WF (2007) Expression and mutation analysis of the discoidin domain receptors 1 and 2 in non-small cell lung carcinoma. Br J Cancer 96:808–814
5. Yang SH, Baek HA, Lee HJ et al (2010) Discoidin domain receptor 1 is associated with poor prognosis of non-small cell lung carcinomas. Oncol Rep 24:311–319
6. Miao L, Zhu S, Wang Y et al (2013) Discoidin domain receptor 1 is associated with poor prognosis of non-small cell lung cancer and promotes cell invasion via epithelial-to-mesenchymal transition. Med Oncol 30:626
7. Valencia K, Ormazabal C, Zandueta C et al (2012) Inhibition of collagen receptor discoidin domain receptor-1 (DDR1) reduces cell survival, homing, and colonization in lung cancer bone metastasis. Clin Cancer Res 18:969–980
8. Payne LS, Huang PH (2014) Discoidin domain receptor 2 signaling networks and therapy in lung cancer. J Thorac Oncol 9:900–904
9. Ren T, Zhang J, Zhang J, Liu X, Yao L (2013) Increased expression of discoidin domain receptor 2 (DDR2): a novel independent prognostic marker of worse outcome in breast cancer patients. Med Oncol 30:397

10. Ren T, Zhang W, Liu X et al (2014) Discoidin domain receptor 2 (DDR2) promotes breast cancer cell metastasis and the mechanism implicates epithelial-mesenchymal transition programme under hypoxia. J Pathol 234:526–537
11. Zhang K, Corsa CA, Ponik SM et al (2013) The collagen receptor discoidin domain receptor 2 stabilizes SNAIL1 to facilitate breast cancer metastasis. Nat Cell Biol 15:677–687
12. Yan Z, Jin S, Wei Z et al (2014) Discoidin domain receptor 2 facilitates prostate cancer bone metastasis via regulating parathyroid hormone-related protein. Biochim Biophys Acta 1842:1350–1363
13. Xu J, Lu W, Zhang S et al (2014) Overexpression of DDR2 contributes to cell invasion and migration in head and neck squamous cell carcinoma. Cancer Biol Ther 15:612–622
14. Sasaki H, Shitara M, Yokota K et al (2012) DDR2 polymorphisms and mRNA expression in lung cancers of Japanese patients. Oncol Lett 4:33–37
15. Miao L, Wang Y, Zhu S et al (2014) Identification of novel driver mutations of the discoidin domain receptor 2 (DDR2) gene in squamous cell lung cancer of Chinese patients. BMC Cancer 14:369
16. Kim D, Ko P, You E, Rhee S (2014) The intracellular juxtamembrane domain of discoidin domain receptor 2 (DDR2) is essential for receptor activation and DDR2-mediated cancer progression. Int J Cancer 135:2547–2557
17. Rikova K, Guo A, Zeng Q et al (2007) Global survey of phosphotyrosine signaling identifies oncogenic kinases in lung cancer. Cell 131:1190–1203
18. Ruiz PA, Jarai G (2011) Collagen I induces discoidin domain receptor (DDR) 1 expression through DDR2 and a JAK2-ERK1/2-mediated mechanism in primary human lung fibroblasts. J Biol Chem 286:12912–12923
19. Iwai LK, Payne LS, Luczynski MT et al (2013) Phosphoproteomics of collagen receptor networks reveals SHP-2 phosphorylation downstream of wild-type DDR2 and its lung cancer mutants. Biochem J 454:501–513
20. Ding L, Getz G, Wheeler DA et al (2008) Somatic mutations affect key pathways in lung adenocarcinoma. Nature 455:1069–1075
21. Davies H, Hunter C, Smith R et al (2005) Somatic mutations of the protein kinase gene family in human lung cancer. Cancer Res 65:7591–7595
22. Hammerman PS, Sos ML, Ramos AH et al (2011) Mutations in the DDR2 kinase gene identify a novel therapeutic target in squamous cell lung cancer. Cancer Discovery 1:78–89
23. Pitini V, Arrigo C, Di Mirto C, Mondello P, Altavilla G (2013) Response to dasatinib in a patient with SQCC of the lung harboring a discoid-receptor-2 and synchronous chronic myelogenous leukemia. Lung Cancer 82:171–172
24. Reungwetwattana T, Dy GK (2013) Targeted therapies in development for non-small cell lung cancer. J Carcinog 12:22
25. Beauchamp EM, Woods BA, Dulak AM et al (2014) Acquired resistance to dasatinib in lung cancer cell lines conferred by DDR2 gatekeeper mutation and NF1 loss. Mol Cancer Ther 13:475–482
26. Yun CH, Mengwasser KE, Toms AV et al (2008) The T790M mutation in EGFR kinase causes drug resistance by increasing the affinity for ATP. Proc Natl Acad Sci U S A 105:2070–2075
27. Bai Y, Kim JY, Watters JM et al (2014) Adaptive responses to dasatinib-treated lung squamous cell cancer cells harboring DDR2 mutations. Cancer Res 74:7217–7228
28. Bollig-Fischer A, Chen W, Gadgeel SM et al (2015) Racial diversity of actionable mutations in non-small cell lung cancer. J Thorac Oncol 10:250–255

# Chapter 9
# DDRs in Healthy and Cancerous Reproductive Systems

**Marta Marco and Paul R. Gill**

## 9.1 Introduction

Many members of the receptor tyrosine kinase (RTK) family have been shown to play key regulatory roles in diverse critical cellular processes including growth, differentiation, metabolism, and migration. Ligand binding by RTK extracellular domains transduces environmental signals to the cell interior resulting in an appropriate cellular response. Ligand binding to specific RTK receptors is the basic mechanism for transduction of a wide range of environmental signals. Ligand-dependent RTK activation leads to intracellular RTK auto-/transphosphorylation resulting in the formation of intracellular adaptor protein binding sites. Upon adaptor protein binding, the signal is further propagated, ultimately leading to a cellular response. Mutation resulting in altered RTK mRNA formation or expression or RTK protein structure that leads to aberrant RTK signaling has been causally linked to progression of numerous inflammatory diseases and cancers as well as bone disorders.

The human genome encodes 58 RTKs, which are divided into 20 subfamilies based on modular architecture of their ectodomains [1]. The discoidin domain receptor (DDR) subfamily includes two members in vertebrates, DDR1 and DDR2 [2]. In

M. Marco, Ph.D. (✉)
Area de Biologia Tumoral, Departamento de Bioquímica Clínica, Facultad de Quimica, Universidad de la República, Av. Gral. Flores 2124, Montevideo, Uruguay

Laboratorio de Biotecnología, Polo Tecnológico de Pando, Facultad de Química, Universidad de la República, Montevideo, Uruguay
e-mail: mmarco@fq.edu.uy

P.R. Gill, Ph.D.
Unidad de Microbiología Molecular, Instituto de Investigaciones Biológicas Clemente Estable, IIBCE, Av. Italia 3318, Montevideo, Uruguay

El Grupo Esmeralda Uruguay, Sierras de Rocha, IIBCE, Av. Italia 3318, Montevideo, Uruguay
e-mail: prgillrivers@gmail.com

 
R. Fridman, P.H. Huang (eds.), *Discoidin Domain Receptors in Health and Disease*, DOI 10.1007/978-1-4939-6383-6_9

humans DDR2 has only one isoform, while DDR1 exists in five isoforms (DDR1a–e) generated through alternative splicing [3]. Prototypical RTKs are activated by small soluble polypeptide growth factors, cytokines and hormones. DDRs are unique among RTKs in that they recognize and are activated by specific peptide sequences in collagen [4, 5], a major component of the vertebrate extracellular matrix (ECM) [6]. DDRs are thus attributed with having a direct role in RTK signaling with ECM [4, 5]. Prior to these studies with DDRs, the ECM had been considered to contribute to RTK signaling only indirectly by capturing and storing growth factors [7]. In this time, a signaling role for collagen had only been described for integrins and the platelet receptor, glycoprotein VI [8–10]. Importantly, DDR activation by collagen was shown to be independent of β1 integrins [11]. Both DDRs are activated by fibrillar collagens (e.g., types I, II, and III), while their preferences for non-fibrillar collagens are distinct, with DDR1 and DDR2 being specific for type IV and type X collagen, respectively [4, 5, 12]. DDR activation strictly requires collagen to be in its native triple-helical conformation; heat-denatured collagen (gelatin) is not recognized by DDRs [4, 5].

Recent reviews discuss downstream signaling pathways and roles for DDRs in human disease [10, 13–15]. The biological outcomes of DDR activation and the roles they play in various human diseases are only partially understood. DDRs are widely expressed in mammalian tissues and regulate cell adhesion, migration, proliferation, and differentiation [10, 13], as well as remodeling of ECM by matrix metalloproteinases (MMPs) [16, 17]. Mutation of DDRs leading to altered expression has been associated with many different kinds of cancers. The role DDRs may play in various cancers is complicated in that they may act as pro-tumorigenic or anti-tumorigenic receptors, and their effect is highly dependent on the type and stage of cancer. DDRs are thus recognized as potentially important targets for development of therapeutic agents in cases where they promote tumorigenesis. This chapter focuses on the role of DDRs in cancers of the male and female reproductive system, with particular attention to ovarian and prostate cancers.

## 9.2 DDR2 in Steroidogenesis and Gametogenesis

We begin with a discussion of the role of DDR2 in normal healthy development of ovaries and testes. DDR2 is constitutively expressed in somatic interstitial cells of ovary and testis during postnatal development from prepuberty to adulthood [18]. A role for DDR2 in normal development of reproductive organs was shown with the isolation and characterization of smallie (*slie*), a spontaneous recessive mutation in mice [18]. The *slie* mutation was mapped within a 2 Mb region in chromosome 1 in which a 150 kb deletion resulted in the absence of expression of a single gene, i.e., *Ddr2*.

Homozygous *slie* mice (lacking DDR2, $Ddr2^{slie/slie}$) are dwarf (30–40 % reduced body weight and shortened long bones), and infertility results in both sexes suggesting that DDR2 is a newly discovered molecular player critical for reproduction [18].

Although the pituitary gland of prepubertal mutant mice was reduced in size compared with that of wild-type mice, gene expression levels for pituitary and hypothalamic-releasing hormones did not differ between *Ddr2*$^{slie/slie}$ and wild-type mice. This is in contrast with other spontaneous mutations in mice resulting in dwarfism in which neuroendocrine function is altered. Dwarf mutants of this type include Snell dwarf (*dw*), dwarf (*df*), and little (*lit*) mice [19–22]. Mutants *dw*, *df*, and *lit* have played valuable roles in understanding the interplay of the body size, reproduction, and neuroendocrine axis including the unraveling of the regulatory network of growth hormone underlying neuroendocrine organization and activation.

The study of Kano et al. with *Ddr2*$^{slie/slie}$ mice, in contrast, indicates that gonad dysfunction in these mice is due to defects in hormone-responsive pathways in peripheral tissue [18]. As noted above, DDR2 is activated by collagens I, II, III, and X (also V), but not by basement membrane collagen IV. Besides the gonadal development discussed below, DDR2 is involved in remodeling the ECM during proliferation, development, and differentiation as well as tissue repair. The shortening of long bones in *Ddr2*$^{slie/slie}$ mice may be due to the absence of DDR2 to interact with collagen X. This is the main collagen in chondrocyte ECM and appears to be involved in signaling these cells to proliferate. Mice that do not have collagen X have reduced chondrocyte proliferation as well as stunted long bones [12]. It is thus speculated that interaction of collagen X and DDR2 may be involved in signaling chondrocytes to proliferate, but in *Ddr2*$^{slie/slie}$ mice, this is not possible and normally longer bones were shorter in these mice.

Collagen I is indicated to affect steroidogenesis in the ovarian follicle [23] and to be an endogenous ligand for DDR2 in ovary [18]. There is an overlap in distribution of collagen I and DDR2 in the ovary. Collagen I is found throughout the ECM of the mouse ovary with higher concentrations in the stroma near surface epithelium and follicular compartments, consistently present during follicle maturation [24]. As the follicle develops, oocytes, the antrum, and the granulosa cell layer express collagen I [24, 25]. DDR2 detected by indirect immunofluorescence in ovaries of adult wild-type mice was found in interstitial and thecal cells but not in cumulus, inner layer granulosa cells or oocytes. FSH (follicle-stimulating hormone) initiates follicular growth, affecting granulosa cells during the initial phase of the menstrual cycle. The subsequent surge in LH (luteinizing hormone) that triggers ovulation supports thecal interna cells, which express LH receptor (LHR), to provide precursor sex steroids to granulosa cells (e.g.,for estrogen production). After rupture of the follicle, these cells differentiate into theca lutein cells that secrete progesterone. In testis of adult wild-type mice, DDR2 receptors were detected primarily in somatic Leydig cells [18]. These cells are responsive to LH for testosterone production. DDR2 is expressed in both ovary and testis including interstitial cells, but not in germ cells. In addition *Ddr2* gene expression in gonads was not different between prepubertal and adult wild-type mice.

Follicular development in female *Ddr2*$^{slie/slie}$ mutants reaches the Graafian stage, but luteinization of thecal cells was incomplete and corpora lutea were not formed [18]. This defect could not be restored with addition of exogenous LH, in which luteinization and steroid secretion from thecal cells remained deficient. In the striking absence of corpora lutea in females and greatly reduced number of spermatids and presence of atrophy of spermatogonia, Sertoli and Leydig cells in males were the most significant differences found between *Ddr2*$^{slie/slie}$ and wild-type mice at 4 months. No gross abnormalities were found in *Ddr2*$^{slie/slie}$ mice in organs and tissues. These data imply that collagen I and DDR2 receptors share a common pathway facilitating steroidogenesis, i.e., steroid production necessary for maturity of structures and functions of the ovulatory cycle.

Ovarian transcriptome microarray analysis of wild-type and *Ddr2*$^{slie/slie}$ mice was compared to identify relative changes in gene expression throughout the mouse genome resulting from loss of DDR2 [23]. Decreases in relative expression of genes in several gene categories that likely regulate or affect ovary development, ovulation, and steroid hormone production were found in *Ddr2*$^{slie/slie}$ mice ovaries compared to those of wild-type mice. Decreased expression of LHR and prostaglandin type E and F receptors was observed in *Ddr2*$^{slie/slie}$ ovaries. Surprisingly, MMP gene expression and protein production, which are typically activated in the canonical signaling pathway for DDRs, were not affected by the loss of DDR2 in *Ddr2*$^{slie/slie}$ ovaries. Results were consistent with DDR2 signaling regulating classical endocrine pathways known to be critical in ovary development and the ovarian cycle, particularly LH-responsive gene expression [23].

In spite of the lack of detailed molecular or cellular mechanistic models for how DDR2 may interact with the LHR, co-regulated receptors, or other molecules affecting LHR expression, it can be speculated that DDR2 may affect somatic cell proliferation in response to the LH surge. It can also be speculated that DDR2 and LHR or associated gene products positively co-regulate each other to gradually induce remodeling of the follicular wall and release of the oocyte. Release of the oocyte during the LH surge is facilitated by contraction of smooth muscle cells of the theca externa (outer layer of theca folliculi) and involves increased levels of cAMP and progesterone PGF2α. In addition, *Ddr2* expression is increased by exogenous addition of LH in wild-type, but not in *Ddr2*$^{slie/slie}$ mutant ovaries further implicating an important interaction between these receptors. Lack of DDR2 signaling in *Ddr2*$^{slie/slie}$ mice may trigger anovulation by altering gene expression of *LHR* and, to a lesser extent, *Ptgfr* and *Ptger2*. Reduced expression of these receptors, in particular *LHR*, may lead to subsequent downregulation of anti-apoptosis genes arising from impaired hormone signaling and in turn drive follicular apoptosis, anovulation, and ultimately infertility in *Ddr2*$^{slie/slie}$ mutants [23].

DDR2 is also essential for maintenance of adult male gametogenesis as indicated above. DDR2 is expressed primarily in Leydig cells, and procollagen I, precursor of collagen I, is produced by interstitial cells on the outside of the seminiferous tubules [26]. In addition as in the ovary, there is an overlap in DDR2 and collagen I localization suggesting a role for DDR2 in spermatogenesis. In a separate study, expression of several receptors, enzymes, and proteins related to spermatogenesis was investigated in wild-type and *Ddr2*$^{slie/slie}$ mice at 10 weeks and 5 months of age [26]. Levels of LHR, StAR, P450c17, Hsd3b6, and Fshr were comparable for wild-type and *Ddr2*$^{slie/slie}$ mice at 10 weeks, but with Fshr levels appearing slightly higher in the mutant. Also, P450scc was somewhat lower in *Ddr2*$^{slie/slie}$ versus wild-type mice at

10 weeks. At 5 months, all of these proteins, except Fshr, were greatly reduced in *Ddr2*$^{slie/slie}$ versus wild-type mice. Fshr levels, in contrast, were significantly higher in *Ddr2*$^{slie/slie}$ mice compared to wild-type mice. At 10 weeks, a lower but sustained steroidogenesis was evident in Leydig cells of *Ddr2*$^{slie/slie}$ mice compared to wild-type mice. At 5 months a great decline in steroidogenesis in *Ddr2*$^{slie/slie}$ mice was evident that paralleled the lost of LHR. Testosterone was significantly reduced in *Ddr2*$^{slie/slie}$ mice at 5 months of age, but LH was similar in both types of mice at both 10 weeks and 5 months of age. Thus, there are interesting similarities with the functions of FSH and Fshr in spermatogenesis and follicle development in *Ddr2*$^{slie/slie}$. Ovarian follicles could develop only to the Graafian stage, and at 5 months of age, spermatogenesis proceeded only to the round spermatid stage in *Ddr2*$^{slie/slie}$ mice.

Atrophy of Leydig cells was also evident in *Ddr2*$^{slie/slie}$ mice at 5 months. Mature sperm could be produced in younger *Ddr2*$^{slie/slie}$ mice but mature sperm production was not maintained at 5 months. There were no differences in the number of cells undergoing programmed cell death in testes of wild-type and *Ddr2*$^{slie/slie}$ mice at 10 days, but a significantly greater number were undergoing this process in *Ddr2*$^{slie/slie}$ mice at 5 months. The number of differentiated spermatozoa in seminal fluid was also significantly lower in *Ddr2*$^{slie/slie}$ mice at 5 months. The absence of DDR2 affects Leydig cells directly or germ cell communication indirectly. The absence of DDR2 expression alters fertility, spermatogenesis, Leydig cell response to LH, and testosterone production in older *Ddr2*$^{slie/slie}$ male mice. DDR2 thus provides essential functions for the maintenance of spermatogenesis [26].

## 9.3 Male Reproductive System Cancers

### *9.3.1 PCa Disease Development and DDR1*

Prostate cancer (PCa) is a very common male malignancy, third in morbidity and sixth in mortality for cancer in males [27]. PCa is also one of the most heritable of all cancers with genetic factors estimated to account for 42 % of risk [28]. The prostate secretes fluid that nourishes and protects the sperm and requires testosterone to function properly. In PCa, cell proliferation becomes androgen independent. As outlined below, *DDR1* has been shown to be associated with genes clearly linked with risk for PCa. The importance of DDR1 in PCa disease development is implicated in that it functionally interacts with the major PCa antigen, PCA1 [29], and thus plays an important or perhaps an essential role in formation of malignant tumors.

High-throughput genome-wide association studies (GWAS) have been highly successful in discovering susceptibility loci for disease [30], and more than 30 loci have been associated with risk for PCa [31]. GWAS of diverse populations representing different ethnicities including Chinese, European, and Japanese have consistently identified loci encoding hepatocyte nuclear factor-1, homeobox B (HNF1B, a POU homeodomain-containing transcription factor), as a risk locus for PCa [32–36]. *HNF1B* variant rs4430796 (chromosome 17q12) was one of the first loci found to be associated with PCa in both European and Japanese men [32, 35]. Another *HNF1B* variant found to be strongly associated with PCa risk, rs11649743 (chromosome 17q12), was separated from variant rs4430796 by a recombination hotspot [37]. Large-scale fine mapping using single nucleotide polymorphism analysis showed that four other *HNF1B* variants together with variant rs4430796 provided the best model for PCa risk in this chromosomal region [31]. This same study also found variant rs11649743 to be associated with risk, as expected. The mechanism(s) by which variants at 17q12 may alter regulation or splicing of HNFB1 transcripts, e.g., to be associated with PCa risk is not known, and detailed structure–function studies are very clearly needed.

HNF1B, also known as transcription factor 2, was initially defined as a liver-specific factor of the homeobox-containing helix–turn–helix family and is produced as one of three isoforms. Isoforms HNF1B(A) and HNF1B(B) seem to be transcriptional activators, and isoform HNF1B(C) a transcriptional repressor [38]. More recently HNF1B was shown to play a regulatory role in nephron and pancreas development [39, 40]. Normally, HNF1B is believed to form homodimers or heterodimers that bind to enhancer and promoter sequences leading to either activation or inhibition of transcription of selected target genes.

A functional role for HNF1B in PCa remains unknown, and there are only vague suggestions on how it may function as a transcription factor. Analysis of expression profiles for 12,625 transcripts in prostate tumors from patients with different types of clinical outcomes showed that differential expression of *HNF1B* was associated with recurrence of the disease [41]. *HNF1B* was one of 14 genes associated with recurrence in this study. In addition, downregulation of *HNF1B* expression has been associated with renal cell cancer progression [42]. There are also reported differences in *HNFB1* isoform expression comparing normal healthy versus cancerous prostate tissue in which isoform *HNFB1(B)* was higher and isoform *HNF1(C)* was lower in cancerous tissue [43]. A recent report complementing GWAS showing *HNFB1* to be a major risk gene in PCa used meta-analysis of gene expression data. This study showed that *DDR1* was one of only 12 genes associated with both *HNFB1* and PCa risk [44].

*PCA1* encodes a predicted DNA alkylation damage repair enzyme having high levels of mRNA expression in prostate carcinoma. PCA1 is a marker for PCa in that it is expressed in a high number of both prostate carcinoma samples and atypical cells in high-grade prostatic intraepithelial neoplasias, but not in benign prostatic hyperplasia or normal adjacent tissues. PCA1 is considered to be the human counterpart of *Escherichia coli* AlkB in that *PCA1*-transfected COS-7 cells had resistance to methyl methanesulfonate-induced cell death [45]. AlkB is a Fe (II)/2-oxoglutarate-dependent dioxygenase that removes methyl groups from 1-methyl adenines and 3-methyl cytosines in DNA. Knockdown of *PCA1* by small interfering RNA transfection induced apoptosis in reducing expression of anti-apoptotic Bcl-xl in androgen-independent PCa cell line PC3. In addition, silencing *PCA1* significantly downregulated *DDR1* expression using in vitro matrigel and in vivo chorioallantoic membrane assays [29]. Transfection with *PCA1* increased levels of both Bcl-xl and DDR1 making cells more invasive with upregulation of MMP9 in cell line DU145 [42]. Long-term culture of androgen-sensitive cancer cell line

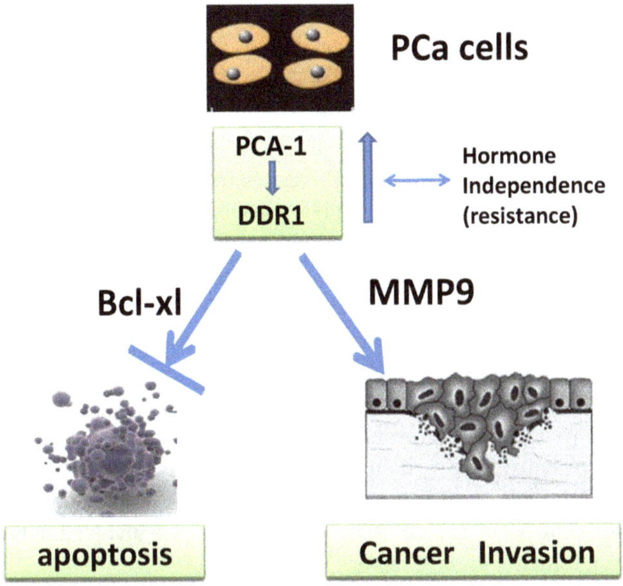


**Fig. 9.1** Schematic presentation of interaction of PCA1 and DDR1 in mediating signals leading to androgen independence (resistance) and cancer progression in PCa. PCA1 and downstream DDR1 and Bcl-xl or MMP9 signals promote cancer invasion and survival. PCA1-mediated signals are amplified and associated with hormone resistance and enhance malignant potential of androgen-independent prostate cancer cells

LNCaP in androgen-free medium resulted in increased levels of PCA1 as well as expression of Bcl-xl and DDR1. PCA1 and DDR1 were both found to be highly expressed in 169 prostate carcinomas, including preneoplastic lesions, but no expression was found in normal epithelium. Interestingly, knockdown of DDR1 resulted in suppressed cancer cell invasion and also reduced cell survival in which the level of Bcl-xl expression was reduced to the same level as in knockdown of PCA1. Overexpression of PCA1 resulted in increased constitutive mRNA expression and gelatinolytic activity of MMP9 as well as enhanced cancer invasion in vitro; however these activities were strongly suppressed in DDR1 knockdown experiments. MMP9 is thus an effector molecule in PCA1–DDR1 signaling. This signaling pathway involving DDR1 is outlined in Fig. 9.1. Together, these results indicate that PCA1 and DDR1 are associated with establishment of a hormone-independent state and malignancy potential in PCa [29].

### 9.3.2 *PCa Bone Metastasis and DDR2*

Dissemination and spread of aggressive PCas most frequently results in bone metastasis and accounts for nearly 80 % of PCa metastases. Metastasis of PCa cells to the bone involves bone–cancer cell cross talk mediated by various factors and cytokines

[46, 47]. TGF-β secretion in the bone metastatic environment can promote active proliferation and dissemination abilities in PCa cells [48, 49]. It should be acknowledged that TGF-β, together with bone morphogenic protein (BMP), is also intimately involved in normal healthy osteoblast differentiation and bone formation [50]. PCa cells also secrete osteolytic factors, very likely the most important of which is parathyroid hormone-related protein (PTHrP), as it is found at elevated levels in 90 % of PCa patients with bone metastasis. PTHrP produced by tumor cells binds to its receptor (PTHR1) on osteoblasts thereby stimulating expression of receptor activator of nuclear factor kB ligand (RANKL), which then induces osteoclast differentiation and activation [51–55]. Bone metastatic PCa cells also highly express bone-related genes that confer osteomimetic properties [56]. Runt-related transcription factor 2 (Runx2), an important transcription factor in osteogenic commitment, significantly contributes to expression of genes conferring osteomimetic characteristics [57, 58]. Expression of osteomimetic characteristics by these cancerous cells is considered to contribute to their survival in the bone matrix environment [56]. Runx plays critical roles in PCa bone metastasis in that it regulates RANKL expression, MMP2, MMP9, and MMP13 that are involved in bone turnover and vascular endothelial growth factor (VEGF) [59, 60].

DDR2 plays an important role in ECM remodeling in that it regulates MMPs and stimulates cell proliferation, adhesion, and migration. DDR2 normally controls osteoblast differentiation and bone formation by regulation of Runx2 phosphorylation and transcription activation dependent on ERK MAPK [61, 62]. Recently, Yan et al. reported that DDR2 was highly expressed in bone metastatic PCa cells compared with normal healthy cells [63]. When these cancer cells were made to constitutively express DDR2, their mobility and invasiveness were significantly increased. Knockdown of DDR2, on the other hand, resulted in corresponding decreases in proliferation and differentiation of osteoblast function. There was also a dramatic stimulation in osteoclast differentiation and bone resorption when DDR2 was highly expressed, whereas knockdown of DDR2 in PCa cells greatly reduced osteoclast differentiation.

Yan et al. proposed that the elevated levels of DDR2 in bone metastatic PCa cells regulate PTHrP expression and secretion as well as gene promoter activity by way of regulating phosphorylation and transcriptional activity of Runx2 to in turn affect RANKL expression in osteoblasts leading to osteoclast activation and resorption of the bone. TGF-β stimulation of DDR2, which binds to type I collagen (most abundant in bone), mediated TGF-β-induced bone resorption. This signaling pathway involving DDR2 is outlined in Fig. 9.2. These authors previously showed that DDR2 activation leads to secretion and activity of MMP2 and MMP13 and may contribute significantly to early stages of bone metastasis. DDR2 expression on these cells should also target them to type I collagen, e.g., to potential bone metastatic sites. In the recent study [63], they indicate that PTHrP at these sites is stimulated by DDR2 activation in turn acting on osteoblasts to stimulate their own differentiation and bone formation as well as osteoclast differentiation and bone resorption (Fig. 9.2). This should result in the release of stored TGF-β from the bone matrix to regulate DDR2 expression. In this way, DD2 can continue to act as mediator in a vicious cycle of bone destruction and metastatic cancer cell proliferation.

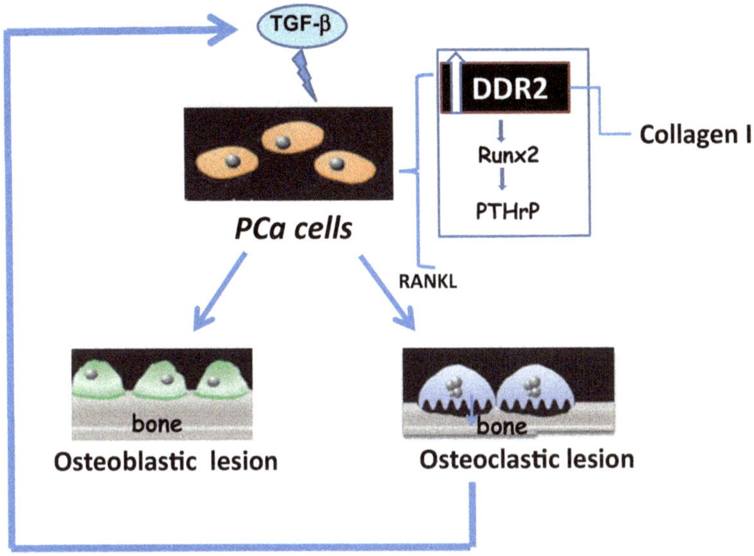


**Fig. 9.2** A schematic model depicting molecular mechanisms underlying DDR2-mediated prostate cancer cell bone metastasis. TGF-β in tumor microenvironment stimulates expression of DDR2, which is responsible for Runx2 activation and subsequently PTHrP secretion. In bone metastasis sites, PTHrP promotes osteoblast proliferation, differentiation, and function; at the same time DDR2 stimulates RANKL expression in osteoblasts, leading to the activation of osteoclasts and enhanced osteoclastogenesis. Furthermore, DDR2 facilitates specific binding of PCa cells to collagen I, which is the main collagen of bone

DDR2 is thus involved at four important levels in progression and establishment of bone metastatic PCa cells, i.e., invasion, homing or targeting, and promotion of both osteoblastic and osteoclastic activities.

### *9.3.3 DDR2 Overexpression in PCa and Testis Carcinomas*

Kim et al. recently reported that DDR2 is overexpressed in several cancerous tissues including those of the lung, bladder, stomach, and kidney as well as from the prostate and testis [64]. The aim of their study was to better understand the potential function of the unusually long intracellular juxtamembrane region (IJR) of DDRs, compared with other RTKs, in collagen-dependent activation. They found that the central part of this region, ca. 50 % of the IJR in DDR2, shared very little similarity with the same region in DDR1a. This distinct centralized 59 aa DDR2 IJR subregion, designated as JM2, was found to be an essential contributor to receptor dimerization, along with the transmembrane region (TM). JM2 was also required for efficient binding of collagen to the extracellular discoidin (DS) domain and to be required for activation after collagen binding. This study was focused experimentally on a lung cancer model rather than prostate or testis models.

## 9.4 Other Cancers of the Male Reproductive System

Studies describing DDRs involved in cancers of the epididymis, vas deferens, and seminal vesicle have not been reported.

## 9.5 Female Reproductive System Cancers

### *9.5.1 Ovarian Cancer*

In contrast with normal healthy ovaries in which DDR2 plays an essential role in follicle development and steroidogenesis, analysis of a cDNA library of SKOV-3, an EOC cell line, showed that DDR1 was expressed [65]. They also reported that in various types of ovarian cancer cell lines, DDR1 mRNA was expressed in moderately or poorly differentiated ovarian tumors, while benign or borderline tumors showed very little expression [65]. Another study showed that DDR1 mRNA expression was localized within cancerous ovarian epithelial tissue [66]. DDR1 protein expression in ovarian tumors was first reported by Heinzelmann-Schwarz et al. [67]. High levels of DDR1 protein were found in all EOC histological types and in ovarian epithelial inclusion cysts, which are thought to be sites of metaplastic change within normal ovarian tissue. There was no significant correlation between DDR1 protein expression and prognosis [67]. DDR1 expression levels in ovarian cancer tissue samples were increased by an average of 6.7-fold compared with normal ovarian tissue. DDR1 protein was highly expressed in 69 % (46/67) of serous ovarian cancer tissue samples, whereas it was undetectable in normal ovarian surface epithelium [68, 69]. Recently, Testuri and Marco found, however, that high levels of DDR1 protein expression are present in both high-grade and borderline EOC human tumors from all of over 50 patients using immunohistochemical methods [unpublished results].

Another study of serous ovarian tumors found high levels of DDR1 protein in high-grade and advanced stage tumors compared with low-grade and early-stage tumors and a significant correlation between DDR1 protein expression and poor outcome [70]. Thus, two independent studies [67, 70] suggest that DDR1 may be a useful target and biomarker for early [67] and advanced [70] ovarian cancer. DDR1 was expressed significantly more frequently in high-grade (79 %) and advanced stage (77 %) tumors compared to low-grade (50 %) and early-stage (43 %) tumors. Although the functional role for DDR1 in EOC remains to be understood, results indicate that DDR1 expression may serve as both a potential biomarker and a molecular target for advanced EOC [70].

Expression of DDR1 in EOC was described as predominantly cytoplasmic [67] suggesting that the receptor had been internalized to further propagate the activating signal [71, 72]. Alternatively, significant internalization may suggest adding another function in tumorigenesis not related to surface-mediated cell adhesion, such as receptor trafficking. These results also suggested that increased levels of DDR1 are correlated with early events in EOC development and have potential application for early detection of disease. Another study used short hairpin RNA (shRNA) interference to knockdown N-myc downstream regulated gene 1 (NDRG1), an important gene that in some cases may promote tumor invasion and in others act as a tumor suppressor. Results showed that DDR2 was one of ten genes upregulated in common by an ovarian cancer cell line (HO-8910PM) and a cervical cancer cell line (CaSki). Interestingly, NDRG1 knockdown also resulted in reduction of MMP7 in both cell lines, which is considered to be a pro-invasion gene [73].

## 9.6 Other Gynecological Cancers

### *9.6.1 Endometrial Cancer*

Endometrial cancer (EC) is the most frequently found of invasive tumors in the female genital tract. Although usually detected in its initial stages, still 20 % of patients are first diagnosed after advanced disease has developed. To date, no characterized molecular marker has been validated for diagnosis of EC. Two different studies on EC reported that DDR1 mRNA was significantly upregulated in tumor tissues and in uterine aspirates compared with normal tissue. This suggests that DDR1 could serve as a novel diagnostic/prognostic marker for this type of cancer [74, 75]. Recently, somatically mutated RTK genes were sequenced from 112 EC tumors and as well were evaluated for copy number alterations. This study showed the occurrence of low-frequency somatic mutations in DDR1 among serous, clear cell, and endometrioid ECs [76].

### *9.6.2 Cervical and Other Cancers*

As discussed above using shRNA methods to suppress NDRG1 expression, the same strategy was used in CaSki cells (a metastatic cervical cancer cell line). Results showed that DDR2 was upregulated as well and was related to increased tumor cell adhesion, migration, and invasion activities but not cell proliferation using in vitro assays [73]. Also, the morphology of the cells was not affected by downregulation of NDRG1. However, there are also reports showing that NDRG1 may be a putative tumor metastasis promoter gene [73]. There are currently no other studies describing DDRs as being involved in cancers of other specific organs of the female reproductive system, i.e., the uterus, fallopian tubes, vulva, or vagina.

## 9.7 EOC Development and Metastasis

### 9.7.1 Pathology of Primary Ovarian Tumor

While all cells within the ovary can give rise to malignancies, EOCs are the most common and lethal. Intraperitoneal dissemination is the primary metastatic route for EOC [77, 78]. Poor prognosis relates, in large part, to rapid progression of peritoneal metastasis compared to that of the hematological metastatic route, and distinct mechanisms involved contribute to devastating outcomes. EOCs are a heterogeneous group of cancers that may be categorized into two major groups [79]. Type I EOCs are low-grade, slow-growing tumors that are thought to arise from benign ovarian lesions and include all four major histotypes: serous, endometrioid, mucinous, and clear cell. In contrast, type II EOCs frequently have p53 mutations and are thus genetically unstable. Histologically these tumors are high-grade serous, mixed epithelial, or undifferentiated carcinomas. These cancers are thought to seed the ovarian surface and pelvic peritoneum concurrently, which explains why they are rarely present as stage I disease [79]. See Table 9.1 for more detail in differences between EOC cancer types.

EOCs generally have an insidious onset. Due to the asymptomatic nature of early-stage disease, most patients are not diagnosed until after their tumors have metastasized intraperitoneally. At this point, their chance of surviving beyond 5 years is only about 25 % [80], and this is largely due to the diffuse peritoneal lesions that impede surgical eradication. In fact, the completeness of surgical debulking is the best predictor of survival. Chemotherapy, while initially effective, ultimately fails to prevent disease progression because patients almost inevitably develop recurrent resistant disease [81, 82]. Intraperitoneal metastases can cause peritoneal organ adhesion and malfunction, massive ascites, and/or pleural effusions [83–85], leading to mortality.

**Table 9.1** Histologic and associated genetic differences between EOC cell types

| | Putative precursor | Most frequent mutation(s) |
|---|---|---|
| *Type I tumors* | | |
| Low-grade serous carcinoma | Serous borderline tumor | *KRAS, BRAF, ERBB2* |
| Low-grade endometrioid carcinoma | Endometrioma | *CTNNB1, PIK3CA, PTEN, ARID1A* |
| Clear cell carcinoma | Endometrioma | *PIK3CA, ARID1A, FBXW7* |
| Mucinous carcinoma | Mucinous borderline tumor | *KRAS* |
| *Type II tumors* | | |
| High-grade serous carcinoma | Fallopian tube epithelium | *TP53, BRCA1/2* |
| High-grade endometrioid carcinoma | Not recognized | *TP53* |
| Undifferentiated carcinoma | Not recognized | Unknown |
| Carcinosarcoma | Not recognized | *TP53* |

A key factor contributing to the poor prognosis of intraperitoneally metastasizing cancers is the rapid, self-perpetuating, feed-forward cycle of seeding and growth that is fueled by inflammation. Since the prognosis of patients with peritoneal metastases is tightly correlated with the completeness of surgical cytoreduction [86, 87], and widespread metastases are not amenable to surgery, the development of novel strategies to arrest metastatic progression is imperative [88]. Steps in the progression of EOC are listed below.

### 9.7.2 *Intraperitoneal Dissemination of Cancer Cells*

Cancer cells can freely disseminate in the peritoneal cavity after exfoliating from exposed primary intraperitoneal tumors: ovarian epithelia for type I EOC or fallopian tube epithelia for type II EOC. In either case, once suspended in the peritoneal fluid, single cells detach and then aggregate to form spheroids and then can establish as new tumors attached to the peritoneal membrane. In this way, cancer cells can resist anoikis, a specialized form of apoptosis triggered by a lack of attachment to other cells or to the ECM, and evade clearance by peritoneal lymphatics. Accordingly, interactions between tumor cells and the peritoneum are key contributors to metastatic progression, which, if successfully blocked, should promote clearance or death of cancer cells.

### 9.7.3 *Cellular and Molecular Properties of the Peritoneum*

The peritoneum is comprised of a single layer of mesothelial cells and its associated underlying ECM, which cover the vast surface of the abdominal and pelvic cavities, as well as visceral organs. Mesothelial cells apically secrete glycosaminoglycans, surfactants, and proteoglycans to provide an anti-adhesive peritoneal surface that ensures the appropriate gliding of the abdominal viscera and prevents intra-abdominal organ fusion. Mesothelial cells regulate entry of leukocytes and inflammatory cells into the peritoneal cavity. In response to injury or insult, these cells release chemokines MCP-1 and IL-8 and upregulate cell surface adhesion molecules ICAM-1 and VCAM-1, to which leukocytes attach [89]. The ECM underlying the peritoneal mesothelial cells is rich in collagen I and fibronectin, with thin deposits of laminin and collagen IV lying directly beneath the mesothelium [90]. The ECM is, for the most part, concealed by the flattened squamous-like mesothelial cell layer; however, it is periodically exposed at lymphatic portals through which the peritoneal fluid drains into the venous circulation. These lymphatic portals are particularly abundant on the omental and subdiaphragmatic peritoneal surfaces and are commonly referred to as "milky spots" because of their whitish appearance that results from the accumulation of resident lymphocytes participating in immune surveillance [89].

Ultrastructural analysis reveals the absence of a basement membrane at milky spots. Instead, the collagen I-rich stromal matrix is exposed. The adjacent mesothelial cells have a cuboidal morphology with disruptions and intercellular gaps that further expose the peritoneal ECM.

### 9.7.4 Cancer Cell Attachment to the Peritoneum

A variety of different adhesion molecules can mediate attachment of cancer cells to mesothelial monolayers. For example, the ovarian cancer biomarker CA125/MUC16 is a transmembrane mucin that binds to mesothelin, a GPI-linked protein expressed by mesothelial cells [91]. Attachment of cancer cells to ECM, on the other hand, is mainly mediated by integrins. The β1 integrin subunit is key as it can pair with a variety of α-integrin subunits to confer binding to most ECM substrata. Blocking β1 integrin inhibits EOC cell attachment and migration on ECM substrata relevant to the peritoneum [92, 93]. β1 integrins also participate in cancer cell attachment to mesothelial monolayers [93–96], which could indicate that cancer cells are binding to mesothelium-associated ECM and/or to mesothelial cell surface VCAM-1.

Many studies have been designed with the assumption that peritoneal metastasis is dependent on cancer cell attachment to mesothelial cells. However, other studies indicate that cancer cells have a much greater affinity for the peritoneal ECM, which is consistent with the clinical pattern of metastatic spread.

### 9.7.5 Metastasis

The clinical pattern of disease progression underscores the concept that cancer cells preferentially attach to areas where the mesothelium is disrupted. During the initial stages of peritoneal metastasis, cancer cells attach to milky spots where the collagen-rich connective tissue matrix is exposed [97–100] (Fig. 9.3). Resident immune cells of milky spots are not able to prevent tumor growth [49, 53, 97, 101]; instead, their production of pro-inflammatory cytokines promotes cancer growth and dissemination. The impact of inflammatory cytokines on peritoneal metastasis is profound and transforms the initial pattern of dissemination, which is limited to milky spots, into widespread peritoneal metastasis [98, 102]. This transformation is triggered by increased exposure of the submesothelial ECM, driven by inflammation, and inflammatory cytokines causing the protective mesothelial cells to retract, exposing the previously obscured underlying ECM. Since cancer cells preferentially attach to the ECM, widespread cancer cell attachment ensues [98, 102] (Fig. 9.3). While activated mesothelial cells may be less efficient in creating a barrier than a quiescent mesothelial monolayer, in either case, these cells protect against cancer cell adhesion by concealing the underlying connective tissue matrix to which cancer cells

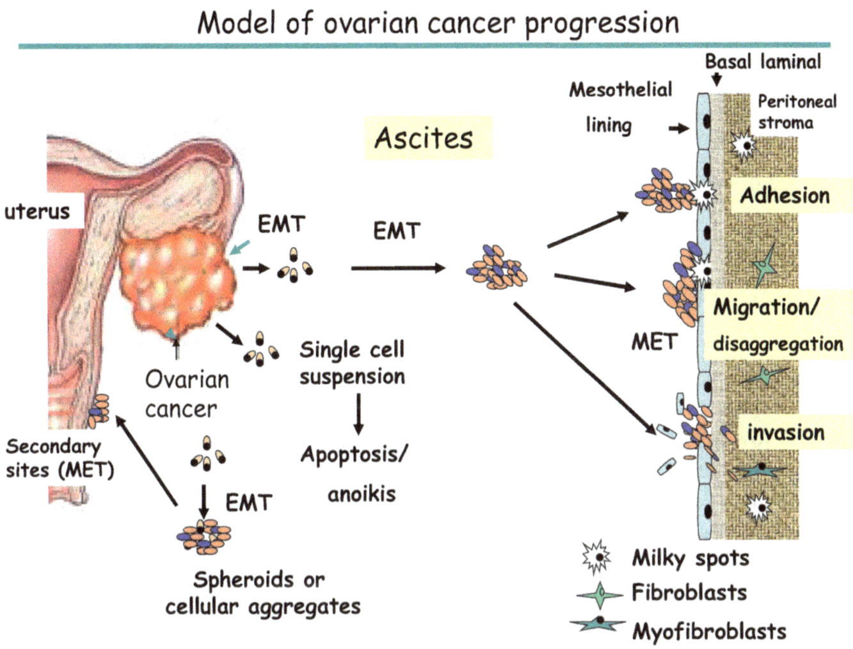


**Fig. 9.3** A working model for metastatic spread in progression of EOC. EOC cells growing on the surface of the ovary undergo EMT to attain motile functions required for cancer metastasis. Rupture of the ovarian tumor results in shedding of tumor cells into the peritoneum where they survive as cellular aggregates/spheroids which secrete cytokines. The surrounding mesothelial and infiltrating blood cells facilitate invasiveness of carcinoma spheroids. In secondary sites, cancer cells initially attach to milky spots where the stromal matrix is exposed, providing direct access to a preferred substrate, collagen I. With disease progression and in response to increasing concentrations of inflammatory mediators, mesothelial cells retract and detach. The resulting exposure of underlying ECM, with a discontinuous basement membrane, facilitates widespread peritoneal metastasis. TGF-β, released by cancer and inflammatory cells, stimulates myofibroblast transdifferentiation. The spheroids attach to and invade the peritoneal matrix. The combination of their contractile and proteolytic capacities remodels the collagen I-rich matrix to facilitate stromal implantation and invasive growth

preferentially attach. Knowing that cancer cells prefer areas where the mesothelium is absent and the peritoneal ECM is exposed highlights the importance of targeting cancer cell interactions with the ECM while simultaneously preventing mesothelial cell retraction.

## 9.8 Suggested Roles for DDRs in EOC

The precise molecular events that occur during development, progression/invasion, and formation of secondary tumors in EOC metastasis are poorly characterized and understood. In addition, contribution to the local microenvironment at secondary sites in tumor progression has been recognized as highly relevant in several metastatic cancers [103–105]; however no information on EOC in this area is available. Establishing a potential role of DDR1 in EOC metastasis is vital to clarifying molecular mechanisms involved in progression of this disease. This in turn may enable identification of novel and more effective treatment strategies against advanced disease.

### *9.8.1 DDR1 in Cancer Metastasis*

DDR1 is widely expressed in epithelial cells of both fetal and adult organs. The highest expression is found in the brain, lungs, placenta, and kidneys. DDR1 is also found at low levels in cells such as melanocytes and those of various adult organs including the heart, liver, skeletal muscle, pancreas, and ovaries [14]. Although the physiological functions of DDR1 are not fully understood, DDR1 signaling is essential for cerebellar granule cell differentiation [106], arterial wound repair [107], and mammary gland development [108]. It is clear that DDR1 is involved in cell interactions with the ECM involving adhesion. DDR1, however, was found to be overexpressed in breast, brain, colon, and lung cancers, thus suggesting that this receptor may play a role in tumorigenesis of these epithelial cancers [109–112]. In breast cancer, DDR1 was overexpressed in both primary breast tumor samples and lymph nodes containing metastatic tumors [113]. DDR1 protein levels were elevated in 100 % of patients with primary and metastatic brain tumors [110], in 61 % of patients with non-small cell lung cancer, and in 64 % of patients with invasive lung adenocarcinoma [114]. Thus, DDR1 expression appears to be elevated in a variety of human cancers. Consistent with studies of DDR1 in these solid tumors, elevated levels of DDR1 were seen in serous ovarian cancer, noted above. Elevated levels of DDR1 were also found for all stages of serous and mucinous ovarian cancers as noted above (Testuri and Marco, unpublished results).

Analysis of DDR1 isoforms in mouse embryonic and adult tissues revealed that DDR1b is the major isoform [109]. However, DDR1a is found in high levels in breast and glioma tumor cell lines [109, 115]. No information is currently available on which DDR1 isoforms are associated with EOC. Additional information on which DDR1 isoforms may contribute to EOC oncogenesis and predicted role in binding collagen I in the submesothelial ECM of the peritoneum remain unknown.

As noted above, integrins are important molecules specifically involved in signaling or adhesion of cancer cells allowing for metastatic establishment. DDRs are similarly implicated to play related roles in some types of cancer, but how they may collaborate in some way with integrins in EOC is also not known. Some hypotheses, however, can be suggested based on current information. The most relevant issue is an apparent regulated "switch" in expression that occurs from that of DDR2 (in normal ovary epithelial cells) to DDR1 (in EOC cells). As discussed above, DDR2 is required for reproductive functions in ovaries, particularly steroid production needed to complete the ovarian cycle.

Our laboratory has evidence that such a switch from DDR2 to DDR1 expression occurs in cells at the beginning of malignant transformation in serous and mucinous borderline ovarian tumors [unpublished results]. It is known that DDRs initiate signaling pathways in a context and cell type-dependent manner. The primary or secondary EOC tumor microenvironment might play a role in switching DDR expression, as the tumor microenvironment has been suggested to play an active role in cancer initiation and progression [103, 116]. Stromal–epithelial cross talk could provide a mechanism for switching since this microenvironment can be very different from that of normal health tissue. This might involve interaction with any number of stromal components including different collagens, of other high molecular weight ECM components as well as growth factors, hormones, cytokines, and chemokines present in the secondary tumor microenvironment that facilitate progression of EOC that may include upregulation of DDR1. Some evidence suggests that the switch to DDR1 could be by the secretion of cytokines by stroma cells [117] or by stimulation by collagen I, normally present in ovary stroma [118].

Detailed studies on signal transduction pathways activated by DDR1 and DDR2 in different cell types and their transcriptional targets have not yet been described. Similarly, very little is known about the regulation of DDR expression, and so far, only a few inflammatory mediators including TNF-α, IL-1β, and LPS have been shown to increase DDR1 expression [119]. Characterization of regulatory mechanisms governing DDR expression involving detailed structural studies has not been done. A recent study in primary human lung fibroblasts suggested a mechanism of induction for DDR1 expression is through collagen I, DDR2, and a JAK2-ERK1/2 pathway [119].

In addition, DDR1 was reported to be able to both promote and inhibit EMT in a ligand- and cell type-dependent manner. In pancreatic cancer cells, DDR1, together with integrin α2β1, promotes cell scattering on collagen I and results in increased expression of mesenchymal marker, N-cadherin [120]. Thus, DDRs can interact with multiple proteins, and these interactions result in complex signaling processes that vary between cell type and can be ligand-based RTK activity dependent or independent. Together, these studies suggest that switching between DDR2 and DDR1 could be associated with EMT and possibly MET described above.

### 9.8.2 DDR Cross Talk with Other Molecules

In addition to mediating direct collagen-dependent signaling, DDRs can also modulate signaling pathways initiated by other matrix receptors (e.g., integrins), cytokines (e.g., TGF-β), and transmembrane receptors (e.g., insulin receptor and Notch1). Cross talk between DDRs and integrin is complex and influences multiple processes including cell adhesion and differentiation. DDR1 can both potentiate and inhibit integrin-mediated signaling. DDR1 cooperates with integrin α2β1 in maintaining mouse embryonic stem cells undifferentiated via activation of selective collagen–DDR and collagen–integrin-mediated signaling pathways that ultimately converge to the self-renewal controlling molecule, Bim-1 [121]. Moreover, overexpression of DDR1 or DDR2 in cells expressing collagen-binding receptor integrins α1β1 and α2β1 results in enhanced integrin-mediated adhesion to collagen due to increased integrin activation rather than increased integrin expression levels [122]. In contrast, DDR1 has also been shown to counteract integrin-mediated signaling when promoting epithelial cell differentiation [123]. Thus DDR1–integrin cross talk is highly dependent on the type of integrins expressed on a given cell type.

DDRs can also modulate signaling initiated by growth factors. Cross talk between DDR1 and TGF-β is critical for proper growth and patterning of mouse mammary gland. TGF-β negatively regulates mammary gland ductal extension and lateral branching by promoting Wnt5a expression and DDR1 phosphorylation [124]. In human mammary cells, Wnt5a acts as an upstream regulator positively affecting collagen-induced DDR1 phosphorylation. In addition, levels of Wnt5a are directly associated with increased cell adhesion and reduced cell migration on collagen [69]. This suggests that Wnt5a might control two important cell functions in regulating the phosphorylation and activation of DDR1. Recently, cross talk between DDR2 and the insulin receptor and between Notch1 and DDR1 was proposed. Stimulation of cells with collagen I and insulin showed that insulin can act to enhance collagen I-dependent signaling by increasing the activity of DDR2 [125]. Finally, it has been proposed that collagen-stimulated DDR1 promotes survival of cancer cells by binding to and activating Notch1 thus promoting the activation of the two transcription factors Hes1 and Hey2 [126]. In conclusion, cross talk of DDRs with various receptors is critical for regulation of cell survival, migration, and differentiation in development as well as in pathological conditions. DDR1 signaling modified by growth factors has not been explored in EOC.

It is not clear why or how DDRs can act as either pro-tumorigenic or anti-tumorigenic receptors. This behavior is highly dependent on the type and stage of cancer. Consistent with a pro-tumorigenic action of DDR2, DDR2-null mice show reduced primary tumor-associated angiogenesis and reduced lung colonization following tail vein injection of melanoma cells [127]. Normally, lung cancers involve DDR1. Studies in breast cancer have shown involvement with either DDR1 or DDR2. A recent study of DDR1 and DDR2 expression in breast cancer patients, e.g., showed a sixfold increase in DDR2 levels in tumor vs. normal tissue, while DDR1 expression was decreased in tumor tissues [128]. In EOC, in contrast, it seems that DDR1 expression contributes to the progression of the disease. This seems to be more similar to what is described for non-small cell lung cancer where DDR1 overexpression is suggested to play a role in this type of cancer [114].

## 9.9 Perspectives

DDR1 overexpression is found in a wide variety of different human cancers as mentioned above suggesting that it has a function in tumor progression. As shown in Fig. 9.3, DDR1 may have a critical role in adhesion, migration, and invasion of

EOC metastasis, but additional study is needed. As a diagnostic tool, elevated DDR1 protein expression levels may be predictive for EOC [67]. However, using DDR1 levels as a marker for EOC is not necessarily specific and may indicate various other malignancies. DDR1 expression levels in body fluid or serum may have clinical prognostic value as a biomarker for cancer in a more general sense. The shedding of the DDR1 ectodomain by membrane-type MMP may add another level to regulate as well as to detect these receptors [129]. These biomarkers could be found in serum of EOC patients. Such understanding is critical for developing selective DDR inhibitors or treatments that can alter activity indirectly to result in novel and safe therapies for the treatment of a broad range of diseases including cancer.

## References


1. Lemmon MA, Schlessinger J (2010) Cell signaling by receptor tyrosine kinases. Cell 141:1117–1134
2. Fu HL, Valiathan RR, Arkwright R, Sohail A, Mihai C, Kumarasiri M, Mahasenan KV, Mobashery S, Huang P, Agarwal G, Fridman R (2013) Discoidin domain receptors: unique receptor tyrosine kinases in collagen-mediated signaling. J Biol Chem 288:7430–7437
3. Alves F, Saupe S, Ledwon M, Schaub F, Hiddemann W, Vogel WF (2001) Identification of two novel, kinase-deficient variants of discoidin domain receptor 1: differential expression in human colon cancer cell lines. FASEB J 15:1321–1323
4. Shrivastava A, Radziejewski C, Campbell E, Kovac L, McGlynn M, Ryan TE, Davis S, Goldfarb MP, Glass DJ, Lemke G, Yancopoulos GD (1997) An orphan receptor tyrosine kinase family whose members serve as nonintegrin collagen receptors. Mol Cell 1:25–34
5. Vogel W, Gish GD, Alves F, Pawson T (1997) The discoidin domain receptor tyrosine kinases are activated by collagen. Mol Cell 1:13–23
6. Kadler KE, Baldock C, Bella J, Boot-Handford RP (2007) Collagens at a glance. J Cell Sci 120:1955–1958
7. Hynes RO (2009) The extracellular matrix: not just pretty fibrils. Science 326:1216–1219
8. Heino J (2007) The collagen family members as cell adhesion proteins. Bioessays 29:1001–1010
9. Leitinger B, Hohenester E (2007) Mammalian collagen receptors. Matrix Biol 26:146–155
10. Leitinger B (2011) Transmembrane collagen receptors. Annu Rev Cell Dev Biol 27:265–290
11. Vogel W, Brakebusch C, Fassler R, Alves F, Ruggiero F, Pawson T (2000) Discoidin domain receptor 1 is activated independently of beta(1) integrin. J Biol Chem 275:5779–5784
12. Leitinger B, Kwan AP (2006) The discoidin domain receptor DDR2 is a receptor for type X collagen. Matrix Biol 25:355–364
13. Vogel WF, Abdulhussein R, Ford CE (2006) Sensing extracellular matrix: an update on discoidin domain receptor function. Cell Signal 18:1108–1116
14. Valiathan RR, Marco M, Leitinger B, Kleer CG, Fridman R (2012) Discoidin domain receptor tyrosine kinases: new players in cancer progression. Cancer Metastasis Rev 31:295–321
15. Borza CM, Pozzi A (2014) Discoidin domain receptors in disease. Matrix Biol 34:185–192
16. Olaso E, Labrador JP, Wang L, Ikeda K, Eng FJ, Klein R, Lovett DH, Lin HC, Friedman SL (2002) Discoidin domain receptor 2 regulates fibroblast proliferation and migration through the extracellular matrix in association with transcriptional activation of matrix metalloproteinase-2. J Biol Chem 277:3606–3613
17. Xu L, Peng H, Wu D, Hu K, Goldring MB, Olsen BR, Li Y (2005) Activation of the discoidin domain receptor 2 induces expression of matrix metalloproteinase 13 associated with osteoarthritis in mice. J Biol Chem 280:548–555

18. Kano K, Marín de Evsikova C, Young J, Wnek C, Maddatu TP, Nishina PM, Naggert JK (2008) A novel dwarfism with gonadal dysfunction due to loss-of-function allele of the collagen receptor gene, *Ddr2*, in the mouse. Mol Endocrinol 22:1866–1880
19. Snell GD (1929) Dwarf, a new Mendelian recessive character of the house mouse. Proc Natl Acad Sci U S A 15:733–734
20. Schaiber RH, Gowen JW (1961) A new dwarf mouse. Genetics 46:896
21. Bartke A (1964) Histology of the anterior hypophysis, thyroid and gonads of two types of dwarf mice. Anat Rec 149:225–235
22. Eicher EM, Beamer WG (1976) Inherited ateliotic dwarfism in mice: characteristics of the mutation, little, on chromosome 6. J Hered 67:87–91
23. Matsumura H, Kano K, Marín de Evsikova C, Young JA, Nishina PM, Naggert JK, Naito K (2009) Transcriptome analysis reveals an unexpected role of a collagen tyrosine kinase receptor gene, *Ddr2*, as a regulator of ovarian function. Physiol Genomics 39:120–129
24. Berkholtz CB, Lai BE, Woodruff TK, Shea LD (2006) Distribution of extracellular matrix proteins type I collagen, type IV collagen, fibronectin, and laminin in mouse folliculogenesis. Histochem Cell Biol 126:583–592
25. Berkholtz CB, Shea LD, Woodruff TK (2006) Extracellular matrix functions in follicle maturation. Semin Reprod Med 24:262–269
26. Kano K, Kitamura A, Matsuwaki T, Morimatsu M, Naito K (2010) Discoidin domain receptor 2 (DDR2) is required for maintenance of spermatogenesis in male mice. Mol Reprod Dev 77:29–37
27. Denmeade SR, Isaacs JT (2004) Development of prostate cancer treatment: the good news. Prostate 58:211–224
28. Lichtenstein P, Holm NV, Verkasalo PK, Iliadou A, Kaprio J, Koskenvuo M, Pukkala E, Skytthe A, Hemminki K (2000) Environmental and heritable factors in the causation of cancer—analyses of cohorts of twins from Sweden, Denmark, and Finland. N Engl J Med 343:78–85
29. Shimada K, Nakamura M, Ishida E, Higuchi T, Yamamoto H, Tsujikawa K, Konishi N (2008) Prostate cancer antigen-1 contributes to cell survival and invasion though discoidin receptor 1 in human prostate cancer. Cancer Sci 99:39–45
30. Manolio TA (2010) Genomewide association studies and assessment of the risk of disease. N Engl J Med 363:166–176
31. Berndt SI, Sampson J, Yeager M, Jacobs KB, Wang Z et al (2011) Large-scale fine mapping of the *HNF1B* locus and prostate cancer risk. Hum Mol Genet 20:3322–3329
32. Gudmundsson J, Sulem P, Steinthorsdottir V, Bergthorsson JT et al (2007) Two variants on chromosome 17 confer prostate cancer risk, and the one in TCF2 protects against type 2 diabetes. Nat Genet 39:977–983
33. Eeles RA, Kote-Jarai Z, Giles GG, Olama AA et al (2008) Multiple newly identified loci associated with prostate cancer susceptibility. Nat Genet 40:316–321
34. Thomas G, Jacobs KB, Yeager M, Kraft P et al (2008) Multiple loci identified in a genome-wide association study of prostate cancer. Nat Genet 40:310–315
35. Takata R, Akamatsu S, Kubo M, Takahashi A et al (2010) Genome-wide association study identifies five new susceptibility loci for prostate cancer in the Japanese population. Nat Genet 42:751–754
36. Liu F, Hsing AW, Wang X, Shao Q et al (2011) Systematic confirmation study of reported prostate cancer risk-associated single nucleotide polymorphisms in Chinese men. Cancer Sci 102:1916–1920
37. Sun J, Zheng SL, Wiklund F, Isaacs SD, Purcell LD, Gao Z, Hsu FC, Kim ST, Liu W, Zhu Y et al (2008) Evidence for two independent prostate cancer risk-associated loci in the HNF1B gene at 17q12. Nat Genet 40:1153–1155
38. Bach I, Yaniv M (1993) More potent transcriptional activators or a transdominant inhibitor of the HNF1 homeoprotein family are generated by alternative RNA processing. EMBO J 12:4229–4242

39. Kato N, Motoyama T (2009) Hepatocyte nuclear factor-1beta (HNF-1beta) in human urogenital organs: its expression and role in embryogenesis and tumorigenesis. Histol Histopathol 24:1479–1486
40. Maestro MA, Cardalda C, Boj SF, Luco RF, Servitja JM, Ferrer J (2007) Distinct roles of HNF1beta, HNF1alpha, and HNF4alpha in regulating pancreas development, beta-cell function and growth. Endocr Dev 12:33–45
41. Glinsky GV, Glinskii AB, Stephenson AJ, Hoffman RM, Gerald WL (2004) Gene expression profiling predicts clinical outcome of prostate cancer. J Clin Invest 113:913–923
42. Buchner A, Castro M, Hennig A, Popp T, Assmann G, Stief CG, Zimmermann W (2010) Downregulation of HNF-1B in renal cell carcinoma is associated with tumor progression and poor prognosis. Urology 76:507–511
43. Harries LW, Perry JR, McCullag P, Crundwell M (2010) Alterations in LMTK2, MSMB and HNF1B gene expression are associated with the development of prostate cancer. BMC Cancer 10:315–326
44. Hu YL, Zhong D, Pang F, Ning QY, Zhang YY, Li G, Wu JZ, Mo ZN (2013) *HNF1b* is involved in prostate cancer risk via modulating androgenic hormone effects and coordination with other genes. Genet Mol Res 12:1327–1335
45. Konishi N, Nakamura M, Ishida E, Shimada K, Mitsui E, Yoshikawa R, Yamamoto H, Tsujikawa K (2005) High expression of a new marker PCA-1 in human prostate carcinoma. Clin Cancer Res 11:5090–5097
46. Casimiro S, Guise TA, Chirgwin J (2009) The critical role of the bone microenvironment in cancer metastases. Mol Cell Endocrinol 310:71–81
47. Yoneda T (2011) Mechanism and strategy for treatment of cancer metastasis to bone. Gan To Kagaku Ryoho 38:877–884
48. Juárez P, Guise TA (2011) TGF-β in cancer and bone: implications for treatment of bone metastases. Bone 48:23–29
49. Buijs JT, Juárez P, Guise TA (2011) Therapeutic strategies to target TGF-β in the treatment of bone metastases. Curr Pharm Biotechnol 12:2121–2137
50. Chen G, Deng C, Li Y-P (2012) TGF-β and BMP signaling in osteoblast differentiation and bone formation. Int J Biol Sci 8:272–288
51. Soki FN, Park SI, McCauley LK (2012) The multifaceted actions of PTHrP in skeletal metastasis. Future Oncol 8:803–817
52. Isowa S, Shimo T, Ibaragi S, Kurio N, Okui T, Matsubara K, Hassan NM, Kishimoto K, Sasaki A (2010) PTHrP regulates angiogenesis and bone resorption via VEGF expression. Anticancer Res 30:2755–2767
53. Liao J, Li X, Koh AJ, Berry JE, Thudi N, Rosol TJ, Pienta KJ, McCauley LK (2008) Tumor expressed PTHrP facilitates prostate cancer-induced osteoblastic lesions. Int J Cancer 123:2267–2278
54. Mak IW, Cowan RW, Turcotte RE, Singh G, Ghert M (2011) PTHrP induces autocrine/paracrine proliferation of bone tumor cells through inhibition of apoptosis. PLoS One 6, e19975
55. Kremer R, Li J, Camirand A, Karaplis AC (2011) Parathyroid hormone related protein (PTHrP) in tumor progression. Adv Exp Med Biol 720:145–160
56. Rucci N, Teti A (2010) Osteomimicry: how tumor cells try to deceive the bone. Front Biosci (Schol Ed) 2:907–915
57. Pratap J, Lian JB, Javed A, Barnes GL, van Wijnen AJ, Stein JL, Stein GS (2006) Regulatory roles of Runx2 in metastatic tumor and cancer cell interactions with bone. Cancer Metastasis Rev 25:589–600
58. Akech J, Wixted JJ, Bedard K, van der Deen M, Hussain S, Guise TA, van Wijnen AJ, Stein JL, Languino LR, Altieri DC, Pratap J, Keller E, Stein GS, Lian JB (2010) Runx2 association with progression of prostate cancer in patients: mechanisms mediating bone osteolysis and osteoblastic metastatic lesions. Oncogene 29:811–821
59. Pratap J, Lian JB, Stein GS (2011) Metastatic bone disease: role of transcription factors and future targets. Bone 48:30–36

60. Baniwal SK, Khalid O, Gabet Y, Shah RR, Purcell DJ, Mav D, Kohn-Gabet AE, Shi Y, Coetzee GA, Frenkel B (2010) Runx2 transcriptome of prostate cancer cells: insights into invasiveness and bone metastasis. Mol Cancer 9:258–275
61. Zhang Y, Jin S, Jiangtian Y, Xin B, Ren T, Yao L (2011) An essential role of discoidin domain receptor 2 (DDR2) in osteoblast differentiation and chondrocyte maturation via modulation of Runx2 activation. J Bone Miner Res 26:604–617
62. Lin KL, Chou CH, Hsieh SC, Hwa SY, Lee MT, Wang FF (2010) Transcriptional upregulation of DDR2 by ATF4 facilitates osteoblastic differentiation through p38 MAPK mediated Runx2 activation. J Bone Miner Res 25:2489–2503
63. Yan Z, Jin S, Wei Z, Huilian H, Zhanhai Y, Yue T, Juan L, Jing L, Libo Y, Xu L (2014) Discoidin domain receptor 2 facilitates prostate cancer bone metastasis via regulating parathyroid hormone-related protein. Biochim Biophys Acta 1842:1350–1363
64. Kim D, Ko P, You E, Rhee S (2014) The intracellular juxtamembrane domain of discoidin domain receptor 2 (DDR2) is essential for receptor activation and DDR2-mediated cancer progression. Int J Cancer 135:2547–2557
65. Laval S, Butler R, Shelling AN, Hanby AM, Poulsom R, Ganesan TS (1994) Isolation and characterization of an epithelial-specific receptor tyrosine kinase from an ovarian cancer cell line. Cell Growth Differ 5:1173–1183
66. Shelling AN, Butler R, Jones T, Laval S, Boyle JM, Ganesan TS (1995) Localization of an epithelial-specific receptor kinase (EDDR1) to chromosome 6q16. Genomics 25:584–587
67. Heinzelmann-Schwarz VA, Gardiner-Garden M, Henshall SM, Scurry J, Scolyer RA, Davies MJ et al (2004) Overexpression of the cell adhesion molecules DDR1, Claudin 3, and Ep-CAM in metaplastic ovarian epithelium and ovarian cancer. Clin Cancer Res 10:4427–4436
68. Hansen C, Greengard P, Nairn AC, Andersson T, Vogel WF (2006) Phosphorylation of DARPP-32 regulates breast cancer cell migration downstream of the receptor tyrosine kinase DDR1. Exp Cell Res 312:4011–4018
69. Jonsson M, Andersson T (2001) Repression of Wnt-5a impairs DDR1 phosphorylation and modifies adhesion and migration of mammary cells. J Cell Sci 114:2043–2053
70. Quan J, Yahata T, Adachi S, Yoshihara K, Tanaka K (2011) Identification of receptor tyrosine kinase, discoidin domain receptor 1 (DDR1), as a potential biomarker for serous ovarian cancer. Int J Mol Sci 12:971–982
71. Christofori G (2003) Changing neighbours, changing behaviour: cell adhesion molecule-mediated signalling during tumour progression. EMBO J 22:2318–2323
72. Vogel WF (2002) Ligand-induced shedding of discoidin domain receptor 1. FEBS Lett 514:175–180
73. Zhao G, Chen J, Deng Y, Gao F, Zhu J, Feng Z, Lv X, Zhao Z (2011) Identification of NDRG1-regulated genes associated with invasive potential in cervical and ovarian cancer cells. Biochem Biophys Res Commun 408:154–159
74. Colas E, Perez C, Cabrera S, Pedrola N, Monge M, Castellvi J et al (2011) Molecular markers of endometrial carcinoma detected in uterine aspirates. Int J Cancer 129:2435–2444
75. Domenyuk VP, Litovkin KV, Verbitskaya TG, Dubinina VG, Bubnov VV (2007) Identification of new DNA markers of endometrial cancer in patients from the Ukrainian population. Exp Oncol 29:152–155
76. Rudd ML, Mohamed H, Price JC, O'Hara AJ, Le Gallo M, Urick ME, NISC Comparative Sequencing Program, Cruz P, Zhang S, Hansen NF, Godwin AK, Sgroi DC, Wolfsberg TG, Mullikin JC, Merino MJ, Bell DW (2014) Mutational analysis of the tyrosine kinome in serous and clear cell endometrial cancer uncovers rare somatic mutations in TNK2 and DDR1. BMC Cancer 26:884–892
77. Lu Z, Wang J, Wientjes MG, Au JL (2010) Intraperitoneal therapy for peritoneal cancer. Future Oncol 6:1625–1641
78. Sadeghi B, Arvieux C, Glehen O, Beaujard AC, Rivoire M, Baulieux J et al (2000) Peritoneal carcinomatosis from non-gynecologic malignancies: results of the EVOCAPE 1 multicentric prospective study. Cancer 88:358–363

79. Karst AM, Drapkin R (2010) Ovarian cancer pathogenesis: a model in evolution. J Oncol 2010:932371
80. Siegel R, Naishadham D, Jemal A (2012) Cancer statistics, 2012. CA Cancer J Clin 62:10–29
81. Agarwal R, Kaye SB (2003) Ovarian cancer: strategies for overcoming resistance to chemotherapy. Nat Rev Cancer 3:502–516
82. Chien JR, Aletti G, Bell DA, Keeney GL, Shridhar V, Hartmann LC (2007) Molecular pathogenesis and therapeutic targets in epithelial ovarian cancer. J Cell Biochem 102:1117–1129
83. Kucukmetin A, Naik R, Galaal K, Bryant A, Dickinson HO (2010) Palliative surgery versus medical management for bowel obstruction in ovarian cancer. Cochrane Database Syst Rev 7, CD007792
84. Stratton JF, Tidy JA, Paterson ME (2001) The surgical management of ovarian cancer. Cancer Treat Rev 27:111–118
85. Tan DS, Agarwal R, Kaye SB (2006) Mechanisms of transcoelomic metastasis in ovarian cancer. Lancet Oncol 7:925–934
86. Deraco M, Baratti D, Laterza B, Balestra MR, Mingrone E, Macri A et al (2011) Advanced cytoreduction as surgical standard of care and hyperthermic intraperitoneal chemotherapy as promising treatment in epithelial ovarian cancer. Eur J Surg Oncol 37:4–9
87. Harmon RL, Sugarbaker PH (2005) Prognostic indicators in peritoneal carcinomatosis from gastrointestinal cancer. Int Semin Surg Oncol 2:3
88. Sodek KL, Murphy KJ, Brown TJ, Ringuette MJ (2012) Cell–cell and cell–matrix dynamics in intraperitoneal cancer metastasis. Cancer Metastasis Rev 31:397–414
89. Mutsaers SE (2002) Mesothelial cells: their structure, function and role in serosal repair. Respirology 7:171–191
90. Witz CA, Montoya-Rodriguez IA, Cho S, Centonze VE, Bonewald LF, Schenken RS (2001) Composition of the extracellular matrix of the peritoneum. J Soc Gynecol Invest 8:299–304
91. Rump A, Morikawa Y, Tanaka M, Minami S, Umesaki N, Takeuchi M et al (2004) Binding of ovarian cancer antigen CA125/MUC16 to mesothelin mediates cell adhesion. J Biol Chem 279:9190–9198
92. Burleson KM, Casey RC, Skubitz KM, Pambuccian SE, Oegema TR Jr, Skubitz AP (2004) Ovarian carcinoma ascites spheroids adhere to extracellular matrix components and mesothelial cell monolayers. Gynecol Oncol 93:170–181
93. Burleson KM, Hansen LK, Skubitz AP (2004) Ovarian carcinoma spheroids disaggregate on type I collagen and invade live human mesothelial cell monolayers. Clin Exp Metastasis 21:685–697
94. Lessan K, Aguiar DJ, Oegema T, Siebenson L, Skubitz AP (1999) CD44 and beta1 integrin mediate ovarian carcinoma cell adhesion to peritoneal mesothelial cells. Am J Pathol 154:1525–1537
95. Casey RC, Burleson KM, Skubitz KM, Pambuccian SE, Oegema TR Jr, Ruff LE et al (2001) Beta 1-integrins regulate the formation and adhesion of ovarian carcinoma multicellular spheroids. Am J Pathol 159:2071–2080
96. Strobel T, Cannistra SA (1999) Beta1-integrins partly mediate binding of ovarian cancer cells to peritoneal mesothelium in vitro. Gynecol Oncol 73:362–367
97. Krist LF, Kerremans M, Broekhuis-Fluitsma DM, Eestermans IL, Meyer S, Beelen RH (1998) Milky spots in the greater omentum are predominant sites of local tumour cell proliferation and accumulation in the peritoneal cavity. Cancer Immunol Immunother 47:205–212
98. Mochizuki Y, Nakanishi H, Kodera Y, Ito S, Yamamura Y, Kato T et al (2004) TNF-alpha promotes progression of peritoneal metastasis as demonstrated using a green fluorescence protein (GFP)-tagged human gastric cancer cell line. Clin Exp Metastasis 21:39–47
99. Sorensen EW, Gerber SA, Sedlacek AL, Rybalko VY, Chan WM, Lord EM (2012) Omental immune aggregates and tumor metastasis within the peritoneal cavity. Immunol Res 45:185–194
100. Tsujimoto H, Takhashi T, Hagiwara A, Shimotsuma M, Sakakura C, Osaki K et al (1995) Site-specific implantation in the milky spots of malignant cells in peritoneal dissemination:

immunohistochemical observation in mice inoculated intraperitoneally with bromodeoxyuridine-labelled cells. Br J Cancer 71:468–472

101. Oosterling SJ, van der Bij GJ, Bogels M, van der Sijp JR, Beelen RH, Meijer S et al (2006) Insufficient ability of omental milky spots to prevent peritoneal tumor outgrowth supports omentectomy in minimal residual disease. Cancer Immunol Immunother 55:1043–1051
102. Stadlmann S, Raffeiner R, Amberger A, Margreiter R, Zeimet AG, Abendstein B et al (2003) Disruption of the integrity of human peritoneal mesothelium by interleukin-1beta and tumor necrosis factor-alpha. Virchows Arch 443:678–685
103. Allen M, Louise Jones J (2011) Jekyll and Hyde: the role of the microenvironment on the progression of cancer. J Pathol 223:162–176
104. Swartz MA, Iida N, Roberts EW, Sangaletti S, Wong MH, Yull FE, Coussens LM, DeClerck YA (2012) Tumor microenvironment complexity: emerging roles in cancer therapy. Cancer Res 72:2473–2480
105. Finley SD, Popel AS (2013) Effect of tumor microenvironment on tumor VEGF during anti-VEGF treatment: systems biology predictions. J Natl Cancer Inst 105:802–811
106. Bhatt RS, Tomoda T, Fang Y, Hatten ME (2000) Discoidin domain receptor 1 functions in axon extension of cerebellar granule neurons. Genes Dev 14:2216–2228
107. Hou G, Vogel W, Bendeck MP (2001) The discoidin domain receptor tyrosine kinase DDR1 in arterial wound repair. J Clin Invest 107:727–735
108. Vogel WF, Aszodi A, Alves F, Pawson T (2001) Discoidin domain receptor 1 tyrosine kinase has an essential role in mammary gland development. Mol Cell Biol 21:2906–2917
109. Perez JL, Jing SQ, Wong TW (1996) Identification of two isoforms of the Cak receptor kinase that are coexpressed in breast tumor cell lines. Oncogene 12:1469–1477
110. Weiner HL, Huang H, Zagzag D, Boyce H, Lichtenbaum R, Ziff EB (2000) Consistent and selective expression of the discoidin domain receptor-1 tyrosine kinase in human brain tumors. Neurosurgery 47:1400–1409
111. Rikova K, Guo A, Zeng Q, Possemato A, Yu J, Haack H, Nardone J, Lee K, Reeves C, Li Y et al (2007) Global survey of phosphotyrosine signaling identifies oncogenic kinases in lung cancer. Cell 131:1190–1203
112. Alves F, Vogel W, Mossie K, Millauer B, Hofler H, Ullrich A (1995) Distinct structural characteristics of discoidin I subfamily receptor tyrosine kinases and complementary expression in human cancer. Oncogene 10:609–618
113. Barker KT, Martindale JE, Mitchell PJ, Kamalati T, Page MJ, Phippard DJ, Dale TC, Gusterson BA, Crompton MR (1995) Expression patterns of the novel receptor-like tyrosine kinase DDR, in human breast tumours. Oncogene 10:569–575
114. Yang SH, Baek HA, Lee HJ, Park HS, Jang KY, Kang MJ, Lee DG, Lee YC, Moon WS, Chung MJ (2010) Discoidin domain receptor 1 is associated with poor prognosis of non-small cell lung carcinomas. Oncol Rep 24:311–319
115. Ram R, Lorente G, Nikolich K, Urfer R, Foehr E, Nagavarapu U (2006) Discoidin domain receptor-1a (DDR1a) promotes glioma cell invasion and adhesion in association with matrix metalloproteinase-2. J Neurooncol 76:239–248
116. Balkwill FR, Capasso M, Hagemann T (2012) The tumor microenvironment at a glance. J Cell Sci 125:5591–5596
117. Kamohara H, Yamashiro S, Galligan C, Yoshimura T (2001) Discoidin domain receptor 1 isoform-a (DDR1alpha) promotes migration of leukocytes in three-dimensional collagen lattices. FASEB J 15:2724–2726
118. Miao L, Zhu S, Wang Y, Li Y, Ding J, Dai J, Cai H, Zhang D, Song Y (2013) Discoidin domain receptor 1 is associated with poor prognosis of non-small cell lung cancer and promotes cell invasion via epithelial-to-mesenchymal transition. Med Oncol 30:626
119. Ruiz PA, Jarai G (2011) Collagen I induces discoidin domain receptor (DDR) 1 expression through DDR2 and a JAK2-ERK1/2-mediated mechanism in primary human lung fibroblasts. J Biol Chem 286:12912–12923

120. Shintani Y, Fukumoto Y, Chaika N, Svoboda R, Wheelock MJ, Johnson KR (2008) Collagen I-mediated up-regulation of N-cadherin requires cooperative signals from integrins and discoidin domain receptor 1. J Cell Biol 180:1277–1289
121. Suh HN, Han HJ (2011) Collagen I regulates the self-renewal of mouse embryonic stem cells through alpha2beta1 integrin- and DDR1-dependent Bmi-1. J Cell Physiol 226:3422–3432
122. Xu H, Bihan D, Chang F, Huang PH, Farndale RW, Leitinger B (2012) Discoidin domain receptors promote alpha1beta1- and alpha2beta1-integrin mediated cell adhesion to collagen by enhancing integrin activation. PLoS One 7, e52209
123. Yeh YC, Lin HH, Tang MJ (2012) A tale of two collagen receptors, integrin β1 and discoidin domain receptor 1, in epithelial cell differentiation. Am J Physiol Cell Physiol 303:C1207–C1217
124. Roarty K, Serra R (2007) Wnt5a is required for proper mammary gland development and TGF-beta-mediated inhibition of ductal growth. Development 134:3929–3939
125. Iwai LK, Chang F, Huang PH (2013) Phosphoproteomic analysis identifies insulin enhancement of discoidin domain receptor 2 phosphorylation. Cell Adhes Migr 7:161–164
126. Kim HG, Hwang SY, Aaronson SA, Mandinova A, Lee SW (2011) DDR1 receptor tyrosine kinase promotes prosurvival pathway through Notch1 activation. J Biol Chem 286:17672–17681
127. Zhang S, Bu X, Zhao H, Yu J, Wang Y, Li D, Zhu C, Zhu T, Ren T, Liu X, Yao L, Su J (2014) A host deficiency of discoidin domain receptor 2 (DDR2) inhibits both tumor angiogenesis and metastasis. J Pathol 232:436–448
128. Ren T, Zhang J, Liu X, Yao L (2013) Increased expression of discoidin domain receptor 2 (DDR2): a novel independent prognostic marker of worse outcome in breast cancer patients. Med Oncol 30:397
129. Fu HL, Sohail A, Valiathan RR, Wasinski BD, Kumarasiri M, Mahasenan KV, Bernardo MM, Tokmina-Roszyk D, Fields GB, Mobashery S, Fridman R (2013) Shedding of discoidin domain receptor 1 by membrane-type matrix metalloproteinases. J Biol Chem 288:12114–12129

# Chapter 10
# Small Molecule Inhibitors of Discoidin Domain Receptors (DDRs)

**Xiaoyun Lu and Ke Ding**

## Abbreviations

| | |
|---|---|
| DDRs | Discoidin domain receptors |
| 2D | Two dimensional |
| DS | Discoidin domain |
| ECM | Extracellular matrix |
| EJXM | Extracellular juxtamembrane |
| FDA | Food and Drug Administration |
| FLiK | Fluorescent labels in kinases |
| FUSION | Functional Signature-Based Ontology |
| IJXM | Intracellular juxtamembrane |
| IR | Insulin receptor |
| KD | Kinase domain |
| MOA | Mechanism of action |
| NMR | Nuclear magnetic resonance |
| NSCLC | Non-small cell lung cancer |
| PK | Pharmacokinetics |
| PMBCl | *para*-Methoxybenzyl chloride |
| QSAR | Quantitative structure-activity relationship |
| RTKs | Receptor tyrosine kinases |
| SQCC | Squamous lung cancer |
| S768R | Serine 768 to arginine 768 |
| TMT | Transmembrane |
| VEGFR | Vascular endothelial growth factor receptor |

X. Lu • K. Ding (✉)
School of Pharmacy, Jinan University, 601 Huangpu Avenue West, Guangzhou 510632, China

Guangzhou Institutes of Biomedicine and Health, Chinese Academy of Sciences, #190 Kaiyuan Avenue, Guangzhou 510530, China
e-mail: ding_ke@gibh.ac.cn



R. Fridman, P.H. Huang (eds.), *Discoidin Domain Receptors in Health and Disease*, DOI 10.1007/978-1-4939-6383-6_10

## 10.1 Introduction

The discoidin domain receptors (DDRs) are a unique class of transmembrane receptor tyrosine kinases (RTKs) which were discovered in the early 1990s by homology cloning [1–3]. Two types of DDRs, i.e., DDR1 and DDR2, have been identified with distinct expression profiles and ligand specificities. DDR1 has five isoforms (i.e., DDR1a, DDR1b, DDR1c, DDR1d, and DDR1e) generated by alternative splicing, while there is only one isoform of DDR2 found to date (Table 10.1) [4, 5]. DDR1 is widely expressed in epithelial cells in the lung, kidney, colon, and brain [3, 4, 6], whereas DDR2 is primarily expressed in mesenchymal cells including fibroblasts, myofibroblasts, smooth muscle, and skeletal in the kidney, skin, lung, heart, and connective tissues [7]. Unlike other RTKs using typical ligands for activation which are mostly soluble polypeptide growth factors, cytokines, hormones, etc., DDRs are activated by collagens—the most abundant components of the extracellular matrix (ECM) [8–11]. Upon collagen binding, DDRs undergo autophosphorylation [12] to regulate fundamental cellular processes, such as proliferation, survival, differentiation, adhesion, and matrix remodeling [13, 14]. However, overexpression and/or "gain-of-function" mutation of DDRs may also elicit unregulated signaling to cause epithelial migration, metastasis, and cellular invasion in diseases [5, 15–22].

Extensive genetic and functional studies have highlighted DDRs as new players in cancer development and progression [10, 16, 23, 24]. For instance, DDR1 has been demonstrated to be associated with the poor prognosis in non-small cell lung cancer (NSCLC) and is also implicated in cell survival and invasiveness in hepatocellular carcinoma, pituitary adenoma, pancreatic cancer, and prostate cancer [24, 25]. Overexpression and/or mutation of DDRs is also frequently detected in many cancer cell lines or tissues [26–28]. A "gain-of-function" serine 768 to arginine 768 (S768R) mutation of DDR2 has been recently identified as the driving force in a portion of squamous lung cancer patients [21]. Besides many types of human cancer, DDRs may also serve as new potential therapeutic targets for drug discovery against inflammation [29, 30], fibrosis [5, 31–33], etc.

Given the accumulating evidences implicating DDRs' crucial roles in various diseases, several classes of DDR small molecular inhibitors have been developed [10, 34]. This chapter provides an overview of recent progress on the discovery of DDR inhibitors as new potential therapeutic agents and/or research tools for further investigation.

**Table 10.1** DDRs subfamily members: isoforms and distribution

| DDR subfamily | Isoforms | Domains contained | Distribution |
|---|---|---|---|
| DDR1 | DDR1a | DS, DS-like, EJXM, TM, IJXM, KD | Epithelial cells |
| | DDR1b | DS, DS-like, EJXM, TM, IJXM, KD | |
| | DDR1c | DS, DS-like, EJXM, TM, IJXM, KD | |
| | DDR1d | DS, DS-like, EJXM, TM, short IJXM | |
| | DDR1e | DS, DS-like, EJXM, TM, IJXM, Inactive KD | |
| DDR2 | DDR2 | DS, DS-like, EJXM, TM, IJXM, KD | Mesenchymal cells |

## 10.2 Structural Basis of DDRs

Structurally, DDRs are comprised of six domains including (1) an N-terminal discoidin domain (DS domain), (2) a discoidin-like domain (DS-like domain), (3) an extracellular juxtamembrane (EJXM) region, (4) a single transmembrane (TM) domain, (5) a large intracellular juxtamembrane (IJXM) region, and (6) the C-terminal intracellular kinase domain (KD) [12]. The DS domain of various DDR isoforms has a highly conserved region and contains the collagen ligand binding site as determined by NMR and X-ray crystallographic technologies [35–38]. Sequence variation in the DS domain defines the specificity and types of the binding collagens [39, 40]. The collagen binding site of DDR2 is composed of the residues of Trp52, Thr56, Asp69, Arg105, Glu113, Cys73, and Cys177 (Fig. 10.1a) [36], which are similar to those in DDR1 (Fig. 10.1b) [41]. The DS-like domain has a very comparable sequence to that of DS domain, but its function is still not fully elucidated [41, 42]. The EJXM region contains an *N*- and/or *O*-glycosylation site for receptor- or ligand-induced activation [43]. The TM domain which links the ectodomain and intracellular domain of DDRs plays a critical role in the mediation of collage-independent receptor dimerization [44]. There are a number of tyrosine residues that are present in the IJXM region which serve as docking sites for essential DDRs regulators [8]. DDR1a, DDR1b, DDR1c, and DDR2 are characterized as full-length functional receptors which are comprised of all six structural domains [12, 45], whereas the other two DDR1 isoforms, DDR1d and DDR1e, are truncated or kinase-inactive receptors without kinase activity (Table 10.1) [45].

Very few crystal structural information about the kinase domain of DDRs is available. Only recently, researchers in the University of Oxford and Dana-Farber Cancer Institute independently disclosed the X-ray structures of DDR1 kinase domain co-crystallized with different small molecular inhibitors [46, 47]. The X-ray

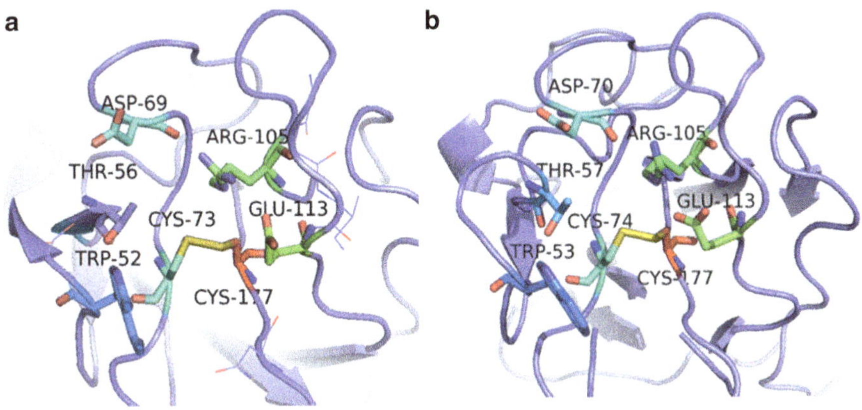


**Fig. 10.1** (**a**) The collagen binding site of DDR2 discoidin domain (PDB code: 2WUH); (**b**) the collagen binding site of DDR1 discoidin domain (PDB code: 4AG4). The binding residues are shown in *stick colored* by element type (carbon in *green*, oxygen in *red*, and nitrogen in *blue*)

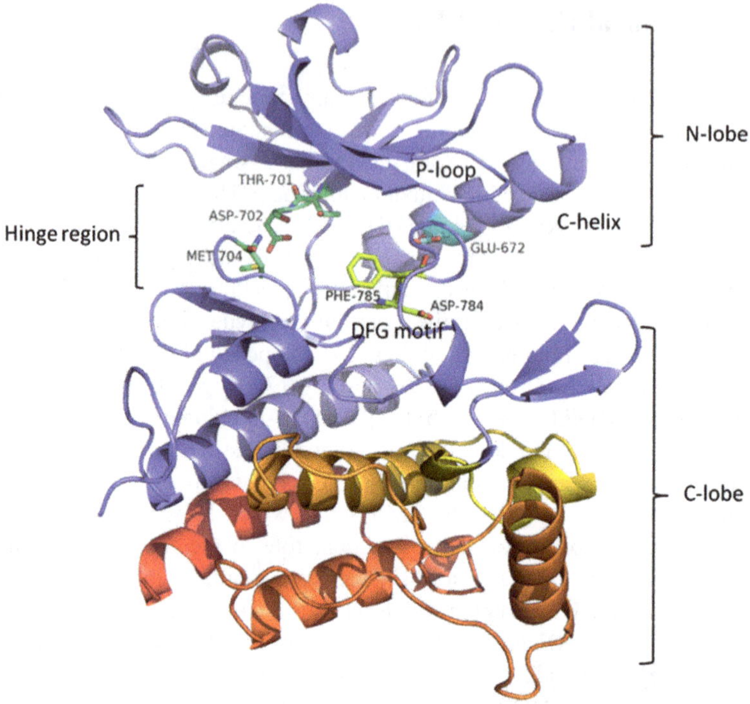


**Fig. 10.2** Overall structure of DDR1 kinase domain (PDB: 3ZOS). The active residues in the ATP binding site are shown in *stick colored* by elemental type (carbon in *green* or light *green* (DFG motif), oxygen in *red*, and nitrogen in *blue*)

structure of DDR2 kinase domain remains unavailable. Similar to that of typical RTKs, DDR1 kinase domain is composed of N-terminal lobes, P-loop, activation loop, C-helix, F-helix, DFG motif, and C-terminal lobes. The key residues of Glu672, Asp784, Thr701, Asp702, and Met704 in the protein form a well-defined pocket for ATP binding (Fig. 10.2) [46, 47]. Canning et al. showed in detail the inhibitor binding differences between DDR1 and Abl in both the shape and sequence of the ATP binding site, which is mainly observed in the DDR1 P-loop accommodated outside the ATP pocket within β-hairpin. The structural information of the DDR1 kinase domain provides a fundamental basis for the discovery of DDR1-specific inhibitors.

## 10.3 Small Molecule Inhibitors of DDRs

Given their great contributions in various human diseases, different approaches have been applied to suppress the activation of DDR signals (Fig. 10.3). It has been well demonstrated that the activation of DDRs is initiated by collagen binding with

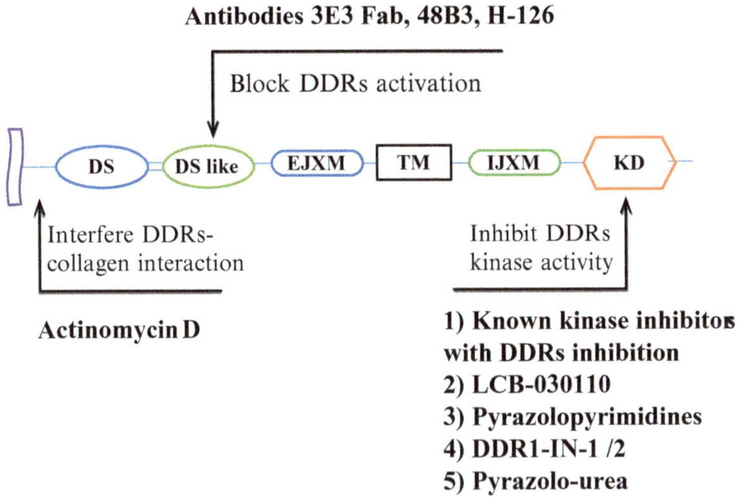


**Fig. 10.3** Key approaches to inhibit the activation of DDR signaling pathway

the DS domain of the receptors. Therefore, one of the most straightforward strategies to inhibit DDR signals is to interrupt the collagen-DDR protein-protein interaction by a small chemical, the so-called DDR-collagen interaction blockers [33]. The advantage of this strategy is that the compounds may achieve great specificity against DDRs without interfering with other kinases because they lack the DS domain. Natural product actinomycin D [48] has been reported to selectively inhibit DDR2 activation through the blockage of DDR-collagen interaction. Similar to other RTKs, ATP binding with the C-terminal kinase domain plays a critical role for the autophosphorylation of DDRs. Consequently, a typical kinase inhibitor targeting the ATP binding site represents another highly powerful strategy to inactivate the DDR signals. However, given the fact that many kinases share similar sequences in the kinase domains, target specificity is one of the most challenging issues for this type of compounds. A number of DDR1/DDR2 kinase inhibitors have been discovered to date with different potency and selectivity profiles. In addition, monoclonal antibodies, i.e., 3E3 Fab [41], 48B3 [49], and H-126 [50], have also developed to inhibit DDR1 activation without interfering with collagen binding by selectively targeting the DS-like domain in the receptor [41]. Herein, we would like to describe the recent advances on the discovery of DDRs inactivators with focus on the small molecule kinase inhibitors.

### *10.3.1 DDR-Collagen Interaction Blocker*

Actinomycin D (**1**, Fig. 10.4), also known generically as actinomycin or dactinomycin, was first isolated by Selman Waksman et al. from soil bacteria of the genus *Streptomyces* in 1940. It was approved as a chemotherapy drug for a variety of

**Fig. 10.4** Chemical structure of actinomycin D

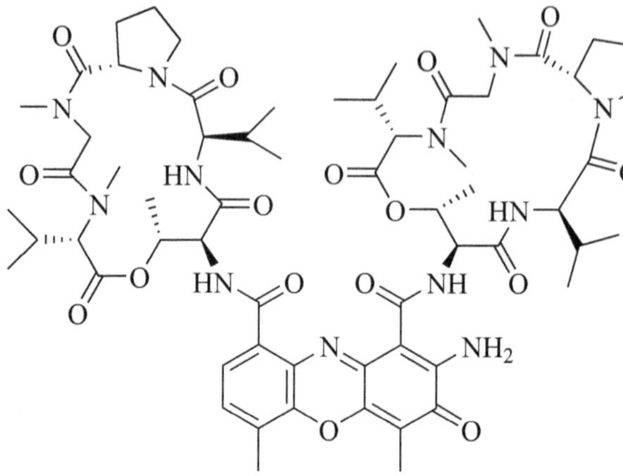

**1 Actinomycin D**

cancers by the US Food and Drug Administration (FDA) in 1964. The primary mechanism of this drug is to prevent elongation of RNA chain by binding at the transcription initiation complex of DNA [51]. The compound was recently identified as a novel selective blocker of DDR2-collagen (type I) interaction with an $IC_{50}$ value of approximately 40 μM by utilizing a *Spodoptera frugiperda* 9 (sf9) insect cell-based screening technology [48]. It was also validated that actinomycin D dose-dependently inhibits the collagen-stimulated activation of DDR2 in HEK293 cells stably expressing DDR2 with an $IC_{50}$ value of 9.0 μM and without altering the DDR2 protein level, whereas it does not show obvious effect on the activation of DDR1b, epidermal growth factor receptor (EGFR), or insulin receptor (IR) in the similar cellular systems, which supporting its selective inhibition against DDR2. However, it is still unclear whether the drug binds directly to DDR2 or type I collagen for its inhibitory activity against the association between DDR2 and collagen. It is only a logical assumption that actinomycin D may target the extracellular domain of DDR2 because it does not inhibit the DDR1 and collagen interaction. Further investigation is necessary to demonstrate the mechanistic details for its blockage against DDR-collagen interaction.

## *10.3.2 Small Molecule Inhibitors Targeting the Kinase Domain of DDRs*

### 10.3.2.1 Nonselective Inhibition Against DDRs by Known Kinase Inhibitors

Human DDR1 and DDR2 (Swiss-Prot ID: Q08345 and Q16832, respectively) display high sequence similarity to many other tyrosine kinases. In particular, approximate 61 % sequence identity is found in a sequence alignment of DDR1 and DDR2

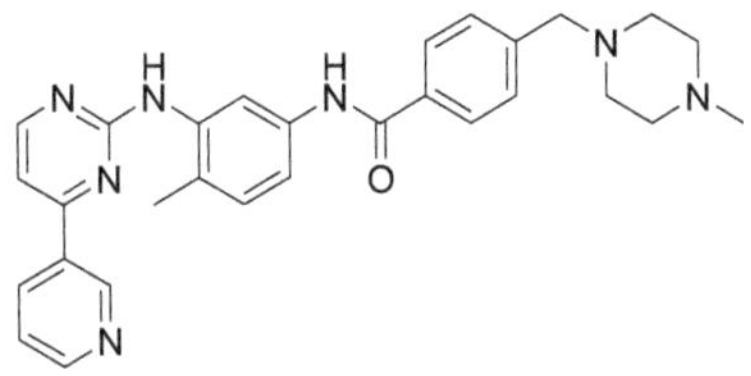

**2 STI-571, Imatinib**

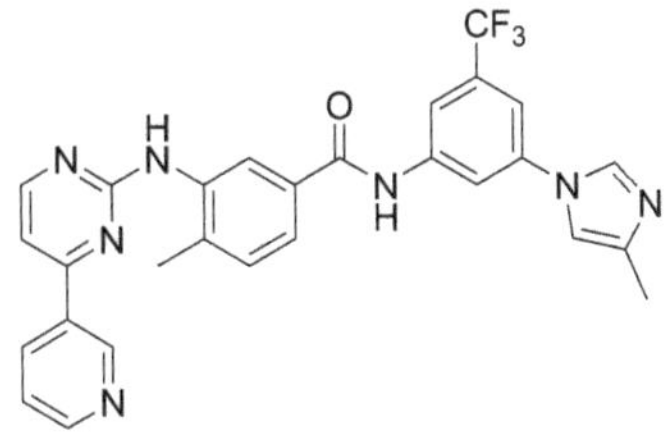

**3 Nilotinib**

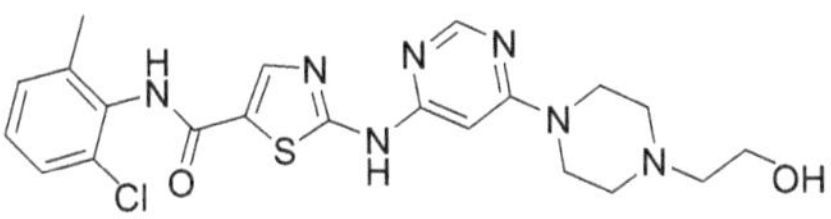

**4 Dasatinib**

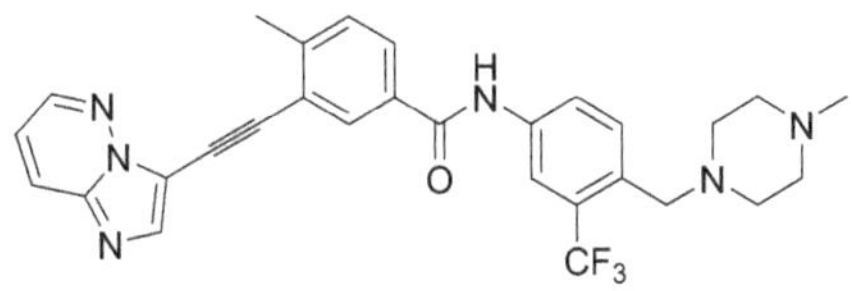

**5 Ponatinib, AP24534**

**Fig. 10.5** Chemical structures of imatinib, nilotinib, dasatinib, and ponatinib

**Table 10.2** The $IC_{50}$ values of imatinib, nilotinib, and dasatinib targeting DDR1 and DDR2

| Inhibitors | Type | $IC_{50}$DDR1 (nM) | $IC_{50}$DDR2 (nM) | Selective |
|---|---|---|---|---|
| Imatinib | Type II | 337 | 675 | No |
| Nilotinib | Type II | 43 | 55 | No |
| Dasatinib | Type I | 0.5 | 1.4 | No |

with Bcr-Abl in the ATP binding domain [34]. Therefore, almost all previously reported Bcr-Abl inhibitors were found to exhibit DDR kinase inhibition. Imatinib (**2**), nilotinib (**3**), dasatinib (**4**), and ponatinib (**5**) (Fig. 10.5), which are FDA-approved Bcr-Abl inhibitors for the clinical management of chronic myelogenous leukemia (CML), have been reported to target DDR1 and DDR2 by chemical proteomic profiling studies [34, 52, 53]. Further biochemical investigation reveals that they strongly suppress the kinase activities of DDRs and prevent collagen-mediated DDRs autophosphorylation [34, 54]. In a cell-free TR-FRET kinase assay, dasatinib was demonstrated as one of the most potent DDR1b and DDR2 inhibitors with $IC_{50}$ values of 0.5 nM and 1.4 nM, respectively. The corresponding $IC_{50}$ values for imatinib and nilotinib are 337 nm, 675 nm, and 43 nm, and 55 nM, respectively (Table 10.2). Imatinib bound in the DFG-out ATP binding pocket of DDR1 with a similar mode to that of Bcr-Abl (Fig. 10.6) [55]. Similar to imatinib, nilotinib is predicted to bind to the DDR1 kinase domain in an inactive DFG-out conformation, whereas dasatinib adopts a type I binding with DDR1 in DFG-in conformation.

Moreover, both imatinib and nilotinib have been shown to display promising therapeutic efficacy for renal, liver, and pulmonary fibrosis in different animal models, which is due to, at least in part, their inhibition against DDR kinases. Given the

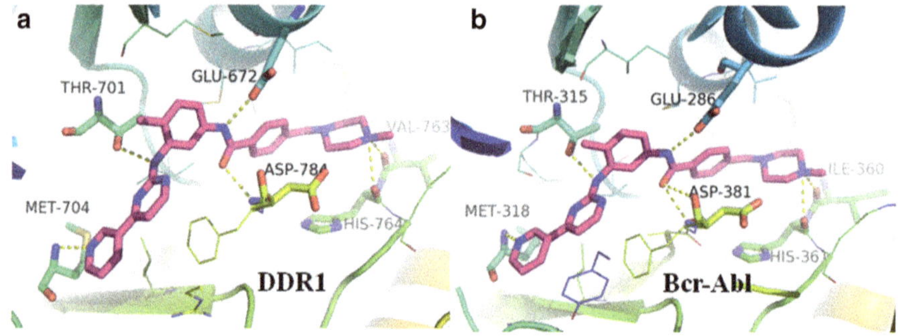


**Fig. 10.6** The crystallographic complex of imatinib bound in the DFG-out ATP binding pocket of (A) DDR1 (PDB code: 4BKJ) and (B) Bcr-Abl (PDB code: 2HYY). The active residues in ATP binding site are shown in *stick colored* by element type (carbon in *green*, oxygen in *red*, and nitrogen in *blue)*. Ponatinib is shown in *stick colored* by carbon in *purple*, nitrogen in *blue*, and oxygen in *red*

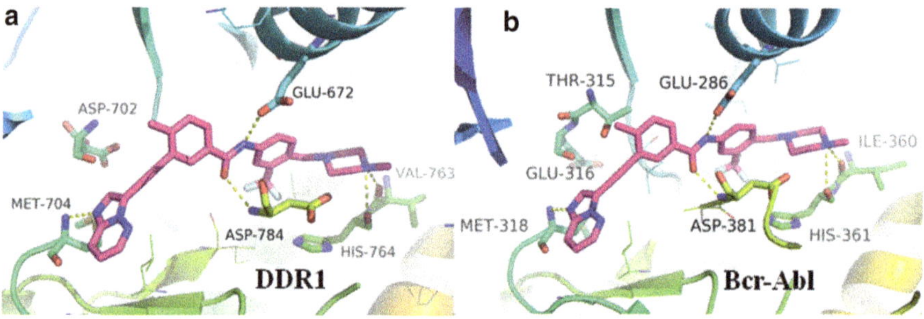


**Fig. 10.7** The crystallographic complex of ponatinib bound in the DFG-out ATP binding pocket of (A) DDR1 (PDB code: 3IK3) and (B) Bcr-Abl (PDB code: 3OXZ). The active residues in ATP binding site are shown in *stick colored* by element type (carbon in *green*, oxygen in *red*, and nitrogen in *blue*). Ponatinib is shown in *stick colored* by carbon in *purple*, nitrogen in *blue*, oxygen in *red*, and fluoride in *cyan*

strong inhibitory effect of dasatinib against DDR2, a phase II clinical investigation was initiated for its therapeutic efficacy in squamous lung cancer (SQCC) patients harboring "gain-of-function" mutation of DDR2 in 2011 [21, 56]. A successful case has been reported to show a significant shrinkage in tumor size for dasatinib-treated SQCC patients with DDR2 S768R mutation [57]. However, the clinical investigation is recently terminated due to toxicity issues related to the multi-target inhibition of the drug [58].

Ponatinib (**5**, AP24534, Iclusig®) (Fig. 10.5) is a newly FDA-approved Bcr-Abl inhibitor for the treatment of resistant or intolerant CML and $Ph^+$ ALL patients against imatinib, especially those that harbor $Bcr\text{-}Abl^{T315I}$ mutation [59, 60]. It also exhibits strong suppression on DDRs with low nanomolar $IC_{50}$ values. More significantly, a co-crystal structure of ponatinib with DDR1 kinase domain is disclosed to demonstrate that the compound binds in the DFG-out ATP binding pocket of DDR1 with a similar mode to that in Bcr-Abl kinase (Fig. 10.7). The imidazo[1,2-*b*]pyridazine core occupies the adenine pocket of the kinase, the amide forms two hydrogen bonds with Glu672 and Asp784, and the trifluoromethylphenyl group bound deeply into the hydrophobic pocket (Fig. 10.7a) [46]. However, the US FDA temporarily suspended the marketing of this drug due to the increasing numbers of blood clots observed in ponatinib-treated patients in October 2013. The drug was then reauthorized for sale after a revised indication statement and a boxed warning were made by the manufacturer in December 2013.

Other Bcr-Abl inhibitors, such as bafetinib (**6**, INNO-406) [61] and GZD824 (**7**) (Fig. 10.8) [62], were also reported to exhibit strong inhibition against DDRs by chemical proteomics and kinase profiling studies. Bafetinib inhibits DDR2 with $IC_{50}$ value of 220 nM, whereas GZD824 tightly binds with DDR1 with a $K_d$ value of 0.9 nM. In addition to the Bcr-Abl inhibitors, a number of other kinase suppressors were reported to display strong inhibition against DDRs with nanomolar activity (Fig. 10.8). For instance, the multikinase inhibitor sorafenib (**8**) has been demonstrated to tightly bind both DDR1 and DDR2 with $K_d$ values of 1.5 nM and 6.6 nM, respectively [63]. A potent VEGFR inhibitor pazopanib (**9**) which was approved and launched for the treatment of advanced or metastatic renal cell carcinoma in 2009 and more recently in soft tissue sarcoma is reported to exhibit strong binding with DDR2 with $K_d$ value of 57 nM [63]. BIRB-796 (**10**), a P38 MAPK inhibitor, has also been identified to be a strong DDR1 and DDR2 inhibitor with $K_d$ values of 1.9 nM and 33 nM, respectively. Moreover, foretinib (**15**), an orally available multikinase inhibitor of c-Met and VEGFR2, showed a strong inhibition against DDR1 with $K_d$ value of 200 picomolar [63].

#### 10.3.2.2 LCB-030110

In order to identify novel DDR2 inhibitors, a research group in Korea conducted a random screen against their in-house compound library. A series of thienopyridine derivatives are found to exhibit good inhibitory activity against the kinase. Further medicinal chemistry efforts enabled them to identify LCB-030110 (**18**, Fig. 10.9) as novel potent DDR2 inhibitor [64, 65]. The compound inhibited both activated and inactivated DDR2 with $IC_{50}$ values of 6.0 nM and 145 nM, respectively. It also potently suppressed the collagen-induced autophosphorylation of DDR1 and DDR2 with $IC_{50}$ values of approximately 164 nM and 171 nM, respectively, in the engineered HEK293 cell models. Further kinase selectivity profiling investigation revealed that the compound **18** displays broad-spectrum inhibition against a panel of tyrosine kinases such as c-Src, Lck, Lyn, Fyn, Yes, Fgr, Hck, Blk, Btk, Syk, VEGFR2, EphA3, FLT3, and PDGFRα with low nM $IC_{50}$ values. LCB-030110 suppressed the proliferation and migration of primary dermal fibroblasts induced by transforming growth factor β1 and type I collagen. It also strongly inhibited cell migration and nitric oxide, inducible nitric oxide synthase, cyclooxygenase 2, and tumor necrosis factor-α synthesis in J774A.1 macrophage cells activated by lipopolysaccharide. Furthermore, LCB-030110 showed strong inhibition against fibroinflammation and hypertrophic scar formation after a tropical treatment in a rabbit ear wound healing model.

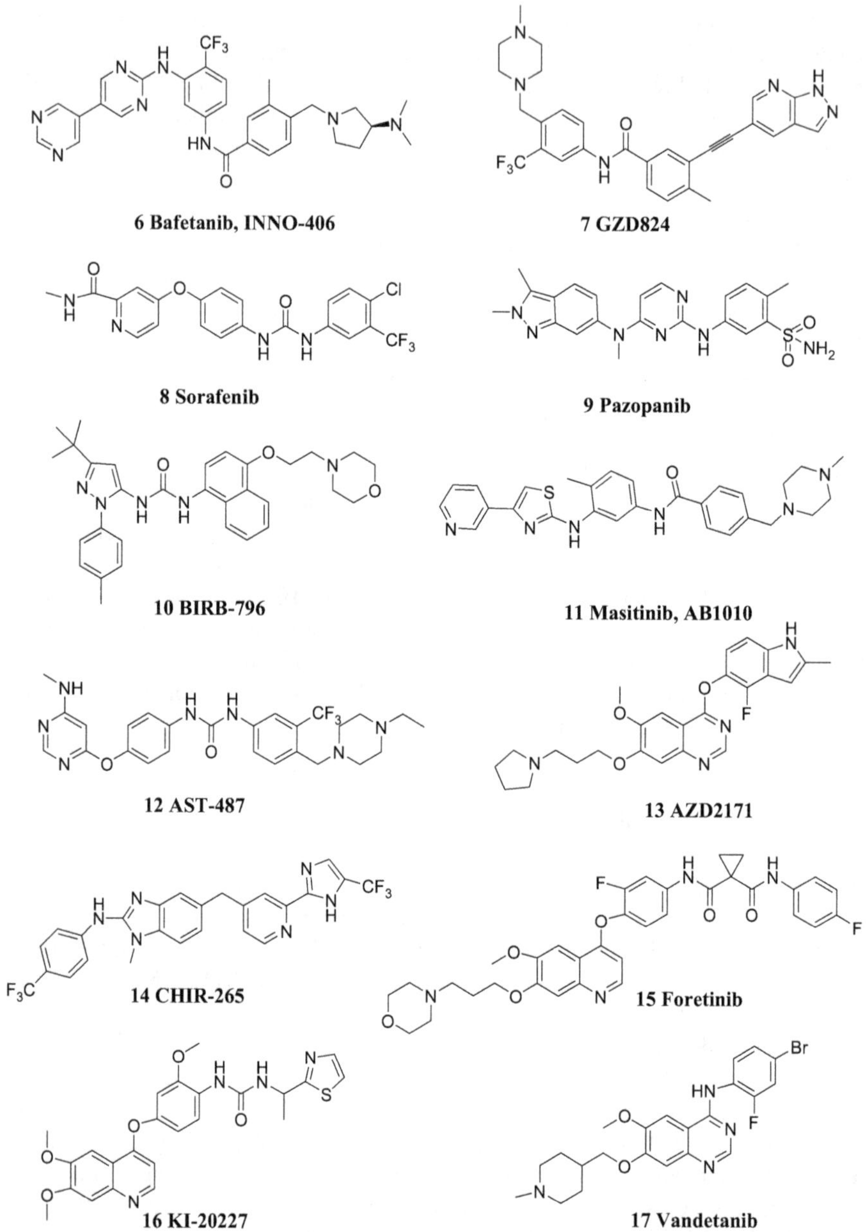


**Fig. 10.8** Chemical structures of some known kinase inhibitors with DDR inhibition

#### 10.3.2.3 Pyrazolopyrimidine Alkyne Derivatives

Although a number of small molecules have been reported to exhibit strong inhibition against DDRs, almost all of them display relatively poor DDR kinase selectivity. In order to identify new DDR inhibitor with good selectivity profile, Ding et al.

**Fig. 10.9** Chemical structure of LCB-030110 (**18**)

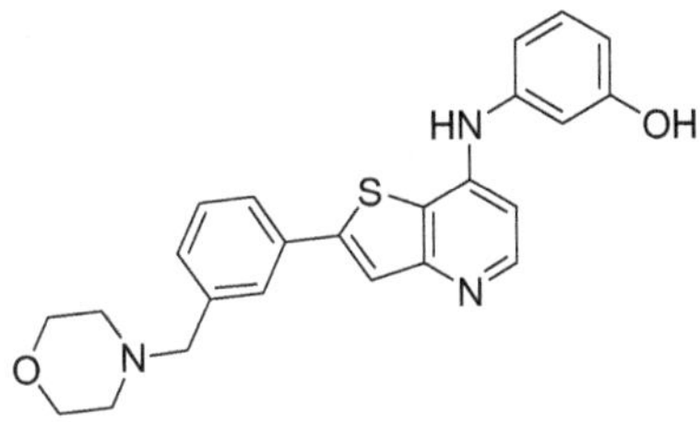


**18 LCB 03-0110**

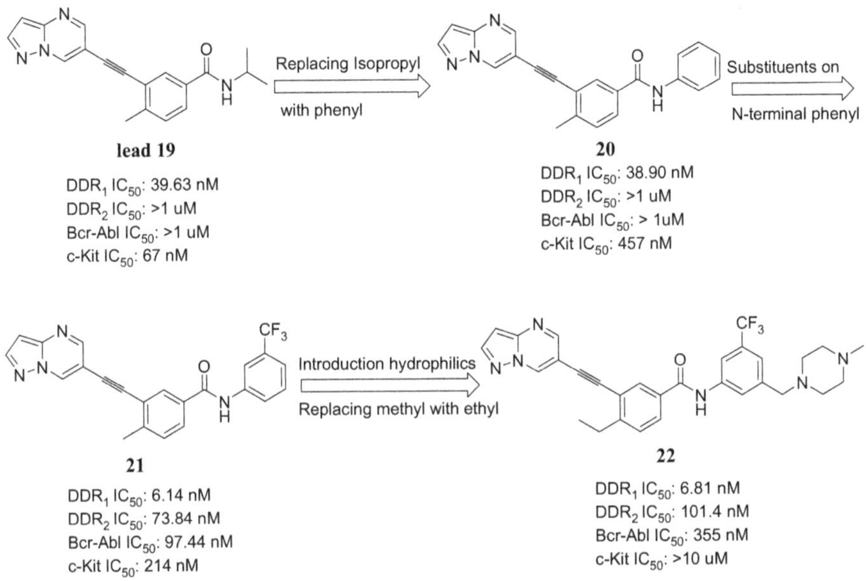


**Fig. 10.10** A discovery process for the highly selective DDR1 inhibitor **22**

screened a self-constructed kinase inhibitor-focused library of approximately 2000 compounds using a well-established LANCE Ultra kinase assay [66]. The DDRs selectivity of the compounds was also monitored by determining their inhibitory effect on Bcr-Abl and c-Kit kinases, which show high sequence similarity to DDRs in kinase domain. *N*-isopropyl-4-methyl-3-(2-(pyrazolo[1,5-*a*]pyrimidin-6-yl)ethynyl) benzamide (**19**) was identified as an initial "hit" potently inhibiting DDR1 with an $IC_{50}$ value of 39.6 nM, while it was significantly less potent against DDR2 and Bcr-Abl with $IC_{50}$ values of over 1.0 μM, respectively [66]. However, compound **19** also potently inhibited c-Kit with $IC_{50}$ value of 67 nM.

Further structure optimization was conducted to improve the compound's selectivity profile (Fig. 10.10) [66]. Replacement of the isopropyl in **19** with phenyl group affords new inhibitor **20**, which maintains the strong DDR1 inhibitory activity and exhibits about 12 times selectivity against DDR1 over c-Kit. Structural modification efforts in the *N*-terminal phenyl region of **20** resulted in the discovery of compound **22** (**7rh**) as a novel DDR1 inhibitor. The compound strongly inhibits DDR1 with an $IC_{50}$ value of 6.81 nM, whereas its activities against DDR2, Bcr-Abl, and c-Kit are significantly less potent, with $IC_{50}$ values of 101.4, 355 nM, and over 1 μM, respectively. A direct binding assay validates that the compound tightly binds to the ATP binding site of DDR1 with a $K_d$ value of 0.6 nM. A kinase profiling study against 465 kinase reveals that compound **22** is a selective DDR1 inhibitor with the *S* (35) and *S* (10) score values of 0.035 and 0.008, respectively, at 100 nM which is about 160 times higher than its $K_d$ value with DDR1.

Compound **22** dose-dependently suppresses the autophosphorylation of DDR1 and downstream signaling as determined by Western blot analysis [66]. It also inhibits the proliferation of a panel of cancer cells highly expressing DDR1, including A549, NCI-H23, NCI-H460 human NSCLC cells, MCF-7, T47D human breast cancer cells; HCT116 colon cancer cells; and K562 CML cells, with low μM or nM $IC_{50}$ values. Further investigation demonstrates that compound **22** strongly inhibits invasion, cell matrix adhesion, and tumorigenicity in NCI-H23 human NSCLC cells. Moreover, in vivo pharmacokinetic studies reveal that compound **22** possesses good PK profile with oral bioavailability of 67.4 % and reasonable half-life of 15.5 h in rats.

#### 10.3.2.4 DDR1-IN-1/DDR1-IN-2

In view that both of imatinib and nilotinib are typical type II DDR inhibitors, a research team led by Prof. Nathanael S. Gray in the Dana-Farber Cancer Institute of Harvard Medical School developed a library of potential type II kinase inhibitors by using a "pharmacophore-based design" strategy in conjunction with the structural information of current inhibitors. Screening across a panel of 451 kinases using the KINOMEscan approach against the small library resulted in the identification of DDR1-IN-1 (**23**) and DDR1-IN-2 (**24**) as novel potent and selective binders of DDR1 (Fig. 10.11) [47]. The compounds are further confirmed to inhibit the kinase activities of DDR1 and DDR2 with $IC_{50}$ values of 105, 413 nM for DDR1-IN-1 and 47 nM, 145 nM for DDR1-IN-2, respectively. They also exhibit strong blockage on collagen-induced autophosphorylation of DDR1 in U2OS cells with $EC_{50}$ values of 86 nM and 9.0 nM, respectively. Further cellular growth inhibition studies reveal that DDR-IN-2 blocks the proliferation of a variety of cancer cell lines expressing DDR1 at sub-micromolar ranges. A KINOMEscan profiling study was also conducted to demonstrate that DDR-IN-1 displays a good selective profile on DDR1 with *S* (1) score of 0.01 against 451 kinases at 1.0 μM, whereas DDR-IN-2 exhibits less selectivity than DDR-IN-1, with the *S* (1) score of 0.07 [47].

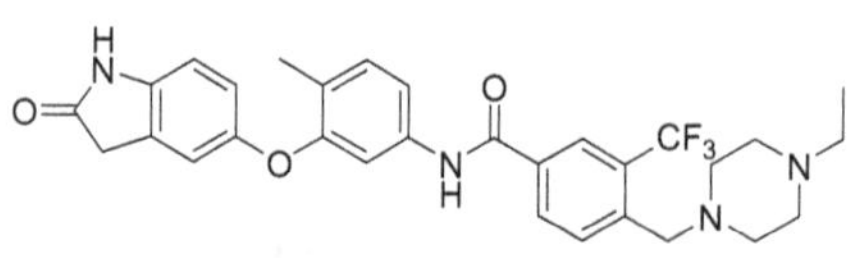

**23 DDR1-IN-1**

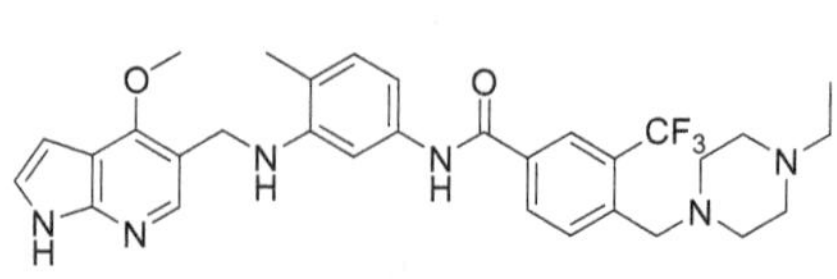

**24 DDR1-IN-2**

**Fig. 10.11** Chemical structures of DDR-IN-1 (**23**) and DDR-IN-2 (**24**)

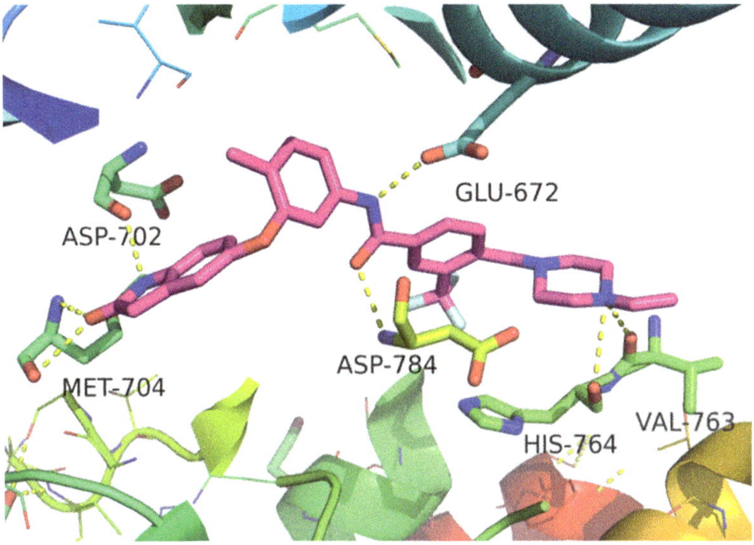


**Fig. 10.12** Co-crystal structure of DDR-IN-1 bound to DDR1 (PDB ID: 4CKR). The active residues in ATP binding site are shown in *stick colored* by element type (carbon in *green*, oxygen in *red*, and nitrogen in *blue*). DDR-IN-1 is shown in *stick colored* by carbon in *purple*, nitrogen in *blue*, oxygen in *red*, and fluorine in *cyan*

Moreover, a co-crystal structure of DDR-IN-1 with DDR1 is also disclosed from the research group (PDB ID: 4CKR) (Fig. 10.12) [47]. It shows that DDR-IN-1 interacts with the kinase domain of DDR1 with inactive DFG-out conformation. The trifluoromethyl group binds deeply into a hydrophobic pocket directly adjacent to the ATP binding site, while the indolinone motif of DDR-IN-1 makes a pair of hydrogen bonds with the hinge region Asp702 and Met704 and the amide group forms two hydrogen bonds with the residue Glu672 and Asp784. The nitrogen of piperazine also forms two hydrogen bonds with Val763 and His764. A computational study also suggests that DDR-IN-2 binds to DDR1 with a similar mode to that of DDR-IN-1. But an additional hydrogen bond is formed between the nitrogen atom in the linker region of DDR-IN-2 with the gatekeeper Thr701, which may contribute to the improved potency, and that of DDR-IN-1.

#### 10.3.2.5 Other DDR Kinase Inhibitors

Most recently, a series of pyrazolo-urea derivatives (**25**–**28**) were identified as novel type II/III binders of DDR2 with low nM $K_d$ values by using a fluorescent labels in kinases (FLiK) assay (Fig. 10.13) [67]. The DDR2 inhibitory activities of the compounds were further validated by using standard kinase suppressive assays. Among all the compounds evaluated, compound **27** tightly binds with DDR2 with a $K_d$ value of 20 nM. It also strongly inhibits the kinase activity of DDR2 with an $IC_{50}$ value of 8 nM [67]. A series of benzamide- and quinazolinedione-based selective

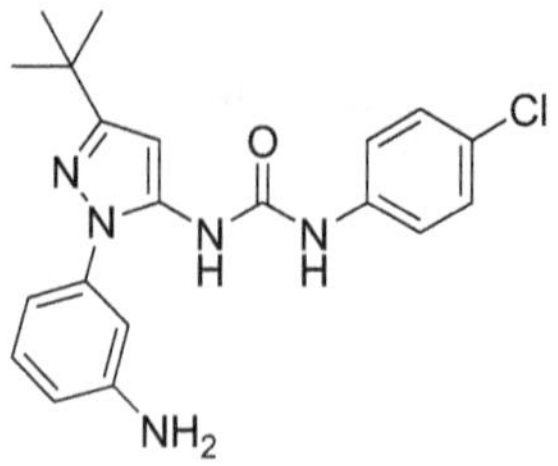

**25**

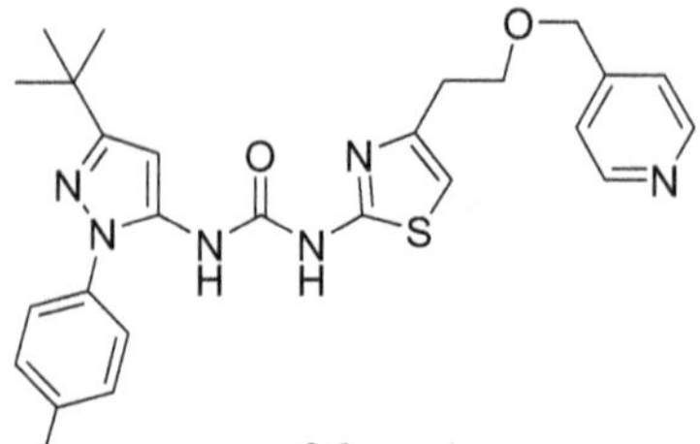

**26**

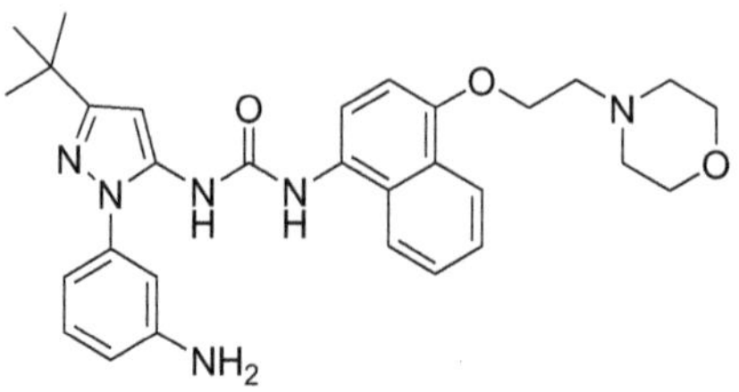

**27**

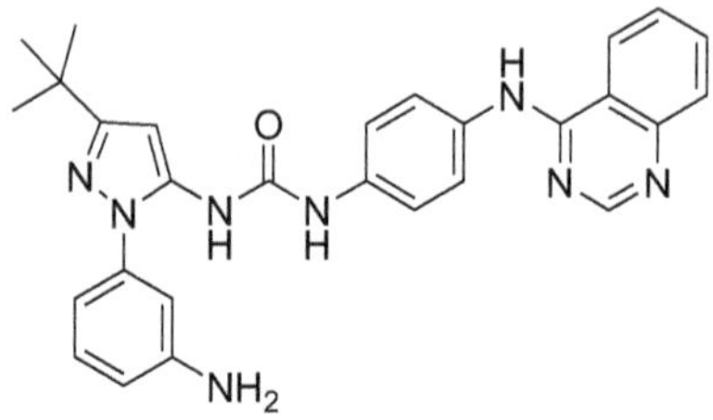

**28**

**Fig. 10.13** Chemical structures of pyrazolo-urea derivatives **25–28**

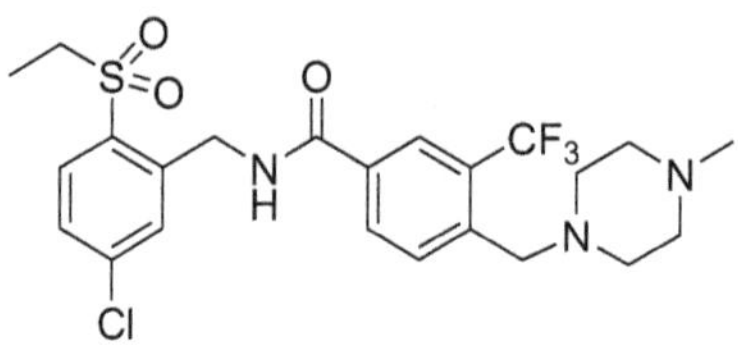

**29**

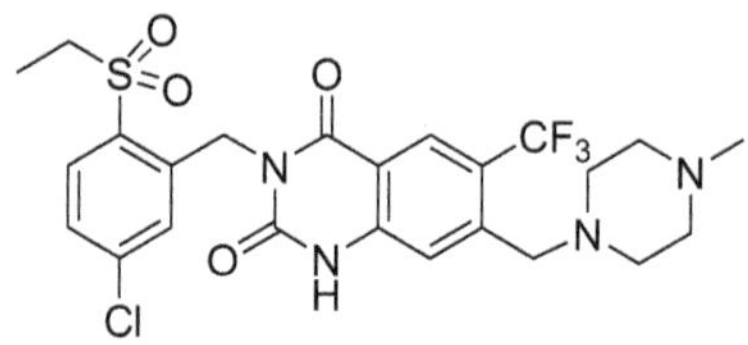

**30**

**Fig. 10.14** Chemical structures of compounds **29** and **30**

DDR1 inhibitors were also discovered in a Japanese group in 2013 (Fig. 10.14) [68, 69]. Representative compounds **29** and **30** exhibit $IC_{50}$ values against DDR1 of 9.7 nM and 21 nM, respectively, in a LanthaScreen Eu kinase binding assay. The compounds are claimed to be potentially useful for the treatment of cancer, fibrosis, and inflammation, but further biological investigation is needed to support their potential therapeutic indications.

### *10.3.3 DDR Signal Pathway Inhibitors with Unknown Mechanism*

In order to obtain new DDR2 signaling pathway inhibitors, MacMillan J. B. and co-workers from the University of Texas Southwestern Medical Center performed a Functional Signature-Based Ontology (FUSION) map screening against a library of

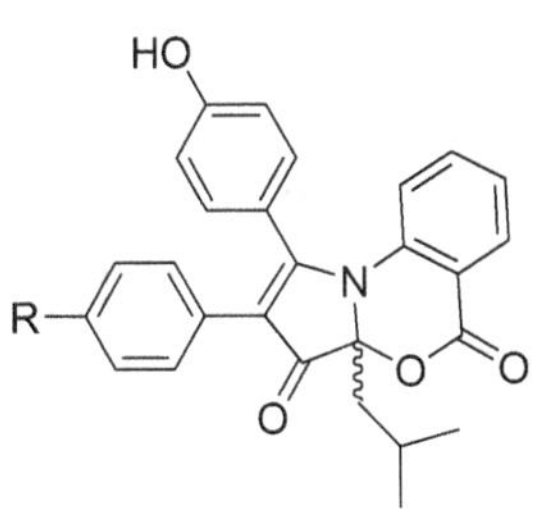

**31 Discoipyrrole A R=OH**
**32 Discoipyrrole B R=H**

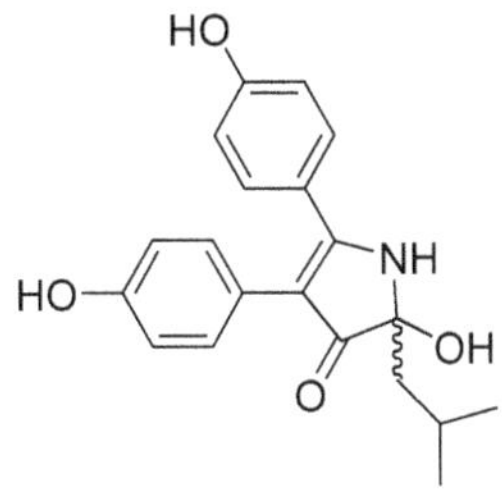

**33 Discoipyrrole C**

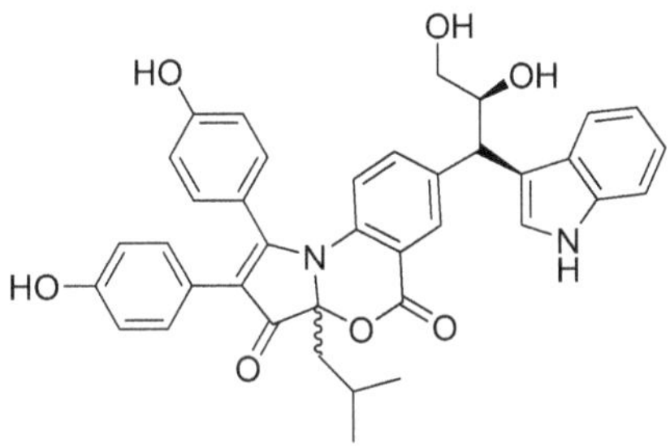

**34 Discoipyrrole D**

**Fig. 10.15** Chemical structures of discoipyrrole derivatives

6500 natural product fractions. A fraction from marine-derived *Bacillus hunanensis* strain SNA-048 was identified to potently inhibit the DDR2 signals in HCT116 colon cancer cells with an identical functional signature to that of DDR2 siRNA. Further investigation reveals that the fraction inhibits DDR2-dependent migration of BR5 fibroblasts through a 2D Matrigel matrix at 1.0 μg/mL. A bioactivity-guided isolation of the fraction led to the identification of discoipyrroles A (**31**), B (**32**), and D (**34**) as novel DDR2 pathway inhibitors with a 3*H*-benzo[*d*]pyrrolo[1,3]oxazine-3,5-dione chemical scaffold (Fig. 10.15) [70]. The compounds potently inhibit BR5 fibroblast migration at 1.0 μM, with discoipyrrole A displaying the most potent inhibition at 95 % followed by discoipyrrole B and discoipyrrole C (**33**) with 85 % inhibition. Interestingly, the structurally simplified compound D also displayed moderate inhibition against the BR5 fibroblast migration at 1.0 μM. The compounds potently inhibited the proliferation of HCC366 non-small cell lung cancer (NSCLC) cells harboring "gain-of-function" DDR2 mutation with $IC_{50}$ values of 0.12, 0.19, 0.70, and 0.275 μM, respectively. But their activities against A549 NSCLC cells with wild-type DDR2 are significantly less potent. Further investigation suggests that the compounds induce a posttranslational modification of DDR2, which is predicted as glycosylation based on electrophoretic mobility. However, further work is necessary to elucidate the mechanistic details for DDR2 inhibition and to identify the molecular target.

## 10.4 Conclusions

Investigations on the biological functions of DDRs suggest them to be novel attractive therapeutic targets for drug discovery against various diseases. Several classes of small molecule inhibitors have been discovered to potently suppress the biological functions of DDRs, which may become new promising candidates for the treatment of cancer, inflammation, and fibrosis. Despite the progress on DDR inhibitor development, medicinal chemistry efforts are still highly worthwhile for further validation of DDRs as drug targets and the development of DDR-based therapies. For instance, it will be of much value to develop novel DDR inhibitors with tissue or pathway specificity because the receptors display distinct expression profiles and diverse functions. Further molecular mechanism study and crystallographic characterization of DDR-DS inhibitor complexes will also be helpful in the understanding of collagen-induced DDR activation and the development of specific DDR inhibitors. Nevertheless, the current progress on DDR inhibitor discovery reviewed in this chapter will provide a solid foundation for further validation of DDRs as potential targets and development of DDR inhibitors as a new therapeutic strategy for various chronic diseases.

## References


1. Johnson JD, Edman JC, Rutter WJ (1993) A receptor tyrosine kinase found in breast carcinoma cells has an extracellular discoidin I-like domain. Proc Natl Acad Sci U S A 90:5677–5681
2. Lai C, Lemke G (1994) Structure and expression of the Tyro 10 receptor tyrosine kinase. Oncogene 9:877–883
3. Alves F, Vogel W, Mossie K, Millauer B, Hofler H, Ullrich A (1995) Distinct structural characteristics of discoidin I subfamily receptor tyrosine kinases and complementary expression in human cancer. Oncogene 10:609–618
4. Lemmon MA, Schlessinger J (2010) Cell signaling by receptor tyrosine kinases. Cell 141:1117–1134
5. Vogel WF, Abdulhussein R, Ford CE (2006) Sensing extracellular matrix: an update on discoidin domain receptor function. Cell Signal 18:1108–1116
6. Perez JL, Jing SQ, Wong TW (1996) Identification of two isoforms of the Cak receptor kinase that are coexpressed in breast tumor cell lines. Oncogene 12:1469–1477
7. Lee JE, Kang CS, Guan XY, Kim BT, Kim SH, Lee YM, Moon WS, Kim DK (2007) Discoidin domain receptor 2 is involved in the activation of bone marrow-derived dendritic cells caused by type I collagen. Biochem Biophys Res Commun 352:244–250
8. Vogel W, Gish GD, Alves F, Pawson T (1997) The discoidin domain receptor tyrosine kinases are activated by collagen. Mol Cell 1:13–23
9. Shrivastava A, Radziejewski C, Campbell E, Kovac L, McGlynn M, Ryan TE, Davis S, Goldfarb MP, Glass DJ, Lemke G, Yancopoulos GD (1997) An orphan receptor tyrosine kinase family whose members serve as non integrin collagen receptors. Mol Cell 1:25–34
10. Valiathan RR, Marco M, Leitinger B, Kleer CG, Fridman R (2012) Discoidin domain receptor tyrosine kinases: new players in cancer progression. Cancer Metastasis Rev 31:295–321
11. Leitinger B (2003) Molecular analysis of collagen binding by the human discoidin domain receptors, DDR1 and DDR2. Identification of collagen binding sites in DDR2. J Biol Chem 278:16761–16769

12. Fu HL, Valiathan RR, Arkwright R, Sohail A, Mihai C, Kumarasiri M, Mahasenan KV, Mobashery S, Huang P, Agarwal G, Fridman R (2013) Discoidin domain receptors: unique receptor tyrosine kinases in collagen-mediated signaling. J Biol Chem 288:7430–7437
13. Hou G, Vogel WF, Bendeck MP (2002) Tyrosine kinase activity of discoidin domain receptor 1 is necessary for smooth muscle cell migration and matrix metalloproteinase expression. Circ Res 90:1147–1149
14. Agarwal G, Mihai C, Iscru DF (2007) Interaction of discoidin domain receptor 1 with collagen type. J Mol Biol 367:443–455
15. Radiola I (2011) Discoidin domain receptors role in human disease. Not Sci Biol 3:7–12
16. Zhang K, Corsa CA, Ponik SM, Prior JL, Piwnica-Worms D, Eliceiri KW, Keely PJ, Longmore GD (2013) The collagen receptor discoidin domain receptor 2 stabilizes SNAIL1 to facilitate breast cancer metastasis. Nat Cell Biol 15:677–687
17. Lu KK, Trcka D, Bendeck MP (2011) Collagen stimulates discoidin domain receptor 1- mediated migration of smooth muscle cells through Src. Cardiovasc Pathol 20:71–76
18. Yeh YC, Lin HH, Tang MJ (2012) A tale of two collagen receptors, integrin beta1 and discoidin domain receptor 1, in epithelial cell differentiation. Am J Physiol Cell Physiol 303:C1207–C1217
19. Vogel WF, Aszódi A, Alves F, Pawson T (2001) Discoidin domain receptor 1 tyrosine kinase has an essential role in mammary gland development. Mol Cell Biol 21:2906–2917
20. Kano K, Kitamura A, Matsuwaki T, Morimatsu M, Naito K (2010) Discoidin domain receptor 2 (DDR2) is required for maintenance of spermatogenesis in male mice. Mol Reprod Dev 77:29–37
21. Hammerman PS, Sos ML, Ramos AH, Xu C, Dutt A, Zhou W, Brace LE, Woods BA, Lin W, Zhang J, Deng X, Lim SM, Heynck S, Peifer M, Simard JR, Lawrence MS, Onofrio RC, Salvesen HB, Seidel D, Zander T, Heuckmann JM, Soltermann A, Moch H, Koker M, Leenders F, Gabler F, Querings S, Ansén S, Brambilla E, Brambilla C, Lorimier P, Brustugun OT, Helland A, Petersen I, Clement JH, Groen H, Timens W, Sietsma H, Stoelben E, Wolf J, Beer DG, Tsao MS, Hanna M, Hatton C, Eck MJ, Janne PA, Johnson BE, Winckler W, Greulich H, Bass AJ, Cho J, Rauh D, Gray NS, Wong KK, Haura EB, Thomas RK, Meyerson M (2011) Mutations in the DDR2 kinase gene identify a novel therapeutic target in squamous cell lung cancer. Cancer Discovery 1:78–89
22. Olaso E, Labrador JP, Wang L, Ikeda K, Eng FJ, Klein R, Lovett DH, Lin HC, Friedman SL (2002) Discoidin domain receptor 2 regulates fibroblast proliferation and migration through the extracellular matrix in association with transcriptional activation of matrix metalloproteinase-2. J Biol Chem 277:3606–3613
23. Yang SH, Baek HA, Lee HJ, Park HS, Jang KY, Kang MJ, Lee DG, Lee YC, Moon WS, Chung MJ (2010) Discoidin domain receptor 1 is associated with poor prognosis of non-small cell lung carcinomas. Oncol Rep 24:311–319
24. Shimada K, Nakamura M, Ishida E, Higuchi T, Yamamoto H, Tsujikawa K, Konishi N (2008) Prostate cancer antigen-1 contributes to cell survival and invasion through discoidin receptor 1 in human prostate cancer. Cancer Sci 99:39–45
25. Couvelard A, Hu J, Steers G, O'Toole D, Sauvanet A, Belghiti J, Bedossa P, Gatter K, Ruszniewski P, Pezzella F (2006) Identification of potential therapeutic targets by gene-expression profiling in pancreatic endocrine tumors. Gastroenterology 131:1597–1610
26. Ford CE, Lau SK, Zhu CQ, Andersson T, Tsao MS, Vogel WF (2007) Expression and mutation analysis of the discoidin domain receptors 1 and 2 in non-small cell lung carcinoma. Br J Cancer 96:808–814
27. Barker KT, Martindale JE, Mitchell PJ, Kamalati T, Page MJ, Phippard DJ, Dale TC, Gusterson BA, Crompton MR (1995) Expression patterns of the novel receptor-like tyrosine kinase, DDR, in human breast tumours. Oncogene 10:569–575
28. Quan J, Yahata T, Adachi S, Yoshihara K, Tanaka K (2011) Identification of receptor tyrosine kinase, discoidin domain receptor 1 (DDR1), as a potential biomarker for serous ovarian cancer. Int J Mol Sci 12:971–982
29. Xu L, Peng H, Wu D, Hu K, Goldring MB, Olsen BR, Li Y (2005) Activation of the discoidin domain receptor 2 induces expression of matrix metalloproteinase 13 associated with osteoarthritis in mice. J Biol Chem 280:548–555

30. Poudel B, Ki HH, Lee YM, Kim DK (2013) Induction of IL-12 production by the activation of discoidin domain receptor 2 via NF-kB and JNK pathway. Biochem Biophys Res Commun 434:584–588
31. Mihai C, Iscru DF, Druhan LJ, Elton TS, Agarwal G (2006) Discoidin domain receptor 2 inhibits fibrillogenesis of collagen type 1. J Mol Biol 361:864–876
32. Olaso E, Arteta B, Benedicto A, Crende O, Friedman SL (2011) Loss of discoidin domain receptor 2 promotes hepatic fibrosis after chronic carbon tetrachloride through altered paracrine interactions between hepatic stellate cells and liver-associated macrophages. Am J Pathol 179:2894–2904
33. Borza CM, Pozzi A (2014) Discoidin domain receptors in disease. Matrix Biol 34:185–192.
34. Day E, Waters B, Spiegel K, Alnadaf T, Manley PW, Buchdunger E, Walker C, Jarai G (2008) Inhibition of collagen-induced discoidin domain receptor 1 and 2 activation by imatinib, nilotinib and dasatinib. Eur J Pharmacol 599:44–53
35. Ichikawa O, Osawa M, Nishida N, Goshima N, Nomura N, Shimada I (2007) Structural basis of the collagen-binding mode of discoidin domain receptor 2. EMBO J 26:4168–4176
36. Carafoli F, Bihan D, Stathopoulos S, Konitsiotis AD, Kvansakul M, Farndale RW, Leitinger B, Hohenester E (2009) Crystallographic insight into collagen recognition by discoidin domain receptor 2. Structure 17:1573–1581
37. Xu H, Raynal N, Stathopoulos S, Myllyharju J, Farndale RW, Leitinger B (2011) Collagen binding specificity of the discoidin domain receptors: binding sites on collagens II and III and molecular determinants for collagen IV recognition by DDR1. Matrix Biol 30:16–26
38. Abdulhussein R, McFadden C, Fuentes-Prior P, Vogel WF (2004) Exploring the collagen-binding site of the DDR1 tyrosine kinase receptor. J Biol Chem 279:31462–31470
39. Leitinger B, Steplewski A, Fertala A (2004) The D2 period of collagen II contains a specific binding site for the human discoidin domain receptor, DDR2. J Mol Biol 344:993–1003
40. Leitinger B, Kwan AP (2006) The discoidin domain receptor DDR2 is a receptor for type X collagen. Matrix Biol 25:355–364
41. Carafoli F, Mayer MC, Shiraishi K, Pecheva MA, Chan LY, Nan R, Leitinger B, Hohenester E (2012) Structure of the discoidin domain receptor 1 extracellular region bound to an inhibitory Fab fragment reveals features important for signaling. Structure 20:688–697
42. Carafoli F, Hohenester E (2013) Collagen recognition and transmembrane signaling by discoidin domain receptors. Biochim Biophys Acta 1834:2187–2194
43. Curat CA, Eck M, Dervillez X, Vogel WF (2001) Mapping of epitopes in discoidin domain receptor 1 critical for collagen binding. J Biol Chem 276:45952–45958
44. Noordeen NA, Carafoli F, Hohenester E, Horton MA, Leitinger B (2006) A transmembrane leucine zipper is required for activation of the dimeric receptor tyrosine kinase DDR1. J Biol Chem 281:22744–22751
45. Alves F, Saupe S, Ledwon M, Schaub F, Hiddemann W, Vogel WF (2001) Identification of two novel, kinase-deficient variants of discoidin domain receptor 1: differential expression in human colon cancer cell lines. FASEB J 15:1321–1323
46. Canning P, Elkins JM, Goubin S, Mahajan P, Bradley A, Coutandin D, Von Delft F, Arrowsmith CH, Edwards AM, Bountra C, Bullock A (2014) Structural mechanisms determining inhibition of the collagen receptor DDR1 by selective and multi-targeted type II kinase inhibitors. J Mol Biol 426:2457–2470
47. Kim HG, Tan L, Weisberg EL, Liu F, Canning P, Choi HG, Ezell SA, Wu H, Zhao Z, Wang J, Mandinova A, Griffin JD, Bullock AN, Liu Q, Lee SW, Gray NS (2013) Discovery of a potent and selective DDR1 receptor tyrosine kinase inhibitor. ACS Chem Biol 8:2145–2150
48. Siddiqui K, Kim GW, Lee DH, Shin HR, Yang EG, Lee NT, Yang BS (2009) Actinomycin D identified as an inhibitor of discoidin domain receptor 2 interaction with collagen through an insect cell based screening of a drug compound library. Biol Pharm Bull 32:136–141
49. Ram R, Lorente G, Nikolich K, Urfer R, Foehr E, Nagavarapu U (2006) Discoidin domain receptor-1a (DDR1a) promotes glioma cell invasion and adhesion in association with matrix metalloproteinase-2. J Neurooncol 76:239–248

50. Castro-Sanchez L, Soto-Guzman A, Navarro-Tito N, Martinez-Orozco R, Salazar EP (2010) Native type IV collagen induces cell migration through a CD9 and DDR1-dependent pathway in MDA-MB-231 breast cancer cells. Eur J Cell Biol 89:843–852
51. Sobell H (1985) Actinomycin and DNA transcription. Proc Natl Acad Sci U S A 82:5328–5331
52. Bantscheff M, Eberhard D, Abraham Y, Bastuck S, Boesche M, Hobson S, Mathieson T, Perrin J, Raida M, Rau C, Reader V, Sweetman G, Bauer A, Bouwmeester T, Hopf C, Kruse U, Neubauer G, Ramsden N, Rick J, Kuster B, Drewes G (2007) Quantitative chemical proteomics reveals mechanisms of action of clinical ABL kinase inhibitors. Nat Biotechnol 25:1035–1044
53. Hantschel O, Rix U, Superti-Furga G (2008) Target spectrum of the BCR-ABL inhibitors imatinib, nilotinib and dasatinib. Leuk Lymphoma 49:615–619
54. Rix U, Hantschel O, Durnberger G, RemsingRix LL, Planyavsky M, Fernbach NV, Kaupe I, Bennett KL, Valent P, Colinge J, Köcher T, Superti-Furga G (2007) Chemical proteomic profiles of the BCR-ABL inhibitors imatinib, nilotinib, and dasatinib reveal novel kinase and nonkinase targets. Blood 110:4055–4063
55. Cowan-Jacob SW, Fendrich G, Floersheimer A, Furet P, Liebetanz J, Rummel G, Rheinberger P, Centeleghe M, Fabbro D, Manley PW (2007) Structural biology contributions to the discovery of drugs to treat chronic myelogenous leukaemia. Acta Crystallogr D Biol Crystallogr 63:80–93
56. Trial of Dasatinib in Subjects With Advanced Cancers Harboring DDR2 Mutation or Inactivating B-RAF Mutation. (2012) http://clinicaltrials.gov/show/NCT01514864
57. Pitini V, Arrigo C, Di Mirto C, Mondello P, Altavilla G (2013) Response to dasatinib in a patient with SQCC of the lung harboring a discoid receptor 2 and synchronous chronic myelogenous leukemia. Lung Cancer 82:171–172
58. Dasatinib in Advanced Squamous Cell Lung Cancer (2011) http://clinicaltrials.gov/ct2/show/NCT01491633.
59. Huang WS, Metcalf CA, Sundaramoorthi R, Wang Y, Zou D, Thomas RM, Zhu X, Cai L, Wen D, Liu S, Romero J, Qi J, Chen I, Banda G, Lentini SP, Das S, Xu Q, Keats J, Wang F, Wardwell S, Ning Y, Snodgrass JT, Broudy MI, Russian K, Zhou T, Commodore L, Narasimhan NI, Mohemmad QK, Iuliucci J, Rivera VM, Dalgarno DC, Sawyer TK, Clackson T, Shakespeare WC (2010) Discovery of 3-[2-(imidazo[1,2-b]pyridazin-3-yl)ethynyl]-4-methyl-N-{4-[(4-methyl-piperazin-1-yl)methyl]-3-(trifluoromethyl)phenyl}benzamide (AP24534), a potent, orally active pan-inhibitor of breakpoint cluster region-abelson (BCR-ABL) kinase including the T315I gatekeeper mutant. J Med Chem 53:4701–4719
60. O'Hare T, Shakespeare WC, Zhu X, Eide CA, Rivera VM, Wang F, Adrian LT, Zhou T, Huang WS, Xu Q, Metcalf CA, Tyner JW, Loriaux MM, Corbin AS, Wardwell S, Ning Y, Keats JA, Wang Y, Sundaramoorthi R, Thomas M, Zhou D, Snodgrass J, Commodore L, Sawyer TK, Dalgarno DC, Deininger MW, Druker BJ, Clackson T (2009) AP24534, a pan-BCR-ABL inhibitor for chronic myeloid leukemia, potently inhibits the T315I mutant and overcomes mutation-based resistance. Cancer Cell 16:401–412
61. Rix U, Remsing Rix LL, Terker AS, Fernbach NV, Hantschel O, Planyavsky M, Breitwieser FP, Herrmann H, Colinge J, Bennett KL, Augustin M, Till JH, Heinrich MC, Valent P, Superti-Furga G (2010) A comprehensive target selectivity survey of the BCR-ABL kinase inhibitor INNO-406 by kinase profiling and chemical proteomics in chronic myeloid leukemia cells. Leukemia 24:44–50
62. Xm R, Pan XF, Zhang Z, Wang DP, Lu XY, Li YP, Wen DH, Long HY, Luo JF, Feng YB, Zhuang XX, Zhang FX, Liu JQ, Leng F, Lang XF, Bai Y, She MQ, Tu ZC, Pan JX, Ding K (2013) Identification of GZD824 as an orally bioavailable inhibitor that targets phosphorylated and nonphosphorylated breakpoint cluster region-Abelson (Bcr-Abl) kinase and overcomes clinically acquired mutation-induced resistance against imatinib. J Med Chem 56:879–894
63. Davis MI, Hunt JP, Herrgard S, Ciceri P, Wodicka LM, Pallares G, Hocker M, Treiber DK, Zarrinkar PP (2011) Comprehensive analysis of kinase inhibitor selectivity. Nat Biotechnol 29:1046–1051

64. Yang BS, Kim YZ, Park TK, Woo SH, Lee HS, Kim SY, Cho JU, Lee HB (2013) Anti-inflammatory compound having inhibitory activity against multiple tyrosine kinases and pharmaceutical composition containing same. US Patent 20130072482, 21 Mar 2013.
65. Sun X, Phan TN, Jung SH, Kim SY, Cho JU, Lee HS, Woo SH, Park TK, Yang BS (2012) LCB 03–0110, a novel pan-discoidin domain receptor/c-Src family tyrosine kinase inhibitor, suppresses scar formation by inhibiting fibroblast and macrophage activation. J Pharmacol Exp Ther 340:510–519
66. Gao MS, Duan L, Luo JF, Zhang LW, Lu XY, Zhang Y, Zhang Z, Tu ZC, Xu Y, Ren XM, Ding K (2013) Discovery and optimization of 3-(2-(pyrazolo[1,5-a]pyrimidin-6-yl)ethynyl) benzamides as novel selective and orally bioavailable discoidin domain receptor 1 (DDR1) inhibitors. J Med Chem 56:3281–3295
67. Rachters A, Nguyen HD, Phan T, Simard JR, Julian CG, Rauh ED (2014) Identification type II and type III DDR2 inhibitors. J Med Chem 57:4252–4262
68. Takeshi M, Satoshi N, Sousuke H, Hatsuo K, Kihito H, Hideaki S, Hiroshi T, Yoshito N (2013) Preparation of benzamide derivatives as discoidin domain receptor 1 (DDR1) inhibitors. WO 2013161851, 31 Oct 2013.
69. Takeshi M, Hatsuo K, Satoshi, N, Sousuke H, Kihito H, Hideaki S, Hiroshi T, Toshiyuki M (2013) Preparation of quinazolinedione derivatives as discoidin domain receptor 1 (DDR1) inhibitors. WO 2013161853. 31 Oct 2013.
70. Hu Y, Potts MB, Colosimo D, Herrera-Herrera ML, Legako AG, Yousufuddin M, White MA, MacMillan JB, Hu Y, Potts MB, Colosimo D, Herrera-Herrera ML, Legako AG, Yousufuddin M, White MA, MacMillan JB (2013) Discoipyrroles A-D: isolation, structure determination, and synthesis of potent migration inhibitors from Bacillus hunanensis. J Am Chem Soc 135:13387–13392

# Chapter 11
# Discoidin Domain Receptor Signalling Networks

**Leo K. Iwai, Leo S. Payne, Dina Allam, and Paul H. Huang**

## 11.1 Introduction

The discoidin domain receptors (DDRs), comprising DDR1 and DDR2, are an atypical class of receptor tyrosine kinases (RTKs) that bind to and are activated by collagen in the extracellular matrix (ECM) [1, 2]. Upon collagen binding, these receptors display unique phosphorylation dynamics that are characterised by delayed and sustained activation, such that activation is initiated in the timescale of hours rather than minutes and is sustained for days [3]. As a result of work from multiple research groups over the past 15 years, we have gained some understanding of the intracellular proteins that interact with and are responsible for propagating signalling downstream of these receptors. The development of global profiling approaches such as transcriptomic and proteomic analyses has also contributed to our increasing collective knowledge regarding the signalling networks activated by the DDRs. While there are still clear gaps in our understanding of these receptors, the experimental data available to date has generated a molecular portrait of some of the critical signalling pathways that mediate the biological functions of these unique receptors [4].

L.K. Iwai, M.Sc., Ph.D.
Laboratório Especial de Toxinologia Aplicada (LETA), Center of Toxins, Immune-response and Cell Signaling (CeTICS), Instituto Butantan, Av Vital Brasil 1500, São Paulo 05503-000, Brazil

L.S. Payne, Ph.D.
School of Biological Sciences, The University of Auckland, Thomas Building, 3A Symonds Street Auckland, New Zealand

D. Allam
UCL School of Public Policy, 29/31 Tavistock Square, London WC1H9QU, UK

P.H. Huang, Ph.D. (✉)
Division of Cancer Biology, Institute of Cancer Research, 237 Fulham Road, London SW3 6JB, UK
e-mail: paul.huang@icr.ac.uk



R. Fridman, P.H. Huang (eds.), *Discoidin Domain Receptors in Health and Disease*, DOI 10.1007/978-1-4939-6383-6_11

In this chapter, we summarise our current knowledge of DDR signalling across multiple biological systems. We describe the specific interactions between phosphotyrosine sites on the DDRs with signalling adaptors, kinases and phosphatases and the experimental approaches that have been employed to map these binding events. Additionally, we detail the signalling pathways that are initiated upon collagen activation of these receptors. We then highlight reported crosstalk between the DDRs and receptor signalling systems such as the integrins and Wnt pathways. An increasing body of evidence suggests that at least some of the signalling functions of the DDRs occur via phosphorylation- and kinase-independent mechanisms, and we elaborate on the findings from these recent studies. It should be noted that in many cases, the reported DDR signalling data is far from unified; in fact conflicting reports are common in the field. Where applicable, we highlight such inconsistencies and the context-dependent nature of DDR signalling.

## 11.2 Discoidin Domain Receptor 1 (DDR1)

Like most RTKs, ligand binding results in kinase activation and phosphorylation of multiple tyrosine residues within the intracellular domains of the DDRs. In the case of DDR1, there are 15 tyrosine residues located in the juxtamembrane region, kinase domain and C-terminal tail that have the potential to be phosphorylated in this manner [4]. In this section, we first detail the critical phosphotyrosine residues on DDR1 and the key interacting proteins that bind to these phosphorylation sites and initiate the propagation of downstream signalling. Known canonical and non-canonical signalling pathways are also discussed in the context of their roles in driving DDR1 biological function.

### *11.2.1 Receptor Phosphorylation and Key Binding Partners*

The interactions between DDR1 and the key binding partners discussed in this section are summarised in Fig. 11.1. The most well-characterised binding partner for DDR1 is the SHC1 (Src homology 2 domain-containing transforming protein 1) adaptor protein which contains an N-terminal phosphotyrosine binding (PTB) domain and a C-terminal Src homology 2 (SH2) domain, both of which are capable of interacting with phosphotyrosine residues on RTKs. Through the use of multiple assays including site-directed mutagenesis, co-immunoprecipitation and quantitative mass spectrometry, it has been rigorously demonstrated that SHC1 binds to the phosphorylated Y513 site in the DDR1 juxtamembrane domain [3, 5]. Recruitment of SHC1 leads to the initiation of signal transduction pathways involving mitogen-activated protein kinase (MAPK) and nuclear factor kB (NFkB) (discussed in more detail in the following sections). In addition to SHC1, other well-characterised SH2 or PTB domain-containing adaptor proteins have also been shown to bind DDR1 in a phosphotyrosine-dependent manner. Using bacterial two-hybrid screens and

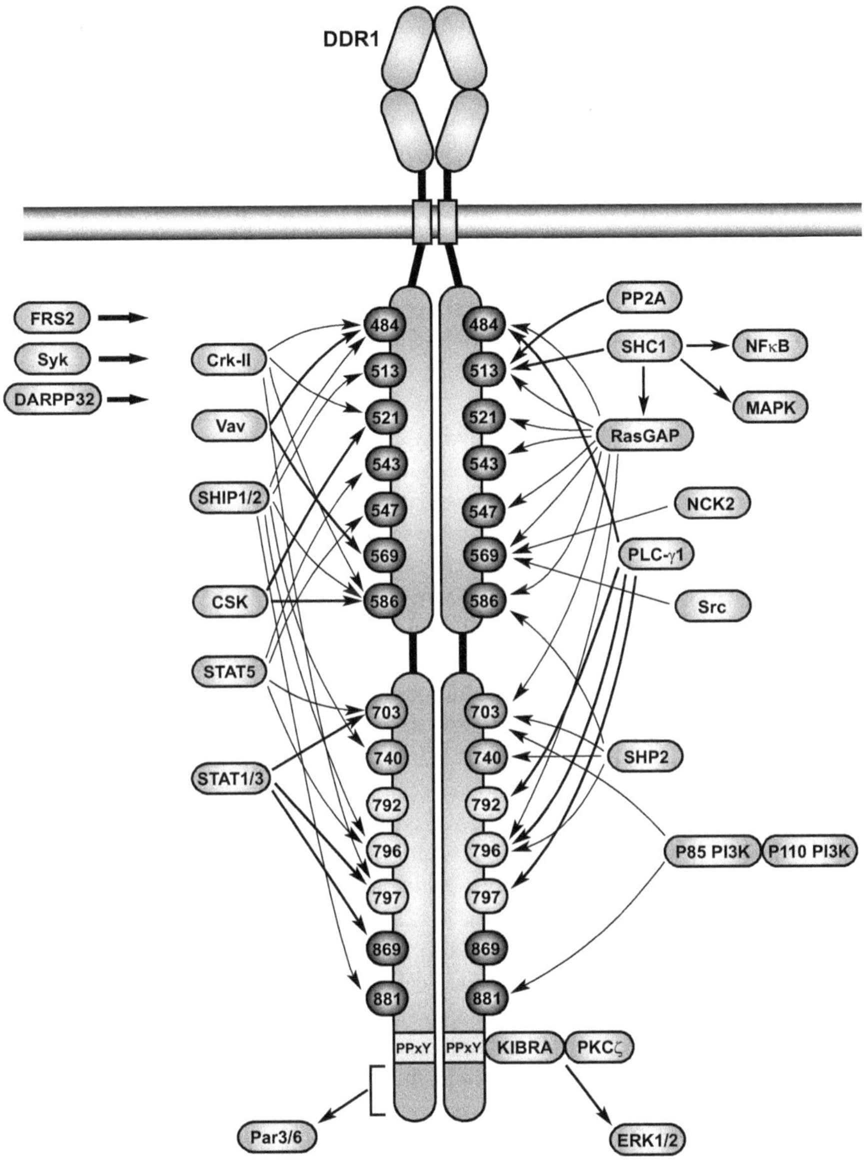


**Fig. 11.1** Reported protein–protein interactions and downstream signalling components of DDR1. The figure depicts known tyrosine phosphorylation sites (*numbered circles*) in the cytoplasmic domains of DDR1 and the downstream effector proteins that have been shown to either interact or be phosphorylated by the receptor. Collagen-dependent but phosphotyrosine-independent interactions (e.g., via the PPxY motif) are also displayed

co-immunoprecipitation in mammalian cell lines, Vogel and colleagues demonstrated that the NCK2 adaptor protein binds to activated DDR1 [6]. Consensus motif analysis of the binding preference for the SH2 domain of NCK2 revealed Y484 in the DDR1 juxtamembrane domain as the potential phosphosite responsible for this interaction. However, subsequent phosphopeptide affinity purification and analysis by mass spectrometry performed by Lemeer et al. showed that in vitro, NCK2 bound preferentially to Y569 in the juxtamembrane domain [5]. Interestingly, in the same study the Y484 site was identified as a binding site for the Crk-II adaptor protein and the Vav2 and Vav3 guanidine nucleotide exchange factors, which can also function as adaptor proteins. Additional in vitro binding sites for Crk-II were also identified at Y521, Y586 and Y797 and for Vav2 at Y569 [5]. Employing engineered chimera receptors composed of distinct domains of the platelet-derived growth factor receptor (PDGFR), DDR1 and tropomyosin receptor kinase A (TrkA) receptors, Foehr et al. found that the juxtamembrane portion of DDR1 was capable of recruiting the fibroblast growth factor receptor substrate 2 (FRS2) adaptor protein [7]. However, it was not established if the interaction is dependent on a specific phosphotyrosine residue in this region.

In addition to adaptor proteins, DDR1 is also capable of recruiting a number of phosphatases, including src homology domain-containing phosphatase 2 (SHP-2) and SH2 domain-containing inositol-5′-phosphatases-1 and inositol-5′-phosphatases-2 (SHIP-1 and SHIP-2). The interaction between DDR1 and the protein tyrosine phosphatase SHP-2 has been described by two groups. In the first study, Koo et al. showed that DDR1 binds to SHP-2 specifically via the Y740 site within the kinase domain [6]. In contrast, studies using Madin Darby canine kidney (MDCK) cells indicated that DDR1 interacts with SH2 and PTP domains of SHP-2 through the Y703 and Y796 sites [8]. A subsequent phosphopeptide affinity purification study found that SHP-2 derived from placenta lysates binds tightly to Y586 and Y740 on DDR1 [5]. Taken together, these studies suggest that DDR1 does indeed bind to SHP-2; however, the exact sites through which this interaction occurs may be context dependent, being influenced by, for example, the specific cell type being studied or the activation state of SHP-2. SHIP-1 and SHIP-2 are inositol phosphatases that are critical regulators of biological signal transduction networks, with the expression of the former being restricted to haematopoietic cells. Lemeer et al. found that these proteins showed a strong preference for the Y740 site, but also bound to Y484, Y513, Y586 and Y796 and Y881 with high affinity [5]. The biological consequence of DDR1 and SHIP-1/-2 interactions is currently unknown and requires further investigation.

DDR1 interacts directly with multiple kinases. For instance, the DDR1 kinase domain contains an YXXM binding motif for association with the SH2 domain of the p85 regulatory subunit of phosphoinositide-3-kinase (PI3K). Interactions between DDR1 and this subunit have been identified upon collagen stimulation in the T47D breast cancer cell line [9]. Phosphopeptide pull-down analysis localised the binding of the p85 subunit in complex with the p110 subunit to the consensus binding motif at Y881 (pYELM) and, to a lesser extent, to the non-consensus motif at Y703 (pYMEN) [5]. Src kinase has been shown to interact with DDR1 in smooth muscle cells [10]. Lu et al. demonstrated that Src co-immunoprecipitates with DDR1, and furthermore, receptor tyrosine phosphorylation is dependent on Src activity. Lameer et al. localised this binding event to Y569 on the juxtamembrane of the receptor [5]. In a proteomic analysis of interactors of CSK (C-terminal Src tyrosine kinase, the negative regulator of Src), DDR1 was found to bind to this kinase via its SH2 domain [11]. In the reciprocal experiment, CSK was found to bind to Y521 and Y586 of DDR1 in a phosphopeptide pull-down analysis [5]. Finally Neuhaus et al. reports that the non-receptor tyrosine kinase Syk (spleen tyrosine kinase) constitutively interacts with, and is activated by, DDR1, which results in the potent inhibition of cell migration [12].

Signal transducers and activators of transcription (STATs) are transcription factors that link receptor tyrosine phosphorylation directly to gene transcription. In mammary epithelial HC11 cells, DDR1 has been shown to transiently interact and phosphorylate STAT5 in conjugation with prolactin-induced Janus kinase 2 (Jak2) activity [13]. Lameer et al. showed that STAT5b interacts with a number of phosphotyrosine sites on the receptor including Y543, Y547, Y703 and Y796 [5]. It should be noted that these interactions were not responsive to phosphopeptide competition experiments, suggesting that the reported interactions with DDR1 may not be specific to STAT5 or that these binding events may be mediated by indirect interactions via another protein. In addition to STAT5, DDR1 also regulates the phosphorylation of STAT1 and STAT3. Wang et al. showed that DDR1 inhibited STAT1/3 activation via SHP-2 binding to Y703 or Y796 [8]. This interaction antagonises collagen-induced α2β1 integrin-mediated cell migration. Mutation at either Y703 or Y796 sites prevented SHP-2 recruitment and restored STAT1/3 tyrosine phosphorylation. Interestingly, phosphopeptide pull-down and competition binding experiments show that STAT1/3 is capable of binding directly to DDR1 at Y703, Y797 and Y869 [5]. This discrepancy between the cellular data and in vitro pull-down experiments may potentially be explained by the fact that the cellular context is often overlooked in in vitro experiments and spatial effects such as subcellular compartmentalisation of signalling effector proteins are not preserved in such studies.

In their proteomic analysis of DDR1 interactions, Lameer et al. showed that Ras GTPase-activating protein (RasGAP) and phospholipase C gamma 1 (PLCγ1), two proteins that have previously not been implicated with DDR1 signalling, were capable of binding directly to phosphotyrosine peptides representing activated sites within the intracellular domain of DDR1 [5]. RasGAP was found to be one of the most abundant proteins that interact with DDR1, with a particular preference for the Y484 site. However, this interaction could not be recapitulated in cellular immunoprecipitation experiments with a DDR1 antibody. It is therefore unclear if RasGAP plays a functional role in DDR1 biology. PLCγ1 is a phospholipase enzyme that propagates signalling downstream of other receptor tyrosine kinases, including the epidermal growth factor receptor (EGFR) and PDGFR. Peptide pull-down assays revealed that PLCγ1 was capable of binding Y484 and to a lesser extent activation loop sites Y792/Y796/Y797.

### 11.2.2 Phosphotyrosine-Independent Interactions

A number of proteins interact with DDR1 in a collagen-dependent but phosphotyrosine-independent manner. For instance, KIBRA (kidney and brain expressed protein) is a scaffold protein that is characterised by two N-terminal WW domains which recognise specific proline-containing sequences and contains a conserved membrane-targeting C2 domain. KIBRA was found to interact with the WW domain binding motif PPxY in DDR1 [14]. This interaction was lost upon activation of DDR1 by collagen I or IV. DDR1 and KIBRA form a trimeric complex with protein kinase C zeta (PKCζ), a protein that regulates fundamental cellular processes including proliferation and cell motility. Collagen activation and subsequent dissociation of the trimeric complex result in activation of the ERK1/2 MAPK pathway.

Dopamine- and cAMP-regulated neuronal phosphoprotein, molecular mass 32 kDa (DARPP-32) acts as a phosphatase inhibitor when phosphorylated at T34 by protein kinase A (PKA) and as a kinase inhibitor upon T75 phosphorylation by cyclin-dependent kinase (Cdk)-1 or Cdk-5. Vogel and colleagues used a yeast two-hybrid screening system to identify DARPP-32 as a DDR1-interacting protein [15]. Validation experiments in human mammary epithelial cells revealed that DARPP-32 constitutively interacts with the juxtamembrane domain of DDR1 prior to collagen stimulation and the interaction is reduced following collagen activation. DARPP-32 was not expressed or poorly expressed in multiple breast cancer cell lines, and forced expression of this protein impaired cancer cell migration downstream of DDR1. These findings suggest that loss of DARPP-32 expression may play a key role in cancer cell motility and invasion.

Hidalgo-Carcedo et al. showed that DDR1 is localised at cell–cell contacts in A431 cells and regulates collective cell migration. Using a candidate-based approach, the authors found that the C-terminal of DDR1 binds specifically to the PDZ domains of the cell polarity regulators protease-activated receptors 3 and 6 (Par3 and Par6) [16]. Intriguingly, this interaction is independent of DDR1 binding to collagen as well as receptor kinase activity consistent with its localisation at cell–cell junctions (where collagen is not present). The authors found that this complex localises RhoE at cell–cell junctions through a yet-to-be determined mechanism which promotes collective cell invasion in 3D, 2D and organotypic models.

In summary, these three studies highlight that in addition to phosphotyrosine-mediated signalling, phosphorylation-independent functions play important roles in mediating the biological functions of DDR1, in particular its regulation of cancer cell migration.

### *11.2.3 Downstream Signalling Pathways*

The downstream signalling pathways activated by DDR1 are poorly characterised, and our knowledge is fragmentary at best. The MAPK pathways are activated downstream of DDR1 as demonstrated by several groups in a series of studies. For instance, in mouse microglial cells, collagen activation of DDR1 increased the production of nitric oxide, Cox-2, CD40 and MMP-9 via the activation of the p38 and JNK MAPK as well as the $NF_{k}B$ pathway [17]. In DDR1 knockout mice, bleomycin-induced lung fibrosis failed to upregulate the p38 MAPK pathway compared to wild-type mice [18]. Smooth muscle cells from DDR1 null mice were defective in ERK1/2 pathway activation when challenged with collagen [10]. In contrast, DDR1 negatively regulated the growth of mouse mesangial cells, through the constitutive suppression of MAPK signalling [19]. Depletion of DDR1 in these cells resulted in the upregulation of ERK1/2 and p38 and subsequently increased cell proliferation. Similarly, in mesenchymal stem cells cultured in 3D collagen gels, DDR1 suppressed the phosphorylation of ERK1/2, accompanied by increased osteoblast differentiation [20]. Intriguingly, a report by L'Hote et al. showed that in T47D breast cancer cells, activation of DDR1 by collagen failed to stimulate the ERK1/2 MAPK pathway [9].

Ongusaha et al. identified DDR1 as a response gene for the p53 tumour suppressor protein [21]. The authors found that p53 induces DDR1 expression and activation in response to DNA damage and that DDR1 is a direct p53 transcription target. The receptor acts in a positive feedback loop where DDR1 activation results in the initiation of the RAS/RAF/MAPK cascade leading to the accumulation of $p19^{ARF}$ and increased levels of p53 and its targets. DDR1 was also shown to protect the cells from p53-induced stress by promoting cellular arrest rather than caspase-dependent apoptosis. Taken together, the studies described above suggest that the regulation of the MAPK pathway downstream of DDR1 is not universal and likely to be cell-type specific and context dependent.

### *11.2.4 Crosstalk with Other Receptor Signalling Systems*

Wnt-5a is a tumour suppressor that belongs to the wingless-related integration site (*Wnt*) gene family. Studies in mammary epithelial cells revealed Wnt-5a expression enhances adhesion and collagen-induced activation of DDR1, implicating it as a downstream component of the Wnt pathway [22]. Transactivation of DDR1 by Wnt-5a is thought to occur through the non-canonical Wnt/$Ca^{2+}$ pathway, potentially involving PKC activation. Dejmek et al. showed that heterotrimeric G protein ($G_{i/o}$) signalling mediates Wnt-5a activation of DDR1 and that Wnt-5a stimulation of Src is essential for DDR1 activation [23] (Fig. 11.2). Furthermore, transforming growth factor β (TGF-β) has been reported to promote DDR1 activation by increasing Wnt-5a expression levels to regulate mammary gland development and branching [24]. It has been suggested that Wnt-5a loss in mammary epithelial cancer cells leads to reduced DDR1 signalling and adhesion, enabling tumour cells to break away from the basement membrane resulting in distant metastasis, although this hypothesis remains to be experimentally proven [25].

Despite binding to different motifs on collagen, DDR1 has been shown to interact with the collagen binding integrins α1β1 and α2β1 through the convergence of downstream signalling pathways [26, 27]. For instance, N-cadherin is overexpressed during epithelial–mesenchymal transition (EMT) in human pancreatic cancer cells in response to collagen I, and increased N-cadherin expression in tumour cells has been reported to increase metastasis and aggressive phenotype. A study on pancreatic cancer BxPC3 cells treated with collagen I activation revealed that N-cadherin upregulation is dependent on both the DDR1 and α2β1 integrin signalling pathways [28]. In

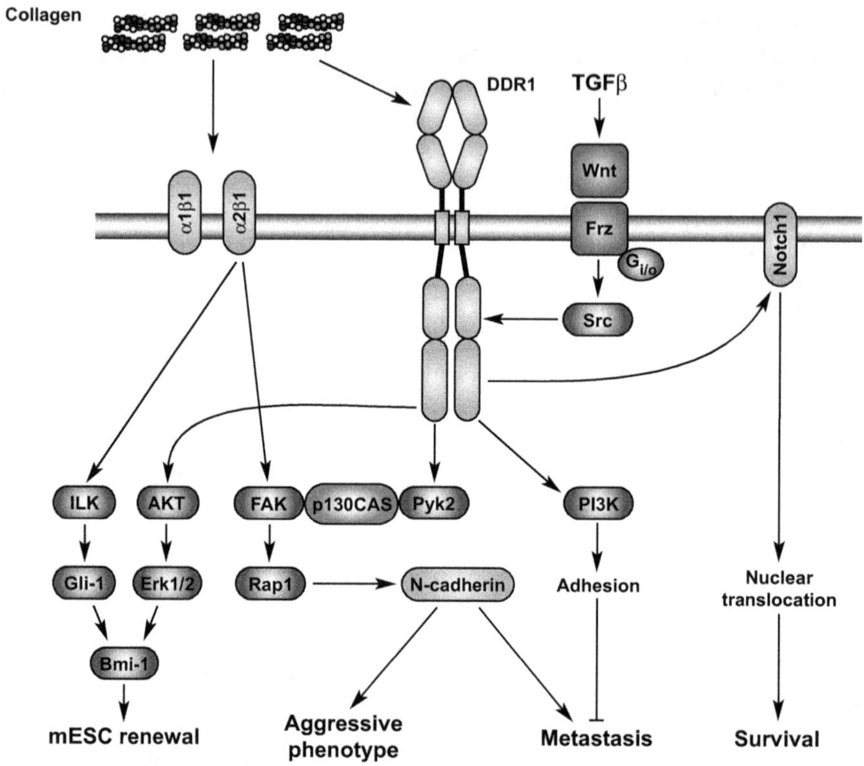


**Fig. 11.2** Reported crosstalk of DDR1 with other receptor signalling systems. DDR1 has been shown to interact with a number of different receptor systems, including the integrins (α1β1), the Wnt-5a signalling system and Notch1. This leads to the activation of multiple downstream signalling pathways including FAK, AKT and Erk1/2, ultimately influencing cellular phenotypes including survival, differentiation and invasion

this signalling scheme, protein tyrosine kinase 2 (Pyk2) was found to interact specifically with DDR1, while α2β1 integrin activated the nonoverlapping focal adhesion kinase (FAK) pathway with p130CAS serving as a scaffold and point of convergence for both classes of receptors (Fig. 11.2). This complex then signals through the Rap1 GTPase, activating the JNK pathway leading to N-cadherin upregulation. Similarly, in mouse embryonic stem cells (mESCs), collagen I binding results in the activation of both α2β1 integrin and DDR1 and is required for maintaining mESCs in an undifferentiated state [29]. In this case, α2β1 integrin activation leads to integrin-linked kinase (ILK) phosphorylation and Gli-1 expression, while DDR1 promotes AKT and ERK1/2 phosphorylation. Importantly, the authors showed that the parallel integrin and DDR1 pathways merge at the level of Bmi-1 via Gli-1 and ERK, respectively, to control mESC self-renewal capacity. There are additionally well-characterised interactions between these two classes of receptors in epithelial cell differentiation which is elaborated in detail in the relevant chapter by Yeh and Tang.

Using tandem affinity purification, Kim et al. showed that the Notch1 receptor interacts with DDR1 [30]. DDR1 activation by collagen results in the nuclear translocation of Notch1 and transcription of Notch1 target genes. The authors showed that Notch1 activation lies downstream of DDR1 and is essential for DDR1-mediated cancer cell survival in response to genotoxic stress. Further studies are required to identify the pathways that regulate Notch 1 activation by DDR1 and the implications of this interaction for targeted therapy in cancer.

## 11.3 Discoidin Domain Receptor 2

The literature available describing DDR2 signalling is sparse, and little is known about the pathways that are activated by this receptor. There are 13 tyrosine residues in the intracellular domains of the DDR2 that have the potential to be phosphorylated upon receptor activation [4]. However, only a subset of these sites has been found to be endogenously phosphorylated in cells in response to collagen activation. In this section, we summarise the tyrosine phosphorylation sites activated on DDR2 and its known signalling pathways (Fig. 11.3). We also highlight deficits in our current knowledge, which would benefit from more intensive future investigation.

### *11.3.1 Receptor Phosphorylation and Key Binding Partners*

Src is required for DDR2 auto-phosphorylation in hepatic stellate cells and skin fibroblasts, promoting receptor association with the SHC1 adaptor and inducing matrix metalloproteinase 2 (MMP-2) promoter activity [31]. Src family kinase (SFK)-deficient cells show reduced DDR2 phosphorylation in response to collagen stimulation. Consistent with this data, inhibition of Src activity by chemical inhibitors result in the ablation of DDR2 phosphorylation in cells. The authors of this study demonstrated that Y471 in the juxtamembrane of DDR2 is essential for the bimolecular interaction with SHC1. In a subsequent study by Yang et al., the authors used in vitro kinase assays to demonstrate that 3 tyrosine residues (Y736, Y740 and Y741) in the activation loop of DDR2 were targets for SRC phosphorylation [32]. Phosphorylation at these sites triggered intrinsic DDR2 kinase activity which led to the auto-phosphorylation of additional tyrosine residues, facilitating subsequent recruitment of SHC1. Intriguingly, the authors discovered that Y740 on DDR2 is an auto-inhibitory site and that mutating this site to phenylalanine results in a collagen-independent constitutively active DDR2 receptor. In addition to Src and SHC1, in silico analysis of DDR2 reveals potential tyrosine containing consensus sequence for the SH2 domains of NCK1 (non-catalytic region of tyrosine kinase adaptor protein 1), GAP (GTPase-activating protein) and p85 subunit of PI3K [33].

Using quantitative mass spectrometry, our laboratory has shown that five tyrosine residues are endogenously phosphorylated upon DDR2 activation by collagen I [34].

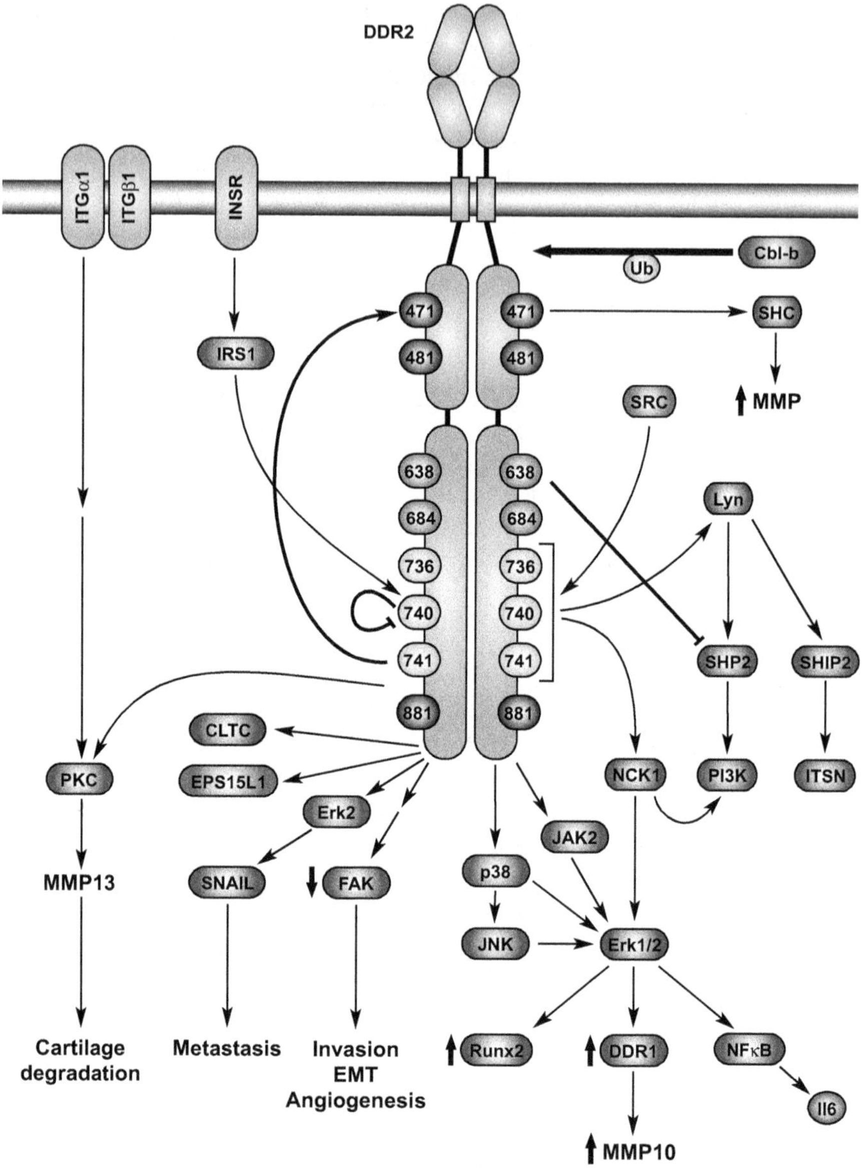


**Fig. 11.3** Known DDR2 protein–protein interactions and downstream signalling. DDR2 initiates its signalling through tyrosine phosphorylation of receptor sites (indicated in *numbered circles*). This leads to the propagation of downstream signalling of multiple pathways as depicted in the figure, resulting in diverse phenotypes including metastasis, angiogenesis and cartilage degradation. In addition, DDR2 degradation is regulated by ubiquitination by Cbl-b. Finally, DDR2 has been shown to interact with the insulin receptor (INSR) system as well as the integrins (ITGα1β1)

In addition to the activation loop sites Y736 and Y740, we demonstrated that Y481 on the juxtamembrane domain and Y684 and Y813 in the kinase domain of DDR2 were endogenously phosphorylated in cells stimulated with collagen. All these tyrosine residues displayed delayed and sustained phosphorylation kinetics that is consistent with previous reports of DDR2 activation [3]. The only exception was Y481 which was found to be constitutively phosphorylated in the absence of collagen. The cellular data was validated with kinase assays with the cytoplasmic domain of DDR2 activated by Src in vitro. Interestingly, in addition to tyrosine sites, several serine residues were also found to be phosphorylated in these in vitro kinase assays, including S446, S461, S674 and S696. Similarly, other phosphoproteomic studies in cancer cell lines have also identified serine-phosphorylated residues including S459, S461, S467, S485, S501, S675 and S737 (www.phosphosite.org). However, the signalling and functional consequences of these serine residues are currently unknown.

### *11.3.2 Downstream Signalling Pathways*

Collagen binding to DDR2 leads to the activation of a number of different signalling pathways with the most well characterised being the MAPK pathway. For instance, this pathway has been studied in the context of osteoblast differentiation and chondrocyte maturation. Two reports showed that silencing of DDR2 by genetic approaches suppresses osteoblast differentiation and matrix mineralisation [35, 36]. Overexpression of DDR2 induced activation and phosphorylation of Runx2, a master regulator in skeletal development. Phosphorylation of Runx2 occurs downstream of DDR2 via the activation of the p38 MAPK and ERK1/2 pathways. This results in the upregulation of bone markers and promotes osteoblast differentiation. In a related study, Klatt et al. showed in primary human chondrocytes that collagen II activation of DDR2 promotes the release of the pro-inflammatory cytokine interleukin 6 (IL-6) [37]. Using chemical inhibitors that target components of the MAPK and PI3K pathways, the authors demonstrated that DDR2 propagates the p38, JNK and ERK1/2 MAPKs leading to NFkB activation and subsequent upregulation of IL-6. The authors suggest that increased levels of collagen II fragments that arise as a result of cartilage matrix degradation may promote a positive feedforward loop that involves activation of DDR2 and release of IL-6. As a consequence of these findings, the authors propose that targeting these pathways therapeutically may be a means to overcome osteoarthritis.

DDR2 activation of the ERK MAPK pathway has also been shown in breast cancer cells. Zhang et al. showed that DDR2 activated the ERK2 kinase which results in the phosphorylation and stabilisation of the SNAIL1 transcription factor [38]. This pathway promotes breast cancer cell invasion through the maintenance of the EMT phenotype, facilitating breast cancer metastasis. This aspect of DDR2 function is elaborated in greater detail in the chapter by Longmore and colleagues. In primary lung fibroblasts, Ruiz and Jarai demonstrated that DDR2 activation by collagen leads to the upregulation of DDR1 expression as well as secretion of

MMP-10 [39]. This pathway is specific to DDR2 and is dependent on the downstream activation of JAK2 and ERK1/2. In a subsequent study by the same group, the authors showed that in normal human lung fibroblasts, both DDR2 and DDR1 were required for cell migration through collagen matrices but only DDR2 was required for fibroblast proliferation, implicating these receptors in a functional role in lung physiology [40].

Focal adhesion kinase (FAK) is a major tyrosine kinase localised to focal adhesions and implicated in tumour invasion, EMT and angiogenesis [41]. FAK down-regulation on fibrillar collagen was reported in vascular smooth muscle cells (VSMCs). Using engineered collagen 1 extracellular matrices, Bhadriraju et al. showed that the downregulation of FAK expression in VSMCs was regulated by DDR2 [42]. It remains unclear if the regulation of FAK expression by DDR2 is a unique event separate from the well-established FAK activation by the collagen binding integrins and if signalling integration between these two receptor classes intersects at the level of FAK upon collagen binding.

Our laboratory has employed phosphoproteomics to determine the global signalling networks activated by DDR2 upon collagen engagement [34]. Using HEK293 cells ectopically expressing DDR2, we found that stimulation of DDR2 resulted in the phosphorylation of a large number of proteins with a delayed and sustained activation profile over 7 time points across 24 h. These proteins include adaptor proteins (NCK1), kinases (Lyn, PI3K and ERK), phosphatases (SHP-2 and SHIP-2) and regulators of endocytosis (CLTC, EPS15L1 and ITSN1). Importantly we demonstrate that activation of a subset of these pathways is independent of collagen binding integrins and specific to DDR2. Using a targeted proteomic approach, we further show that DDR2 activation of SHP-2 is dependent on intact kinase activity of the receptor. Extending this work to DDR2 squamous cell lung cancer mutations identified from the exon sequencing efforts of the TCGA (The Cancer Genome Atlas) consortium, we show that the I638F kinase domain mutation of DDR2 is kinase inactivating [43]. This results in the downregulation of DDR2 receptor phosphorylation and inhibition of SHP-2 activation. Phenotypically, wild-type DDR2 inhibits cell growth in cancer cells when challenged with fibrillar collagen, and the I683F mutation by virtue of its loss of function relieves this growth suppression [34, 44]. Since DDR2 is currently being evaluated as a target for squamous cell lung cancer patients [45], our data suggests that ascertaining the activity of DDR2 mutants in patients is critical in stratifying patients who are likely to benefit from targeted therapy against this receptor.

In a recent study by Yu et al., it was shown that Cbl-b but not Cbl was responsible for ubiquitination of the DDR2 receptor [46]. Ubiquitination of DDR2 and interaction of Cbl-b with the receptor were enhanced upon activation by collagen II. The interaction and subsequent ubiquitination by Cbl-b promote receptor degradation by the proteasome. Using pre-osteoblastic MC3T3-E1 cells and mouse primary synovial fibroblasts, the authors demonstrated that Cbl-b was critical for regulating the collagen II-DDR2-MMP-13 axis important in promoting cartilage degradation in arthritis. Depletion of Cbl-b enhances the stability of the DDR2 protein which results in the upregulation of its downstream target MMP-13. This finding is consistent with the phenotype of Cbl-b-/- mice which display significant erosion of cartilage [47]. This important study is the first demonstration that DRR2, like other classical growth factor receptors such as EGFR and PDGFR, is also negatively regulated by ubiquitination and that Cbl-b is a key player in facilitating the degradation and termination of collagen-mediated signalling by DDR2.

### *11.3.3 Crosstalk with Other Receptor Signalling Systems*

Similar to DDR1, the DDR2 receptor has been shown to interact with the collagen binding integrins. Using primary goat chondrocytes cultured on collagen, Vonk et al. found that both integrin α1 (ITGα1) and DDR2 are required to modulate MMP-13 expression [48]. Interestingly, the study also revealed the expression levels of both receptors were linked, as silencing ITGα1 leads to the reduction in DDR2 expression. Downstream of integrin and DDR2 activation, PKC played an important role in promoting MMP-13 expression. Similar to silencing of ITGα1, inhibition of PKC led to a decrease in DDR2 expression and a corresponding downregulation of MMP-13.

DDR2 also appears to interact with components of the insulin signalling pathway. A report from our group has shown there is crosstalk at the receptor level between the insulin signalling pathway and DDR2 [49]. While stimulation of DDR2 expressing HEK293 cells with collagen I resulted in the phosphorylation of DDR2 in the Y740 site in the activation loop as expected, co-stimulation of these cells with insulin and collagen led to an unanticipated increase in the phosphorylation levels in this site. This finding suggests that the insulin pathway may cooperate with collagen signalling by enhancing DDR2 phosphorylation and activity. In another study by Zurakowski et al., overexpression of DDR2 in 3T3-L1 pre-adipocytes inhibited insulin receptor substrate 1 (IRS-1) tyrosine phosphorylation by insulin treatment [50]. Correspondingly, this impairment of IRS-1 signalling resulted in reduced glucose transport and cell proliferation. Taken together, these data suggest that expression of DDR2 may be important for insulin response and could play a pathological role in insulin resistance in diabetes patients.

## 11.4 Conclusion

The summary provided in this chapter is reassuring in that the DDR field is starting to piece together the global signalling profile of these important RTKs. However, it also highlights the daunting task ahead in reconciling the conflicting experimental data that is present in the literature. A number of outstanding questions remain unanswered. These include: (1) What is the functional role of the delayed and sustained activation of the DDRs? (2) What are the mechanisms that direct delayed receptor activation and negative regulation of receptor phosphorylation? (3) Can we assign a canonical signalling network to the DDRs that encompass the majority of signalling components that are reproducibly regulated in multiple biological contexts? (4) How does kinase-independent and kinase-dependent signalling cooperate to mediate the biological functions of the DDRs? (5) How prevalent is DDR crosstalk with other receptor systems and does this crosstalk occur naturally in biology?

In particular, we assert that future work on the DDRs should focus on determining a unified canonical signalling profile of these receptors. Much like classical growth factor receptors such as EGFR, PDGFR and c-Met where much of the canonical signalling networks have been rigorously defined, it is critical as a field that we move towards characterising the pathways that are reproducibly activated in different biological systems. It is only in this manner can we begin to assign function to specific intracellular signalling components and establish a mechanistic understanding of DDR biology.

## References


1. Payne LS, Huang PH (2013) The pathobiology of collagens in glioma. Mol Cancer Res 11(10):1129–1140
2. Fu HL, Valiathan RR, Arkwright R, Sohail A, Mihai C, Kumarasiri M, Mahasenan KV, Mobashery S, Huang P, Agarwal G, Fridman R (2013) Discoidin domain receptors: unique receptor tyrosine kinases in collagen-mediated signaling. J Biol Chem 288(11):7430–7437
3. Vogel W, Gish GD, Alves F, Pawson T (1997) The discoidin domain receptor tyrosine kinases are activated by collagen. Mol Cell 1(1):13–23
4. Iwai LK, Luczynski MT, Huang PH (2014) Discoidin domain receptors: a proteomic portrait. Cell Mol Life Sci 71(17):3269–3279
5. Lemeer S, Bluwstein A, Wu Z, Leberfinger J, Muller K, Kramer K, Kuster B (2012) Phosphotyrosine mediated protein interactions of the discoidin domain receptor 1. J Proteomics 75(12):3465–3477
6. Koo DH, McFadden C, Huang Y, Abdulhussein R, Friese-Hamim M, Vogel WF (2006) Pinpointing phosphotyrosine-dependent interactions downstream of the collagen receptor DDR1. FEBS Lett 580(1):15–22
7. Foehr ED, Tatavos A, Tanabe E, Raffioni S, Goetz S, Dimarco E, De Luca M, Bradshaw RA (2000) Discoidin domain receptor 1 (DDR1) signaling in PC12 cells: activation of juxtamembrane domains in PDGFR/DDR/TrkA chimeric receptors. FASEB J 14(7):973–981
8. Wang CZ, Su HW, Hsu YC, Shen MR, Tang MJ (2006) A discoidin domain receptor 1/SHP-2 signaling complex inhibits alpha2beta1-integrin-mediated signal transducers and activators of transcription 1/3 activation and cell migration. Mol Biol Cell 17(6):2839–2852
9. L'Hote CG, Thomas PH, Ganesan TS (2002) Functional analysis of discoidin domain receptor 1: effect of adhesion on DDR1 phosphorylation. FASEB J 16(2):234–236
10. Lu KK, Trcka D, Bendeck MP (2011) Collagen stimulates discoidin domain receptor 1-mediated migration of smooth muscle cells through Src. Cardiovasc Pathol 20(2):71–76
11. Yang G, Li Q, Ren S, Lu X, Fang L, Zhou W, Zhang F, Xu F, Zhang Z, Zeng R, Lottspeich F, Chen Z (2009) Proteomic, functional and motif-based analysis of C-terminal Src kinase-interacting proteins. Proteomics 9(21):4944–4961
12. Neuhaus B, Buhren S, Bock B, Alves F, Vogel WF, Kiefer F (2011) Migration inhibition of mammary epithelial cells by Syk is blocked in the presence of DDR1 receptors. Cell Mol Life Sci 68(22):3757–3770
13. Faraci-Orf E, McFadden C, Vogel WF (2006) DDR1 signaling is essential to sustain Stat5 function during lactogenesis. J Cell Biochem 97(1):109–121

14. Hilton HN, Stanford PM, Harris J, Oakes SR, Kaplan W, Daly RJ, Ormandy CJ (2008) KIBRA interacts with discoidin domain receptor 1 to modulate collagen-induced signalling. Biochim Biophys Acta 1783(3):383–393
15. Hansen C, Greengard P, Nairn AC, Andersson T, Vogel WF (2006) Phosphorylation of DARPP-32 regulates breast cancer cell migration downstream of the receptor tyrosine kinase DDR1. Exp Cell Res 312(20):4011–4018
16. Hidalgo-Carcedo C, Hooper S, Chaudhry SI, Williamson P, Harrington K, Leitinger B, Sahai E (2011) Collective cell migration requires suppression of actomyosin at cell-cell contacts mediated by DDR1 and the cell polarity regulators Par3 and Par6. Nat Cell Biol 13(1):49–58
17. Seo MC, Kim S, Kim SH, Zheng LT, Park EK, Lee WH, Suk K (2008) Discoidin domain receptor 1 mediates collagen-induced inflammatory activation of microglia in culture. J Neurosci Res 86(5):1087–1095
18. Avivi-Green C, Singal M, Vogel WF (2006) Discoidin domain receptor 1-deficient mice are resistant to bleomycin-induced lung fibrosis. Am J Respir Crit Care Med 174(4):420–427
19. Curat CA, Vogel WF (2002) Discoidin domain receptor 1 controls growth and adhesion of mesangial cells. J Am Soc Nephrol 13(11):2648–2656
20. Lund AW, Stegemann JP, Plopper GE (2009) Mesenchymal stem cells sense three dimensional type I collagen through discoidin domain receptor 1. Open Stem Cell J 1:40–53
21. Ongusaha PP, Kim JI, Fang L, Wong TW, Yancopoulos GD, Aaronson SA, Lee SW (2003) p53 induction and activation of DDR1 kinase counteract p53-mediated apoptosis and influence p53 regulation through a positive feedback loop. EMBO J 22(6):1289–1301
22. Jonsson M, Andersson T (2001) Repression of Wnt-5a impairs DDR1 phosphorylation and modifies adhesion and migration of mammary cells. J Cell Sci 114(Pt 11):2043–2053
23. Dejmek J, Dib K, Jonsson M, Andersson T (2003) Wnt-5a and G-protein signaling are required for collagen-induced DDR1 receptor activation and normal mammary cell adhesion. Int J Cancer 103(3):344–351
24. Roarty K, Serra R (2007) Wnt5a is required for proper mammary gland development and TGF-beta-mediated inhibition of ductal growth. Development 134(21):3929–3939
25. Dejmek J, Leandersson K, Manjer J, Bjartell A, Emdin SO, Vogel WF, Landberg G, Andersson T (2005) Expression and signaling activity of Wnt-5a/discoidin domain receptor-1 and Syk plays distinct but decisive roles in breast cancer patient survival. Clin Cancer Res 11(2 Pt 1):520–528
26. Staudinger LA, Spano SJ, Lee W, Coelho N, Rajshankar D, Bendeck MP, Moriarty T, McCulloch CA (2013) Interactions between the discoidin domain receptor 1 and beta1 integrin regulate attachment to collagen. Biol Open 2(11):1148–1159
27. Xu H, Bihan D, Chang F, Huang PH, Farndale RW, Leitinger B (2012) Discoidin domain receptors promote alpha1beta1- and alpha2beta1-integrin mediated cell adhesion to collagen by enhancing integrin activation. PLoS One 7(12), e52209
28. Shintani Y, Fukumoto Y, Chaika N, Svoboda R, Wheelock MJ, Johnson KR (2008) Collagen I-mediated up-regulation of N-cadherin requires cooperative signals from integrins and discoidin domain receptor 1. J Cell Biol 180(6):1277–1289
29. Suh HN, Han HJ (2011) Collagen I regulates the self-renewal of mouse embryonic stem cells through alpha2beta1 integrin- and DDR1-dependent Bmi-1. J Cell Physiol 226(12): 3422–3432
30. Kim HG, Hwang SY, Aaronson SA, Mandinova A, Lee SW (2011) DDR1 receptor tyrosine kinase promotes prosurvival pathway through Notch1 activation. J Biol Chem 286(20): 17672–17681
31. Ikeda K, Wang LH, Torres R, Zhao H, Olaso E, Eng FJ, Labrador P, Klein R, Lovett D, Yancopoulos GD, Friedman SL, Lin HC (2002) Discoidin domain receptor 2 interacts with Src and Shc following its activation by type I collagen. J Biol Chem 277(21):19206–19212
32. Yang K, Kim JH, Kim HJ, Park IS, Kim IY, Yang BS (2005) Tyrosine 740 phosphorylation of discoidin domain receptor 2 by Src stimulates intramolecular autophosphorylation and Shc signaling complex formation. J Biol Chem 280(47):39058–39066
33. Vogel W (1999) Discoidin domain receptors: structural relations and functional implications. FASEB J 13:S77–S82

34. Iwai LK, Payne LS, Luczynski MT, Chang F, Xu H, Clinton RW, Paul A, Esposito EA, Gridley S, Leitinger B, Naegle KM, Huang PH (2013) Phosphoproteomics of collagen receptor networks reveals SHP-2 phosphorylation downstream of wild-type DDR2 and its lung cancer mutants. Biochem J 454(3):501–513
35. Lin KL, Chou CH, Hsieh SC, Hwa SY, Lee MT, Wang FF (2010) Transcriptional upregulation of DDR2 by ATF4 facilitates osteoblastic differentiation through p38 MAPK-mediated Runx2 activation. J Bone Miner Res 25(11):2489–2503
36. Zhang Y, Su J, Yu J, Bu X, Ren T, Liu X, Yao L (2011) An essential role of discoidin domain receptor 2 (DDR2) in osteoblast differentiation and chondrocyte maturation via modulation of Runx2 activation. J Bone Miner Res 26(3):604–617
37. Klatt AR, Zech D, Kuhn G, Paul-Klausch B, Klinger G, Renno JH, Schmidt J, Malchau G, Wielckens K (2009) Discoidin domain receptor 2 mediates the collagen II-dependent release of interleukin-6 in primary human chondrocytes. J Pathol 218(2):241–247
38. Zhang K, Corsa CA, Ponik SM, Prior JL, Piwnica-Worms D, Eliceiri KW, Keely PJ, Longmore GD (2013) The collagen receptor discoidin domain receptor 2 stabilizes SNAIL1 to facilitate breast cancer metastasis. Nat Cell Biol 15(6):677–687
39. Ruiz PA, Jarai G (2011) Collagen I induces discoidin domain receptor (DDR) 1 expression through DDR2 and a JAK2-ERK1/2-mediated mechanism in primary human lung fibroblasts. J Biol Chem 286(15):12912–12923
40. Ruiz PA, Jarai G (2012) Discoidin domain receptors regulate the migration of primary human lung fibroblasts through collagen matrices. Fibrogenesis Tissue Repair 5:3
41. Zhao J, Guan JL (2009) Signal transduction by focal adhesion kinase in cancer. Cancer Metastasis Rev 28(1–2):35–49
42. Bhadriraju K, Chung KH, Spurlin TA, Haynes RJ, Elliott JT, Plant AL (2009) The relative roles of collagen adhesive receptor DDR2 activation and matrix stiffness on the downregulation of focal adhesion kinase in vascular smooth muscle cells. Biomaterials 30(35):6687–6694
43. Hammerman PS, Sos ML, Ramos AH, Xu C, Dutt A, Zhou W, Brace LE, Woods BA, Lin W, Zhang J, Deng X, Lim SM, Heynck S, Peifer M, Simard JR, Lawrence MS, Onofrio RC, Salvesen HB, Seidel D, Zander T, Heuckmann JM, Soltermann A, Moch H, Koker M, Leenders F, Gabler F, Querings S, Ansen S, Brambilla E, Brambilla C, Lorimier P, Brustugun OT, Helland A, Petersen I, Clement JH, Groen H, Timens W, Sietsma H, Stoelben E, Wolf J, Beer DG, Tsao MS, Hanna M, Hatton C, Eck MJ, Janne PA, Johnson BE, Winckler W, Greulich H, Bass AJ, Cho J, Rauh D, Gray NS, Wong KK, Haura EB, Thomas RK, Meyerson M (2011) Mutations in the DDR2 kinase gene identify a novel therapeutic target in squamous cell lung cancer. Cancer Discovery 1(1):78–89
44. Wall SJ, Werner E, Werb Z, DeClerck YA (2005) Discoidin domain receptor 2 mediates tumor cell cycle arrest induced by fibrillar collagen. J Biol Chem 280(48):40187–40194
45. Payne LS, Huang PH (2014) Discoidin domain receptor 2 signaling networks and therapy in lung cancer. J Thorac Oncol 9(6):900–904
46. Yu J, Zhao H, Zhang Y, Liu YC, Yao L, Li X, Su J (2014) Ubiquitin ligase Cbl-b acts as a negative regulator in discoidin domain receptor 2 signaling via modulation of its stability. FEBS Lett 588(9):1509–1514
47. Jeon MS, Atfield A, Venuprasad K, Krawczyk C, Sarao R, Elly C, Yang C, Arya S, Bachmaier K, Su L, Bouchard D, Jones R, Gronski M, Ohashi P, Wada T, Bloom D, Fathman CG, Liu YC, Penninger JM (2004) Essential role of the E3 ubiquitin ligase Cbl-b in T cell anergy induction. Immunity 21(2):167–177
48. Vonk LA, Doulabi BZ, Huang C, Helder MN, Everts V, Bank RA (2011) Collagen-induced expression of collagenase-3 by primary chondrocytes is mediated by integrin α1 and discoidin domain receptor 2: a protein kinase C-dependent pathway. Rheumatology (Oxford) 50(3):463–472
49. Iwai LK, Chang F, Huang PH (2013) Phosphoproteomic analysis identifies insulin enhancement of discoidin domain receptor 2 phosphorylation. Cell Adhes Migr 7(2):161–164
50. Zurakowski H, Gagnon A, Landry A, Layne MD, Sorisky A (2007) Discoidin domain receptor 2 impairs insulin-stimulated insulin receptor substrate-1 tyrosine phosphorylation and glucose uptake in 3T3-L1 adipocytes. Horm Metab Res 39(8):575–581

# Chapter 12
# Discoidin Domain Receptor Signaling and Pharmacological Inhibitors

**Pedro A. Ruiz-Castro, Duncan Shaw, and Gabor Jarai**

## 12.1 Introduction

Discoidin domain receptors (DDRs) are members of the receptor tyrosine kinase (RTK) superfamily and are distinguished by the presence of a discoidin-homology region in their extracellular domain [1, 2]. Unlike typical RTKs that use soluble, peptide-like growth factors as ligands, DDRs are activated by various types of triple-helical collagens, major components of the extracellular matrix (ECM) [3, 4]. DDRs from a subfamily among the 58 known human receptor tyrosine kinases consisting of two related members, DDR1 and DDR2, have been identified with distinct expression profiles and ligand specificities. DDR1 is widely expressed in epithelial

P.A. Ruiz-Castro
Novartis Institutes for Biomedical Research, Respiratory Diseases, Wimblehurst Road, Horsham RH12 5AB, UK

Sanford-Burnham Medical Research Institute, 10901 N Torrey Pines Rd, La Jolla, CA 92037, USA

Division of Gastroenterology and Hepatology, Novartis Institutes for Biomedical Research, Rämistrasse 100, Zurich 8091, Switzerland

D. Shaw, Ph.D.
Novartis Institutes for Biomedical Research, Respiratory Diseases, Wimblehurst Road, Horsham RH12 5AB, UK

Global Discovery Chemistry, Novartis Institutes for Biomedical Research, 100 Technology Square, Cambridge, MA 02039, USA

G. Jarai, Ph.D. (✉)
Novartis Institutes for Biomedical Research, Respiratory Diseases, Wimblehurst Road, Horsham RH12 5AB, UK

Fibrotic Diseases, Disease Sciences and Biology, Bristol-Myers Squibb, 311 Pennington-Rocky Hill Road, Pennington, NJ 08534, USA
e-mail: gabor.jarai@novartis.com

 
R. Fridman, P.H. Huang (eds.), *Discoidin Domain Receptors in Health and Disease*, DOI 10.1007/978-1-4939-6383-6_12

cells, particularly of the lung, kidney, mammary gland, and gastrointestinal tract, whereas DDR2 is confined to cells of mesenchymal origin, including fibroblasts, myofibroblasts, and smooth and skeletal muscle cells [5, 6]. DDRs are activated by many different collagens, with fibrillar collagens acting as ligands for both receptors. DDR1 is primarily activated by collagens I–V, VIII, and XI, while DDR2 responds to collagens I, II, III, V, and non-fibrillar collagen X, but not the basement membrane collagen IV [3, 4, 7].

Both DDR1 and DDR2 are single-pass type I transmembrane glycoproteins characterized by an N-terminal discoidin domain followed by another globular domain predicted to belong to the same superfamily and unique to DDRs [8]. The extracellular discoidin domains are connected by an extracellular juxtamembrane region to a transmembrane helix that is linked to the C-terminal tyrosine kinase domain by a large cytosolic juxtamembrane region. Within the intracellular region, DDR1 and DDR2 contain 15 and 14 tyrosine residues, respectively, that may serve as potential phosphorylation sites. DDR2 has one known isoform, whereas DDR1 has five isoforms generated by alternative splicing: DDR1a, b, c, d, and e. DDR1c is the longest with 919 amino acids, whereas DDR1a and DDR1b lack 37 or 6 amino acids in the juxtamembrane and kinase regions, respectively. DDR1d and DDR1e appear to be non-signaling variants as they lack the kinase domain or the adenosine triphosphate (ATP)-binding site within the kinase domain, respectively [1, 9]; however they may have alternative functions, by serving as decoy receptors or modulating the activity of other DDR1 splice variants via receptor dimerization.

Similar to collagen-binding integrins, DDRs recognize discrete amino acid motifs in collagen, with the DDR collagen-binding sites entirely contained within their discoidin domains [10]. To date, four DDR2-binding sites [11] and one DDR1-binding site [12] have been identified in collagens II and III. The primary binding site for both DDR1 and DDR2 contains the GVMGFO motif (where O is hydroxyproline), which has been shown to be present in the fibrillar collagens I–III [13–15]. The identification of GVMGFO as a DDR-binding motif enabled the determination of the crystal structure of the discoidin domain of DDR2 bound to a GVMGFO-containing triple-helical peptide. The central collagen-binding residues found in DDR2 are strictly conserved in DDR1, which is consistent with the affinity of both receptors to fibrillar collagens. Interestingly, several collagen-binding residues outside the GVMGFO peptide-binding interface are not conserved in DDR1, suggesting a role for these amino acids in the collagen-binding specificity of DDRs [16]. Furthermore, a recent report has shown that replacing these residues in DDR2 with the corresponding DDR1 residues enabled the receptor construct to bind collagen IV [15]. DDRs bind to the native triple-helical form of collagen and not to individual collagen α-chains or heat-denatured collagen [4, 10]; however, triple-helical peptides containing the collagen-binding motif induce DDR2 activation indicating that the supramolecular assembly of collagen into long fibrils or fibers is not required for DDR signaling [14].

The amount of DDR1 on the cell surface has been shown to be regulated by ectodomain shedding, a protease-mediated process that causes the release of the extracellular domain of membrane proteins. DDR1 is shed both constitutively [1] and in a collagen-dependent [17, 18] manner in a process that requires a metalloprotease, which cleaves DDR1 within the extracellular juxtamembrane region generating a membrane-anchored C-terminal fragment [18]. Both membrane-anchored matrix metalloproteases (MMPs) and ADAMs (a disintegrin and metalloproteinase) families have been shown to be capable of DDR1 shedding; however the exact identity of the proteinases regulating DDR expression under physiological conditions is not known. Interestingly, there is also evidence for the role of SRC, a tyrosine kinase involved in DDR activation, in collagen I-induced shedding of DDR1 [17, 19].

While the majority of RTKs are single-chain receptors that generate dimeric complexes upon ligand binding, it is believed that DDRs exist as preformed dimers at the cell surface. Supporting this notion, recent data suggest that DDRs have the highest potential for transmembrane helix self-interactions in the entire RTK superfamily via the action of a leucine-zipper motif and that residues within the transmembrane helix are required for DDR signaling [20, 21]. Furthermore, dimerization of the DDR ectodomains has been shown to be required for ligand binding for both DDR1 and DDR2 [10, 22, 23]. How collagen binding to the DDR discoidin domain results in the activation of the cytosolic kinase domain is still unclear, but most likely involves conformational changes along the entire receptor that triggers phosphorylation of tyrosine residues in the intracellular region, similarly to what occurs in many RTKs [8]. Nonetheless, the structural and cellular mechanisms regulating these changes and whether one or both discoidin domains within the preformed DDR dimer need to be occupied by collagen remain to be elucidated.

Similar to other RTKs, DDRs have been shown to play crucial roles in the regulation of a wide range of cellular processes, such as proliferation, survival, differentiation, adhesion, migration, wound repair, and ECM remodeling [5, 6, 24, 25]. Dysregulation of DDRs has been linked to several human disorders, including atherosclerosis [26], arthritis [27, 28], inflammation [29], cancer [30], and a variety of fibrotic disorders [6]. For instance, both DDR1 and DDR2 have been shown to be upregulated in kidney and liver fibrosis, respectively [31, 32]. DDR1$^{-/-}$ mice have been shown to be resistant to bleomycin-induced lung inflammation and fibrosis [33], and DDR2$^{-/-}$ mice present delayed wound healing and slow fibroblast proliferation [5, 6]. In cancer, overexpression of DDR1 has been implicated in cell survival and invasiveness in hepatocellular carcinoma, pituitary adenoma, and prostate cancer [34–36]. Elevated expression levels of DDRs have also been found in the lung [37], esophageal [38], breast [39, 40], ovary [1, 41], liver [42], pancreas [43], prostate [35], and brain [44] cancers. Inhibition of DDR1 by small interfering RNAs (siRNAs) has been shown to inhibit tumorigenicity and bone metastasis in lung cancer and to reduce cancer cell chemoresistance [45–47]. Selective suppression of DDR2 is also linked to antitumor activity in mouse xenografts of squamous lung cancer cells [48–50].

Given the accumulating evidence implicating DDRs in a wide range of diseases, DDR1 and DDR2 are considered as potential novel targets for therapeutic intervention [30, 51]. Most targeted therapies interfering with tyrosine kinases involve small-molecule inhibitors designed to inhibit intracellular kinase activity or monoclonal antibodies designed to interfere with ectodomain function. The recent determination of the crystal structure of the complex between DDR2 and a triple-helical collagen peptide could also facilitate the design of small peptide inhibitors [16]. Small-molecule tyrosine kinase inhibitors are mostly ATP-competitive inhibitors, which interfere with the transfer of the terminal phosphate from ATP onto proteins containing a tyrosine residue. Type I inhibitors constitute the majority of ATP-competitive inhibitors and recognize the active conformation of the kinase. By contrast, type II inhibitors recognize the inactive conformation of the kinase, which is sometimes referred to as DFG-out due to the conformational rearrangement of this Asp–Phe–Gly motif [52]. The structure of the kinase domain is highly conserved, making it very challenging to design selective small-molecule inhibitors. As a result, many of the inhibitors used in the clinic target multiple kinases, including DDRs [53, 54]. In addition to the substrate-binding site and the ATP-binding site of the tyrosine kinase domain, the juxtamembrane region, which can play an autoinhibitory role in the catalytic activity of RTKs, has been suggested to be an attractive target due to the lack of sequence similarity among juxtamembrane regions between different RTK families [55]. The soluble ectodomains of DDR1 and DDR2 have also been shown to have an important role in modulating endogenous receptor activation and function [23, 56, 57]. As a consequence, targeting the extracellular domain of DDRs with chemical compounds has been suggested as a strategy for the inhibition of receptor function. For example, a recent screening with DDR2-expressing insect cells identified the anticancer drug actinomycin D as an inhibitor of DDR2. Interestingly, this antibiotic blocked the interaction between DDR2 and collagen I in HEK293 cells, but did not interfere with DDR1 activation. The mechanisms underlying actinomycin D-mediated inhibition of DDR2 remain to be elucidated [58].

## 12.2 DDR Activation and Binding Partners

Like all RTKs, DDRs undergo autophosphorylation on intracellular tyrosine residues upon ligand binding, but compared to the rapid response of RTKs, collagen-induced DDR phosphorylation is unusually slow and sustained [3, 4]. Several models of receptor activation have been proposed to explain the slow kinetic profile of DDR1 activation. In a recent study, collagen I has been demonstrated to induce rapid aggregation of preformed DDR1b dimers followed by receptor internalization into early endosomal vesicles and posterior recycling of the receptor to the cell surface. Because DDR1 endocytosis appears to precede receptor phosphorylation, full receptor activation is likely to take place within the endocytic vesicles [59]. Interestingly, DDR1b and DDR1c, but not DDR1a, present a motif in the intracellular juxtamembrane region that has been shown to regulate integrin-/clathrin-dependent endocytosis and recycling [60], suggesting that DDR internalization may be an isoform-dependent regulatory mechanism. Given the fact that collagen is typically an abundant ligand in the tissue microenvironment, the delayed activation of DDRs may constitute an evolutionary mechanism to avoid excessive receptor activation until a threshold of ligand concentration or exposure time is reached.

Moreover, some studies suggest that the kinetic profile of DDR phosphorylation might serve as a delayed negative mechanism to downregulate early activation of other receptors and signaling pathways. For instance, DDR1 has been shown to antagonize α2β1 integrin signaling by reducing the activation of signal transducers and activators of transcription (STAT)1/3 and Cdc42 [61, 62]. Since α2β1 integrin activation and formation of focal adhesion signaling complexes take place within minutes of collagen binding, it is plausible that DDRs have evolved to provide a late negative feedback mechanism for the suppression of integrin-mediated cell responses. The slow accumulation of late effectors for negative regulation has been postulated to facilitate the transition to a new cellular state [63], which is consistent with the crucial role of DDRs in cellular processes such as differentiation, epithelial-to-mesenchymal transition (EMT), survival, and migration [5, 6, 24, 25]. Because RTKs often use the same adaptor proteins and signal through common signaling pathways to induce cellular responses [64], it has been suggested that the sustained phosphorylation of DDRs and subsequent recruitment of adaptor and signaling molecules in the proximity of RTKs may be a mechanism for the rapid activation of some RTKs. According to this model, ligand binding-induced RTK phosphorylation would result in the competitive recruitment of adaptor and signaling proteins, such as phosphoinositide 3-kinase (PI3K) or the tyrosine kinase SYK from complexes with the phosphorylated DDRs to these RTKs, thereby facilitating rapid activation of RTK-dependent signaling pathways [65]. Interestingly, collagen-independent DDR1 hyperphosphorylation observed in the presence of the phosphatase inhibitor pervanadate suggests that the DDR phosphorylation profile is also regulated by tyrosine phosphatases. This may indicate that DDR1 rapidly alternates between phosphorylated and non-phosphorylated states [66]; however, the molecular mechanisms driving this DDR phosphorylation dynamic and how it exactly relates to intracellular signaling remain unclear.

The binding of collagen to DDRs leads to receptor autophosphorylation at distinct tyrosine residues within the cytosolic juxtamembrane and kinase domains, which serve as docking sites for adaptor signaling molecules containing SRC homology-2/3 (SH2/3) and phosphotyrosine-binding (PTB) domains (Fig. 12.1). Activated DDR1 has been shown to bind to a variety of adaptor proteins and signaling molecules, including ShcA [4], the non-muscle myosin heavy chain (NMHC)-IIA [67], the focal adhesion kinase (FAK)-related protein tyrosine kinase (PYK2) [68], the phosphatase SHP-2 [61, 69], and the SH2 domain-containing adaptor proteins NCK1/2 [69] and C-terminal SRC tyrosine kinase (CSK) [70]. The structural subunit p85α of PI3K has also been reported to directly bind DDR1 in normal and cancer mammary epithelial cells [66]. The inhibition of PI3K with the nonselective PI3K inhibitor wortmannin resulting in a significant decrease of mammary epithelial cell adhesion to collagen may suggest a role for PI3K-mediated DDR1 signaling in cell adhesion to the ECM [71]; however, these data require further confirmation as wortmannin is a nonselective inhibitor that targets other signaling pathways as well. In a recent proteomic study analyzing phosphotyrosine-mediated protein–protein interactions of DDR1, Lemeer et al. have also identified Ras GTPase-activating protein (RasGAP), the adaptor protein CRKII and the guanine nucleotide exchange

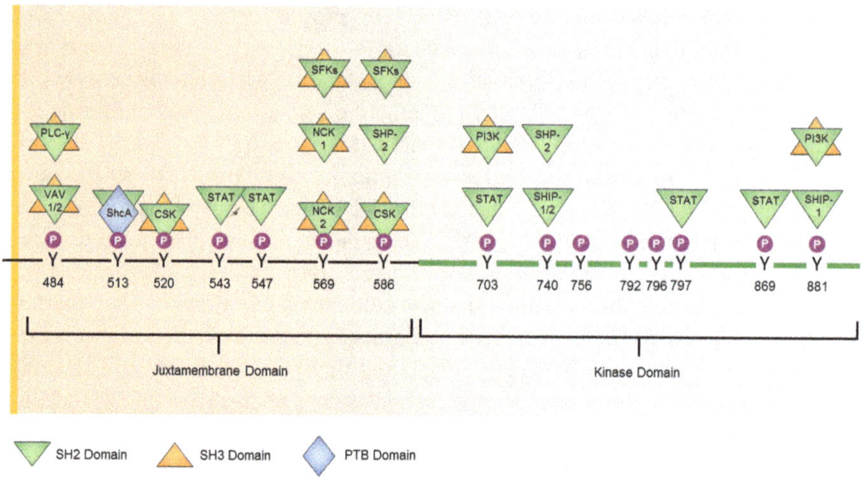


**Fig. 12.1** DDR1 intracellular binding partners. Collagen binding triggers the autophosphorylation DDRs at multiple tyrosine residues within the cytosolic juxtamembrane and kinase domains, which generate docking sites for adaptor molecules containing SH2/3 and PTB domains. Activated DDR1 recruits a variety of adaptor proteins and signaling molecules, including ShcA, NCK1/2, p85α PI3K, VAV2/3, the protein tyrosine kinase CSK, the phosphatases SHP-2 and SHIP-1/2, and several members of the STAT family. The PTB domain in ShcA and the SH2 domain in CSK and NCK2 directly bind phospho-DDR1. *PLC-γ* phospholipase C γ, *SFK* SRC family kinase, *SHIP-1/2* inositol 5′-phosphatase-1/2. The indicated amino acid annotations refer to the DDR1b isoform [30]

factors VAV3 and VAV2 [72], which has also been shown to bind other RTKs including epidermal growth factor receptor (EGFR) and platelet-derived growth factor receptor (PDGFR) through the SH2 domain of VAV2. Interestingly, VAV2 has recently been shown to play a role in RTK endocytosis pointing to a possible role of VAV2 in DDR1 internalization [73]. Several members of the STAT family are also recruited to DDR1 by a SH2 domain-mediated mechanism, including STAT-1α/β, STAT-3, and STAT-5β [72]. Of note, several reports have shown that collagen-induced DDR1 activation triggers STAT-5β, but not STAT-5α phosphorylation and transcriptional activity in mammary epithelial cancer cells [74]. Constitutive phosphorylation of STAT-5β appears to correlate with increased carcinogenesis in several human tumors supporting the proposed role of DDR1 in tumor progression [75]; however, the evidence for this remains indirect. While the intracellular juxtamembrane domain of DDR1 seems to be the primary binding site for adaptor proteins such as ShcA, the kinase domain appears to bind PI3Ks and STATs. Interestingly, the only binding partners showing no preference for specific tyrosine residues were RasGAP and tyrosine phosphatases, which were recruited at various locations in both domains. This suggests that DDR1 may regulate signaling by bringing the phosphatases and their substrates in close proximity, pointing to a highly compartmentalized and fine-tuned regulation of DDR1-induced signaling [72].

While the aforementioned interacting partners are only recruited by DDR1 in a phosphotyrosine-dependent manner, several signaling proteins appear to bind

DDR1 in the absence of collagen stimulation. For example, inactive tyrosine kinase SYK has been observed to be constitutively bound to DDR1 via its SH2 domain regardless of the phosphorylation status of DDR1 [76]. Furthermore, a recent study indicates that dopamine- and cAMP-regulated neuronal phosphoprotein (DARPP)-32 binds to the juxtamembrane region of DDR1 and dissociates from it upon activation of DDR1 by collagen. Binding of DARPP-32 to inactive DDR1 was shown to impair migration in human epithelial breast cancer cells. Interestingly, the anti-migratory function of DARPP-32 requires the presence of DDR1, suggesting a role for inactive DDR1 in sequestrating DARPP-32 at the plasma membrane in order to inhibit cell migration [77]. Similarly to DARPP-32, the interaction between the cytoplasmatic phosphoprotein KIBRA and DDR1 was shown to be stronger in the absence of collagen in human breast cancer cells. KIBRA binds DDR1 via the PPxY motif of DDR1, interacting in a tripartite complex with the protein kinase PKCζ. Interestingly, KIBRA was shown to block collagen-/DDR1-induced extracellular signal-regulated kinase (ERK)1/2 mitogen-activated protein kinase (MAPK) activation, suggesting that unstimulated DDR1 interacts with KIBRA to inhibit ERK1/2 MAPK signaling and subsequent migration of human mammary cells [78]. Since both DARPP-32 and KIBRA dissociate DDR1 in the presence of collagen, it is plausible that collagen induces an intramolecular rearrangement of the DDR1 intracellular region, which results in masking of the binding sequences of these proteins. Phosphotyrosine-dependent DDR1-binding partners, such as ShcA, NCK2, or SHP-2, might also competitively displace DARPP-32 or KIBRA upon collagen-mediated DDR1 activation [77, 78]. Recent reports indicate that DDR1 forms a complex with E-cadherin through interaction of their extracellular domains independently of collagen stimulation and DDR1 activation in polarized epithelial cells. This interaction has been suggested to negatively regulate DDR1 activity and collagen-mediated cell migration by sequestering DDR1 to adherent junctions, which prevents its interaction with collagen [79]. E-cadherins play a crucial role in the maintenance of epithelial polarity and tissue architecture, and the membrane levels of E-cadherin are regulated by the balance between exocytosis and endocytosis [80, 81]. Interestingly, the presence of DDR1 in cell–cell junctions is also required to stabilize cell surface E-cadherin by preventing collagen-induced E-cadherin endocytosis and suggests a role for inactive DDR1 in the maintenance of both cell junctions and cell polarity in epithelia [82]. Furthermore, the localization of DDR1 to cell–cell junctions has been shown to suppress actomyosin contractility through the recruitment of the cell polarity regulators PAR3 and PAR6 to the intracellular domain of DDR1 [40]. PAR3 and PAR6 activation and Cdc42-dependent actomyosin contractility are required for collective cancer cell invasion [83–85]. This result further supports the role for active DDR1 in cell migration and cancer progression.

Similarly to DDR1, DDR2 has been shown to form a complex with SRC in COS-7 cells and activated rat hepatic stellate cells [19]. SRC phosphorylates three tyrosine residues, Y736, Y740, and Y741, in the DDR2 kinase domain activation loop, which in turn promotes the formation of DDR2 cytosolic domain-Shc signaling complexes. ShcA binds in the juxtamembrane region of DDR2 at Y471, but unlike in the case of DDR1, this interaction is mediated by the SH2 domain rather than the

PTB domain [86]. Recruitment of both the tyrosine kinase SRC and the adaptor protein ShcA appears to be required for full receptor autophosphorylation upon collagen stimulation [19]. Interestingly, substitution of Y740 for phenylalanine stimulates collagen-independent autophosphorylation and activation of DDR2, suggesting that Y740 plays a role in the inhibition of DDR2 phosphorylation [86].

## 12.3 DDR Signaling Pathways and DDR–Integrin Cross Talk

The identification of signaling pathways and molecules that are specifically activated by DDRs is essential for the understanding of the contribution of DDRs to disease pathogenesis and for the identification of therapeutic targets for the treatment of DDR-related diseases such as fibrosis and cancer. Nonetheless, depicting the signal transduction pathways linked to DDR activation is proving to be a challenging task, due to several factors. DDR signaling is cell- and collagen-type specific, and therefore different collagens elicit different cellular responses depending on tissue and cellular context [3, 4, 6]. Additionally, DDRs appear to engage in different molecular complexes and trigger different signaling cascades and cellular responses, depending on whether they are involved in cell–matrix interactions or cell–cell interactions independent of collagen [40, 79, 82]. Moreover, an increasing body of evidence suggests that DDRs interact with several other receptors, including the developmental Notch1 and Frizzled5 receptors [46, 71] and the insulin receptor [87]. Finally, DDR signaling can also interact with signals from other collagen receptors such as integrins, with which DDRs have been shown to synergize or antagonize cell responses depending on the cell system studied [61, 62, 68, 88, 89]. In spite of all the aforementioned, considerable progress has been made in recent years in understanding the molecular mechanisms underlying the activation of signaling events downstream of DDRs (Fig. 12.2).

Collagen IV-activated DDR1 has been linked to an increase in the expression of cyclooxygenase (COX)2 through activation of the survival nuclear factor (NF)-kB in human breast cancer cells [45]. DDR1 is a transcriptional target for the p53 tumor suppressor protein, and several studies have shown that DNA damage induced either by $\gamma$-irradiation or chemotherapeutic agents triggers p53-mediated DDR1 expression in cancer cells [45, 90, 91]. Together, these results suggest that COX2 is an important mediator for DDR1/NF-kB-mediated suppression of p53-induced apoptosis, suggesting a role for DDR1 in chemoresistance and cell survival [45]. In macrophages, both NF-kB and c-jun N-terminal kinase (JNK) MAPK have been found to play a critical role in DDR1-induced expression of the immune response regulator of inducible nitric oxide synthase and subsequent production of nitric oxide. Collagen-/DDR1-mediated activation of NF-kB was shown to be regulated through the recruitment of Shc to the receptor [92]. DDR1 also signals via JNK MAPK to induce the upregulation of N-cadherin and EMT in pancreatic cancer cells [68]. While unstimulated DDR1 has been shown to suppress ERK1/2 MAPK phosphorylation [78], collagen-stimulated DDR1 induces ERK1/2 MAPK activation and cell

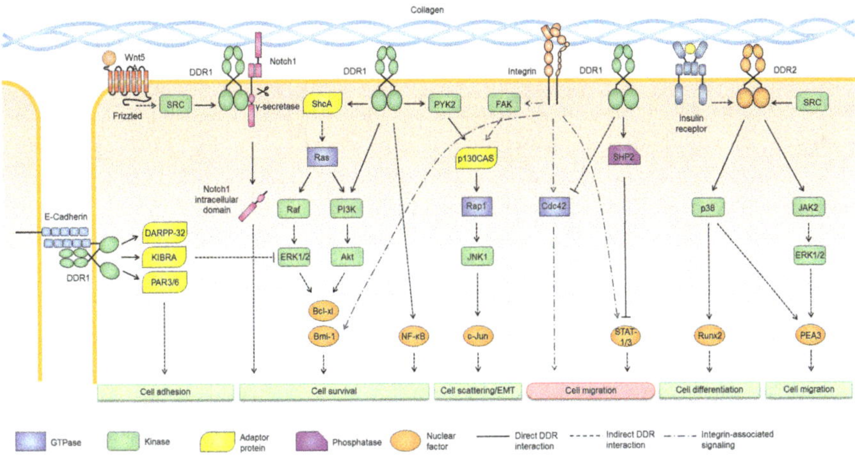


**Fig. 12.2** DDR-associated signaling pathways. DDRs participate in different complexes and induce different signaling pathways, depending on whether they are involved in cell–matrix interactions or collagen-independent cell–cell interactions. Collagen-activated DDR1 induces cell survival through the activation of NF-kB and several MAPKs, including the Ras/Raf/ERK and Ras/PI3K/ERK signaling axes, while DDR2 signals through p38 and the JAK2/ERK MAPK cascades. In a collagen-independent mechanism, binding of inactive DDR1 to E-cadherin in cell–cell junctions appears to block collagen-induced cell migration. DDR1 interacts with several survival receptors, including Notch1 and Frizzled5, while DDR2 interacts with the insulin receptor to promote cell migration. DDR1 has been shown to synergize or antagonize collagen-binding integrin signaling cascades. Thus DDR1-activated PYK2 and α2β1 integrin-activated FAK cooperate to induce JNK1 activity and cell scattering. Signals from DDR1 and α2β1 integrin also converge to enhance Bmi-1 activation. Conversely, DDR1 has been shown to suppress cell migration by blocking integrin-mediated Cdc42 and STAT1/3 activity. Arrows pointing to DDRs indicate pathways involved in DDR activation, and arrows pointing away from DDRs indicate pathways induced by DDRs. The squared green boxes indicate processes induced by DDRs, and the rounded red box indicates a process suppressed by DDRs

adhesion and proliferation in mesangial cells and human breast cancer cells [93]. Similarly, DDR1-induced ERK1/2 MAPK phosphorylation mediates human vascular smooth muscle cells (SMCs) migration through an SRC-mediated mechanism [94]. Recent studies have shown that collagen I-induced DDR1 activation triggers Ras/Raf/ERK1/2 signaling pathway activation and Ras-dependent Akt phosphorylation in colon carcinoma cells and mouse embryonic stem cells (mESCs) [88, 90]. Interestingly, several studies have implicated the Ras/Raf/ERK MAPK signaling axis in the transcriptional regulation of DDR1 expression in a variety of normal and cancer cells. For example, activation of primary human T cells via their T-cell receptor upregulates DDR1 expression through the Ras/Raf/ERK MAPK and the protein kinase PKC pathways, thereby promoting T-cell migration in three-dimensional collagen matrices [57, 95]. Since the activation of DDR1 also triggers Ras/Raf/ERK signaling pathway, a positive feedback loop has been suggested in which DDR1 could positively regulate its own expression [90].

Cross talk between DDR2 and MAPK signaling pathways has also been demonstrated. For example, DDR2 induces SMC migration through p38 MAPK activation [96], and several studies have shown that DDR2 plays a crucial role in osteoblast differentiation and chondrocyte maturation through ERK1/2- and p38 MAPK-mediated activation of the transcription factor Runx2 [97, 98]. DDR2-mediated ERK1/2 MAPK phosphorylation has also been shown to induce MMP-13 expression through activator protein (AP)-1 and Runx2 activation in synovial fibroblasts in the presence of collagen II [99]. A recent report indicates that DDR2-mediated NF-kB and JNK activation is responsible for collagen I-induced IL-2 expression in dendritic cells [100]. DDR2 has also been shown to mediate the induction of IL-2 and CD86 and allogeneic T-cell reaction indicating a role for DDR2 in mediating dendritic cell-regulated immune responses induced by collagen [101, 102]. In primary human fibroblasts, collagen I-activated DDR2 has been shown to induce DDR1 and MMP-10 expression through a Janus kinase (JAK)2- and ERK1/2 MAPK-mediated mechanism, which also involves the recruitment of polyoma enhancer A-binding protein 3 to the DDR1 promoter [103]. Furthermore, DDR2-/JAK2-mediated DDR1 and collagen IV-degrading MMP expression promoted fibroblast migration through collagen IV, suggesting a physiological role for DDR2 in fibroblast transmigration through basement membranes. These results suggest a coordinated regulation and function of both DDRs in response to fibrillar collagen [104].

There is accumulating evidence that DDRs interact with other receptors that regulate or are regulated by DDR signaling. For example, DDR1- and collagen-induced survival has been shown to be Notch1 dependent. DDR1 directly binds the Notch1 receptor in the presence of collagen I and triggers Notch1 cleavage by γ-secretase, thereby generating the Notch 1 intracellular domain, which induces the expression of cell survival genes such as *Hes1* and *Hey2* [46]. Conversely, collagen activation of DDR1 is stimulated by other signaling molecules. For instance, transforming growth factor (TGF)-β-regulated Wnt5a, a member of the wingless-related protein family, has been shown to mediate collagen-induced mammary cell migration and adhesion via SRC tyrosine kinase-mediated activation of DDR1 [71, 105, 106].

A recent study using a mass spectrometry-based phosphoproteomic approach points to a direct cross talk between DDR2 and the insulin receptor by which insulin promotes DDR2 phosphorylation and signaling in the presence of collagen I [87]. Specifically, insulin stimulation enhanced collagen I-induced DDR2 phosphorylation at Y740, which is required for full receptor activation [86]. Clustering analysis showed that the SH2 domain-containing inositol 5'-phosphatase (SHIP)-2 and the cytoskeletal kinase SgK269/PEAK1 present the same phosphorylation profile as DDR2, suggesting a cooperative relationship between DDR2 and the insulin receptor in cell spreading and migration [107–109].

Integrins are the main cell adhesion molecules for cellular binding to ECMs, and four β1 integrins (α1β1, α2β1, α10β1, and α11β1) function primarily as collagen receptors [110]. Since both β1 integrins and DDRs bind to collagen, several studies have investigated whether DDR signaling is linked to integrin-mediated cellular responses. Integrins and RTKs have been shown to interact to regulate many biological processes. Their cooperation involves mutual transactivation, regulation of their cellular compartmentalization, and signaling cross talk [111]. Although DDR tyrosine phosphorylation and activation are independent of collagen-binding integrins [112], some of the DDR1-activated signaling pathways have been shown to converge with downstream integrin-dependent signaling and facilitate or suppress integrin-mediated cellular responses. For example, in epithelial Madin-Darby canine kidney (MDCK) cells, collagen I-activated DDR1 negatively regulates integrin signal transduction by inhibiting α2β1-induced tyrosine phosphorylation of STAT1 and STAT3 and cell migration through the recruitment and activation of SHP-2. Interestingly, neither the SHP-2N-SH2 nor C-SH2 domain alone seems to interact with DDR1, but both SH2 domains and the phosphotyrosyl phosphatase (PTP) domain of SHP-2 are required for binding to DDR1. This supports previous reports suggesting that both the SHP-2N-SH2 and C-SH2 domains are important for binding with phosphotyrosyl-containing proteins and for C-terminal PTP regulatory function [113]. Mutation of tyrosine residues Y703 and Y796 in DDR1 abolished the recruitment of SHP-2 to the receptor, thereby restoring STAT1 and STAT3 activity and collagen-induced cell migration [61]. Consistent with its role in stabilization of cell-surface E-cadherin in cell–cell junctions, collagen-activated DDR1 has also been shown to suppress α2β1 integrin-FAK-Cdc42-mediated E-cadherin destabilization and cell spreading in MDCK cells [62, 79, 114]. E-cadherin-mediated adhesion is critical in maintaining tissue architecture during development and tissue repair, and several reports have shown that the cell membrane pool of E-cadherin plays a critical role during cell migration [115]. The role of DDR1 in increased junctional stability suggests that in the presence of proliferative signals, DDR1 is released from cell–cell junctions and exposed to collagen. Subsequent DDR1 stimulation results in the inhibition of the β1 integrin signaling and Cdc42 activation. Because Cdc42 destabilizes E-cadherin/catenin complexes, DDR1-mediated inhibition of Cdc42 results in increased stability of the complexes. These results indicate opposing effects of DDR1 and β1 integrins in cell migration and differentiation, suggesting that the signaling balance between DDR1 and β1 integrins may contribute to cell fate decisions during differentiation [114]. Some studies suggest that cooperation between DDR1 and β1 integrins also promotes cell adhesion to collagen [89], and in a recent report, Staudinger et al. have shown that DDR1 expression levels positively regulate β1 integrin functional interactions with fibrillar collagen by regulating posttranslational modifications of the β1 integrin subunit [116]. In contrast, the cooperation of DDR1 and integrin α2β1 is required to mediate collagen I-induced EMT and cell invasion in pancreatic cancer cells. In these cells, DDR1-activated PYK2 and α2β1 integrin-activated FAK converge to induce the activation of the small GTPase Rap1 and JNK1, which in turn upregulates N-cadherin expression through c-Jun, thereby promoting cell scattering. Interestingly, p130CAS, a scaffold protein that binds to FAK and PYK2, also serves as a scaffold for both collagen receptors [68]. Recently, it has been shown that in the presence of collagen I, the DDR1-induced Ras-PI3K/Akt-ERK1/2 MAPK pathway and the α2β1-induced integrin-linked kinase (ILK)-Notch-Gli-1 pathway merge and cooperate to enhance Bmi-1 activation and promote mESC self-renewal [88].

Taken together, these results suggest that DDRs and integrins can play cooperating or antagonizing roles in the regulation of many fundamental cellular processes, highlighting a critical balance between DDRs and integrins in the regulation of cell function. It is important to understand the potential interplay between both receptor types and the specific contribution of each of them to collagen-mediated functional responses in order to effectively design novel therapeutic approaches targeting DDRs.

The rapidly increasing evidence that links DDR function and signaling to fundamental cellular processes in a wide range of immune, inflammatory, remodeling, and malignant disorders highlights the potential novel therapeutic opportunity that selective inhibitors of these unique receptors may represent.

## 12.4 Pharmacological Inhibitors of DDR1 and DDR2

From a pharmacological perspective, DDR1 and DDR2 offer few surprises in terms of the molecules known to be direct inhibitors of the ATP site. The general picture was quite typical of that seen for many RTKs. The profile of molecules identified in an internal high-throughput screen was reasonably typical of that seen in other RTK screens. The hit set identified showed a strong preference compounds which showed a DFG-out (type II) binding mode, with significant representation of DFG in (type I) binders. This binding pocket profile presents a significant challenge to the identification of selective inhibitors of DDR1/DDR2, although excellent progress has been made toward this goal in recent years.

The identification by a chemical proteomic approach of DDR1 as one of the key interaction partners of the clinical RTK inhibitors dasatinib [54], nilotinib [54], and imatinib [117] was nevertheless a striking result for the field and has inspired further developments (Fig. 12.3).

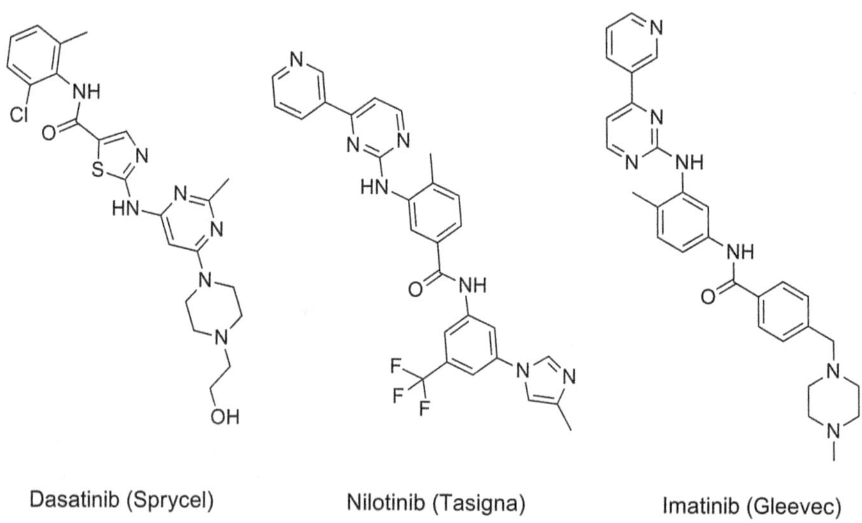


**Fig. 12.3** The chemical structures of RTK inhibitors identified to interact with DDR1 and DDR2

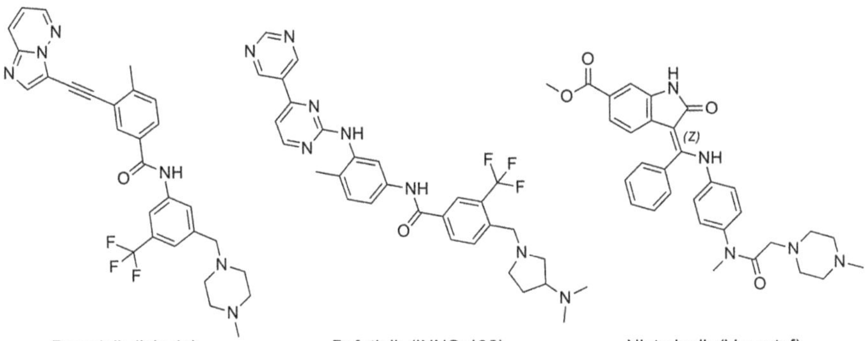


**Fig. 12.4** The chemical structures of additional clinical phase the multi-kinase inhibitors with DDR1 and DDR2 activity

Subsequently, it was also confirmed that RTK inhibitors dasatinib, nilotinib, and imatinib were indeed potent ATP-competitive binders to DDR1 and DDR2 in a TR FRET assay; furthermore, it was shown that the compounds were able to inhibit both collagen-induced autophosphorylation in human embryonic kidney (HEK)293 cells and MCP-1 release in monocytic cells. The consideration of binding modes and homology modeling suggested that despite high homology within the active sites of RTKs, it could be possible to obtain selective inhibitors of DDR1 and DDR2 based on significant differences seen in the ATP phosphate-binding loop (p-loop) of DDR1 and DDR2 [118].

An even wider range of clinical compounds including bafetinib [119], ponatinib [120], and nintedanib (Shaw et al., unpublished data) have now been identified as inhibitors of DDR1 and DDR2 (Fig. 12.4); however these molecules are to a greater or lesser extent multi-kinase inhibitors; by extension the in vivo and in vitro effects of such molecules cannot easily be ascribed to DDR1/DDR2 inhibition alone. The search for more selective tools is however going on in earnest and has been fruitful in elucidating further some of the biology of the DDR family.

The selective DDR1/DDR2 inhibitor DDR1-IN-1 is a type II inhibitor that shows excellent kinase selectivity (Fig. 12.5). This molecule was designed by a library approach based around the known DDR inhibitor nilotinib [121]. The closely related molecule DDR1-IN-2, derived from imatinib, showed a somewhat lower in vitro selectivity. In a range of cancer cell lines, DDR-IN-1 showed a markedly reduced antiproliferative effect relative to DDR-IN-2. This suggests that the inhibition of DDR1/DDR2 in itself is not sufficient to exert pharmacological effects in these cell lines and that any DDR derived effects of DDR-IN-2 are working in synergy with the inhibition of one or more of the kinase targets of DDR-IN-2. Combinatorial kinase inhibitor screening identified a synergistic effect between the selective DDR1/2 inhibitor DDR1-IN-1 and GSK2126458, a PI3K/mTOR inhibitor [121].

The identification of selective DDR1 and/or DDR2 inhibitors however continues to present a very exciting opportunity to understand further the biology of these proteins. Ding et al. have identified a potent and selective inhibitor of DDR1, compound 1 (Fig. 12.6), which is much less potent against BCR-abl, c-Kit, and importantly

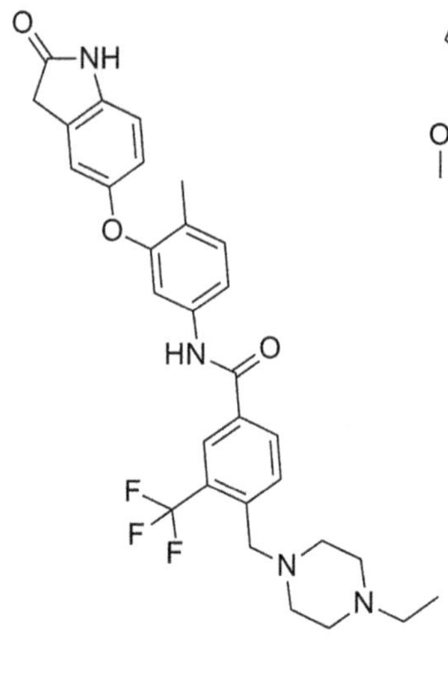

DDR1-IN-1

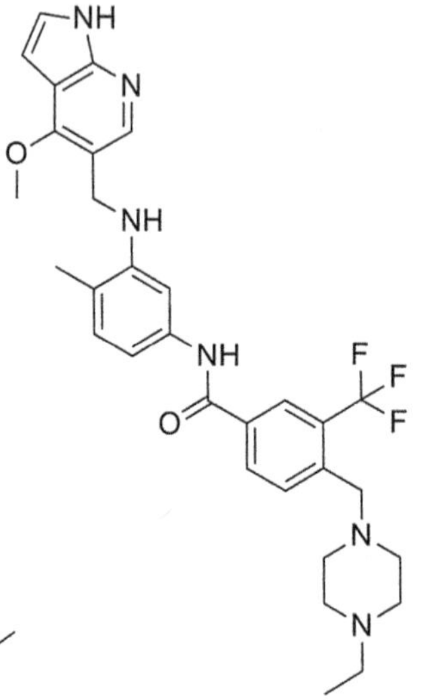

DDR1-IN-2

**Fig. 12.5** The chemical structures of selective DDR1/DDR2 inhibitors

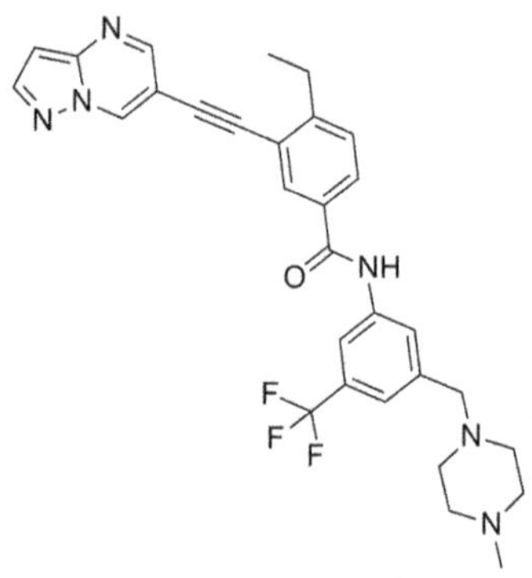

1

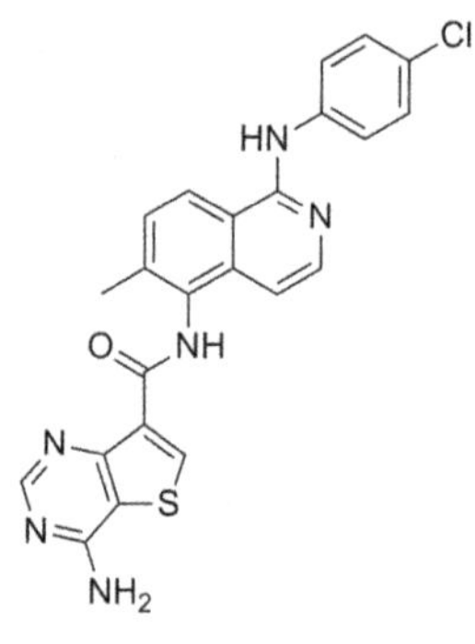

2

**Fig. 12.6** The chemical structure of a compound with good DDR1 selectivity (1) and of a novel RAF/DDR inhibitor (2)

DDR2 [51]. Taking the structure of the known DDR1/DDR2 inhibitor ponatinib as a starting point, and pursuing an extensive medicinal chemistry optimization program, selective inhibitor 1 was discovered. The compound shows around 20-fold selectivity for DDR1 over DDR2 and 60-fold selectivity over Bcr-Abl. Screening against a wider selection of kinases revealed that compound 1 inhibits only 3 % of the kinases tested by more than 35 % (at a concentration of 0.1 μM). Compound 1 was also able to inhibit the downstream signaling of DDR1 in NCI-H23 non-small cell lung cancer

(NSCLC) cells. Further investigation showed that at low micromolar concentrations, the compound could inhibit invasiveness, cell–matrix adhesion, and tumorigenicity in human NSCLC cells. This result contrasts somewhat with the overall conclusions from the study of DDR-IN-1, but the difference is likely to be due to the role of the other kinase activities still present in both molecules.

The promising Rat pharmacokinetics of inhibitor 1 is also of great interest and suggests that this would be a useful tool compound to investigate the in vivo properties of a potent and selective DDR inhibitor.

Pharmaceutical companies also continue to develop DDR inhibitors, and a recent patent publication claims compounds of a chemo type illustrated by compound 2 (Fig. 12.6) as inhibitors of RAF and DDR kinases [122].

In summary, although recent developments have led to the identification of very promising pharmacological tool inhibitors of DDR1 and DDR2, there still remains a need for more selective tools to further elucidate the biology of these important kinases.

## References


1. Alves F, Vogel W, Mossie K, Millauer B, Hofler H, Ullrich A (1995) Distinct structural characteristics of discoidin I subfamily receptor tyrosine kinases and complementary expression in human cancer. Oncogene 10(3):609–618
2. Vogel W (1999) Discoidin domain receptors: structural relations and functional implications. FASEB J 13:S77–S82
3. Shrivastava A, Radziejewski C, Campbell E, Kovac L, McGlynn M, Ryan TE, Davis S, Goldfarb MP, Glass DJ, Lemke G, Yancopoulos GD (1997) An orphan receptor tyrosine kinase family whose members serve as nonintegrin collagen receptors. Mol Cell 1(1):25–34
4. Vogel W, Gish GD, Alves F, Pawson T (1997) The discoidin domain receptor tyrosine kinases are activated by collagen. Mol Cell 1(1):13–23
5. Labrador JP, Azcoitia V, Tuckermann J, Lin C, Olaso E, Manes S, Bruckner K, Goergen JL, Lemke G, Yancopoulos G, Angel P, Martinez C, Klein R (2001) The collagen receptor DDR2 regulates proliferation and its elimination leads to dwarfism. EMBO Rep 2(5):446–452
6. Vogel WF, Abdulhussein R, Ford CE (2006) Sensing extracellular matrix: an update on discoidin domain receptor function. Cell Signal 18(8):1108–1116
7. Leitinger B, Kwan AP (2006) The discoidin domain receptor DDR2 is a receptor for type X collagen. Matrix Biol 25(6):355–364. doi:10.1016/j.matbio.2006.05.006
8. Lemmon MA, Schlessinger J (2010) Cell signaling by receptor tyrosine kinases. Cell 141(7):1117–1134. doi:10.1016/j.cell.2010.06.011
9. Alves F, Saupe S, Ledwon M, Schaub F, Hiddemann W, Vogel WF (2001) Identification of two novel, kinase-deficient variants of discoidin domain receptor 1: differential expression in human colon cancer cell lines. FASEB J 15(7):1321–1323
10. Leitinger B (2003) Molecular analysis of collagen binding by the human discoidin domain receptors, DDR1 and DDR2. Identification of collagen binding sites in DDR2. J Biol Chem 278(19):16761–16769. doi:10.1074/jbc.M301370200
11. Farndale RW, Lisman T, Bihan D, Hamaia S, Smerling CS, Pugh N, Konitsiotis A, Leitinger B, de Groot PG, Jarvis GE, Raynal N (2008) Cell-collagen interactions: the use of peptide toolkits to investigate collagen-receptor interactions. Biochem Soc Trans 36(Pt 2):241–250. doi:10.1042/BST0360241

12. Gu TL, Deng X, Huang F, Tucker M, Crosby K, Rimkunas V, Wang Y, Deng G, Zhu L, Tan Z, Hu Y, Wu C, Nardone J, MacNeill J, Ren J, Reeves C, Innocenti G, Norris B, Yuan J, Yu J, Haack H, Shen B, Peng C, Li H, Zhou X, Liu X, Rush J, Comb MJ (2011) Survey of tyrosine kinase signaling reveals ROS kinase fusions in human cholangiocarcinoma. PLoS One 6(1), e15640. doi:10.1371/journal.pone.0015640
13. Ichikawa O, Osawa M, Nishida N, Goshima N, Nomura N, Shimada I (2007) Structural basis of the collagen-binding mode of discoidin domain receptor 2. EMBO J 26(18):4168–4176
14. Konitsiotis AD, Raynal N, Bihan D, Hohenester E, Farndale RW, Leitinger B (2008) Characterization of high affinity binding motifs for the discoidin domain receptor DDR2 in collagen. J Biol Chem 283(11):6861–6868
15. Xu H, Raynal N, Stathopoulos S, Myllyharju J, Farndale RW, Leitinger B (2011) Collagen binding specificity of the discoidin domain receptors: binding sites on collagens II and III and molecular determinants for collagen IV recognition by DDR1. Matrix Biol 30(1):16–26. doi:10.1016/j.matbio.2010.10.004
16. Carafoli F, Bihan D, Stathopoulos S, Konitsiotis AD, Kvansakul M, Farndale RW, Leitinger B, Hohenester E (2009) Crystallographic insight into collagen recognition by discoidin domain receptor 2. Structure 17(12):1573–1581. doi:10.1016/j.str.2009.10.012
17. Slack BE, Siniaia MS, Blusztajn JK (2006) Collagen type I selectively activates ectodomain shedding of the discoidin domain receptor 1: involvement of Src tyrosine kinase. J Cell Biochem 98(3):672–684. doi:10.1002/jcb.20812
18. Vogel WF (2002) Ligand-induced shedding of discoidin domain receptor 1. FEBS Lett 514(2–3):175–180
19. Ikeda K, Wang LH, Torres R, Zhao H, Olaso E, Eng FJ, Labrador P, Klein R, Lovett D, Yancopoulos GD, Friedman SL, Lin HC (2002) Discoidin domain receptor 2 interacts with Src and Shc following its activation by type I collagen. J Biol Chem 277(21):19206–19212
20. Finger C, Escher C, Schneider D (2009) The single transmembrane domains of human receptor tyrosine kinases encode self-interactions. Sci Signal 2(89), ra56. doi:10.1126/scisignal.2000547
21. Noordeen NA, Carafoli F, Hohenester E, Horton MA, Leitinger B (2006) A transmembrane leucine zipper is required for activation of the dimeric receptor tyrosine kinase DDR1. J Biol Chem 281(32):22744–22751. doi:10.1074/jbc.M603233200
22. Agarwal G, Kovac L, Radziejewski C, Samuelsson SJ (2002) Binding of discoidin domain receptor 2 to collagen I: an atomic force microscopy investigation. Biochemistry 41(37):11091–11098
23. Agarwal G, Mihai C, Iscru DF (2007) Interaction of discoidin domain receptor 1 with collagen type 1. J Mol Biol 367(2):443–455. doi:10.1016/j.jmb.2006.12.073
24. Chin GS, Lee S, Hsu M, Liu W, Kim WJ, Levinson H, Longaker MT (2001) Discoidin domain receptors and their ligand, collagen, are temporally regulated in fetal rat fibroblasts in vitro. Plast Reconstr Surg 107(3):769–776
25. Hou G, Vogel WF, Bendeck MP (2002) Tyrosine kinase activity of discoidin domain receptor 1 is necessary for smooth muscle cell migration and matrix metalloproteinase expression. Circ Res 90(11):1147–1149
26. Ferri N, Carragher NO, Raines EW (2004) Role of discoidin domain receptors 1 and 2 in human smooth muscle cell-mediated collagen remodeling: potential implications in atherosclerosis and lymphangioleiomyomatosis. Am J Pathol 164(5):1575–1585
27. Li W, Zhang YQ, Liu XP, Yao LB, Sun L (2005) Regular expression of discoidin domain receptor 2 in the improved adjuvant-induced animal model for rheumatoid arthritis. Chin Acad Med Sci 20(2):133–137
28. Xu L, Peng H, Wu D, Hu K, Goldring MB, Olsen BR, Li Y (2005) Activation of the discoidin domain receptor 2 induces expression of matrix metalloproteinase 13 associated with osteoarthritis in mice. J Biol Chem 280(1):548–555. doi:10.1074/jbc.M411036200
29. Kamohara H, Yamashiro S, Galligan C, Yoshimura T (2001) Discoidin domain receptor 1 isoform-a (DDR1alpha) promotes migration of leukocytes in three-dimensional collagen lattices. FASEB J 15(14):2724–2726

30. Valiathan RR, Marco M, Leitinger B, Kleer CG, Fridman R (2012) Discoidin domain receptor tyrosine kinases: new players in cancer progression. Cancer Metastasis Rev 31(1–2):295–321. doi:10.1007/s10555-012-9346-z
31. Lee R, Eidman KE, Kren SM, Hostetter TH, Segal Y (2004) Localization of discoidin domain receptors in rat kidney. Nephron Exp Nephrol 97(2):e62–e70. doi:10.1159/000078407
32. Olaso E, Ikeda K, Eng FJ, Xu L, Wang LH, Lin HC, Friedman SL (2001) DDR2 receptor promotes MMP-2-mediated proliferation and invasion by hepatic stellate cells. J Clin Invest 108(9):1369–1378
33. Avivi-Green C, Singal M, Vogel WF (2006) Discoidin domain receptor 1-deficient mice are resistant to bleomycin-induced lung fibrosis. Am J Respir Crit Care Med 174(4):420–427
34. Park HS, Kim KR, Lee HJ, Choi HN, Kim DK, Kim BT, Moon WS (2007) Overexpression of discoidin domain receptor 1 increases the migration and invasion of hepatocellular carcinoma cells in association with matrix metalloproteinase. Oncol Rep 18(6):1435–1441
35. Shimada K, Nakamura M, Ishida E, Higuchi T, Yamamoto H, Tsujikawa K, Konishi N (2008) Prostate cancer antigen-1 contributes to cell survival and invasion though discoidin receptor 1 in human prostate cancer. Cancer Sci 99(1):39–45. doi:10.1111/j.1349-7006.2007.00655.x
36. Yoshida D, Teramoto A (2007) Enhancement of pituitary adenoma cell invasion and adhesion is mediated by discoidin domain receptor-1. J Neurooncol 82(1):29–40. doi:10.1007/s11060-006-9246-6
37. Yang SH, Baek HA, Lee HJ, Park HS, Jang KY, Kang MJ, Lee DG, Lee YC, Moon WS, Chung MJ (2010) Discoidin domain receptor 1 is associated with poor prognosis of non-small cell lung carcinomas. Oncol Rep 24(2):311–319
38. Nemoto T, Ohashi K, Akashi T, Johnson JD, Hirokawa K (1997) Overexpression of protein tyrosine kinases in human esophageal cancer. Pathobiology 65(4):195–203
39. Barker KT, Martindale JE, Mitchell PJ, Kamalati T, Page MJ, Phippard DJ, Dale TC, Gusterson BA, Crompton MR (1995) Expression patterns of the novel receptor-like tyrosine kinase, DDR, in human breast tumours. Oncogene 10(3):569–575
40. Hidalgo-Carcedo C, Hooper S, Chaudhry SI, Williamson P, Harrington K, Leitinger B, Sahai E (2011) Collective cell migration requires suppression of actomyosin at cell-cell contacts mediated by DDR1 and the cell polarity regulators Par3 and Par6. Nat Cell Biol 13(1):49–58. doi:10.1038/ncb2133
41. Laval S, Butler R, Shelling AN, Hanby AM, Poulsom R, Ganesan TS (1994) Isolation and characterization of an epithelial-specific receptor tyrosine kinase from an ovarian cancer cell line. Cell Growth Differ 5(11):1173–1183
42. Shen Q, Cicinnati VR, Zhang X, Iacob S, Weber F, Sotiropoulos GC, Radtke A, Lu M, Paul A, Gerken G, Beckebaum S (2010) Role of microRNA-199a-5p and discoidin domain receptor 1 in human hepatocellular carcinoma invasion. Mol Cancer 9:227. doi:10.1186/1476-4598-9-227
43. Couvelard A, Hu J, Steers G, O'Toole D, Sauvanet A, Belghiti J, Bedossa P, Gatter K, Ruszniewski P, Pezzella F (2006) Identification of potential therapeutic targets by gene-expression profiling in pancreatic endocrine tumors. Gastroenterology 131(5):1597–1610. doi:10.1053/j.gastro.2006.09.007
44. Ram R, Lorente G, Nikolich K, Urfer R, Foehr E, Nagavarapu U (2006) Discoidin domain receptor-1a (DDR1a) promotes glioma cell invasion and adhesion in association with matrix metalloproteinase-2. J Neurooncol 76(3):239–248
45. Das S, Ongusaha PP, Yang YS, Park JM, Aaronson SA, Lee SW (2006) Discoidin domain receptor 1 receptor tyrosine kinase induces cyclooxygenase-2 and promotes chemoresistance through nuclear factor-kappaB pathway activation. Cancer Res 66(16):8123–8130
46. Kim HG, Hwang SY, Aaronson SA, Mandinova A, Lee SW (2011) DDR1 receptor tyrosine kinase promotes prosurvival pathway through Notch1 activation. J Biol Chem 286(20):17672–17681. doi:10.1074/jbc.M111.236612
47. Valencia K, Ormazabal C, Zandueta C, Luis-Ravelo D, Anton I, Pajares MJ, Agorreta J, Montuenga LM, Martinez-Canarias S, Leitinger B, Lecanda F (2012) Inhibition of collagen receptor discoidin domain receptor-1 (DDR1) reduces cell survival, homing, and colonization in lung cancer bone metastasis. Clin Cancer Res 18(4):969–980. doi:10.1158/1078-0432.CCR-11-1686

48. Drilon A, Rekhtman N, Ladanyi M, Paik P (2012) Squamous-cell carcinomas of the lung: emerging biology, controversies, and the promise of targeted therapy. Lancet Oncol 13(10):e418–e426. doi:10.1016/S1470-2045(12)70291-7
49. Hammerman PS, Sos ML, Ramos AH, Xu C, Dutt A, Zhou W, Brace LE, Woods BA, Lin W, Zhang J, Deng X, Lim SM, Heynck S, Peifer M, Simard JR, Lawrence MS, Onofrio RC, Salvesen HB, Seidel D, Zander T, Heuckmann JM, Soltermann A, Moch H, Koker M, Leenders F, Gabler F, Querings S, Ansen S, Brambilla E, Brambilla C, Lorimier P, Brustugun OT, Helland A, Petersen I, Clement JH, Groen H, Timens W, Sietsma H, Stoelben E, Wolf J, Beer DG, Tsao MS, Hanna M, Hatton C, Eck MJ, Janne PA, Johnson BE, Winckler W, Greulich H, Bass AJ, Cho J, Rauh D, Gray NS, Wong KK, Haura EB, Thomas RK, Meyerson M (2011) Mutations in the DDR2 kinase gene identify a novel therapeutic target in squamous cell lung cancer. Cancer Discovery 1(1):78–89. doi:10.1158/2159-8274.CD-11-0005
50. Ohashi K, Pao W (2011) A new target for therapy in squamous cell carcinoma of the lung. Cancer Discovery 1(1):23–24. doi:10.1158/2159-8274.CD-11-0069
51. Gao M, Duan L, Luo J, Zhang L, Lu X, Zhang Y, Zhang Z, Tu Z, Xu Y, Ren X, Ding K (2013) Discovery and optimization of 3-(2-(Pyrazolo[1,5-a]pyrimidin-6-yl)ethynyl)benzamides as novel selective and orally bioavailable discoidin domain receptor 1 (DDR1) inhibitors. J Med Chem 56(8):3281–3295. doi:10.1021/jm301824k
52. Zhang J, Yang PL, Gray NS (2009) Targeting cancer with small molecule kinase inhibitors. Nat Rev Cancer 9(1):28–39. doi:10.1038/nrc2559
53. Broekman F, Giovannetti E, Peters GJ (2011) Tyrosine kinase inhibitors: Multi-targeted or single-targeted? World J Clin Oncol 2(2):80–93. doi:10.5306/wjco.v2.i2.80
54. Rix U, Hantschel O, Durnberger G, Remsing Rix LL, Planyavsky M, Fernbach NV, Kaupe I, Bennett KL, Valent P, Colinge J, Kocher T, Superti-Furga G (2007) Chemical proteomic profiles of the BCR-ABL inhibitors imatinib, nilotinib, and dasatinib reveal novel kinase and nonkinase targets. Blood 110(12):4055–4063. doi:10.1182/blood-2007-07-102061
55. Hubbard SR (2004) Juxtamembrane autoinhibition in receptor tyrosine kinases. Nat Rev Mol Cell Biol 5(6):464–471
56. Flynn LA, Blissett AR, Calomeni EP, Agarwal G (2010) Inhibition of collagen fibrillogenesis by cells expressing soluble extracellular domains of DDR1 and DDR2. J Mol Biol 395(3):533–543. doi:10.1016/j.jmb.2009.10.073
57. Hachehouche LN, Chetoui N, Aoudjit F (2010) Implication of discoidin domain receptor 1 in T cell migration in three-dimensional collagen. Mol Immunol 47(9):1866–1869. doi:10.1016/j.molimm.2010.02.023
58. Siddiqui K, Kim GW, Lee DH, Shin HR, Yang EG, Lee NT, Yang BS (2009) Actinomycin D identified as an inhibitor of discoidin domain receptor 2 interaction with collagen through an insect cell based screening of a drug compound library. Biol Pharm Bull 32(1):136–141
59. Mihai C, Chotani M, Elton TS, Agarwal G (2009) Mapping of DDR1 distribution and oligomerization on the cell surface by FRET microscopy. J Mol Biol 385(2):432–445
60. Margadant C, Monsuur HN, Norman JC, Sonnenberg A (2011) Mechanisms of integrin activation and trafficking. Curr Opin Cell Biol 23(5):607–614. doi:10.1016/j.ceb.2011.08.005
61. Wang CZ, Su HW, Hsu YC, Shen MR, Tang MJ (2006) A discoidin domain receptor 1/SHP-2 signaling complex inhibits alpha2beta1-integrin-mediated signal transducers and activators of transcription 1/3 activation and cell migration. Mol Biol Cell 17(6):2839–2852. doi:10.1091/mbc.E05-11-1068
62. Yeh YC, Wang CZ, Tang MJ (2009) Discoidin domain receptor 1 activation suppresses alpha-2beta1 integrin-dependent cell spreading through inhibition of Cdc42 activity. J Cell Physiol 218(1):146–156. doi:10.1002/jcp.21578
63. Amit I, Citri A, Shay T, Lu Y, Katz M, Zhang F, Tarcic G, Siwak D, Lahad J, Jacob-Hirsch J, Amariglio N, Vaisman N, Segal E, Rechavi G, Alon U, Mills GB, Domany E, Yarden Y (2007) A module of negative feedback regulators defines growth factor signaling. Nat Genet 39(4):503–512. doi:10.1038/ng1987
64. Gordus A, Krall JA, Beyer EM, Kaushansky A, Wolf-Yadlin A, Sevecka M, Chang BH, Rush J, MacBeath G (2009) Linear combinations of docking affinities explain quantitative differences in RTK signaling. Mol Syst Biol 5:235. doi:10.1038/msb.2008.72

65. Fu HL, Valiathan RR, Arkwright R, Sohail A, Mihai C, Kumarasiri M, Mahasenan KV, Mobashery S, Huang P, Agarwal G, Fridman R (2013) Discoidin domain receptors: unique receptor tyrosine kinases in collagen-mediated signaling. J Biol Chem 288(11):7430–7437. doi:10.1074/jbc.R112.444158
66. L'Hote CG, Thomas PH, Ganesan TS (2002) Functional analysis of discoidin domain receptor 1: effect of adhesion on DDR1 phosphorylation. FASEB J 16(2):234–236
67. Huang Y, Arora P, McCulloch CA, Vogel WF (2009) The collagen receptor DDR1 regulates cell spreading and motility by associating with myosin IIA. J Cell Sci 122(Pt 10):1637–1646. doi:10.1242/jcs.046219
68. Shintani Y, Fukumoto Y, Chaika N, Svoboda R, Wheelock MJ, Johnson KR (2008) Collagen I-mediated up-regulation of N-cadherin requires cooperative signals from integrins and discoidin domain receptor 1. J Cell Biol 180(6):1277–1289
69. Koo DH, McFadden C, Huang Y, Abdulhussein R, Friese-Hamim M, Vogel WF (2006) Pinpointing phosphotyrosine-dependent interactions downstream of the collagen receptor DDR1. FEBS Lett 580(1):15–22
70. Yang G, Li Q, Ren S, Lu X, Fang L, Zhou W, Zhang F, Xu F, Zhang Z, Zeng R, Lottspeich F, Chen Z (2009) Proteomic, functional and motif-based analysis of C-terminal Src kinase-interacting proteins. Proteomics 9(21):4944–4961. doi:10.1002/pmic.200800762
71. Dejmek J, Dib K, Jonsson M, Andersson T (2003) Wnt-5a and G-protein signaling are required for collagen-induced DDR1 receptor activation and normal mammary cell adhesion. Int J Cancer 103(3):344–351. doi:10.1002/ijc.10752
72. Lemeer S, Bluwstein A, Wu Z, Leberfinger J, Muller K, Kramer K, Kuster B (2012) Phosphotyrosine mediated protein interactions of the discoidin domain receptor 1. J Proteome 75(12):3465–3477. doi:10.1016/j.jprot.2011.10.007
73. Thalappilly S, Soubeyran P, Iovanna JL, Dusetti NJ (2010) VAV2 regulates epidermal growth factor receptor endocytosis and degradation. Oncogene 29(17):2528–2539. doi:10.1038/onc.2010.1
74. Faraci-Orf E, McFadden C, Vogel WF (2006) DDR1 signaling is essential to sustain Stat5 function during lactogenesis. J Cell Biochem 97(1):109–121
75. Xi S, Zhang Q, Gooding WE, Smithgall TE, Grandis JR (2003) Constitutive activation of Stat5b contributes to carcinogenesis in vivo. Cancer Res 63(20):6763–6771
76. Dejmek J, Leandersson K, Manjer J, Bjartell A, Emdin SO, Vogel WF, Landberg G, Andersson T (2005) Expression and signaling activity of Wnt-5a/discoidin domain receptor-1 and Syk plays distinct but decisive roles in breast cancer patient survival. Clin Cancer Res 11(2 Pt 1):520–528
77. Hansen C, Greengard P, Nairn AC, Andersson T, Vogel WF (2006) Phosphorylation of DARPP-32 regulates breast cancer cell migration downstream of the receptor tyrosine kinase DDR1. Exp Cell Res 312(20):4011–4018. doi:10.1016/j.yexcr.2006.09.003
78. Hilton HN, Stanford PM, Harris J, Oakes SR, Kaplan W, Daly RJ, Ormandy CJ (2008) KIBRA interacts with discoidin domain receptor 1 to modulate collagen-induced signalling. Biochim Biophys Acta 1783(3):383–393. doi:10.1016/j.bbamcr.2007.12.007
79. Wang CZ, Yeh YC, Tang MJ (2009) DDR1/E-cadherin complex regulates the activation of DDR1 and cell spreading. Am J Physiol Cell Physiol 297(2):C419–C429. doi:10.1152/ajpcell.00101.2009
80. Bryant DM, Stow JL (2004) The ins and outs of E-cadherin trafficking. Trends Cell Biol 14(8):427–434. doi:10.1016/j.tcb.2004.07.007
81. Takeichi M (1991) Cadherin cell adhesion receptors as a morphogenetic regulator. Science 251(5000):1451–1455
82. Eswaramoorthy R, Wang CK, Chen WC, Tang MJ, Ho ML, Hwang CC, Wang HM, Wang CZ (2010) DDR1 regulates the stabilization of cell surface E-cadherin and E-cadherin-mediated cell aggregation. J Cell Physiol 224(2):387–397. doi:10.1002/jcp.22134
83. Benink HA, Bement WM (2005) Concentric zones of active RhoA and Cdc42 around single cell wounds. J Cell Biol 168(3):429–439. doi:10.1083/jcb.200411109
84. Etienne-Manneville S, Manneville JB, Nicholls S, Ferenczi MA, Hall A (2005) Cdc42 and Par6-PKC zeta regulate the spatially localized association of Dlg1 and APC to control cell polarization. J Cell Biol 170(6):895–901. doi:10.1083/jcb.200412172

85. Wood W, Jacinto A, Grose R, Woolner S, Gale J, Wilson C, Martin P (2002) Wound healing recapitulates morphogenesis in Drosophila embryos. Nat Cell Biol 4(11):907–912. doi:10.1038/ncb875
86. Yang K, Kim JH, Kim HJ, Park IS, Kim IY, Yang BS (2005) Tyrosine 740 phosphorylation of discoidin domain receptor 2 by Src stimulates intramolecular autophosphorylation and Shc signaling complex formation. J Biol Chem 280(47):39058–39066. doi:10.1074/jbc.M506921200
87. Iwai LK, Chang F, Huang PH (2013) Phosphoproteomic analysis identifies insulin enhancement of discoidin domain receptor 2 phosphorylation. Cell Adhes Migr 7(2):161–164. doi:10.4161/cam.22572
88. Suh HN, Han HJ (2011) Collagen I regulates the self-renewal of mouse embryonic stem cells through alpha2beta1 integrin- and DDR1-dependent Bmi-1. J Cell Physiol 226(12):3422–3432. doi:10.1002/jcp.22697
89. Xu H, Bihan D, Chang F, Huang PH, Farndale RW, Leitinger B (2012) Discoidin domain receptors promote alpha1beta1- and alpha2beta1-integrin mediated cell adhesion to collagen by enhancing integrin activation. PLoS One 7(12), e52209. doi:10.1371/journal.pone.0052209
90. Ongusaha PP, Kim JI, Fang L, Wong TW, Yancopoulos GD, Aaronson SA, Lee SW (2003) p53 induction and activation of DDR1 kinase counteract p53-mediated apoptosis and influence p53 regulation through a positive feedback loop. EMBO J 22(6):1289–1301
91. Sakuma S, Saya H, Tada M, Nakao M, Fujiwara T, Roth JA, Sawamura Y, Shinohe Y, Abe H (1996) Receptor protein tyrosine kinase DDR is up-regulated by p53 protein. FEBS Lett 398(2–3):165–169
92. Kim SH, Lee S, Suk K, Bark H, Jun CD, Kim DK, Choi CH, Yoshimura T (2007) Discoidin domain receptor 1 mediates collagen-induced nitric oxide production in J774A.1 murine macrophages. Free Radic Biol Med 42(3):343–352. doi:10.1016/j.freeradbiomed.2006.10.052
93. Curat CA, Vogel WF (2002) Discoidin domain receptor 1 controls growth and adhesion of mesangial cells. J Am Soc Nephrol 13(11):2648–2656
94. Lu KK, Trcka D, Bendeck MP (2011) Collagen stimulates discoidin domain receptor 1-mediated migration of smooth muscle cells through Src. Cardiovasc Pathol 20(2):71–76. doi:10.1016/j.carpath.2009.12.006
95. Chetoui N, El Azreq MA, Boisvert M, Bergeron ME, Aoudjit F (2011) Discoidin domain receptor 1 expression in activated T cells is regulated by the ERK MAP kinase signaling pathway. J Cell Biochem 112(12):3666–3674. doi:10.1002/jcb.23300
96. Chen SC, Wang BW, Wang DL, Shyu KG (2008) Hypoxia induces discoidin domain receptor-2 expression via the p38 pathway in vascular smooth muscle cells to increase their migration. Biochem Biophys Res Commun 374(4):662–667
97. Lin KL, Chou CH, Hsieh SC, Hwa SY, Lee MT, Wang FF (2010) Transcriptional upregulation of DDR2 by ATF4 facilitates osteoblastic differentiation through p38 MAPK-mediated Runx2 activation. J Bone Miner Res 25(11):2489–2503. doi:10.1002/jbmr.159
98. Zhang Y, Su J, Yu J, Bu X, Ren T, Liu X, Yao L (2011) An essential role of discoidin domain receptor 2 (DDR2) in osteoblast differentiation and chondrocyte maturation via modulation of Runx2 activation. J Bone Miner Res 26(3):604–617
99. Su J, Yu J, Ren T, Zhang W, Zhang Y, Liu X, Sun T, Lu H, Miyazawa K, Yao L (2009) Discoidin domain receptor 2 is associated with the increased expression of matrix metalloproteinase-13 in synovial fibroblasts of rheumatoid arthritis. Mol Cell Biochem 330(1–2):141–152
100. Poudel B, Ki HH, Lee YM, Kim DK (2013) Induction of IL-12 production by the activation of discoidin domain receptor 2 via NF-kappaB and JNK pathway. Biochem Biophys Res Commun 434(3):584–588. doi:10.1016/j.bbrc.2013.03.118
101. Lee JE, Kang CS, Guan XY, Kim BT, Kim SH, Lee YM, Moon WS, Kim DK (2007) Discoidin domain receptor 2 is involved in the activation of bone marrow-derived dendritic cells caused by type I collagen. Biochem Biophys Res Commun 352(1):244–250. doi:10.1016/j.bbrc.2006.11.010

102. Poudel B, Yoon DS, Lee JH, Lee YM, Kim DK (2012) Collagen I enhances functional activities of human monocyte-derived dendritic cells via discoidin domain receptor 2. Cell Immunol 278(1–2):95–102. doi:10.1016/j.cellimm.2012.07.004
103. Ruiz PA, Jarai G (2011) Collagen I induces discoidin domain receptor (DDR) 1 expression through DDR2 and a JAK2-ERK1/2-mediated mechanism in primary human lung fibroblasts. J Biol Chem 286(15):12912–12923
104. Ruiz PA, Jarai G (2012) Discoidin domain receptors regulate the migration of primary human lung fibroblasts through collagen matrices. Fibrogenesis Tissue Repair 5:3. doi:10.1186/1755-1536-5-3
105. Jonsson M, Andersson T (2001) Repression of Wnt-5a impairs DDR1 phosphorylation and modifies adhesion and migration of mammary cells. J Cell Sci 114(Pt 11):2043–2053
106. Roarty K, Serra R (2007) Wnt5a is required for proper mammary gland development and TGF-beta-mediated inhibition of ductal growth. Development 134(21):3929–3939. doi:10.1242/dev.008250
107. Kato K, Yazawa T, Taki K, Mori K, Wang S, Nishioka T, Hamaguchi T, Itoh T, Takenawa T, Kataoka C, Matsuura Y, Amano M, Murohara T, Kaibuchi K (2012) The inositol 5-phosphatase SHIP2 is an effector of RhoA and is involved in cell polarity and migration. Mol Biol Cell 23(13):2593–2604. doi:10.1091/mbc.E11-11-0958
108. Kelber JA, Klemke RL (2010) PEAK1, a novel kinase target in the fight against cancer. Oncotarget 1(3):219–223
109. Prasad N, Topping RS, Decker SJ (2001) SH2-containing inositol 5'-phosphatase SHIP2 associates with the p130(Cas) adapter protein and regulates cellular adhesion and spreading. Mol Cell Biol 21(4):1416–1428. doi:10.1128/MCB.21.4.1416-1428.2001
110. Leitinger B, Hohenester E (2007) Mammalian collagen receptors. Matrix Biol 26(3):146–155
111. Comoglio PM, Boccaccio C, Trusolino L (2003) Interactions between growth factor receptors and adhesion molecules: breaking the rules. Curr Opin Cell Biol 15(5):565–571
112. Vogel W, Brakebusch C, Fassler R, Alves F, Ruggiero F, Pawson T (2000) Discoidin domain receptor 1 is activated independently of beta(1) integrin. J Biol Chem 275(8):5779–5784
113. Neel BG, Gu H, Pao L (2003) The 'Shp'ing news: SH2 domain-containing tyrosine phosphatases in cell signaling. Trends Biochem Sci 28(6):284–293. doi:10.1016/S0968-0004(03)00091-4
114. Yeh YC, Wu CC, Wang YK, Tang MJ (2011) DDR1 triggers epithelial cell differentiation by promoting cell adhesion through stabilization of E-cadherin. Mol Biol Cell 22(7):940–953. doi:10.1091/mbc.E10-08-0678
115. Palacios F, Schweitzer JK, Boshans RL, D'Souza-Schorey C (2002) ARF6-GTP recruits Nm23-H1 to facilitate dynamin-mediated endocytosis during adherens junctions disassembly. Nat Cell Biol 4(12):929–936. doi:10.1038/ncb881
116. Staudinger LA, Spano SJ, Lee W, Coelho N, Rajshankar D, Bendeck MP, Moriarty T, McCulloch CA (2013) Interactions between the discoidin domain receptor 1 and beta1 integrin regulate attachment to collagen. Biol Open 2(11):1148–1159. doi:10.1242/bio.20135090
117. Bantscheff M, Eberhard D, Abraham Y, Bastuck S, Boesche M, Hobson S, Mathieson T, Perrin J, Raida M, Rau C, Reader V, Sweetman G, Bauer A, Bouwmeester T, Hopf C, Kruse U, Neubauer G, Ramsden N, Rick J, Kuster B, Drewes G (2007) Quantitative chemical proteomics reveals mechanisms of action of clinical ABL kinase inhibitors. Nat Biotechnol 25(9):1035–1044. doi:10.1038/nbt1328
118. Day E, Waters B, Spiegel K, Alnadaf T, Manley PW, Buchdunger E, Walker C, Jarai G (2008) Inhibition of collagen-induced discoidin domain receptor 1 and 2 activation by imatinib, nilotinib and dasatinib. Eur J Pharmacol 599(1–3):44–53
119. Rix U, Remsing Rix LL, Terker AS, Fernbach NV, Hantschel O, Planyavsky M, Breitwieser FP, Herrmann H, Colinge J, Bennett KL, Augustin M, Till JH, Heinrich MC, Valent P, Superti-Furga G (2010) A comprehensive target selectivity survey of the BCR-ABL kinase inhibitor INNO-406 by kinase profiling and chemical proteomics in chronic myeloid leukemia cells. Leukemia 24(1):44–50. doi:10.1038/leu.2009.228
120. Kotz J (2011) The DDR is in. SciBX 4(20):1–2. doi:10.1038/scibx.2011.559

121. Kim HG, Tan L, Weisberg EL, Liu F, Canning P, Choi HG, Ezell SA, Wu H, Zhao Z, Wang J, Mandinova A, Griffin JD, Bullock AN, Liu Q, Lee SW, Gray NS (2013) Discovery of a Potent and Selective DDR1 Receptor Tyrosine Kinase Inhibitor. ACS Chem Biol 8(10):2145–2150. doi:10.1021/cb400430t
122. Bae I-H, Son J-B, Han S, Kwak E-J, Kim H-S, Song J-Y, Byun E-Y, Jun S-A, Ahn Y-G, Suh, K-H WO 2013/100632 A1

# Chapter 13
# Functions of DDR1 in Epithelial Cell Differentiation

**Yi-Chun Yeh and Ming-Jer Tang**

## 13.1 Collagen in Epithelial Cell Differentiation

The assembly of cell–cell adherens junctions is key to determining epithelial cell differentiation. However, the switching point in controlling epithelial cell differentiation as well as adherens junction assembly is largely dependent on extracellular signals, such as growth factors, cytokines, and the extracellular matrix (ECM). For instance, growth factors such as transforming growth factor β1 (TGF-β1) and hepatocyte growth factor (HGF) induce epithelial cell dedifferentiation via either suppression of E-cadherin gene expression, a key molecule in adherens junctions, or disruption of E-cadherin-dependent cell–cell adhesion [1–5]. The bone morphogenetic protein 7 (BMP 7) exhibits opposite effects in regulating epithelial cell differentiation and also counteracts TGF-β1-induced renal fibrosis [6].

Accumulation of fibrillar collagen is commonly seen in many diseases, including carcinogenesis and tissue fibrosis, in which normal epithelium are replaced by poorly organized fibroblast or mesenchymal cells [7–9]. Under normal physiological conditions, epithelial cells reside atop a basement membrane sheet, which forms a natural barrier to the underlying fibrillar collagen. However, during tissue fibrosis or cancer progression, the integrity of the basement membrane is interrupted due to remodeling by active fibroblasts or transformed epithelial cells [9]. This causes the epithelial cells to be exposed to the underlying fibrillar collagen. In vitro studies have demonstrated that the fibrillar collagen coating reduces epithelial plasticity, including cell polarity, aggregation, and junctional stability, whereas it increases

Y.-C. Yeh • M.-J. Tang, M.D., Ph.D. (✉)
Department of Physiology, National Cheng Kung University Medicine College, Tainan 70101, Taiwan
e-mail: mjtang1@mail.ncku.edu.tw



R. Fridman, P.H. Huang (eds.), *Discoidin Domain Receptors in Health and Disease*, DOI 10.1007/978-1-4939-6383-6_13

mesenchymal properties, such as cell migration, invasion, and morphological changes [10–12]. Critically, the expression of fibrillar collagen, particularly type I, facilitates TGF-β1-, TGF-β3-, and fibroblast growth factor-1 (FGF-1)-induced loss of epithelial plasticity [13–15]. These results demonstrate the negative consequences of fibrillar collagen in terms of regulating epithelial cell differentiation.

Two major types of collagen receptors have been identified in epithelial cells: integrins and discoidin domain receptors (DDRs). Integrins are composed of an α and β subunit, which are connected by noncovalent interactions. Integrins α1β1, α2β1, α10β1, and α11β1 are the major receptors for collagen binding [16]. Among these receptors, the functions of α2β1 integrin in collagen- and TGF-β1-mediated epithelial cell dedifferentiation have been extensively studied [14, 17, 18]. However, as a collagen receptor, DDR1 shows the opposite effects in regulating epithelial cell differentiation in comparison with integrin β1.

DDRs contain the homolog domain identified in Discoidin I protein, which is expressed in the slime mold *Dictyostelium discoideum* with the functions of cell adhesion and migration [19, 20]. DDR1 and DDR2 are two distinct members of the DDRs family [21–23]. DDR1 is widely expressed during embryonic development and is predominantly expressed in the epithelium, particularly in the skin, lung, kidney, gut, and brain of adult mice [24]. In contrast, DDR2 is mainly expressed in the connective tissue, muscle, and heart [21, 25, 26]. Although DDR1 has been discovered with its major expression pattern in epithelia cells [24, 27, 28], its functions in epithelial cell differentiation have only recently been discovered. Before considering in detail how DDR1 regulates epithelial cell differentiation, it is vital to understand the process of epithelial cell differentiation as well as how the assembly of adherens junctions contributes to it.

## 13.2 Epithelial Cell Differentiation

Epithelial cells form tightly cohesive sheets that line the interior and exterior surfaces of our body. This structure is essential for the integrity of tissues during embryogenesis and prohibits the invasion of pathogens throughout the life cycle of the body. Differentiated epithelial cells are revealed during the early embryonic stage, known as the epidermis, in which epithelial cells are tightly connected to one another by cell–cell adhesions. The epidermis develops into ectoderm and endoderm after gastrulation. Ectoderm cells then further differentiate into the skin and nervous system, while endoderm cells develop into the gut, lungs, liver, etc. This process is also known as organogenesis, in which epithelial cells eventually acquire the unique functions of different organs. These long-term processes of embryogenesis are therefore accomplished by unquantifiable cell division, migration, and even apoptosis. Once cells enter the cell cycle, or become involved in migratory movement, or the loss of neighboring cells, epithelial cells temporally reduce or lose their properties, for example, cell–cell adhesions. After the removal of stimuli, such as growth factors or cytokines, cells then regain cell–cell contacts and re-establish

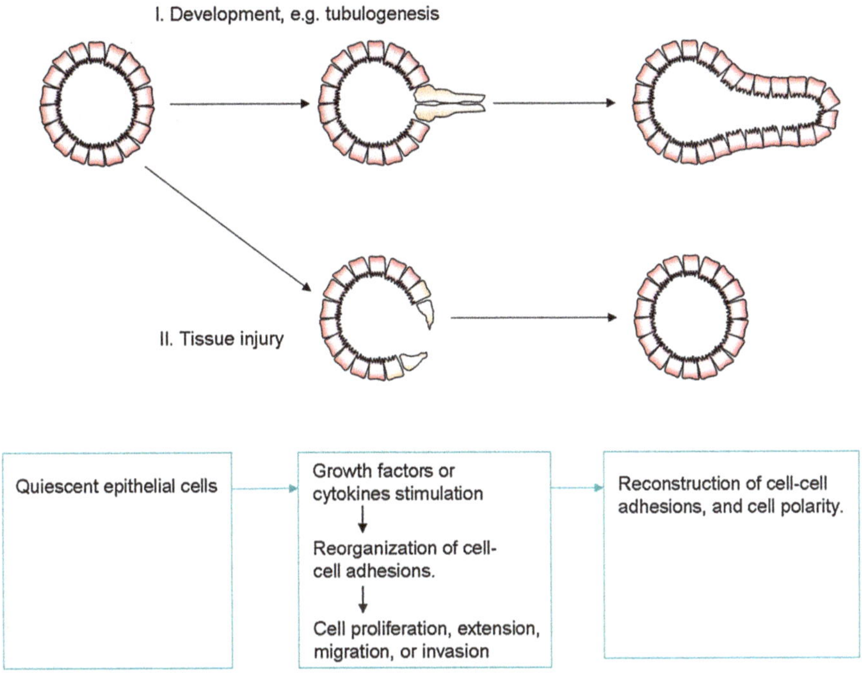


**Fig. 13.1** Epithelial cell differentiation. During development, microenvironmental cues, including growth factors, trigger the rearrangement of cell–cell adhesions, and promote the subsequent events, including cell proliferation, extension, migration, and invasion, in specific regions of epithelial cells. During these processes, many cellular events occur, the most important of which are the reconstruction of cell–cell junctions, the establishment of cell polarity, and the formation of specialized membrane compartments. Epithelial cell differentiation also occurs in adult tissues, for example, after tissue injury. The increase in growth factors or cytokines signals, together with the loss of cell–cell contacts, causes epithelial cells around the injury site to begin extension, and proliferation. Eventually, the growth cells occupy the space and repair the epithelial cell sheets through the formation of cell–cell contacts. Differentiated epithelial cells can also be derived from mesenchymal cells, although this is not included in the above figure

their epithelial properties. As a result, tightly controlled epithelial cell differentiation and dedifferentiation are critical during development (Fig. 13.1). Other than originating from the epidermis, epithelial cells can also develop from mesenchymal cells. For instance, part of the kidney's tubular epithelial cells is derived from metanephrogenic mesenchyme, which originates in the intermediate mesoderm [29–31]. This process involves more complex genetic regulation than simply changes in cell morphology and is known as mesenchymal–epithelial transition [32, 33].

In adult organs, the process of epithelial cell differentiation is still necessary for maintaining the homeostasis of epithelial cells. For example, the epithelial cells in the mouth and intestines have a high renewal rate, whereas the epithelial cells in the skin and respiratory tract feature a high regenerative ability to overcome infection- or injury-induced cell loss. In the case of cell loss during injury, epithelial cells around the injured area lose cell–cell contacts and contact-inhibiting cell growth. Hence, cells change their morphology from cuboidal or squamous cell phenotypes to a flattened fibroblast-like morphology, following which the cells then start to proliferate and recover the injured space. Afterwards, the cells regain cell–cell adhesions and build up the epithelial architecture. This process is delicately controlled, and so loss of this control may lead to disoriented tissue structure or even impair their normal functions [34, 35] (Fig. 13.1).

According to the description above, epithelial cells can be derived from preexisting epithelial cells through dedifferentiation and proliferation. On the other hand, they can also differentiate from mesenchymal cells. Via either means, the construction of cell–cell adhesion and the formation of polarizing membrane compartments are the foremost processes leading to epithelial cell differentiation.

## 13.3 Adherens Junctions in Epithelial Cell Differentiation

Epithelial cells are polarized cells that are connected via different interactions, including tight junctions, adherens junctions, desmosomes, and gap junctions. Among these, both tight junctions and adherens junctions are essential for the establishment of cell polarity. Tight junctions are unique structures found in epithelial cells, which seal beneath the apical membrane and separate the membrane into two compartments, namely the apical and basolateral membrane. The apical membrane develops organ-specific functions or adaptive morphology, such as the secretion of mammary gland epithelial cells, active transport and absorption of intestine or kidney epithelial cells, or the brush border of simple columnar intestine epithelial cells. The basolateral membrane connects to the surrounding cells and transmits signals from the basement membrane. In other words, to accomplish the specific functions of different organs, epithelial cells have developed specialized membrane compartments and tight junctions, which represent the key components for ensuring the identity of different membrane domains. Therefore, the formation of tight junctions is crucial for the functional differentiation of epithelial cells.

Adherens junctions localize on the basolateral side of tight junctions and are considered to be the principle structures in determining the onset of epithelial cell differentiation. The knockdown of the key adherens junctional molecule, E-cadherin, not only disrupts the formation of adherens junctions, but also impairs the assembly of tight junctions [36–38]. However, the downregulation of E-cadherin is an essential step in neural as well as mesodermal differentiation during the early developmental stage, a process known as epithelial–mesenchymal transdifferentiation [39–41]. Evidence suggests that the loss of E-cadherin is harmful in epithelial differentiation but favorable in mesenchymal differentiation, as the loss of E-cadherin is usually accompanied by the increased expression of mesenchymal specific genes, such as fibroblast-specific protein 1 (FSP1) and α-smooth muscle actin (α-SMA) [42–45]. The characteristics of cell migratory and invasive ability are also increased by the loss of E-cadherin function. Therefore, E-cadherin downregulation is closely associated with cancer progression [46–48].

The assembly of adherens junctions is a highly organized and hierarchical process that is initiated by the nectin-based cell–cell adhesions [49, 50]. The transdimerization of nectin activates the small G protein Cdc42 and Rac1 to promote actin organization and recruit non*trans*-interacting E-cadherin to the nectin-based cell–cell adhesion sites [51–53]. Herein, *trans*-homodimer E-cadherin assembles via the $Ca^{2+}$-dependent cadherin extracellular domains, and then activates Rac1 and Cdc42 via the activation of phosphatidylinositol 3-kinase to recruit more non*trans*-interacting E-cadherin [52]. The aggregation of E-cadherin together with the downstream molecules, β-catenin, p120-catenin, α-catenin, vinculin, and actin cytoskeleton, forms a stable *trans*-homocluster complex between neighboring cells [54, 55]. Among these downstream molecules, α-catenin and vinculin directly bind the actin cytoskeleton and control actin belt organization, which is necessary for the stabilization of cadherin–catenin complex and cell–cell adherens junctions [56, 57].

The membrane bound E-cadherin therefore separates into two groups. One is the *cis*- and *trans*-homodimer E-cadherin featuring properties, such a weak adhesive force, fast dynamic rate, and fast turnover rate [55, 58–60]. The other group is the *trans*-homocluster E-cadherin, which forms a stable complex on cell–cell junctions and provides a strong force in maintaining cell–cell contacts and cell architecture [61–63].

## 13.4 Functions of Discoidin Domain Receptor 1 (DDR1) in Epithelial Cell Differentiation

### *13.4.1 Evidence of DDR1-Induced Epithelial Cell Differentiation In Vitro*

The first study to address the function of DDR1 in epithelial cell differentiation was performed on a mammary epithelial cell line [64]. In the study, the authors demonstrated that overexpression of Wnt-5a enhanced cell-to-collagen adhesion and abolished HGF-induced branching morphogenesis in collagen gels, and that an increase in DDR1 activation could be the underlying mechanism. The ectopic expression of Wnt-5a in MCF-7 breast cancer cells, which do not express endogenous Wnt-5a, spontaneously triggered DDR1 phosphorylation and cell aggregation. Although there was no direct evidence showing that activation of DDR1 was required for Wnt-5a-reduced branching morphogenesis upon HGF treatment or Wnt-5a-induced cell phenotypic change, later studies identified the functions of DDR1 in branching tubulogenesis and epithelial morphological changes by using different cell lines.

Branching tubulogenesis is an important developmental feature of many organs, including the lungs, mammary gland, kidneys, etc. This process can be mimicked using Madin–Darby canine kidney (MDCK) cells in a well-established model system. The MDCK cells grown in three-dimensional collagen gels form a ball-like structure in the absence of HGF treatment, and upon HGF treatment the cells develop branching tubules with distinct apical and basolateral membranes and the lumenal structures [65, 66]. The process is completed by the coordination of numerous cellular events, including cell proliferation, reorganization of cell–cell junctions, loss of cell polarity, increase in invasive ability, cell apoptosis, and reconstruction of the cell–cell junctions and lumen structure [67–69].

Studies involving MDCK cells overexpressing wild-type or kinase domain truncated DDR1 indicated that the overexpression of DDR1 suppressed HGF-induced branching tubulogenesis, whereas expression of kinase domain truncated DDR1 suppressed cell aggregation and polarized epithelial architecture but increased long cell protrusion [70]. Further studies indicated that neither the reduction in cell proliferation nor the increase in cell apoptosis was the cause of the reduced number and shorter tubular structures in DDR1 overexpressing cells, but that a reduction in cell migratory and invasive ability was the main reason [70]. Further, a study has now identified the detailed mechanism by which DDR1 suppresses cell migration [71]. The activation of DDR1 by fibrillar collagen suppresses the signal transducer and activator of transcription 1 and 3 (STAT1 and 3) activation through direct association and activation of SHP-2, a Src homology 2 (SH2) domain containing nontransmembrane protein tyrosine phosphatase. Interestingly, STAT1/3 activation is mediated by an integrin β1-dependent adhesion signal [71]. This indicates that collagen induces two counteractive signals through two different membrane receptors, and that the alteration in the expression levels of DDR1 and integrin β1 would determine different cell fates.

Additional evidence shows the counteractive functions of DDR1 and integrin β1 on their functions in regulating epithelial cell spreading, which is also an indicator for epithelial cell differentiation. MDCK cells overexpressed with DDR1, or treated with an integrin β1 blocking antibody, reduces the cell extension area; however, cells overexpressed with dominant negative DDR1 increases the cell extension areas on collagen-coated dishes [70, 72]. This opposite effect on cell spreading regulation is due to the different regulation of Cdc42 activation. Active integrin β1 triggers Cdc42 activation, whereas active DDR1 suppresses Cdc42 activation upon collagen stimulation [72]. The decrease in cell migration, invasion, and spreading suggests the functional role of DDR1 is triggering epithelial differentiation and so provides the mechanism by which DDR1 reduces HGF-induced tubulogenesis in collagen gels.

Actin cytoskeleton organization is an important index for distinguishing the differentiation of epithelial cells [73, 74]. In epithelial cells, junctional complexes, including adherens junctions and tight junctions, coordinates the organization of the actin cytoskeleton. In so doing, actin forms a belt-like structure, also known as cortical actin, around the cell borders. In fibroblast or mesenchymal cells, which have a higher migratory ability compared with epithelial cells, the organization of the actin cytoskeleton is dominated by the integrins-dependent adhesion signal. As a consequence, the actin cytoskeleton is arranged into stress fibers tethered on integrin-mediated focal adhesions [73, 74]. Cells expressing DDR1 exhibit cortical actin around the cell borders. In contrast, cells expressing dominant negative DDR1 or DDR1 shRNA trigger stress fiber assembly [75]. This result suggests that the expression of DDR1 affects actin cytoskeleton organization in epithelial cells.

In addition to actin cytoskeleton organization, the increase in microvilli structure is further evidence supporting the role of DDR1 in promoting epithelial cell differentiation. The microvilli, the small cell membrane protrusion on the apical surface of epithelial cells, perform many cellular functions, such as secretion, absorption, and mechanotransduction. This structure represents polarized and well-differentiated epithelial cells. In DDR1 overexpressing epithelial cells, the number of microvilli structures is significantly increased; while the number of microvilli is significantly decreased in dominant negative DDR1 expressing cells when compared with wild-type cells [75] (Fig. 13.2). Cells grown on a fibrillar collagen-coated dish diminish microvilli structure; however, overexpression of DDR1 further enriches the number of microvilli [75]. This again indicates that activation of DDR1 by fibrillar collagen promotes epithelial cell differentiation and counteracts integrin β1-mediated epithelial cell dedifferentiation.

The discovery of the functions of DDR1 in increasing E-cadherin-mediated cell–cell contacts provides strong evidence to support the role of DDR1 in regulating epithelial cell differentiation and also reveals its underlying mechanism. Overexpression of DDR1 in both a kidney proximal tubule epithelial cell line and a mammary epithelial cell line triggers cell aggregation and shows clear cell boundary. Immunofluorescence studies indicate that expression of DDR1 triggers the membrane localization of E-cadherin, whereas expression of dominant negative DDR1 or DDR1 shRNA promotes cell scattering and cytosolic accumulation of E-cadherin [75] (Fig. 13.2). Adherens junctions provide a strong force for maintaining cell–cell contacts and cell integrity even during cell migration. As a result, cells with strong cell–cell contacts should display less morphological changes and less difference in cell orientations during cell migration. A time-lapse recording indicated that expression of DDR1 not only reduces cell migratory ability but also the morphological changes during cell migration. In contrast, the DDR1 knockdown cells displayed high flexibility in cell–cell association during cell migration [75]. This result suggests that the increase in membrane localization of E-cadherin is correlated with the strong cell–cell adhesive force in DDR1 expressing cells.

An increase in the E-cadherin-mediated adhesion signal would suppress mesenchymal gene expression. The downregulation of mesenchymal marker proteins, including integrin β1, fibronectin, and α-SMA was identified in DDR1 expressing cells. In contrast, cells expressing dominant negative DDR1 or DDR1 shRNA contained higher levels of integrin β1, fibronectin, and α-SMA as compared to wild-type cells [75]. These results offer important evidence in support of the functions of DDR in promoting epithelial cell differentiation (Fig. 13.2).

In summary, expression of DDR1 not only increases epithelial characteristics, as manifested by an increase in microvilli structure, cortical actin assembly, E-cadherin-mediated cell–cell adhesion, but also decreases mesenchymal properties, such as cell migration, invasion, extension, fibronectin deposition, and the protein levels of integrin β1 and α-SMA.

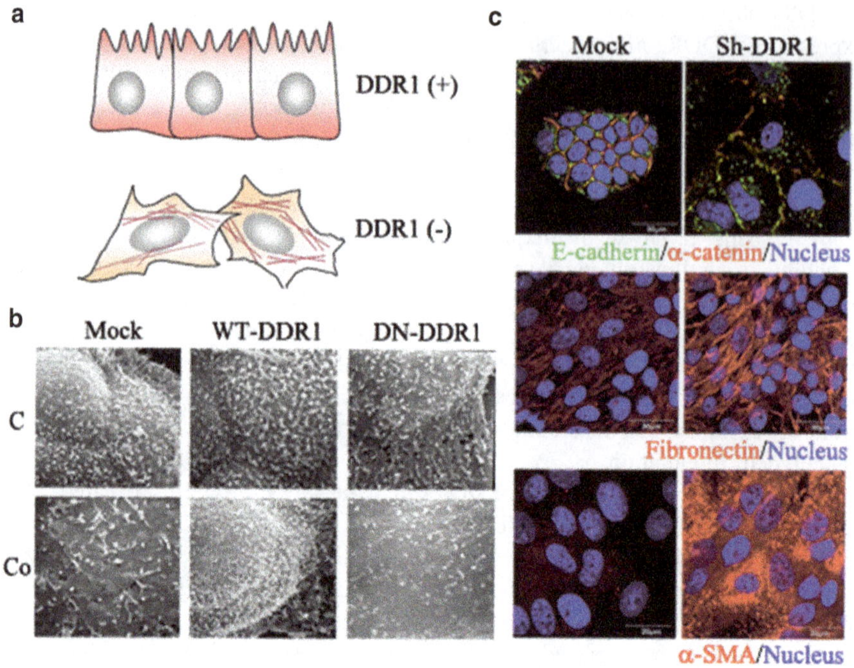


**Fig. 13.2** DDR1 promotes epithelial cell differentiation. (A) Expression of DDR1 increases epithelial cell differentiation with the characteristics of polarized membrane compartments, well-arranged cell–cell junctions, and an actin-belt cytoskeleton beneath cell–cell junctions. Loss of DDR1 promotes a fibroblast phenotypic change, as categorized by a more scattered and flattened cell shape, the cell membrane lack of specialized compartments, and the assembly of stress fibers replacing the actin-belt structure. (B) Various DDR1 expression clones of LLC-PK1 cells, including control (Mock), wild-type DDR1 overexpression clones (WT-DDR1), and dominant negative DDR1 overexpression clones (DN-DDR1) were incubated on coverslips with (Co) or without (C) the type I collagen coating for 48 h. The cell surface microvilli structure was then examined by scanning electron microscopy (SEM). (C) The protein localization of E-cadherin (green) and α-catenin (red), and the protein expression of fibronectin and α-SMA in control (Mock) and DDR1 knockdown cells (Sh-DDR1) were examined with specific antibodies. B and C were reproduced with the permission from Ref. [75]

## *13.4.2 Mechanisms of DDR1-Induced Epithelial Cell Differentiation*

In addition to the junctional localization of E-cadherin, expression of DDR1 increases the protein abundance of E-cadherin. As the mRNA levels and promoter activity of E-cadherin are not different in different DDR1 expression clones, it suggests that transcriptional regulation is not involved in DDR1-augmented E-cadherin protein levels. The measurement of the protein turnover rate suggests that expression of DDR1 reduces the protein turnover rate of E-cadherin, and that the increase in junctional stability of the E-cadherin/catenin complex could be the main cause

[75]. Ample evidence supports this hypothesis. First, the protein complex of E-cadherin and β-catenin is increased in DDR1 expressing cells. Second, overexpression of E-cadherin in DDR1 knockdown cells cannot rescue the membrane expression of E-cadherin and mesenchymal phenotypic changes [75], which suggests that the functional units of the E-cadherin/catenin complex is the key. Finally, the live cell images in cells transfected with mEos conjugated E-cadherin suggest that the membrane mobility of photoconverted mEos-E-cadherin is much lower in DDR1 expressing cells compared with DDR1 knockdown cells. As the membrane fluidity and turnover rate are very different from the dimer form of E-cadherin to *trans*-homo clustering E-cadherin, it suggests that the increase in protein stability of E-cadherin in DDR1 expressing cells is achieved through increasing the protein complex of *trans*-homoclustering E-cadherin.

The immunoprecipitation result suggests the physical interactions between DDR1 and E-cadherin in their extracellular domain [76]. However, the lack of cytosolic domain of DDR1, that is dominant negative DDR1, cannot maintain the membrane stability of E-cadherin [75]. This result suggests that the physical interactions between DDR1 and E-cadherin are not essential for the membrane stability of the E-cadherin/catenin complex, but that downstream signaling mediated by DDR1 is necessary.

Actin cytoskeleton organization is important in regulating the membrane stability of E-cadherin [77–79]. In general, cortical actin maintains adherens junctional stability, whereas stress fibers impair junctional stability. DDR1 has been shown to be involved in the regulation of the Rho family GTPases, including Cdc42 and Rac1 [72, 75], and the main function of these small GTPases is to control the organization of the actin cytoskeleton. As DDR1 knockdown increases the activation level of Cdc42, and either knockdown Cdc42 or overexpression of dominant inactive Cdc42 rescues DDR1 knockdown-triggered E-cadherin endocytosis and stress fiber assembly, this suggests that Cdc42 activation is involved in DDR1-regulated E-cadherin membrane stability [75] (Fig. 13.3). Exactly how DDR1 regulates Cdc42 activation is not clear, but a predicted GTPase-activating protein (GAP) binding element is found in the cytosolic domain of DDR1 [26].

Although Cdc42 activation is required for membrane recruitment of E-cadherin in the initiation of cell–cell contacts, the protein complex of Cdc42-Par6-aPKC is involved in E-cadherin endocytosis via regulating Wasp- and Arp2/3-dependent actin organization [80, 81]. Moreover, through activating myotonic dystrophy kinase-related Cdc42-binding kinase (MRCK), Cdc42 induces actomyosin assembly, which contributes to cell migration, invasion [82, 83], and disturbs the stability of cell–cell contacts [84].

Like Cdc42-MRCK, Rho-ROCK is also involved in actomyosin assembly through regulating myosin-II light chain (MLC2) activity [82, 85], and activation of Rho-ROCK signaling is also involved in the deregulation of E-cadherin-mediated cell–cell contacts [86]. A study using the A431 squamous cell carcinoma cell line showed that DDR1 forms a physical association with E-cadherin, Par3, and Par6 at cell–cell junctions. The complex of DDR1-Par3-Par6 recruits RhoE at cell–cell junctions and suppresses Rho-ROCK-induced actomyosin contractility, which then helps to maintain the cell–cell adhesions [84] (Fig. 13.3). In summary, DDR1

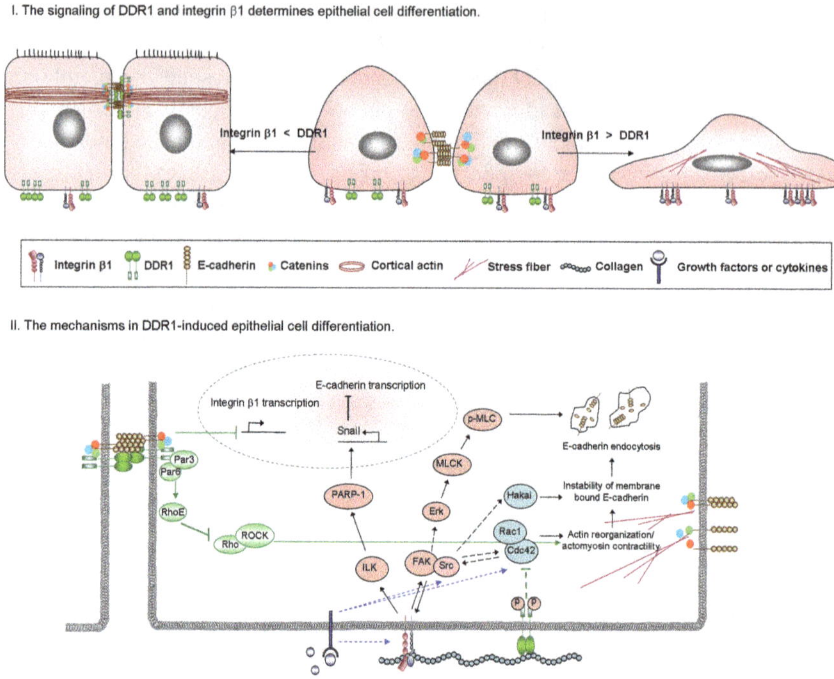


**Fig. 13.3** Mechanisms in DDR1-triggered epithelial cell differentiation. The collagen receptors integrin β1 and DDR1 have different effects on epithelial cell differentiation. An increase in integrin β1 signaling induces the dedifferentiation of epithelial cells, whereas increased DDR1 signaling promotes epithelial cell differentiation. The counter effects on E-cadherin stability explain the opposing actions of integrin β1 and DDR1 in epithelial cell differentiation. Activation of integrin β1 induces loss of E-cadherin-mediated adherens junctions through two different downstream pathways: FAK/Src and ILK. Activation of the FAK/Src pathway triggers signaling through extracellular-regulated kinase (ERK), myosin-light chain kinase (MLCK), and myosin light chain (MLC), which provides contractile force and induces the instability and endocytosis of E-cadherin. On the other hand, Src activation through growth factor signaling triggers the activation of integrin β1 and FAK, which then induces the endocytosis of E-cadherin, suggesting the bidirectional regulation between integrin β1 and Src. Cdc42 and Rac1 activation is another downstream pathway of FAK/Src-induced instability in E-cadherin. The activation of Cdc42 and Rac1-mediated actin cytoskeleton reorganization is necessary for the membrane stability and endocytosis of E-cadherin. Conversely, the activation of Cdc42 positively regulates Src activation and Hakai-mediated endocytosis and degradation of E-cadherin. ILK, another downstream target of integrin β1, increases the transcriptional activity of Poly (ADP-ribose) polymerase-1 (PARP-1), which mediates downregulation of E-cadherin through the induction of the E-cadherin transcriptional repressor, Snail. The different downstream pathways are conducted by microenvironmental cues, for example, the growth factors or cytokines could induce protein instability in E-cadherin through an increase in the integrin β1-mediated adhesion signal. The counteractive function of DDR1 shows an increase in E-cadherin-mediated adherens junctions via either the inactivation of Cdc42 or the inactivation of Rho-ROCK-mediated actomyosin contractility through the association of Par3-Par6. Activation of E-cadherin negatively regulates the gene expression of integrin β1 and the ligand-induced activation of DDR1, suggesting the negative regulation of adherens junctions on cell adhesions. This figure was reproduced with the permission from [99]

triggers epithelial cell differentiation through increasing the membrane stability of the E-cadherin/catenin complex, and inactivation of Cdc42- or Rho-dependent actin organization could be the important mechanism.

The increase of integrin α2β1-dependent adhesion signals is involved in collagen-, TGF-β1-, and unilateral ureteral obstruction (UUO)-induced epithelial cell dedifferentiation and renal fibrosis [13, 17, 18]. The expression of DDR1 suppresses integrin β1-mediated adhesion signals as well as the protein abundance of integrin β1 [72, 75]. An increase in activation or expression of focal adhesion kinase (FAK), Src [87–89] and integrin-linked kinase (ILK) [17, 90–92], the major downstream protein kinases of integrin β1, has been demonstrated to be associated with integrin β1-induced epithelial cell dedifferentiation and disease progression, especially in carcinogenesis and organ fibrosis. The blocking of the FAK/Src signaling axis suppresses integrin β1-induced E-cadherin endocytosis by stabilizing the membrane bound E-cadherin, and switches single-cell migration into collective-cell migration in squamous cell carcinoma cells [89]. Other studies have shown that the signaling cascade, including integrin β1, FAK, extracellular-regulated kinase (ERK), myosin light chain kinase (MLCK), and myosin light chain (MLC), is involved in the junctional localization of E-cadherin [88, 93].

Cdc42 is the downstream molecule involved in the integrin β1/FAK signaling axis-induced cell extension [72], and a reduction of Cdc42 activation decreases the endocytosis of E-cadherin in normal epithelial cells [75]. This suggests that the Cdc42 activation is involved in integrin β1-induced endocytosis of E-cadherin (Fig. 13.3). Interestingly, activation of Cdc42 triggers the ubiquitination and endocytosis of E-cadherin through augmentation of Src activation [94, 95]. These results suggest that a complex feedback loop exists in integrin β1-mediated E-cadherin junctional stability. The bidirectional regulation of integrin β1 and Src, as well as Src and Cdc42, contributes to integrin β1-induced E-cadherin endocytosis and epithelial cell dedifferentiation (Fig. 13.3).

The intercellular and cell-ECM adhesion are key components for maintaining cell tensegrity. The signals which imbalance the force generated by intercellular or cell-ECM adhesion would lead to the changes in cell mobility, invasion capability, cell differentiation, and the cell traction force [96–98]. From this point of view, the counteractive effects on E-cadherin stability explain the opposing actions of integrin β1 and DDR1 in epithelial cell differentiation. As the expression of DDR1 decreases the integrin β1-mediated adhesion signal, it is reasonable to propose that DDR1 increases the E-cadherin-mediated adhesion signal via suppression of the integrin β1-dependent adhesion signal.

During development, the functions of integrin β1 and DDR1 in cell adhesion, migration, tubulogenesis, proliferation, and growth may be either cooperative or opposing. However, after maturation, the functions of integrin β1 and DDR1 in cell differentiation are consistently opposing [99]. The evidence provided here suggests that the counter effects of E-cadherin regulation determine the opposite functions of integrin β1 and DDR1 in epithelial cell differentiation. In other words, the signaling balance between these two collagen receptors, DDR1 and integrin β1, will be crucial in determining epithelial cell differentiation or dedifferentiation (Fig. 13.3).

### *13.4.3 Activation of E-cadherin Negatively Regulates DDR1 Activation*

Although DDR1 positively regulates E-cadherin membrane stability, activation of E-cadherin negatively regulates collagen-induced DDR1 phosphorylation. In various epithelial cells, DDR1 forms physical interactions with E-cadherin, independent of the DDR1 or E-cadherin activation level. However, the activation of E-cadherin actually suppressed collagen-induced DDR1 phosphorylation and limited cell spreading [100]. It is therefore conceivable that stable cell–cell junctions in epithelial cells negatively regulate the activation of DDR1 by trapping DDR1 in the cell–cell junctions. Nevertheless, it is still not clear whether the physical interaction between DDR1 and E-cadherin is required for DDR1-regulated cell–cell adhesion.

This inhibitory effect also occurs on integrin β1. Overexpression of E-cadherin suppresses the gene expression of integrin α5 and β1 in MDA-MB-231 cells [101]. In addition, the $Ca^{2+}$-dependent E-cadherin dimerization suppresses focal adhesion formation and lamellipodia extension and dynamics [102, 103]. These results suggest that the E-cadherin-mediated adherens junctions negatively regulate the integrin β1-mediated adhesion signal (Fig. 13.3). One may therefore speculate that the downregulation of the integrin β1-mediated adhesion signal in DDR1 expressing cells is due to an increase in E-cadherin-mediated cell–cell adhesions.

## 13.5 In vivo Evidence Supporting the Role of DDR1 in Epithelial Cell Differentiation

### *13.5.1 DDR1 Knockout Mice*

DDR1 knockout mice survived gestation but are smaller in size than their control littermates. These mice showed defects in the development of certain organs, such as impaired mammary gland development, a poorly calcified fibula bone, and a narrower pelvis [104]. The DDR2 knockout mice showed dwarfism and a shortening of long bones caused by the decrease in chondrocyte proliferation [105]. A recent study showed that mutation of the DDR2 gene in humans led to similar characteristics, such as shortened limbs, narrowed chest, and chondral calcification [106]. This rare autosomal-recessive disease was first identified in 1993 and is known as spondylo-*meta*-epiphyseal dysplasia (SMED). These results suggest that, despite the different expression patterns of DDR1 and DDR2 in epithelial and stroma cells, overlapping expression and function can exist. Since DDR1 is mainly expressed in epithelium, we focused on the functions of DDR1 in epithelial cell differentiation.

DDR1 knockout females are unable to lactate due to the failure of the alveolar epithelium to secrete milk proteins [104]. During pregnancy, DDR1-null female mice showed hyperproliferation and aberrant differentiation of lobuloalveolar epithelial cells. At birth, the alveoli showed intracellular lipid production and deposition, but failed to secrete milk into the central lumen. In the early pubertal stage, the mammary gland development defect is manifested as a delay in mammary duct outgrowth, and enlargement in the primary ducts and terminal end buds due to a marked increase in the cell proliferation rate in the mutant mice [104]. In addition, a substantial deposition of collagen was shown around the mammary gland epithelial cells in DDR1 knockout mice. The consequence of the alveolar epithelial cells developmental defect is that DDR1 knockout female mice are unable to lactate. In view of the findings that normal mammary epithelia grow and differentiate on the collagen-enriched ECM, these observations indicate that DDR1 has two important functions: inhibition of epithelial cell proliferation and regulation of collagen deposition. Both of the functions play important roles in the regulation of epithelial cell differentiation.

Adult DDR1 knockout mice exhibit proteinuria and urinary acanthocytes. Results from electron microscopy demonstrate a thickening of the subepithelial glomerular basement membrane (GBM) as well as a focal loss of the podocyte slit diaphragms. This data suggests that the loss of cell–matrix communication in DDR1-deficient podocytes appears to result in excess accumulation of basement membrane proteins, which leads to disturbed anchorage of foot processes and disruption of the slit diaphragm. In other words, the interaction between type IV collagen and DDR1 plays an important role in maintaining the structural integrity of the GBM [107].

The loss of DDR1 results in hearing loss due to the substantial structural alteration and the deterioration of the supporting cells, specifically the outer hair cells, which included Deiters', Hensen's, and bordering cells. The loss of DDR1 likewise alters the morphology of basal cells and increases ECM accumulation in the stria vascularis. Severe ultrastructure alterations can be seen in cells proximate to the basilar membrane [108]. The similarities between the kidneys and the inner ear are exhibited in cell types, highly related transporter proteins, ion channels, protein composition in the basement membrane, and most importantly in the disease relevance, such as Alport syndrome, suggesting that the functions of DDR1 could be similar in these organs.

### *13.5.2 Disease*

Upregulation of DDR1 in breast cancer is negatively correlated with cancer cell invasive ability, because the interaction of DDR1 and dopamine and cAMP-regulated neuronal phosphoprotein (DARPP-32) in cancer cells negatively regulates cancer cell migration [109]. A study examining invasive ductal and lobular carcinomas (IDC and ILC) showed that the EMT-related genes, or genes correlated with TGF-β and Wnt signaling were pronounced in ILC. On the other hand, both E-cadherin and DDR1 show prominent expression in IDC instead of ILC [110]. One interesting result showed that DDR1 was coexpressed with E-cadherin in less aggressive breast cancer cell lines, but not in MDA-MB-231, which shows enrichment in EMT-related markers [75]. An increase in DDR1 expression was also observed in borderline ovarian tumors and in low-grade and stage ovarian cancer with the increase of claudin 3, an epithelial cell adhesion molecule [111]. These results suggest that DDR1 positively correlates with cancer cell differentiation, as estimated by the expression of E-cadherin or other cell–cell adhesion molecules, while it negatively correlates with cancer invasion or metastasis.

DDR1 is highly expressed in human bronchial epithelium cells [24]. The immunohistochemistry study indicated that DDR1 is predominantly expressed in the basolateral surface of the bronchial epithelium, and shows an overlap with type IV collagen staining [24]. In the in vitro model system, DDR1 is downregulated in the human bronchial epithelial cells during TGF-β1-induced EMT [112]. DDR1 is also expressed in kidney epithelial cells [70, 75, 107, 113]. In the injury kidney, DDR1 expression shows nonoverlapping with alpha smooth muscle actin (α-SMA) staining, and only expressed in the epithelial cells remaining well cell–cell contact and cuboidal cell shape [99, 113]. Maeyama et al. showed that the reduction of epithelial cell characteristic by overexpression of Slug in MDCK cells reduced the expression of DDR1 [114]. These results suggest that loss of epithelial cell differentiation is accompanied by the downregulation of DDR1.

## 13.6 Conclusion

Preserving the delicate control of epithelial cell differentiation is critical for maintaining various organ functions and tissue integrity. In the present chapter, we do not discuss in detail the distinctive functions or gene regulation of epithelial cells from different organs, for example, the differing properties of simple and stratified squamous epithelial cells, or the specific expression of unique proteins, such as keratin, transporters, or ion channels, but instead focus on the fundamental questions of how cell–cell adhesion, particularly adherens junctions, contribute to epithelial cell differentiation and how DDR1 regulates epithelial cell differentiation through regulating these adherens junctions. We consider how the foremost evidence of DDR1 involvement in growth factor-induced cellular phenotypic changes led to the discovery of the functional role of DDR1 in promoting epithelial cell differentiation. The underlying mechanism in DDR1-regulated epithelial cell differentiation involves stabilizing the E-cadherin-mediated adherens junctions, though more research is needed to provide the exact mechanisms. The in vitro and in vivo evidence supporting the idea that DDR1 triggers epithelial cell differentiation was summarized; however, several important issues still need to be addressed. First, an increase in DDR1 expression was observed in some malignant tumors, such as nonsmall cell lung carcinoma, and the knockdown of DDR1 reduced hypertension-induced renal fibrosis. It is therefore interesting to consider whether the functions of DDR1 in epithelial cell differentiation are cell type specific, and if there are various functions of DDR1 in different types of cells in disease progression? Second, is there a functional compensation between DDR1 and DDR2 during development or disease progression? Finally, whether DDR1-regulated E-cadherin junctional stability is a direct or indirect effect, for example, through suppression of the integrin β1-mediated adhesion signal? These issues provide significant avenues for further study.

## References

1. Oft M, Peli J, Rudaz C et al (1996) TGF-beta1 and Ha-Ras collaborate inmodulating the phenotypic plasticity and invasiveness of epithelial tumor cells. Genes Dev 10(19):2462–2477
2. Sato M, Muragaki Y, Saika S et al (2003) Targeted disruption of TGF-beta1/Smad3 signaling protects against renal tubulointerstitial fibrosis induced by unilateral ureteral obstruction. J Clin Invest 112(10):1486–1494
3. Grotegut S, von Schweinitz D, Christofori G et al (2006) Hepatocyte growth factor induces cell scattering through MAPK/Egr-1-mediated upregulation of snail. EMBO J 25(15): 3534–3545
4. Wang Z, Sandiford S, Wu C et al (2009) Numb regulates cell-cell adhesion and polarity in response to tyrosine kinase signaling. EMBO J 28(16):2360–2373
5. Xu J, Lamouille S, Derynck R (2009) TGF-beta-induced epithelial to mesenchymal transition. Cell Res 19(2):156–172
6. Zeisberg M, Hanai J, Sugimoto H et al (2003) BMP-7 counteracts TGF-beta1-induced epithelial-to-mesenchymal transition and reverses chronic renal injury. Nat Med 9(7):964–968
7. Kivirikko KI (1993) Collagens and their abnormalities in a wide spectrum of diseases. Ann Med 25(2):113–126
8. Zeisberg M, Strutz F, Müller GA (2000) Role of fibroblast activation in inducing interstitial fibrosis. J Nephrol 13(Suppl 3):S111–S120
9. Zeisberg M, Kalluri R (2013) Cellular mechanisms of tissue fibrosis. 1. Common and organ-specific mechanisms associated with tissue fibrosis. Am J Physiol Cell Physiol 304(3): C216–C225
10. Hay ED, Zuk A (1995) Transformations between epithelium and mesenchyme: normal, pathological, and experimentally induced. Am J Kidney Dis 26(4):678–690
11. Shintani Y, Wheelock MJ, Johnson KR (2006) PI3K-rac1-jnk signaling mediates collage I-induced cell scattering and up-regulation of N-cadherin expression in mouse mammary epithelial cells. Mol Biol Cell 17(7):2963–2975
12. Medici D, Nawshad A (2010) Type I collagen promotes epithelial-mesenchymal transition through ILK-dependent activation of NF-kappaB and LEF-1. Matrix Biol 29(3):161–165
13. Vallés AM, Boyer B, Tarone G et al (1996) Alpha 2 beta 1 integrin is required for the collagen and FGF-1 induced cell dispersion in a rat bladder carcinoma cell line. Cell Adhes Commun 4(3):187–199
14. Fan JM, Ng YY, Hill PA et al (1999) Transforming growth factor-beta regulates tubular epithelial-myofibroblast transdifferentiation in vitro. Kidney Int 56(4):1455–1467
15. Shintani Y, Maeda M, Chaika N et al (2008) Collagen I promotes epithelial-to-mesenchymal transition in lung cancer cells via transforming growth factor-beta signaling. Am J Respir Cell Mol Biol 38(1):95–104
16. Hynes RO (2002) Integrins: bidirectional, allosteric signaling machines. Cell 110(6):673–687
17. Li Y, Tan X, Dai C et al (2009) Inhibition of integrin-linked kinase attenuates renal interstitial fibrosis. J Am Soc Nephrol 20(9):1907–1918
18. Yeh YC, Wei WC, Wang YK et al (2010) Transforming growth factor-β1 induces Smad3-dependent β1 integrin gene expression in epithelial-to-mesenchymal transition during chronic tubulointerstitial fibrosis. Am J Pathol 177(4):1743–1754
19. Reitherman RW, Rosen SD, Frasier WA et al (1975) Cell surface species-specific high affinity receptors for discoidin: developmental regulation in Dictyostelium discoideum. Proc Natl Acad Sci U S A 72(9):3541–3545
20. Springer WR, Cooper DN, Barondes SH (1984) Discoidin I is implicated in cell-substratum attachment and ordered cell migration of Dictyostelium discoideum and resembles fibronectin. Cell 39(3 Pt 2):557–564
21. Alves F, Vogel W, Mossie K et al (1995) Distinct structural characteristics of discoidin I subfamily receptor tyrosine kinases and complementary expression in human cancer. Oncogene 10(3):609–618

22. Shelling AN, Butler R, Jones T et al (1995) Localization of an epithelial-specific receptor kinase (EDDR1) to chromosome 6q16. Genomics 25(2):584–587
23. Vogel W, Gish GD, Alves F et al (1997) The discoidin domain receptor tyrosine kinases are activated by collagen. Mol Cell 1(1):13–23
24. Zerlin M, Julius MA, Goldfarb M (1993) NEP: a novel receptor-like tyrosine kinase expressed in proliferating neuroepithelia. Oncogene 8(10):2731–2739
25. Lai C, Lemke G (1991) An extended family of protein-tyrosine kinase genes differentially expressed in the vertebrate nervous system. Neuron 6(5):691–704
26. Vogel W (1999) Discoidin domain receptors: structural relations and functional implications. FASEB J 13(Suppl):S77–S82
27. Di Marco E, Cutuli N, Guerra L et al (1993) Molecular cloning of trkE, a novel trk-related putative tyrosine kinase receptor isolated from normal human keratinocytes and widely expressed by normal human tissues. J Biol Chem 268(32):24290–24295
28. Sakamoto O, Suga M, Suda T et al (2001) Expression of discoidin domain receptor 1 tyrosine kinase on the human bronchial epithelium. Eur Respir J 17(5):969–974
29. Sainio K, Raatikainen-Ahokas A (1999) Mesonephric kidney: a stem cell factory. Int J Dev Biol 43(5):435–439
30. Vainio S, Lehtonen E, Jalkanen M et al (1989) Epithelial-mesenchymal interactions regulate the stage-specific expression of a cell surface proteoglycan, syndecan, in the developing kidney. Dev Biol 134(2):382–391
31. Vainio S, Jalkanen M, Bernfield M et al (1992) Transient expression of syndecan in mesenchymal cell aggregates of the embryonic kidney. Dev Biol 152(2):221–232
32. Esteban MA, Bao X, Zuhang Q et al (2012) The mesenchymal-to-epithelial transition in somatic cell reprogramming. Curr Opin Genet Dev 22(5):423–428
33. Li R, Liang J, Ni S et al (2010) A mesenchymal-to-epithelial transition initiates and is required for the nuclear reprogramming of mouse fibroblasts. Cell Stem Cell 7(1):51–63
34. Nakamura M, Tokura Y (2011) Epithelial-mesenchymal transition in the skin. J Dermatol Sci 61(1):7–13
35. Nieto MA (2013) Epithelial plasticity: a common theme in embryonic and cancer cells. Science 342(6159):1234850
36. Gumbiner B, Stevenson B, Grimaldi A (1988) The role of the cell adhesion molecule uvomorulin in the formation and maintenance of the epithelial junctional complex. J Cell Biol 107(4):1575–1587
37. Gumbiner B (1990) Generation and maintenance of epithelial cell polarity. Curr Opin Cell Biol 2(5):881–887
38. Capaldo CT, Macara IG (2007) Depletion of E-cadherin disrupts establishment but not maintenance of cell junctions in Madin-Darby canine kidney epithelial cells. Mol Biol Cell 18(1):189–200
39. Larue L, Antos C, Butz S (1996) A role for cadherins in tissue formation. Development 122(10):3185–3194
40. Malaguti M, Nistor PA, Blin G et al (2013) Bone morphogenic protein signaling suppresses differentiation of pluripotent cells by maintaining expression of E-cadherin. Elife 2, e01197
41. Schäfer G, Narasimha M, Vogelsang E et al (2014) Cadherin switching during the formation and differentiation of the Drosophila mesoderm: implications for epithelial mesenchymal transitions. J Cell Sci 127(Pt 7):1511–1522
42. Strutz F, Okada H, Lo CW et al (1995) Identification and characterization of a fibroblast marker: FSP1. J Cell Biol 130(2):393–405
43. Okada H, Inoue T, Suzuki H et al (2000) Epithelial-mesenchymal transformation of renal tubular epithelial cells in vitro and in vivo. Nephrol Dial Transplant 15(Suppl 6):44–46
44. Kallui R, Neilson EG (2003) Epithelial-mesenchymal transition and its implications for fibrosis. J Clin Invest 112(12):1776–1784
45. Liu Y (2004) Epithelial to mesenchymal transition in renal fibrogenesis: pathologic significance, molecular mechanism, and therapeutic intervention. J Am Soc Nephrol 15(1):1–12

46. Hirohashi S (1998) Inactivation of the E-cadherin-mediated cell adhesion system in human cancers. Am J Pathol 153(2):333–339
47. Thiery JP (2002) Epithelial-mesenchymal transitions in tumour progression. Nat Rev Cancer 2(6):442–454
48. Guarino M, Rubino B, Ballabio G (2007) The role of epithelial-mesenchymal transition in cancer pathology. Pathology 39(3):305–318
49. Takai Y, Nakanishi H (2003) Nectin and afadin: novel organizers of intercellular junctions. J Cell Sci 116(Pt 1):17–27
50. Takai Y, Irie K, Shimizu K et al (2003) Nectins and nectin-like molecules: roles in cell adhesion, migration, and polarization. Cancer Sci 94(8):655–667
51. Fukuhara A, Irie K, Yamada A et al (2002) Role of nectin in organization of tight junctions in epithelial cells. Genes Cells 7(10):1059–1072
52. Yap AS, Kovacs EM (2003) Direct cadherin-activated cell signaling: a view from the plasma membrane. J Cell Biol 160(1):11–16
53. Sato T, Fujita N, Yamada A et al (2006) Regulation of the assembly and adhesion activity of E-cadherin by nectin and afadin for the formation of adherens junctions in Madin-Darby canine kidney cells. J Biol Chem 281(8):5288–5299
54. Tsukita S, Tsukita S, Nagafuchi A et al (1992) Molecular linkage between cadherins and actin filaments in cell-cell adherens junctions. Curr Opin Cell Biol 4(5):834–839
55. Leckband D, Prakasam A (2006) Mechanism and dynamics of cadherin adhesion. Annu Rev Biomed Eng 8:259–287
56. Vasioukjin V, Fuchs E (2001) Actin dynamics and cell-cell adhesion in epithelia. Curr Opin Cell Biol 13(1):76–84
57. Shen K, Turner JR (2005) Actin depolymerization disrupts tight junctions via caveolae-mediated endocytosis. Mol Biol Cell 16(9):3919–3936
58. Troyanovsky S (2005) Cadherin dimers in cell-cell adhesion. Eur J Cell Biol 84(2–3): 225–233
59. Troyanovsky RB, Sokolov EP, Troyanovsky SM (2006) Endocytosis of cadherin form intracellular junctions is the driving force for cadherin adhesive dimer disassembly. Mol Biol Cell 17(8):3484–4393
60. Pokutta S, Weis WI (2007) Structure and mechanism of cadherins and catenins in cell-cell contacts. Annu Rev Cell Dev Biol 23:237–261
61. Adams CL, Chen YT, Smith SJ et al (1998) Mechanisms of epithelial cell-cell adhesion and cell compaction revealed by high-resolution tracking of E-cadherin-green fluorescent protein. J Cell Biol 142(4):1105–1119
62. Cavey M, Rauzi M, Lenne PF et al (2008) A two-tiered mechanism for stabilization and immobilization of E-cadherin. Nature 453(7196):751–756
63. Nelson WT (2008) Regulation of cell-cell adhesion by the cadherin-catenin complex. Biochem Soc Trans 36(Pt 2):149–155
64. Jönsson M, Andersson T (2001) Repression of Wnt-5a impairs DDR1 phosphorylation and modifies adhesion and migration of mammary cells. J Cell Sci 114(Pt 11):2043–2053
65. Montesano R, Schaller G, Orci L (1991) Induction of epithelial tubular morphogenesis in vitro by fibroblast-derived soluble factors. Cell 66(4):697–711
66. Montesano R, Matsumoto K, Nakamura T et al (1991) Identification of a fibroblast-derived epithelial morphogen as hepatocyte growth factor. Cell 67(5):901–908
67. Pollack AL, Runyan RB, Mostov KE (1998) Morphogenetic mechanisms of epithelial tubulogenesis: MDCK cell polarity is transiently rearranged without loss of cell-cell contact during scatter factor/hepatocyte growth factor-induced tubulogenesis. Dev Biol 204(1):64–79
68. Hogan BL, Kolodzieh PA (2002) Organogenesis: molecular mechanisms of tubulogenesis. Nat Rev Genet 3(7):513–523
69. Marrs JA (2010) Branching morphogenesis: Rac signaling "PIX" tubulogenesis. Focus on "Pak1 regulates branching morphogenesis in 3D MDCK cell culture by a PIX and beta1-integrin-dependent mechanism". Am J Physiol Cell Physiol 299(1):C7–C10

70. Wang CZ, Hsu YM, Tang MJ (2005) Function of discoidin domain receptor HGF-induced branching tubulogenesis MDCK cells in collagen gel. J Cell Physiol 203(1):295–304
71. Wang CZ, Su HW, Hsu YC et al (2006) Discoidin domain receptor 1/SHP-2 signaling complex inhibits α2β1-integrin-mediated signal transducers and activators of transcription 1/3 activation and cell migration. Mol Biol Cell 17(6):2839–2852
72. Yeh YC, Wang CZ, Tang MJ (2009) Discoidin domain receptor 1 activation suppresses alpha-2beta1 integrin-dependent cell spreading through inhibition of Cdc42 activity. J Cell Physiol 218(1):146–156
73. Marceau N, Goyette R, Deschênes J et al (1980) Morphological differences between epithelial and fibroblast cells in rat liver cultures, and the role of cell surface fibronectin and cytoskeletal element organization in cell shape. Ann N Y Acad Sci 349:138–152
74. Yonemura S, Itoh M, Nagafuchi A et al (1995) Cell-to-cell adherens junction formation and actin filament organization: similarities and differences between non-polarized fibroblasts and polarized epithelial cells. J Cell Sci 108(Pt 1):127–142
75. Yeh YC, Wu CC, Wang YK et al (2011) DDR1 triggers epithelial cell differentiation by promoting cell adhesion through stabilization of E-cadherin. Mol Biol Cell 22(7):940–953
76. Eswaramoorthy R, Wang CK, Chen WC et al (2010) DDR1 regulates the stabilization of cell surface E-cadherin and E-cadherin-mediated cell aggregation. J Cell Physiol 224(2): 387–397
77. Hirano S, Nose A, Hatta K et al (1987) Calcium-dependent cell-cell adhesion molecules (cadherins): subclass specificities and possible involvement of actin bundles. J Cell Biol 105(6 Pt 1):2501–2510
78. Smutny M, Cox HL, Leerberg JM et al (2010) Myosin II isoforms identify distinct functional modules that support integrity of the epithelial zonula adherens. Nat Cell Biol 12(7):696–702
79. Mangold S, Wu SK, Norwood SJ et al (2011) Hepatocyte growth factor acutely perturbs actin filament anchorage at the epithelial zonula adherens. Curr Biol 21(6):503–537
80. Georgiou M, Marinari E, Burden J et al (2008) Cdc42, Par6, and aPKC regulate Arp2/3-mediated endocytosis to control local adherens junction stability. Curr Biol 18(21): 1631–1638
81. Leibfried A, Fricke R, Morgan MJ et al (2008) Drosophila Cip4 and WASp define a branch of the Cdc42-Par6-aPKC pathway regulating E-cadherin. Curr Biol 18(21):1639–1648
82. Leung T, Chen XQ, Tan I et al (1998) Myotonic dystrophy kinase-related Cdc42-binding kinase acts as a Cdc42 effector in promoting cytoskeletal reorganization. Mol Cell Biol 18(1):130–134
83. Wilkinson S, Paterson HF, Marchall CJ (2005) Cdc42-MRCK and Rho-ROCK signaling cooperate in myosin phosphorylation and cell invasion. Nat Cell Biol 7(3):255–261
84. Hidalgo-Carcedo C, Hooper S, Chaudhry SI et al (2011) Collective cell migration requires suppression of actomyosin at cell-cell contacts mediated by DDR1 and the cell polarity regulators Par3 and Par6. Nat Cell Biol 13(1):49–58
85. Amano M, Ito M, Kimura K et al (1996) Phosphorylation and activation of myosin by Rho-associated kinase (Rho-kinase). J Biol Chem 271(34):20246–20249
86. Carothers AM, Javid SH, Moran AE et al (2006) Deficient E-cadherin adhesion in C57BL/6J-Min/+ mice is associated with increased tyrosine kinase activity and RhoA-dependent actomyosin contractility. Exp Cell Res 312(4):387–400
87. Avizienyte E, Wyke AW, Jones RJ et al (2002) Src-induced de-regulation of E-cadherin in colon cancer cells requires integrin signaling. Nat Cell Biol 4(8):632–638
88. Avizienyte E, Fincham VJ, Brunton VG et al (2004) Src SH3/2 domain-mediated peripheral accumulation of Src and phosphor-myosin is linked to deregulation of E-cadherin and the epithelial-mesenchymal transition. Mol Biol Cell 15(6):2794–2803
89. Canel M, Serrels A, Miller D et al (2010) Quantitative in vivo imaging of the effects of inhibiting integrin signaling via Src and FAK on cancer cell movement: effects on E-cadherin dynamics. Cancer Res 70(22):9413–9422

90. Li Y, Yang J, Dai C et al (2003) Role for integrin-linked kinase in mediating tubular epithelial to mesenchymal transition and renal interstitial fibrogenesis. J Clin Invest 112(4):503–516
91. Tan C, Costello P, Sanghera J et al (2001) Inhibition of integrin-linked kinase (ILK) suppresses beta-catenin-Lef/Tcf-dependent transcription and expression of the E-cadherin repressor, snail, in APC–/– human colon carcinoma cells. Oncogene 20(1):133–140
92. McPhee TR, McDonald PC, Oloumi A et al (2008) Integrin-linked kinase regulates E-cadherin expression through PARP-1. Dev Dyn 237(10):2737–2747
93. de Rooij J, Kerstens A, Danuser G et al (2005) Integrin-dependent actomyosin contraction regulates epithelial cell scattering. J Cell Biol 171(1):153–164
94. Fujita Y, Krause G, Scheffner M et al (2002) Hakai, a c-Cbl-like protein, ubiquitinates and induces endocytosis of the E-cadherin complex. Nat Cell Biol 4(3):222–231
95. Shen Y, Hirsch DS, Sasiela CA et al (2008) Cdc42 regulates E-cadherin ubiquitination and degradation through an epidermal growth factor receptor to Src-mediated pathway. J Biol Chem 283(8):5127–5137
96. Levental KR, Yu H, Kass L et al (2009) Matrix crosslinking forces tumor progression by enhancing integrin signaling. Cell 139(5):891–906
97. Samuel MS, Lopez JI, McGhee EJ et al (2011) Actomyosin-mediated cellular tension drives increased tissue stiffness and β-catenin activation to induced epidermal hyperplasia and tumor growth. Cancer Cell 19(9):776–791
98. DuFort CC, Paszek MJ, Weaver VM (2011) Balancing forces: architectural control of mechanotransduction. Nat Rev Mol Cell Biol 12(5):308–319
99. Yeh YC, Lin HH, Tang MJ (2012) A tale of two collagen receptors, integrin β1 and discoidin domain receptor 1, in epithelial cell differentiation. Am J Physiol Cell Physiol 303(12):C1207–C1217
100. Wang CZ, Yeh YC, Tang MJ (2009) DDR1/E-cadherin complex regulates the activation of DDR1 and cell spreading. Am J Physiol Cell Physiol 297(2):C419–C429
101. Wu H, Liang YL, Li Z et al (2006) Positive expression of E-cadherin suppresses cell adhesion to fibronectin via reduction of α5β1 integrin in human breast carcinoma cells. J Cancer Res Clin Oncol 132(12):795–803
102. Borghi N, Lowndes M, Maruthamuthu V et al (2009) Regulation of cell motile behavior by crosstalk between cadherin- and integrin-mediated adhesion. Proc Natl Acad Sci U S A 107(30):13324–13329
103. Yamada S, Nelson WJ (2007) Localized zones of Rho and Rac activities drive initiation and expansion of epithelial cell-cell adhesion. J Cell Biol 178(3):517–527
104. Vogel W, Aszódi A, Alves F et al (2001) Discoidin domain receptor 1 tyrosine kinase has an essential role in mammary gland development. Mol Cell Biol 21(8):2906–2917
105. Labrador JP, Azcoitia V, Tuckermann J et al (2001) The collagen receptor DDR2 regulates proliferation and its elimination leads to dwarfism. EMBO Rep 2(5):446–452
106. Bargal R, Cormier-Daire V, Ben-Neriah Z et al (2009) Mutations in DDR2 gene cause SMED with short limbs and abnormal calcifications. Am J Hum Genet 84(1):80–84
107. Gross O, Beirowski B, Harvey SJ et al (2004) DDR1-deficienct mice show localized subepithelial GBM thickening with focal loss of slit diaphragms and proteinuria. Kidney Int 66(1):102–111
108. Meyer zum Gottesberge AM, Gross O, Becker-Lendzian U et al (2008) Inner ear defects and hearing loss in mice lacking the collagen receptor DDR1. Lab Invest 88(1):27–37
109. Hansen C, Greengard P, Nairn AC et al (2006) Phosphorylation of DARPP-32 regulates breast cancer cell migration downstream of the receptor tyrosine kinase DDR1. Exp Cell Res 321(20):4011–4018
110. Turashvili G, Bouchal J, Baumforth K et al (2007) Novel markers for differentiation of lobular and ductal invasive breast carcinomas by laser microdissection and microassay analysis. BMC Cancer 7:55

111. Heinzelmann-Schwarz VA, Gardiner-Garden M, Henshall SM et al (2004) Overexpression of the cell adhesion molecules DDR1, Claudin 3, and Ep-CAM in metaplastic ovarian epithelium and ovarian cancer. Clin Cancer Res 10(13):4427–4436
112. Câmara J, Jarai G (2010) Epithelial-mesenchymal transition in primary human bronchial epithelial cells is Smad-dependent and enhanced by fibronectin and TNF-alpha. Fibrogenesis Tissue Repair 3(1):2
113. Lee R, Eidman KE, Kren SM et al (2004) Localization of discoidin domain receptors in rat kidney. Nephron Exp Nephrol 97(2), e62
114. Maeyama M, Koga H, Selvendiran K et al (2008) Switching in discoidin domain receptor expressions in SLUG-induced epithelial-mesenchymal transition. Cancer 113(10):2823

# Part III
# Discoidin Domain Receptors in Other Pathologies

# Chapter 14
# Discoidin Domain Receptor 2 in Development of Osteoarthritis

**Lin Xu, Peter L. Lee, and Yefu Li**

## 14.1 Feature of Osteoarthritis

Osteoarthritis (OA) is the most common form of arthritis [1] and has been recently defined as "a progressive disease of synovial joints that represents failed repair of joint damage that results from stresses that may be initiated by an abnormality in any of the synovial joint tissues." This ultimately results in the breakdown of cartilage and bone, leading to symptoms of pain, stiffness, and functional disability [2]. Although the precise etiology of OA in most cases is unknown, it is the general consensus that OA is a multifactorial disorder involving both genetic and environmental components [1]. Genetic factors are mutations in genes that result in defects in articular cartilage matrix and chondrocyte metabolism. In the case of cartilage defects, age-dependent OA may occur as a result of even normal mechanical stresses on a defective joint. Environmental factors include obesity, overloading on joints, repetitive injury involving ligaments and menisci, loss of muscle strength, and joint malalignment. These conditions can result in abnormal mechanical stresses on a normal joint, which eventually leads to OA. It is obvious that the initiation of OA can involve multiple joint tissues with one, or a combination of, genetic and environmental factors. Thus, investigation of the detailed pathogenic mechanisms in all of these conditions remains a formidable challenge. However, regardless of the complexity of the etiology for OA, the ultimate consequence of articular cartilage breakdown in the development of OA follows a consistent pathological pattern [3–6].

L. Xu, M.D., Ph.D. (✉) • P.L. Lee • Y. Li, M.D., Ph.D.
Department of Developmental Biology, Harvard School of Dental Medicine, 188 Longwood Avenue, Boston, MA 02115, USA

Faculty of Medicine, Harvard Medical School, 188 Longwood Avenue, Boston, MA 02115, USA
e-mail: lin_xu@hms.harvard.edu



R. Fridman, P.H. Huang (eds.), *Discoidin Domain Receptors in Health and Disease*, DOI 10.1007/978-1-4939-6383-6_14

Results from studies using animal models of OA and human OA tissues indicate that in the process of articular cartilage degeneration, the earliest indication of change is chondrocyte clustering and a general upregulation in synthetic activity of chondrocytes. For example, in our mouse models of OA, we found that the chondrocyte clusters and the overproduction of proteoglycans appeared in the uncalcified (superficial) layer of the articular cartilage prior to any overt pathologic sign of articular cartilage degeneration [7]. With regard to human OA cases, it is extremely difficult to obtain articular cartilage from joints at the earliest stage of articular cartilage degeneration. However, the chondrocyte clusters and the expansion of the pericellular matrix (enlarged lacunae) are seen in the remaining articular cartilage tissue of joints at late stages of OA [4]. Immediately following the upregulation in synthetic activity of chondrocytes, chondrocytes synthesize and release extracellular matrix (ECM)-degrading enzymes. The gradual loss of proteoglycans on the surface region of the articular cartilage follows; this is concurrent with type II collagen degradation. Cracks develop along the articular surface, producing the histological image termed fibrillation. At later stages of the degenerative process, fibrocartilage and osteophytes are formed. It is worth mentioning that two recent studies report an unexpected revelation that the disappearance of aggrecans in articular cartilage of knee joints in adult mice, due to the genetic inactivation of a transcription factor, sex-determining region Y box 9 (Sox9), or the conditional upregulated expression of bone morphogenetic protein 2 in the articular cartilage, does not initiate or accelerate the progression of cartilage degeneration [8, 9]. This completely opposes the prevailing notion that proteoglycans are one of the basic elements and are indispensible for the ability of articular cartilage to resist compressive pressure. One plausible explanation for this observation is that the loss of proteoglycans alone may not be sufficient to initiate or accelerate articular cartilage degeneration. Instead, the degradation of both proteoglycans and type II collagen may be required in the development of OA.

The similar pathological progressions of articular cartilage degeneration suggest that there may be multiple initiating pathways toward the common targets that contribute to articular cartilage degeneration. This eventually leads to OA. The similarities in the pathologic progression also indicate that studies of pathogenetic changes, even in cases of more rare forms of OA, are likely to provide significant information about disease mechanisms and therapeutic targets for treating common forms of the disease. Therefore, the identification of a sequence(s) of molecular events underlying the progression of articular cartilage degeneration may not only help us understand the disease better, but may also provide information for the design of new therapeutic strategies in the prevention and treatment of OA.

We also need to consider that pathogenetic mechanisms by which juvenile OA develops may be different from mechanisms that are responsible for development of OA in adults. It is conceivable that OA joints in juveniles result from immature (developing) joints. Under this condition, fully mature joints are never formed. However, OA joints in adults are the ultimate consequence of the degeneration of the mature articular cartilages of joints. In the remaining part of this chapter, we discuss the roles of DDR2 and other molecules in the development of OA in adults.

## 14.2 Matrix Metalloproteinase 13 in the Development of OA

Matrix metalloproteinase 13 (MMP-13) is one of the major matrix-degrading enzymes that are involved in the breakdown of articular cartilage, which eventually leads to OA. Proteoglycans and type II collagen are two major ECM components in articular cartilage. The degradation of these molecules inevitably leads to a normal joint becoming an OA joint; thus, it would be critical to determine which enzyme/ enzymes are responsible for the degradation of proteoglycans and type II collagen. An ideal candidate would be an enzyme(s) that can degrade both components and should not be expressed in normal articular cartilage. Matrix metalloproteinase 13 (MMP-13) fits this profile exactly.

MMP-13, also known as collagenase-3, was originally cloned from a cDNA library derived from a breast tumor in 1994 [10]. This enzyme can degrade both proteoglycans and type II collagen [11, 12]. Moreover, MMP13 degrades type II collagen more efficiently, in fact 10 times more, than MMP1. Expression of MMP13 is hardly detected in normal mature articular cartilage, but the activity and expression of this enzyme are increased in human OA cartilage and in joints of OA mouse models [13–16]. This is consistent with the observation that the constitutive expression of Mmp-13 in mouse cartilage results in OA-like changes in knee joints [17], and removal of this enzyme prevents articular cartilage erosion in a knee joint instability mouse model of OA [18]. The significant effects of MMP-13 in the development of OA are also highlighted by several other studies. A study by Verzijl et al. indicates that the half-life of type II collagen in humans is 117 years [19]. The long half-life of type II collagen indicates a slow turnover of the collagen in mature articular cartilages. This also suggests that it may not be needed for chondrocytes to continually produce type II collagen in mature articular cartilages. Thus, chondrocytes may possess limited potential for type II collagen replenishment once the collagen is degraded. Results from another study demonstrate that articular cartilage degradation is completely irreversible after induction of the MMP-mediated degradation of aggrecan and type II collagen, but is reversible in the presence of the aggrecanase-mediated aggrecan degradation [20]. If the nature of chondrocytes in mature articular cartilage limits their ability to produce type II collagen, and if the degradation of type II collagen is irreversible, then we must inhibit the activity and expression of MMP-13 in articular cartilage in order to intervene in the development of OA. In fact, numerous pharmacological companies have attempted to inhibit the activity of MMP-13 as a means of delaying the development of OA. However, the broad activity of MMP-13 precludes this enzyme to be a promising target for inhibitory drugs in the treatment of OA [21]. While this may be the case for inhibitors of enzymatic activity, this concern does not apply to molecules involved in the regulation of MMP-13 expression in articular chondrocytes.

Transcriptional analysis of MMP-13 shows that the human MMP-13 promoter contains a TATA box and binding sites for transcription factors such as AP-1 (e.g., c-Fos/c-Jun), PEA-3 (polyoma enhancer activator 3), Runx2 (Runt-related transcription factor 2), and C/EBPβ (CCAAT/enhancer binding protein) [22–26].

Mutations in these binding sites can partially or completely abolish the activity of the human MMP-13 promoter. Numerous transcriptional factors, such as E74-like factor 3, FosA family members, and HIF-2-alpha, have been shown to be involved in induction of MMP-13 in chondrocytes [27–29]. Questions still remain as to how these transcriptional factors are induced during the development of OA, and whether or not the removal of one or a combination of these transcriptional factors is sufficient to significantly attenuate or prevent the progression of OA.

Signaling pathways activated by inflammatory cytokines, such as TNF-α and IL-1β, have been shown to be involved in the transcriptional regulation of MMP-13 expression in chondrocytes [30–33]. It has also been shown that fibronectin fragments (FN-f) can stimulate chondrocytes to synthesize and release MMP-13 via integrin activation [34, 35]. Although these signaling pathways may play a significant role in transcriptional regulation of MMP-13 expression in chondrocytes, a certain degree of articular cartilage damage is required to stimulate the release of the inflammatory cytokines or generate the FN-f. One important question is whether or not a mechanism exists to induce MMP-13 in chondrocytes prior to any significant degradation of articular cartilage. Results from our recent investigations demonstrate that discoidin domain receptor 2 (DDR2) may be a molecule that induces expression of MMP-13 in chondrocytes prior to overt damage of articular cartilage.

## 14.3 DDR2 in the Pathogenesis of OA

What led us to investigate DDR2 in the pathogenesis of OA? The answer is the classical biochemistry event in living systems known as enzyme induction. One example is the inducible lactose-metabolizing enzyme by which the enzyme is induced by its own substrates, see Fig. 14.1.

Bacterial *Escherichia coli* are not able to directly utilize disaccharide lactose for consumption. However, under the condition in which lactose is solely present, the enzyme β-galactosidase is induced in *E. coli*. The enzyme breaks down lactose into monosaccharides (galactose and glucose) that are then metabolized by *E. coli*. We believe that the induction of MMP-13 in chondrocytes may represent this similar situation. For example, MMP-13 is required for tissue turnover to remove ECM molecules such as collagens and proteoglycans, see Fig. 14.2.

The native type II collagen cannot enter chondrocytes to physically induce MMP-13 by itself. Thus, the collagen has to interact with a cell surface molecule(s) and send signals into chondrocytes to induce and release MMP-13. However, if this occurs in relatively quiescent tissues in which there is little tissue turnover, such as mature (adult) articular cartilage, the induction of MMP-13 will result in articular cartilage degradation. If this is the case, the question becomes which cell surface molecule(s) transduce the signal into chondrocytes? Results from our recent investigations suggest that DDR2 may be such a molecule.

DDR2 was originally cloned as a cell surface receptor tyrosine kinase (RTKs) in 1993 [36–39]. The mRNA transcripts of these receptors are detected in several

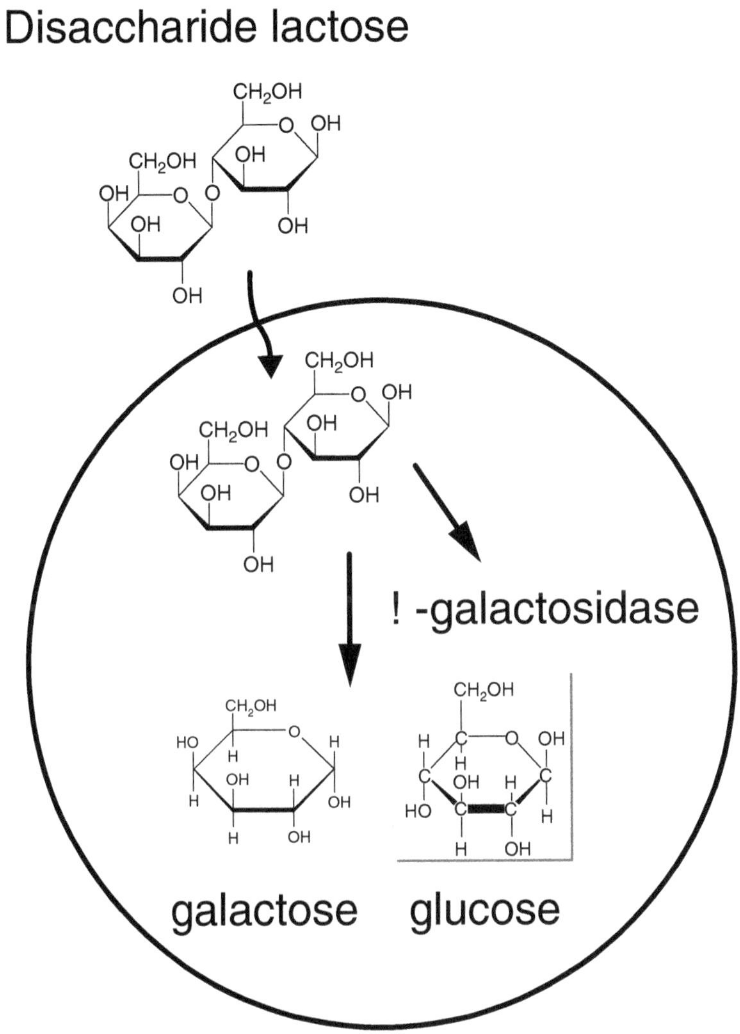


**Fig. 14.1** Induction of β-galactosidase in bacteria *E. coli* by disaccharide lactose. In bacteria, β-galactosidase can convert disaccharide into monosaccharide, for example, from lactose to glucose and galactose. Then, bacteria can consume monosaccharide as energy resource

human and mouse tissues, mainly in cartilage, skeletal muscle, skin, and adipose. In 1997, two research groups reported native fibrillar collagens are ligands for DDR2 [40, 41]. Furthermore, two other independent research groups have identified specific amino acids on DDR2 and type II collagen, which are critical for the interaction of the receptor with type II collagen, see Fig. 14.3 [42–44]. A changc in any of these amino acids will dramatically affect the affinity (interaction) of DDR2 with type II collagen. This suggests that the interaction of the receptor with type II collagen can be interrupted by the alteration of this specific amino acid peptide.

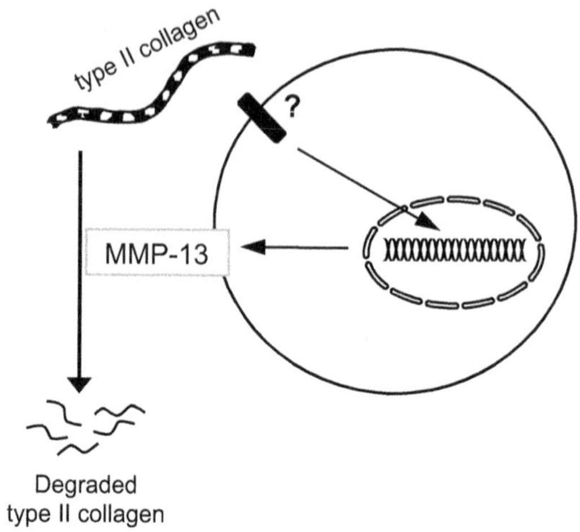


**Fig. 14.2** Induction of the MMP-13 in chondrocytes by native type II collagen. Native type II collagen is too big to move into chondrocyte. Thus, the collagen needs to bind a cell surface receptor and send signal to induce MMP-13 in chondrocytes. MMP-13, in turn, degrades native type II collagen. This indicates that native type II collagen controls its own proteolysis

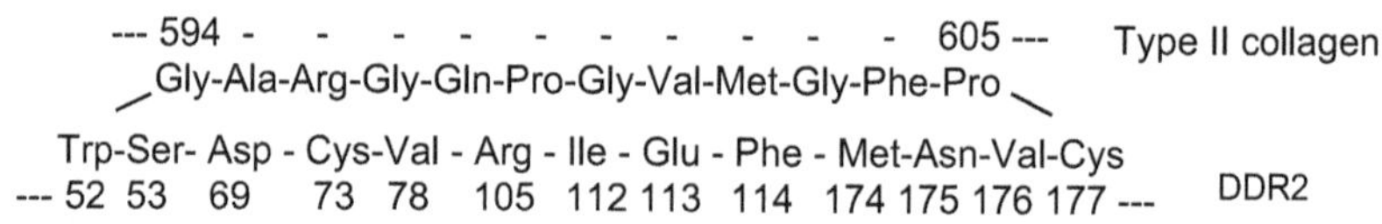


**Fig. 14.3** Critical amino acids for the interaction of DDR2 with type II collagen. An alteration of any of the amino acids in type II collagen and DDR2 shown in this figure will significantly affect the interaction of the two molecules

We examined the expression of DDR2 in human osteoarthritic tissues and mouse models of OA. We found that DDR2 protein was hardly detectable in normal articular cartilage. However, the expression of DDR2 was increased in human osteoarthritic tissues and mouse models of OA [45–49]. Moreover, DDR2 was colocalized with the increased activity and expression of MMP-13 in degenerative articular cartilage [50]. To understand whether or not the activation of DDR2 induced MMP-13 in chondrocytes, we performed a series of in vitro experiments. Results from these experiments showed that (1) the expression of MMP-13 was elevated in chondrocytes cultured on type II collagen. Surprisingly, we found that the expression of DDR2 was also increased in the chondrocytes. This indicated that chondrocytes exposed to native type II collagen were induced to express MMP-13 and the receptor itself. (2) When chondrocytes were cultured on denatured type II collagen (gelatin), the expression of MMP-13 and DDR2 were not induced, suggesting that the chondrocytes responded in a specific manner to triple-helical type II collagen. (3) The overexpression of full-length DDR2 cDNA resulted in increased expression of MMP-13, whereas the overexpression of a truncated DDR2 cDNA (lacking the protein tyrosine kinase) inhibited the increased expression of MMP-13. In addition, DDR2 lacking type II collagen-binding domain (discoidin domain) of the receptor had no effect on the expression of MMP-13 and the receptor itself. These results indicate that increased expression of MMP-13 in chondrocytes requires the interaction of DDR2 with type II collagen. (4) Ras/Raf/MEK/ERK and p38 signaling pathways were involved in the increased expression of MMP-13 in chondrocytes by type II collagen–DDR2 interaction. Based on the aforementioned finds, we carried out in vivo experiments to determine whether a reduction in expression of *Ddr2* in heterozygous $Ddr2^{+/-}$ condition could delay progression of articular cartilage degeneration, caused by either a deficiency in expression of collagen type XI or destabilization of the medial meniscus, in mice. We found that the decrease in the expression of Ddr2 indeed attenuated the progression of articular cartilage degeneration in mouse models of OA [51]. More importantly, three other independent research groups have confirmed the result from our experiments [52–54]. A question remains: how do chondrocytes silence DDR2 in normal mature articular cartilage?

Chondrocytes and their pericellular matrix are considered to be the primary structural and functional units, termed chondrons, of articular cartilage [55–61]. The concept of the chondron was first proposed by Benninghoff in 1925. Approximately 40 years later, Szirmai further evaluated the structure of the chondron by more systematic analysis. At the time, however, the chondron was not widely recognized as a functional unit. Some 20 years later, C.A. Poole's research group completed additional experiments to physically isolate chondrons from cartilage and showed that chondrons are true anatomic and functional entities. Furthermore, Poole's group demonstrated that the structure of the pericellular matrix of chondrocytes is disturbed in human OA cartilage. Chondrons consist of chondrocytes, the pericellular matrix, and a capsule surrounding the pericellular matrix. The pericellular matrix contains laminin, fibronectin, biglycan, decorin, fibromodulin, matrilin 3, and cartilage oligo matrix protein (COMP). The pericellular capsule is mostly composed of type VI and IX collagen, and proteoglycans. The capsule and the pericellular matrix separate chondrocytes from the adjacent interterritorial or territorial matrices containing type II collagen, see Fig. 14.4. Clearly, under normal conditions type II collagen is not exposed to chondrocytes. It is conceivable that the disruption or disappearance of the pericellular matrix enhances the exposure of chondrocytes to type II collagen. This, in turn, elicits the interaction of collagen type II with DDR2, resulting in induction of MMP-13, which leads to the destruction of articular cartilage.

Results from human and mouse genetic studies indicate the significant role of the pericellular matrix of chondrocytes in protecting articular cartilage against the development of OA. For example, the deficiency of one or a combination of two components of the pericellular matrix, such as type VI collagen, type IX collagen, matrilin 3, decorin, biglycan, and fibromodulin, result in early onset of OA in mice [46, 62–65]. In human genetic studies, mutations in type IX collagen and COMP are associated with OA [66–71].

We utilized a tetracycline-controlled gene expression system (Tet-Off system) in mice to investigate whether the pericellular matrix is indeed a critical structure to silence Ddr2. The system provides the ability to conditionally induce expression of Ddr2 in mature articular cartilage. This will prevent any potential abnormalities resulting from

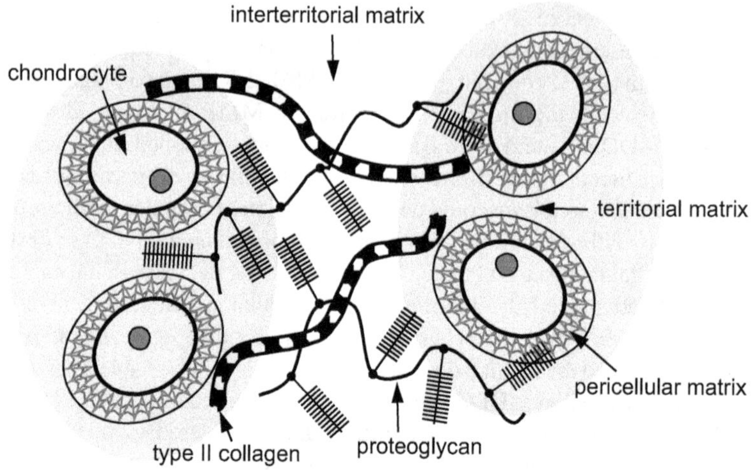


**Fig. 14.4** A schematic illustration of articular cartilage. A chondrocyte is surrounded with the pericellular matrix to form a chondron. Matrix between two chondrons is termed as territorial matrix and matrix among more than three chondrons is termed as interterritorial matrix. Native type II collagens in the territorial and interterritorial matrices are separated from chondrocytes by the pericellular matrix. Thus, the type II collagen does not interact with its receptor, such as DDR2, in normal articular cartilage

increased expression of the receptor during embryonic or early postnatal developmental stages. In the experiment, we induced the overexpression of Ddr2 in the articular cartilage of mouse knee joints at the age of 8 weeks old. It takes, in general, about 6 to 8 weeks for mice to become mature. Results from our experiments showed that the increased expression of Ddr2 did not induce expression of Mmp-13, and there were no OA-like pathologic changes observed in knee joints of the transgenic mice [50]. This indicates that a high-level expression of the inactive DDR2 does not cause the initiation/progression of OA. Our explanation is that although expression of Ddr2 was induced in articular cartilage of knee joints in the transgenic mice, the upregulated Ddr2 was surrounded by pericellular matrix. The pericellular matrix prevents Ddr2 from binding to type II collagen. Thus, Ddr2 was not activated (phosphorylated) and Mmp-13 would not be induced. Our data also suggested that the overexpressed Ddr2, without the interaction of the receptor with its ligand type II collagen, could not lead to autophosphorylation of the receptor itself.

In line with results from our experiments with the transgenic mice, another independent study by Vonk et al. demonstrate that DDR2 is silenced when chondrons are cultured on the collagen-coated plates [53]. It is clearly shown that DDR2 and MMP-13 are induced in chondrocytes in type II collagen-coated plates, whereas there is no induction of DDR2 and MMP-13 in chondrons when the chondrons are used in the type II collagen-coated plates. The results from these investigations support our hypothesis that the pericellular matrix of chondrocytes silences DDR2.

If disruption of the pericellular matrix is required for binding of DDR2 to type II collagen, then an enzyme/enzymes that are capable of degrading the pericellular molecules must play a critical role in development of OA. In fact, a study by Lee and Loeser suggests that proteases, not matrix metalloproteinases, may involve in the degradation of the pericellular matrix [72]. Numerous independent research groups have reported that a serine protease, high temperature requirement A1 (HTRA1), is highly expressed in human OA cartilage and in mouse models of OA. This suggests that HTRA1 may be involved in the development of OA.

## 14.4 HTRA1 in the Pathogenesis of OA

HTRA1 is one of four HTRA family members in human and mouse genomes [73, 74]. Mammalian HTRA1 was originally identified as a gene that was downregulated in SV40 transformed fibroblasts. Substrates of HTRA1 have been identified and include decorin, biglycan, fibromodulin, aggrecans, and fibronectin [75, 76]. All of these molecules are pericellular components of chondrocytes in articular cartilage. It is conceivable that upregulated expression and activity of this enzyme may cause degradation of the pericellular matrix of chondrocytes, which eventually leads to the destruction of a joint. In fact, HTRA1 has been implicated in rheumatoid arthritis (RA) and osteoarthritis (OA), and the level of HTRA1 is found to be higher in synovial fluids obtained from RA and OA patients [77–79]. HTRA1 is the most abundant protease in human OA cartilage. A study with a mouse model of RA demonstrates that increased expression of HtrA1 causes arthritic joints in the mice [80].

Results from our previous study also indicated that the protein level and activity of HtrA1 were increased in knee and temporomandibular joints of OA mouse models [81], including surgically induced models. This suggests that the altered mechanical stress in the surgical models can directly or indirectly induce HTRA1 in chondrocytes. More importantly, elevated expression of the enzyme was associated with disruption of the pericellular matrix of chondrocytes in the OA mouse models. We also found that the expression of this enzyme was increased prior to upregulation of Ddr2 and Mmp-13 in the mouse models. This indicates that HTRA1 may contribute to the development of OA through the degradation of the pericellular network, resulting in enhanced exposure of chondrocytes to type II collagen and activation of DDR2. Data from our experiments with the conditionally induced Ddr2 transgenic mice supports the pericellular matrix-degrading role of HTRA1 in the development of OA. We found that following a surgical destabilization of the medial meniscus (DMM), the progression of articular cartilage degeneration was accelerated in the upregulated Ddr2 transgenic mice, compared to that in wild-type littermates. Our explanation is that DMM surgery induces expression of HtrA1 that degrades the pericellular matrix of chondrocytes. Thus, in the Ddr2-induced transgenic mice, the disruption of the pericellular matrix coupled with the upregulated expression of Ddr2 accelerates the progression of articular cartilage degeneration. The data also indicates that DDR2 is able to accelerate the progression of the cartilage degeneration, but is not able to initiate the cartilage degeneration in the absence of biomechanical and/or biochemical disruption of the pericellular matrix. Therefore, maintaining the integrity of the pericellular matrix is one of the key factors against OA. Obviously, more in vivo experiments are needed to verify the role of HTRA1 in the pathogenesis of OA.

A critical question concerns which factor(s) cause induction of HTRA1 in chondrocytes. We recently investigated whether numerous factors, such as heating at 42 °C, hydrostatic pressure, tumor necrosis factor (TNFα), and transforming growth factor beta 1 (TGF-β1), were involved in the induction of HTRA1 in human primary articular chondrocytes [82]. We found that 42 °C, hydrostatic pressure, and TNFα treatment did not induce HTRA1 in the chondrocytes. In contrast, TGF-β1 induced HTRA1 in these cells.

## 14.5 TGF-β1 in the Pathogenesis of OA

Results from our experiments demonstrate that DDR2 may play a critical role in the development of OA. However, we consider that molecules that eventually lead to a condition resulting in activation of DDR2 are also crucial in the development of OA, such as HTRA1. In the following section, we also discuss about TGF-β1 as one of the molecules involved in the activation of DDR2.

TGF-β1 has been considered an anabolic factor to articular chondrocytes, based largely on results from in vitro and ex vivo experiments in which TGF-β1 can stimulate chondrocytes to synthesize and release extracellular matrix molecules, including proteoglycans and type II collagens [83–86]. In addition, results from two studies indicate that the genetic inactivation of *Smad-3* or disruption of the interaction of Tgf-β1 with its receptor, Tgf-β type II receptor (Tgfbr2), in germline cells causes OA-like knee joints in mice [87, 88]. Moreover, a human genetic study reports that a two-nucleotide deletion, 741-742del AT (nonsense mutation), in SMAD-3 causes early-onset OA in a human family [89]. This is consistent with the results from the animal models, indicating that the lack of Tgf-β1 signaling in the germline cell results in OA. However, observations from other studies also suggest that the increased TGF-β1 signaling may initiate and accelerate articular cartilage degeneration in mature joints.

First, the enhanced production of extracellular matrix molecules, due to an increase in the synthetic activity of chondrocytes, is not necessarily beneficial or physiological in maintenance of the homeostasis of mature articular cartilage. For instance, one of the earliest pathological signs in articular cartilage degeneration is the overproduction of proteoglycans in mouse models of OA [5, 6]. A study by Van den Berg et al. reports that the constitutive overexpression of active TGF-β1 in adult mouse knee joints results in OA associated with increase in the production of proteoglycans in articular cartilage, hyperplasia of synovium and chondro-osteophyte formation [90]. Thus, the overproduction of proteoglycans in articular cartilage could be a pathologic response of the chondrocyte in the early stages of articular cartilage degeneration. This raises an interesting question: does TGF-β1 disrupt homeostasis of articular cartilage instead of repairing the damaged cartilage in mature joints? Second, the human genetic study reports that a nucleotide change, 859C>*T* or 782C>*T* in SMAD-3, increases the level of TGF-β1 and activity of the TGF-β1 signaling pathway in two human families associated with early-onset OA [89]. This is in agreement with the observation from two other studies indicating that the level of TGF-β1 is significantly higher in human osteoarthritic tissues than in healthy articular cartilages [91, 92]. Studies with animal models, by Itayem et al., suggest that intra-articular injections of TGF-β1 into adult rat knee joints may cause early onset of OA [93, 94]. Third, to confirm our in vitro observation that TGF-β1 induced HTRA1 in human chondrocytes, we investigated whether Tgf-β1 induced HtrA1 in articular chondrocytes of two mouse models of OA [7]. We found increases in the expression of Tgf-β1, p-Smad2/3, and HtrA1 in articular chondrocytes of knee joints in the mouse models and increased expressions of p-Smad2/3 and HtrA1 were colocalized in the chondrocyte. In addition, TGF-β1-induced expression of HTRA1 was inhibited by an ALK-5 inhibitor, SB431542, in human and mouse chondrocytes. This suggests that Tgf-β1 canonical signaling may be activated to induce HtrA1 in articular chondrocytes of the mouse models of OA. More importantly, results from another independent research group demonstrate that TGF-β1 induces HTRA1 in human primary chondrocytes [95].

Data from aforementioned investigations raise the question as to what the exact role of TGF-β1 is in the development of OA. Our explanations for this "conflicting" role of TGF-β1 in the pathogenesis of OA are (1) effective TGF-β1 signaling acts in a dose-dependent manner. In this scenario, an appropriate level of TGF-β1 is required for the development and maintenance of articular cartilages. Therefore, TGF-β1 below or above this level results in articular cartilage degeneration. (2) Effective TGF-β1 signaling acts in a developmental stage-dependent manner. In this scenario, TGF-β1 is required for the development of articular cartilage; however, once a joint is formed, TGF-β1 is no longer needed. Therefore, induction of TGF-β1 in an adult joint causes articular cartilage degeneration.

Recently, we started to investigate whether inhibition of Tgf-β1 signaling prevents degeneration of mature knee joints in mouse models of OA [96]. (1) By use of conditional knockout techniques in mice, we removed *Tgfbr2* from articular cartilage of knee joints in mice at 8 weeks of age. We found neither the initiation and acceleration of articular cartilage degeneration in knee joints of mice at the age of 9 months old, nor hypertrophic chondrocytes in the articular cartilage of the mice. (2) We treated a spontaneous mouse model of OA, type XI collagen-haploinsufficiency (*Col11a1*$^{+/-}$) mice, with a neutralizing TGF-β1 antibody by intra-articular injection into the knee joints at 2 months of age. We found that the intra-articular injection of the neutralizing TGF-β1 antibody delayed articular cartilage degeneration approximately 3 months, compared to that in *Col11a1*$^{+/-}$ mice without injection of the antibody. Approximately 15 months is required for *Col11a1*$^{+/-}$ mice to develop OA knee joints (an average life span of the laboratory mice is about 30 months). (3) We treated surgically (DMM) induced OA mice with Losartan for 4 weeks immediately following surgery. Studies indicate that Losartan inhibits the activity of TGF-β1 signaling [96]. We found that treatment with Losartan delayed articular cartilage degeneration compared to that in control littermates. (4) We removed *Tgfbr2* from knee joints of mice at 2 months of age. We then performed DMM surgery on knee joints of the mice. We found that the removal of the *Tgfbr2* delayed articular cartilage degeneration at least 6 weeks, compared to that in wild-type littermates. Approximately 16 weeks is required for mice with DMM to develop OA knee joints. Based on the results from these experiments, we suggest that inhibition of TGF-β1 signaling attenuates articular cartilage degeneration in mature knee joints. Therefore, inhibition activity of TGF-β1, not application of TGF-β1, may be considered in treatment of OA in mature joints.

Data from our experiments demonstrated that the complete removal of Tgf-β1 signaling could, but not completely, delay the progression of articular cartilage degeneration in mice. This suggests that TGF-β1 may not be the only factor that induces HTRA1 in chondrocytes. In fact, data from several other studies indicate that other molecules, such as Wnt/β-catenin and LPS/TLR4 (toll-like receptor), can induce HTRA1 [80, 97].

## 14.6 A Molecular Pathway Underlying Articular Cartilage Degeneration

Based upon the data from our investigations and others' studies, we propose a molecular pathway (see Fig. 14.5) underlying articular cartilage degeneration as follows: Excessive mechanical stresses induced by either a normal mechanical loading of defective joints or an overloading of normal joints can stimulate chondrocytes and other joint tissues to synthesize and release latent TGF-β1 into synovial fluid. This suggestion is supported by studies (1) a study by Lee et al. reports that mechanical injury of bovine cartilage explants causes a significant increase in TGF-β1 gene expression [98]. (2) Results from our study indicated that hydrostatic pressure on human articular chondrocytes in culture increased expression of TGF-β1 in the cells. (3) The latent TGF-β1 is then activated by mechanical shearing of synovial fluid [99, 100]. The active TGF-β1 binds to its cognate receptor, TGFβ receptor II, which induces expression of HTRA1 in chondrocytes. We notice that HTRA1 may be able to inhibit TGF-β1 signaling based on the structural and functional similarity of the Insulin-like growth factor-binding protein domain of HTRA1 with follistatin, a potent antagonist of activin (a TGF-β family protein) and the capability to cleave proTGF-β1 [73, 101]. This suggests that there may be a negative feedback loop between TGF-β1 and HTRA1, which regulates the activity and expression of these two molecules in chondrocytes. Consequences of induction of HTRA1 are degradation of the pericellular matrix and enhanced exposure of chondrocytes to type II collagen. Interaction of chondrocytes with type II collagen results in enhanced signaling through DDR2, which induces the expression of MMP-13 as well as expression of DDR2 itself. MMP-13 degradation of type II collagen and aggrecan results

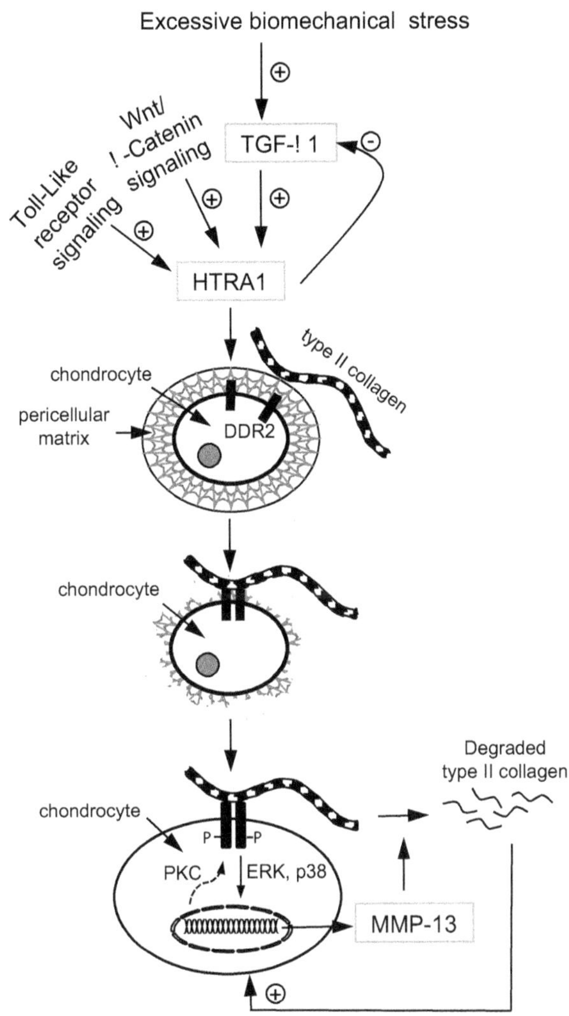


**Fig. 14.5** A schematic illustration of a molecular pathway underlying articular cartilage degeneration

in type II collagen and aggrecan fragments, which in turn may activate signals that further increase the synthesis of MMP-13 [102]. The end result is a feedback amplification loop that causes irreversible articular cartilage degeneration. One legitimate question, from an evolutionary standpoint, is if this molecular pathway leads to irreversible articular cartilage degradation, why do we still retain it? One plausible explanation may be that this molecular pathway is needed for proper tissue turnover during development.

## References

1. Felson DT, Lawrence RC, Dieppe PA et al (2000) Osteoarthritis: new insights. Part 1: the disease and its risk factors. Ann Intern Med 133:635–646
2. Lane NE, Brandt K, Hawker G et al (2011) OARSI-FDA initiative: defining the disease state of osteoarthritis. Osteoarthr Cartil 19(5):478–482. doi:10.1016/j.joca.2010.09.013
3. Hamerman D (1989) The biology of osteoarthritis. N Engl J Med 320(20):1322–1330
4. Lotz MK, Otsuki S, Grogan SP et al (2010) Cartilage cell clusters. Arthritis Rheum 62(8):2006–2018. doi:10.1002/art.27528
5. Xu L, Flahiff CM, Waldman BA et al (2003) Osteoarthritis-like changes and decreased mechanical function of articular cartilage in the joints of chondrodysplasia (cho) mice. Arthritis Rheum 48(9):2509–2518
6. Xu L, Polur I, Lim C et al (2009) Early-onset osteoarthritis of mouse temporomandibular joint induced by partial discectomy. Osteoarthr Cartil 17(7):917–922. doi:10.1016/j.joca.2009.01.002
7. Xu L, Golshirazian I, Asbury BJ et al (2014) Induction of high temperature requirement A1, a serine protease, by TGF-beta1 in articular chondrocytes of mouse models of OA. Histol Histopathol 29(5):609–618
8. Henry SP, Liang S, Akdemir KC et al (2012) The postnatal role of Sox9 in cartilage. J Bone Miner Res 27(12):2511–2525. doi:10.1002/jbmr.1696
9. Blaney Davidson EN, Vitters EL, Bennink MB et al (2015) Inducible chondrocyte-specific overexpression of BMP2 in young mice results in severe aggravation of osteophyte formation in experimental OA without altering cartilage damage. Ann Rheum Dis 74(6):1257–1264. doi:10.1136/annrheumdis-2013-204528
10. Freije JM, Díez-Itza I, Balbín M et al (1994) Molecular cloning and expression of collagenase-3, a novel human matrix metalloproteinase produced by breast carcinomas. J Biol Chem 269(24):16766–16773
11. Mitchell PG, Magna HA, Reeves LM et al (1996) Cloning, expression, and type II collagenolytic activity of matrix metalloproteinase-13 from human osteoarthritic cartilage. J Clin Invest 97(3):761–768
12. Fosang AJ, Last K, Knäuper V et al (1996) Degradation of cartilage aggrecan by collagenase-3 (MMP-13). FEBS Lett 380(1–2):17–20
13. Reboul P, Pelletier JP, Tardif G et al (1996) The new collagenase, collagenase-3, is expressed and synthesized by human chondrocytes but not by synoviocytes. A role in osteoarthritis. J Clin Invest 97(9):2011–2019
14. Shlopov BV, Lie WR, Mainardi CL et al (1997) Osteoarthritic lesions: involvement of three different collagenases. Arthritis Rheum 40(11):2065–2074
15. Tetlow LC, Adlam DJ, Woolley DE (2001) Matrix metalloproteinase and proinflammatory cytokine production by chondrocytes of human osteoarthritic cartilage: associations with degenerative changes. Arthritis Rheum 44(3):585–594
16. Bau B, Gebhard PM, Haag J et al (2002) Relative messenger RNA expression profiling of collagenases and aggrecanases in human articular chondrocytes in vivo and in vitro. Arthritis Rheum 46(10):2648–2657
17. Neuhold LA, Killar L, Zhao W et al (2001) Postnatal expression in hyaline cartilage of constitutively active human collagenase-3 (MMP-13) induces osteoarthritis in mice. J Clin Invest 107(1):35–44
18. Little CB, Barai A, Burkhardt D et al (2009) Matrix metalloproteinase 13-deficient mice are resistant to osteoarthritic cartilage erosion but not chondrocyte hypertrophy or osteophyte development. Arthritis Rheum 60(12):3723–3733. doi:10.1002/art.25002
19. Verzijl N, DeGroot J, Thorpe SR et al (2000) Effect of collagen turnover on the accumulation of advanced glycation end products. J Biol Chem 275(50):39027–39031
20. Karsdal MA, Madsen SH, Christiansen C et al (2008) Cartilage degradation is fully reversible in the presence of aggrecanase but not matrix metalloproteinase activity. Arthritis Res Ther 10(3):R63. doi:10.1186/ar2434

21. Hellio Le Graverand-Gastineau MP (2009) OA clinical trials: current targets and trials for OA. Choosing molecular targets: what have we learned and where we are headed? Osteoarthr Cartil 17(11):1393–1401. doi:10.1016/j.joca.2009.04.009
22. Pendas AM, Balbin M, Llano E et al (1997) Structural analysis and promoter characterization of the human collagenase-3 gene (MMP13). Genomics 40(2):222–233
23. Benbow U, Brinckerhoff CE (1997) The AP-1 site and MMP gene regulation: what is all the fuss about? Matrix Biol 15(8–9):519–526
24. Porte D, Tuckermann J, Becker M et al (1999) Both AP-1 and Cbfa1-like factors are required for the induction of interstitial collagenase by parathyroid hormone. Oncogene 18(3):667–678
25. Hess J, Porte D, Munz C (2001) AP-1 and Cbfa/runt physically interact and regulate parathyroid hormone-dependent MMP13 expression in osteoblasts through a new osteoblast specific element 2/AP-1 composite element. J Biol Chem 276(23):20029–20038
26. Hayashida M, Okazaki K, Fukushi J et al (2009) CCAAT/enhancer binding protein beta mediates expression of matrix metalloproteinase 13 in human articular chondrocytes in inflammatory arthritis. Arthritis Rheum 60(3):708–716. doi:10.1002/art.24332
27. Otero M, Plumb DA, Tsuchimochi K et al (2012) E74-like factor 3 (ELF3) impacts on matrix metalloproteinase 13 (MMP13) transcriptional control in articular chondrocytes under pro-inflammatory stress. J Biol Chem 287(5):3559–3572. doi:10.1074/jbc.M111.265744
28. Ionescu A, Kozhemyakina E, Nicolae C et al (2012) FoxA family members are crucial regulators of the hypertrophic chondrocyte differentiation program. Dev Cell 22(5):927–939. doi:10.1016/j.devcel.2012.03.011
29. Saito T, Kawaguchi H (2010) HIF-2α as a possible therapeutic target of osteoarthritis. Osteoarthr Cartil 18(12):1552–1556. doi:10.1016/j.joca.2010.10.006
30. Mengshol JA, Vincenti MP, Coon CI et al (2000) Interleukin-1 induction of collagenase 3 (matrix metalloproteinase 13) gene expression in chondrocytes requires p38, c-Jun N-terminal kinase, and nuclear factor kappaB: differential regulation of collagenase 1 and collagenase 3. Arthritis Rheum 43(4):801–811
31. Mengshol JA, Vincenti MP, Brinckerhoff CE (2001) IL-1 induces collagenase-3 (MMP-13) promoter activity in stably transfected chondrocytic cells: requirement for Runx-2 and activation by p38 MAPK and JNK pathways. Nucleic Acids Res 29(21):4361–4372
32. Liacini A, Sylvester J, Li WQ et al (2002) Inhibition of interleukin-1-stimulated MAP kinases, activating protein-1 (AP-1) and nuclear factor kappa B (NF-kappa B) transcription factors down-regulates matrix metalloproteinase gene expression in articular chondrocytes. Matrix Biol 21(3):251–262
33. Vincenti MP, Brinckerhoff CE (2002) Transcriptional regulation of collagenase (MMP-1, MMP-13) genes in arthritis: integration of complex signaling pathways for the recruitment of gene-specific transcription factors. Arthritis Res 4(3):157–164
34. Forsyth CB, Pulai J, Loeser RF (2002) Fibronectin fragments and blocking antibodies to alpha2beta1 and alpha5beta1 integrins stimulate mitogen-activated protein kinase signaling and increase collagenase 3 (matrix metalloproteinase 13) production by human articular chondrocytes. Arthritis Rheum 46(9):2368–2376
35. Loeser RF, Forsyth CB, Samarel AM et al (2003) Fibronectin fragment activation of proline-rich tyrosine kinase PYK2 mediates integrin signals regulating collagenase-3 expression by human chondrocytes through a protein kinase C-dependent pathway. J Biol Chem 278(27):24577–24585
36. Zerlin M, Julius MA, Goldfarb M (1993) NEP: a novel receptor-like tyrosine kinase expressed in proliferating neuroepithelia. Oncogene 8(10):2731–2739
37. Di Marco E, Cutuli N, Guerra L et al (1993) Molecular cloning of trkE, a novel trk-related putative tyrosine kinase receptor isolated from normal human keratinocytes and widely expressed by normal human tissues. J Biol Chem 268(32):24290–24295
38. Karn T, Holtrich U, Bräuninger A et al (1993) Structure, expression and chromosomal mapping of TKT from man and mouse: a new subclass of receptor tyrosine kinases with a factor VIII-like domain. Oncogene 8(12):3433–3440

39. Lai C, Gore M, Lemke G (1994) Structure, expression, and activity of Tyro 3, a neural adhesion-related receptor tyrosine kinase. Oncogene 9(3):877–883
40. Vogel W, Gish GD, Alves F et al (1997) The discoidin domain receptor tyrosine kinases are activated by collagen. Mol Cell 1(1):13–23
41. Shrivastava A, Radziejewski C, Campbell E et al (1997) An orphan receptor tyrosine kinase family whose members serve as nonintegrin collagen receptors. Mol Cell 1(1):25–34
42. Leitinger B, Steplewski A, Fertala A (2004) The D2 period of collagen II contains a specific binding site for the human discoidin domain receptor, DDR2. J Mol Biol 344(4):993–1003
43. Konitsiotis AD, Raynal N, Bihan D et al (2008) Characterization of high affinity binding motifs for the discoidin domain receptor DDR2 in collagen. J Biol Chem 283(11):6861–6868. doi:10.1074/jbc.M709290200
44. Osamu I, Masanori O, Noritaka N et al (2007) Structural basis of the collagen-binding mode of discoidin domain receptor 2. EMBO J 26:4168–4176. doi:10.1038/sj.emboj.7601833
45. Xu L, Peng H, Wu D et al (2005) Activation of the discoidin domain receptor 2 induces expression of matrix metalloproteinase 13 associated with osteoarthritis in mice. J Biol Chem 280(1):548–555
46. Hu K, Xu L, Cao L et al (2006) Pathogenesis of osteoarthritis-like changes in joints of type IX collagen-deficient mice. Arthritis Rheum 54(9):2891–2900
47. Xu L, Peng H, Glasson S et al (2007) Increased expression of a collagen receptor discoidin domain receptor 2 in articular cartilage as a key event in the pathogenesis of osteoarthritis. Arthritis Rheum 56(8):2663–2673
48. Lam NP, Li Y, Waldman AB et al (2007) Age-dependent increase of discoidin domain receptor 2 and matrix metalloproteinase 13 expression in temporomandibular joint cartilage of type IX and type XI collagen-deficient mice. Arch Oral Biol 52(6):579–584
49. Sunk IG, Bobacz K, Hofstaetter JG et al (2007) Increased expression of discoidin domain receptor 2 is linked to the degree of cartilage damage in human knee joints: a potential role in osteoarthritis pathogenesis. Arthritis Rheum 56(11):3685–3692
50. Xu L, Polur I, Servais JM et al (2011) Intact pericellular matrix of articular cartilage is required for un-activated discoidin domain receptor 2 in the mouse model. Am J Pathol 179(3):1338–1346. doi:10.1016/j.ajpath.2011.05.023
51. Xu L, Servais J, Polur I et al (2010) Attenuation of osteoarthritis progression by reduction of the discoidin domain receptor 2 in mice. Arthritis Rheum 62(9):2736–2744. doi:10.1002/art.27582
52. Klatt AR, Zech D, Kühn G et al (2009) Discoidin domain receptor 2 mediates the collagen II-dependent release of interleukin-6 in primary human chondrocytes. J Pathol 218(2):241–247. doi:10.1002/path.2529.]
53. Vonk LA, Doulabi BZ, Huang C et al (2011) Collagen-induced expression of collagenase-3 by primary chondrocytes is mediated by integrin α1 and discoidin domain receptor 2: a protein kinase C-dependent pathway. Rheumatology (Oxford) 50(3):463–472. doi:10.1093/rheumatology/keq305
54. Holt DW, Henderson ML, Stockdale CE et al (2012) Osteoarthritis-like changes in the heterozygous sedc mouse associated with the HtrA1-Ddr2-Mmp-13 degradative pathway: a new model of osteoarthritis. Osteoarthr Cartil 20(5):430–439. doi:10.1016/j.joca.2011.11.008
55. Poole CA, Flint MH, Beaumont BW (1988) Chondrons extracted from canine tibial cartilage: preliminary report on their isolation and structure. J Orthop Res 6(3):408–419
56. Poole CA, Ayad S, Schofield JR (1988) Chondrons from articular cartilage: I. Immunolocalization of type VI collagen in the pericellular capsule of isolated canine tibial chondrons. J Cell Sci 90(Pt 4):635–643
57. Poole CA, Wotton SF, Duance VC (1988) Localization of type IX collagen in chondrons isolated from porcine articular cartilage and rat chondrosarcoma. Histochem J 20(10):567–574
58. Poole CA, Honda T, Skinner SJ et al (1990) Chondrons from articular cartilage (II): analysis of the glycosaminoglycans in the cellular microenvironment of isolated canine chondrons. Connect Tissue Res 24(3–4):319–330

59. Poole CA, Matsuoka A, Schofield JR (1991) Chondrons from articular cartilage. III. Morphologic changes in the cellular microenvironment of chondrons isolated from osteoarthritic cartilage. Arthritis Rheum 34(1):22–35
60. Poole CA (1997) Articular cartilage chondrons: form, function and failure. J Anat 191(Pt1):1–13
61. Hunziker EB, Michel M, Studer D (1997) Ultrastructure of adult human articular cartilage matrix after cryotechnical processing. Microsc Res Tech 37(4):271–284
62. Alexopoulos LG, Youn I, Bonaldo P et al (2009) Developmental and osteoarthritic changes in Col6a1-knockout mice. Arthritis Rheum 60(3):771–779. doi:10.1002/art.24293
63. Wadhwa S, Embree M, Ameye L et al (2005) Mice deficient in biglycan and fibromodulin as a model for temporomandibular joint osteoarthritis. Cells Tissues Organs 181(3–4):136–143
64. Wadhwa S, Embree MC, Kilts T et al (2005) Accelerated osteoarthritis in the temporomandibular joint of biglycan/fibromodulin double-deficient mice. Osteoarthr Cartil 13(9): 817–827
65. van der Weyden L, Wei L, Luo J et al (2006) Functional, knockout of the matrilin-3 gene causes premature chondrocyte maturation to hypertrophy and increases bone mineral density and osteoarthritis. Am J Pathol 169(2):515–527
66. Borochowitz ZU, Scheffer D, Adir V et al (2004) Spondylo-epi-metaphyseal dysplasia (SEMD) matrilin 3 type: homozygote matrilin 3 mutation in a novel form of SEMD. J Med Genet 41(5):366–372
67. Hecht JT, Nelson LD, Crowder E et al (1995) Mutations in exon 17B of cartilage oligomeric matrix protein (COMP) cause pseudoachondroplasia. Nat Genet 10(3):325–329
68. Muragaki Y, Mariman EC, van Beersum SE et al (1996) A mutation in the gene encoding the alpha 2 chain of the fibril-associated collagen IX, COL9A2, causes multiple epiphyseal dysplasia (EDM2). Nat Genet 12(1):103–105
69. Mustafa Z, Chapman K, Irven C et al (2000) Linkage analysis of candidate genes as susceptibility loci for osteoarthritis-suggestive linkage of COL9A1 to female hip osteoarthritis. Rheumatology (Oxford) 39(3):299–306
70. Bönnemann CG, Cox GF, Shapiro F et al (2000) A mutation in the alpha 3 chain of type IX collagen causes autosomal dominant multiple epiphyseal dysplasia with mild myopathy. PNAS 97(3):1212–1217
71. Czarny-Ratajczak M, Lohiniva J, Rogala P et al (2001) A mutation in COL9A1 causes multiple epiphyseal dysplasia: further evidence for locus heterogeneity. Am J Hum Genet 69(5):969–980
72. Lee GM, Loeser RF (1998) Interactions of the chondrocyte with its pericellular matrix. Cell Mater 8:135–149
73. Zumbrunn J, Trueb B (1996) Primary structure of a putative serine protease specific for IGF-binding proteins. FEBS Lett 398(2–3):187–192
74. Clausen T, Southan C, Ehrmann M (2002) The HtrA family of proteases: implications for protein composition and cell fate. Mol Cell 10(3):443–455
75. Oka C, Tsujimoto R, Kajikawa M et al (2004) HtrA1 serine protease inhibits signaling mediated by Tgfbeta family proteins. Development 131(5):1041–1053
76. Tsuchiya A, Yano M, Tocharus J et al (2005) Expression of mouse HtrA1 serine protease in normal bone and cartilage and its upregulation in joint cartilage damaged by experimental arthritis. Bone 37(3):323–336
77. Hu SI, Carozza M, Klein M et al (1998) Human HtrA, an evolutionarily conserved serine protease identified as a differentially expressed gene product in osteoarthritic cartilage. J Biol Chem 273(51):34406–34412
78. Grau S, Richards PJ, Kerr B et al (2006) The role of human HtrA1 in arthritic disease. J Biol Chem 281(10):6124–6129
79. Wu J, Liu W, Bemis A et al (2007) Comparative proteomic characterization of articular cartilage tissue from normal donors and patients with osteoarthritis. Arthritis Rheum 56(11):3675–3684

80. Hou Y, Lin H, Zhu L et al (2013) Lipopolysaccharide increases the incidence of collagen-induced arthritis in mice through induction of protease HTRA-1 expression. Arthritis Rheum 65(11):2835–2846. doi:10.1002/art.38124
81. Polur I, Lee PL, Servais JM et al (2010) Role of HTRA1, a serine protease, in the progression of articular cartilage degeneration. Histol Histopathol 25(5):599–608
82. Li YF, Li Y, Xu L et al (2011) Mechanical loading-induced TGF-ß1 mediates cartilage degradation caused by upregulation of HTRA1/DDR2. In: Abstracts of orthopaedic research society annual meeting, Long Beach, CA, January 13–16, 2011
83. Harvey AK, Hrubey PS, Chandrasekhar S (1991) Transforming growth factor-beta inhibition of interleukin-1 activity involves down-regulation of interleukin-1 receptors on chondrocytes. Exp Cell Res 195(2):376–385
84. Galéra P, Vivien D, Pronost S et al (1992) Transforming growth factor-beta 1 (TGF-beta 1) up-regulation of collagen type II in primary cultures of rabbit articular chondrocytes (RAC) involves increased mRNA levels without affecting mRNA stability and procollagen processing. J Cell Physiol 153(3):596–606
85. Rédini F, Mauviel A, Pronost S et al (1993) Transforming growth factor beta exerts opposite effects from interleukin-1 beta on cultured rabbit articular chondrocytes through reduction of interleukin-1 receptor expression. Arthritis Rheum 36(1):44–50
86. van Beuningen HM, van der Kraan PM, Arntz OJ et al (1994) Transforming growth factor-beta 1 stimulates articular chondrocyte proteoglycan synthesis and induces osteophyte formation in the murine knee joint. Lab Invest 71(2):279–290
87. Serra R, Johnson M, Filvaroff EH et al (1997) Expression of a truncated, kinase-defective TGF-beta type II receptor in mouse skeletal tissue promotes terminal chondrocyte differentiation and osteoarthritis. J Cell Biol 139(2):541–552
88. Yang X, Chen L, Xu X et al (2001) TGF-beta/Smad3 signals repress chondrocyte hypertrophic differentiation and are required for maintaining articular cartilage. J Cell Biol 153(1):35–46
89. van de Laar IM, Oldenburg RA, Pals G et al (2011) Mutations in SMAD3 cause a syndromic form of aortic aneurysms and dissections with early-onset osteoarthritis. Nat Genet 43(2):121–126. doi:10.1038/ng.744
90. Bakker AC, van de Loo FA, van Beuningen HM et al (2001) Overexpression of active TGF-beta-1 in the murine knee joint: evidence for synovial-layer-dependent chondro-osteophyte formation. Osteoarthr Cartil 9(2):128–136
91. Schlaak JF, Pfers I, Meyer Zum Büschenfelde KH et al (1996) Different cytokine profiles in the synovial fluid of patients with osteoarthritis, rheumatoid arthritis and seronegative spondylarthropathies. Clin Exp Rheumatol 14(2):155–162
92. Kawamura I, Maeda S, Imamura K et al (2012) SnoN suppresses maturation of chondrocytes by mediating signal cross-talk between transforming growth factor-β and bone morphogenetic protein pathways. J Biol Chem 287(34):29101–29113. doi:10.1074/jbc.M112.349415
93. Itayem R, Mengarelli-Widholm S, Hulth A et al (1997) Ultrastructural studies on the effect of transforming growth factor-β1 on rat articular cartilage. APMIS 105(3):221–228
94. Itayem R, Mengarelli-Widholml S, Reinholt F (1999) The lone-term effect of a short course of transforming growth factorβ1 on rat articular cartilage. APMIS 107(2):183–192
95. Urano T, Narusawa K, Kobayashi S et al (2010) Association of HTRA1 promoter polymorphism with spinal disc degeneration in Japanese women. J Bone Miner Metab 28(2):220–226. doi:10.1007/s00774-009-0124-0
96. Li Y, Chen R, Mian M et al (2014) Inhibition of Tgf-β1 attenuates articular cartilage degeneration in mature knee joints of mouse models of osteoarthritis. Paper presented at the world congress on osteoarthritis, Paris, France, April 24–27, 2014
97. Miclea RL, Siebelt M, Finos L et al (2011) Inhibition of Gsk3β in cartilage induces osteoarthritic features through activation of the canonical Wnt signaling pathway. Osteoarthr Cartil 19(11):1363–1372. doi:10.1016/j.joca.2011.07.014

98. Lee JH, Fitzgerald JB, Dimicco MA et al (2005) Mechanical injury of cartilage explants causes specific time-dependent changes in chondrocyte gene expression. Arthritis Rheum 52(8):2386–2395
99. Albro MB, Cigan AD, Nims RJ et al (2012) Shearing of synovial fluid activates latent TGF-β. Osteoarthr Cartil 20(11):1374–1382. doi:10.1016/j.joca.2012.07.006
100. Albro MB, Nims RJ, Cigan AD et al (2013) Accumulation of exogenous activated TGF-β in the superficial zone of articular cartilage. Biophys J 104(8):1794–1804. doi:10.1016/j.bpj.2013.02.052
101. Atsushi S, Hiroaki N, Akio Y et al (2011) Cerebral small-vessel disease protein HTRA1 controls the amount of TGF-β1 via cleavage of proTGF-β1. Hum Mol Genet 20(9):1800–1810. doi:10.1093/hmg/ddr063
102. Fichter M, Körner U, Schömburg J et al (2006) Collagen degradation products modulate matrix metalloproteinase expression in cultured articular chondrocytes. J Orthop Res 24(1):63–70

# Chapter 15
# DDR1 in Renal Function and Disease

**Christos Chatziantoniou, Aude Dorison, and Jean-Claude Dussaule**

## 15.1 Background: The Long Road for Therapy of Chronic Kidney Disease

Even though significant progress has been made and many aspects of the complex mechanisms orchestrating Chronic Kidney Disease (CKD) have been identified, CKD represents a major problem for public health. It affects up to 10 % of the general population with a prevalence and incidence which has increased worldwide over the past 25 years and has almost doubled both in the United States and Europe [1]. Severe CKD requires renal replacement therapy, which may involve a form of dialysis or renal transplantation [2]. Diabetes and high blood pressure are the major causes of CKD, followed by glomerulonephritis. Independently of the underlying cause, pathogenesis of CKD is characterized by a continuous crosstalk between inflammation and fibrosis leading with time to the progressive impairment of glomerular, tubulointerstitial, and vascular compartments. Today, there are no drugs capable of reducing, arresting, or (even better) reversing the decline of renal function.

Our lab was among the first to demonstrate that CKD can be reversed in animal models by inhibiting the deleterious effects of the renin–angiotensin system [3–6]. However, inhibitors or antagonists of the renin–angiotensin system do not exhibit the same efficiency in arresting or reversing chronic renal disease progression in humans [7, 8]. Therefore, detection of the disease in early stages and discovery of novel therapeutic targets are of crucial importance for an efficient treatment against CKD. As shown in other chapters, DDR1 is a tyrosine kinase receptor that binds directly to collagens. It is expressed in a variety of cells, including vascular smooth

C. Chatziantoniou, Ph.D. (✉) • A. Dorison, Ph.D. • J.-C. Dussaule, Ph.D. M.D
Inserm UMR S 1155 and Sorbonne University, Pierre and Marie Curie University (University of Paris 6), 4, rue de la Chine, 75020 Paris, France
e-mail: christos.chatziantoniou@upmc.fr



R. Fridman, P.H. Huang (eds.), *Discoidin Domain Receptors in Health and Disease*, DOI 10.1007/978-1-4939-6383-6_15

muscle and mesangial cells [9, 10], and its phosphorylation by collagens (all tested types of collagen) leads to a tyrosine kinase-MAPK/p38 cascade activation. For these reasons, it appeared as an interesting candidate to be investigated in connection to the pathogenesis of renal fibrosis. Toward this direction, our team showed that DDR1 is indeed a major mediator of renal inflammation and fibrosis and a promising target for therapy.

## 15.2 DDR1 in Normal Kidneys

DDR1 is expressed in normal kidneys in low levels in vessels and distal tubules, as shown by immunochemistry studies. Its expression has also been reported in podocytes [11]. These investigators reported a slightly higher urinary protein excretion in mice lacking DDR1 expression, but these values were at the normal range. Uremia values were also normal. They have observed that adult mice lacking DDR1 displayed a focal thickening of the glomerular basement membrane without any alteration of the proteins controlling podocyte structure. In a subsequent study, this group reported that this glomerular abnormality was confined to less than 2 % of the total glomerular basement membrane and appeared only in aged mice [12]. Our group, with more than 10 years experience looking at the renal function of mice deficient in DDR1 expression, never observed any structural or functional renal abnormalities under control conditions in these mice until the age of 8 month old.

## 15.3 DDR1 in Renal Disease

### *15.3.1 DDR1 in Hypertensive Nephropathy*

Continuous infusion of angiotensin II through subcutaneous osmotic minipumps or inhibition of endogenous production of NO by NO-synthase inhibitors such as L-NAME is classical model of hypertension-associated renal disease. Both models of hypertension are accompanied by severe renal vascular and glomerular lesions that extend with time to renal interstitium and lead to gradual decline of renal function and structure as evidenced by increasing creatinine, proteinuria, and fibrosis. The initial hypothesis was that if DDR1 is involved in the fibrogenic process acting as a collagen receptor, then mice lacking functional DDR1 should be protected against renal fibrosis compared to wild-type animals and this protection should be independent of blood pressure levels [13]. Indeed, systolic pressure increased to a similar way in wild-type and DDR1-deficient mice after chronic administration of angiotensin II or L-NAME. DDR1 expression was low in the kidney of wild-type mice under normal conditions, whereas it was upregulated during angiotensin II or L-NAME treatment. Interestingly, the increased expression of DDR1 was specific to renal vessels and glomeruli, the two main renal compartments that the abnormal

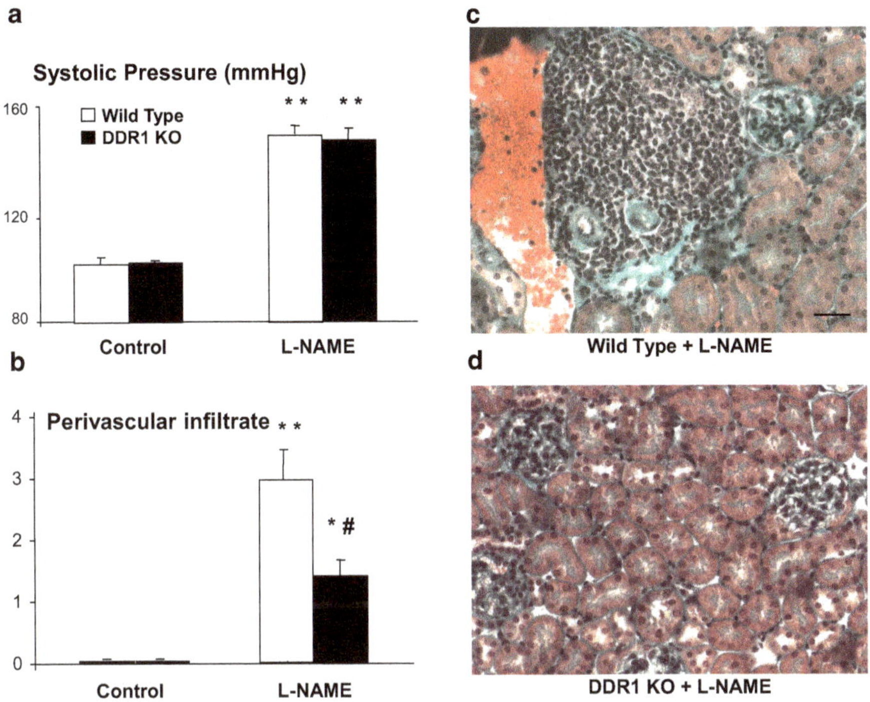


**Fig. 15.1** Systolic blood pressure increase (**a**) and perivascular infiltrate score (**b**) in wild-type and DDR1-deficient mice after 20 weeks of NO inhibition. Values are means ± SE; $n = 10$; * $P < 0.05$ vs. control; ** $P < 0.01$ vs. control; # $P < 0.05$ vs. wild type. Representative examples of extracellular matrix staining (Masson's trichromic solution, bar = 20 μm)) in renal cortex of wild-type (**c**) and DDR1 null mice (**d**) treated for 20 weeks with the NO synthase inhibitor L-NAME. Note the absence of inflammatory cells and the decreased degree of structural alterations in the DDR1 null mice

collagen formation and the subsequent fibrosis are originating in these experimental models. A major finding was that contrary to the wild-type mice, several indexes of renal function and structure such as proteinuria, microalbuminuria, collagen I or IV expressions, and neoformation did not increase and remained close to normal levels indicating prevention against angiotensin II or L-NAME-induced renal damage even when the disease was prolonged to chronic (for the mice) conditions (Fig. 15.1).

An additional striking strain difference was the degree of the inflammatory response. The prolonged action of angiotensin II or L-NAME on vessels is activating inflammatory pathways resulting recruitment of infiltrating cells and induction of renal vascular inflammation [14, 15]. These models are accompanied by important accumulation of infiltrating cells around renal vessels and glomeruli (Fig. 15.1b and c). In sharp contrast, DDR1-deficient mice showed an almost complete absence of inflammatory cell infiltration in the renal vasculature (Fig. 15.1b, d). This impressive protection against inflammation was not due to a phenotype strain difference because basal values of leukocytes were similar between wild-type and DDR1 KO mice.

In addition, mice lacking DDR1expression showed a blunted response of chemokines to sepsis, induced by LPS, and survived at doses that are lethal for wild-type mice. Since DDR1 was not expressed in lymphocytes or macrophages in the hypertensive models of nephropathy, the mechanism of action that appears the most probable is the following: angiotensin II or NO inhibition increase DDR1 expression in renal vessels and mesangial cells (probably through a cellular mechanism involving endothelial cell stress, a common deleterious action of angiotensin II, and L-NAME in the vessel wall). In parallel, angiotensin II and L-NAME directly induce collagen I, III, and IV neoexpression in renal microvessels and glomeruli [16, 17]. As a result, collagens bind DDR1 and induce phosphorylation of its intracellular part. This activation of DDR1 stimulates proinflammatory pathways (such as P38 pathway and/or NFkB), which in turn lead to cytokines synthesis. Cytokine release facilitates inflammatory cell infiltration and further stimulates collagen synthesis and DDR1 expression. Under this scheme, DDR1 is an amplifier of the initial renal vascular lesion by creating a positive deleterious crosstalk between inflammation and extracellular matrix synthesis, which is independent from the initiating cause and which ultimately leads to the development and progression of chronic kidney disease.

### *15.3.2 DDR1 in a Genetic Model of Hereditary Kidney Disease*

Other investigators studied the role of DDR1 in kidney of mice mimicking the Alport syndrome [12]. Alport syndrome is a hereditary disease due to mutations in the chains of collagen IV (mainly a3, a4, and a5 chains). As a result, the mutations in any of these chains alter the proper assembly of the collagen IV network, affecting thus tissues in which collagen IV is an important structural component of basement membranes such as the kidney, ear, and eye. The abnormalities in the structure of glomerular basement membrane cause proteinuria and gradual scarring of glomeruli. Renal function is severely affected from early childhood and usually the patients progress toward End Stage Renal Failure during adolescence. Mice lacking the a3 chain of collagen IV present several histological renal abnormalities similar to those observed in humans and are considered as a model mimicking well the renal pathology of this genetic disorder.

Crossbreeding Alport mice with mice lacking DDR1 expression substantially improved the loss of renal function (decreases in proteinuria and blood urea nitrogen) and increased the lifespan. The genetic invalidation of DDR1 in the Alport mice was also accompanied by reduced levels of several proinflammatory and profibrotic pathways such as TGFbeta, CTGF, NFkB, and IL-6. Of interest, the expression of DDR1 in the Alport mice occurred in podocytes which are the cells involved in this renal disease.

### 15.3.3 DDR1 in Tubular Obstructive Nephropathy

Unilateral Ureteral Obstruction (ligation of the one ureter while the other remains untouched) is a classical model of rapid decomposition of the renal tubular structure. The pathophysiological mechanism is mainly a combination of tubular inflammation, fibrogenesis, and epithelial apoptosis. Although it is not very relevant to human CKD, it is popular among investigators because it represents several advantages: it represents a form of primitive tubulointerstitial injury, is a simple and rapid chirurgical intervention, the contralateral kidney can serve as control, it does not depend on the genetic background of the mice, and the structural decay is relatively fast. There is although the limitation that only the histopathology indexes can be used for analysis. The classical indexes of overall renal function such as proteinuria or creatinine remain normal since the contralateral kidney compensates for the loss of the obstructed kidney. Since interstitial injury and the cellular infiltration and extracellular matrix deposition within the interstitium are considered as key steps in the progression of CKD, the role of DDR1 was studied also in the UUO model using again mice deficient in DDR1 expression [18]. DDR1 was several fold upregulated after UUO at both the mRNA and the protein expression levels. Interestingly, the major induction of DDR1 expression did not occur in the injured tubular epithelial cells. The major cells responsible for the de novo induction of DDR1 expression in the kidney were interstitial cells, and specifically in cells positive to F4/80. Colocalization experiments indicated that the interstitial cells expressing DDR1 were mainly macrophages and to a lesser extent dendritic cells. The mechanism producing renal injury following UUO is characterized by an early onset of renal interstitial inflammation due to the production of inflammatory cytokines, such as MCP-1, by resident interstitial and/or epithelial cells [19, 20]. Infiltration by activated macrophages is a central step in the inflammatory response to UUO and in the progression of the subsequent renal injury [21–24]. This observation indicates that DDR1 is induced locally in cell types that are predominantly activated by the specific physiopathological process and involved in the response to the renal injury: renal vessels and mesangium in hypertension [13], podocytes in a model of Alport syndrome [12], and macrophages in the UUO model [18]. This implies that the absence of DDR1 expression in macrophages was protective against UUO-induced lesions in DDR1-deficient mice. Experiments performed with Boyden chambers, coated with collagen IV or not [18, 25], showed that macrophages freshly isolated from mice lacking DDR1 expression exhibited a deficient capacity of migration compared with macrophages isolated from wild-type animals. This observation suggests that DDR1 expression and activation in macrophages is essential for an efficient migration and indicates the role of DDR1 in interstitial lesions associated to macrophage infiltration in kidneys.

### 15.3.4 DDR1 in Crescentic Glomerulonephritis

Crescentic glomerulonephritis is a lethal rapidly progressive renal pathology. It is an autoimmune disease in which circulating antibodies are directed toward antigens present in the chains of type IV collagen of the glomerular basement membrane. The experimental alloimmune antiglomerular basement membrane (anti-GBM) nephritis referred also as nephrotoxic serum nephritis (NTS) is a model commonly used to study mechanisms of crescentic glomerulonephritis. This model is based on injecting mice (or rats) with a sheep (or goat) serum immunized against glomerular extracts of the mice. This serum obviously contains additional glomerular target antigens compared to human pathology. The injection of NTS induces an immediate major vasoconstriction followed by an inflammatory response characterized with the infiltration of polymorphonuclear cells. This first wave of the innate immune response is followed by T and B-cell activation, which leads to progressive infiltration by CD4+ T cells and macrophages. After few days, significant proteinuria, GFR decrease, renal inflammation, and tubular injury can be observed, associated with glomerular fibrin deposits, the formation of glomerular crescents, and tubulointerstitial fibrosis. After few weeks, animals progress to End Stage Renal Failure and die [26].

As was the case with the previous models, NTS administration was followed by a progressive increase of DDR1 mRNA and protein expressions predominantly in the initial sites of aggression, glomeruli and proximal tubular cells, the two compartments that are the main targets of NTS [26]. As with the hypertensive model was not expressed in the infiltrating cells (macrophages or lymphocytes). It appears thus, that the disease-driven expression of DDR1 is cell/tissue specific, strongly depending on the underlying mechanisms. It can expressed in infiltrating cells in models such as the UUO [18], or in resident cells in response to the inflammatory process in models such as the hypertensive nephropathy [13] or crescentic glomerulonephritis [26], or in a combination of infiltrating and resident cells: DDR1 is presented in smooth muscle cells and macrophages in atherosclerosis [25, 27], and in epithelial cells and macrophages in the model of bleomycin-induced lung fibrosis [28].

As with the previous models of renal disease, mice lacking expression of DDR1 were strongly protected. They showed less proteinuria, uremia, and fibrosis, decreased fibrin deposits and crescentic formation, and blunted inflammatory influx. As a result, these mice displayed a better preservation of renal function and structure and a sevenfold increase in lifespan [26]. In addition, a positive feedback interaction was observed between DDR1 and IL-1 beta: IL-1 beta-induced mRNA and protein expression of DDR1 in cultured podocytes in vitro, and inversely genetic deletion of DDR1 strongly diminished expression of IL-1 beta. Since cytokine secretion such as IL-1 beta by macrophages is a major mechanism to induce podocyte injury in crescentic glomerulonephritis [29, 30], it is probable that the above-described positive interaction between IL-1 beta and DDR1 is responsible for the deleterious effects of DDR1 activation in the renal function in this model.

When antisense oligonucleotides directed against DDR1 were continuously administered in wild-type mice during the development of the disease, an important renoprotection was observed, similar to that seen with the mice lacking DDR1 expression [26]. The experimental approach using antisense oligonucleotides avoided any potential renal or vascular consequence of DDR1 gene deletion during the development of mice and provided a first proof of concept that DDR1 can be a potential target for therapy in renal disease.

## 15.4 Molecular and Cellular Mechanisms of DDR1 in CKD

Still, several questions regarding the mechanisms of action of DDR1 remain unanswered such as the identity of the signal inducing DDR1 neoexpression and/or activation, whether DDR1 is activated only by collagen binding or if it can be transactivated as other tyrosine kinase receptors do, the existence and consequences of DDR1 shedding in vivo, the fate of DDR1 receptor after its phosphorylation, and if the neoexpression and stimulation of DDR1 is enough to initiate and promote renal disease by itself without the synergy of other surrounding factors such as hemodynamic alterations, electrolyte changes, or cell stress.

From a point of view of renal physiopathology, the questions regarding the identification of the signal inducing DDR1 expression in kidneys and the direct promotion of renal disease by DDR1 activation are essential. Since a feature common to the above-presented models are important hemodynamic alterations early after the induction of the disease, it is probable that DDR1 activation is due to a cell event associated with these hemodynamic alterations such as endothelial stress, hypoxia, or shear stress. We are currently investigating these directions and the first results confirm the above-mentioned hypotheses.

Regarding the capacity of DDR1 activation to initiate and promote renal disease, we are currently investigating whether the time conditional and local overexpression of DDR1 in the renal epithelium is a turning point to induce renal damage independently of the external cause. To address this objective, the Paxillin 8 (Pax8) promoter approach was used [31–33] to create conditional strains overexpressing DDR1 in the renal tubular epithelium (Fig. 15.2). This approach consists to create and crossbreed Pax8-rtTA/LC-1

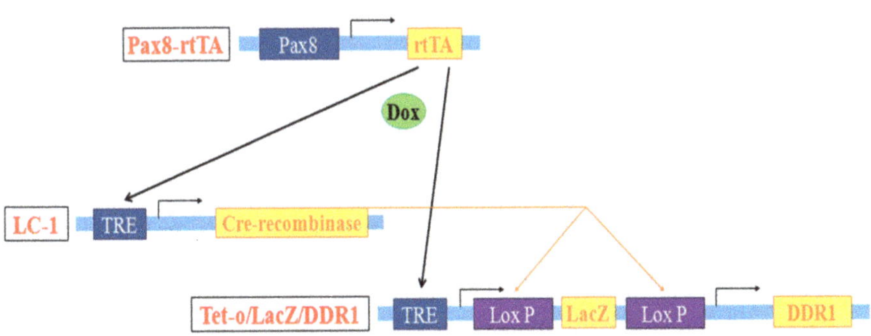


**Fig. 15.2** Experimental design for the creation of a strain of transgenic mice expressing DDR1 in a time and tissue conditional fashion. In this case, the Pax8 promoter approach was used for the creation and activation DDR1 in the renal tubular epithelium

mice with the Tet-o/LacZ/DDR1. Under basal conditions, animals will not express DDR1, but LacZ instead (allowing thus a control of the promoter-tissue specificity). After administration of Dox, rtTA will activate TRE and Cre-recombinase, which will excise LacZ and thus will approach TRE to DDR1. As a result, DDR1 will be overexpressed in the same tissue as LacZ was expressed before. This approach has been already validated in renal physiopathology. The first results are supportive of DDR1 being capable to initiate and promote renal disease.

## 15.5 DDR1 in Human CKD

Clinical data regarding the role of DDR-1 in human nephropathies and the potential effects of its inhibition for therapy are scarce. In few biopsies from focal segmental glomerulosclerosis, diabetic nephropathy, and Goodpasture's syndrome, DDR1 was expressed in glomeruli, parietal epithelial cells, proximal and distal tubules, and the interstitium [26]. Some inhibitors of tyrosine kinase receptors with a selectivity also for DDR1, such as imatinib or nilotinib, have been tested with mixed results. Imatinib is a nonselective inhibitor of several tyrosine kinase receptors, including DDR1. Imatinib is currently used in patients with hematologic disorders such as chronic and acute myeloid leukemia and gastrointestinal stromal tumors. It has been tested in experimental studies with encouraging results in diabetic and lupus nephropathy models. However, it displays several unpleasant secondary effects (such as nausea, diarrhea, periorbital edema, or muscle cramps), probably due to nonselective inhibition of tyrosine kinase receptors which results in epithelial toxicity spread in several tissues. Nilotinib was tested in a model of chronic nephropathy with promising results. However, its tolerance and toxicity in humans has not being tested, and due to lack of selectivity it will probably result in similar to imatinib secondary effects.

## 15.6 DDR1 as a Target for Therapy of CKD

Several elements support the hypothesis that a specific blockade of DDR1 can be promising target for therapy of CKD:

1. DDR1 expression in healthy kidney is low, but it is overexpressed in several models of renal disease.
2. DDR1 activation is accompanying a tissue aggression and is focal to the damaged tissue.
3. DDR1 amplifies renal injury by enhancing the interaction between renal inflammation and fibrosis the two pillars of CKD.
4. Genetic deletion of DDR1 expression impairs the inflammatory response and preserves renal function and structure.

5. Silencing of DDR1 gene with antisense oligonucleotides preserves renal function.
6. Inhibition of DDR1 neoexpression/activation does not affect physiological functions
7. As membrane receptors, DDR1 are easily accessible for detection and therapy intervention.
8. Preliminary experiments show DDR1 overexpression in human renal pathologies.

Unpublished data from our lab with the NTS model further support the hypothesis that DDR1 can be a target for therapy. DDR1 inhibition by antisense after the establishment of the disease (severe proteinuria) was accompanied by an arrest in the progression of renal disease. For the moment, tools do not exist to allow an in depth investigation in human CKD (Elisa kits and specific blockers). Thus, studies about DDR1's levels in renal biopsies, or correlating its expression with the decline of renal function will still wait for a while. Hopefully, the waiting time will not be so long, because several groups of investigators are currently pursuing these objectives and these tools will become available in a nearby future. Already, recent investigations report the synthesis of specific DDR1 inhibitors, in in vitro settings [34–36]. It remains to test their efficiency and selectivity in in vivo studies.

## 15.7 Perspectives: Do DDR1 Inhibitors Have the Potential of Transfer in Human CKD?

Our recent investigations, using mainly experimental models, lead us to identify DDR1 as promising target of therapy. The observations supporting such a role are listed above. The most important limitation is the current lack of widely validated tools ready for use in humans, but this limitation will not exist in a near future as the field is rapidly expanding and several antibodies or blocking agents are expected to be generated or are currently under testing.

## References

1. United States Renal Data System (2003) Annual data report: incidence and prevalence of ESRD. Am J Kidney Dis 42:S37–S173
2. Lozano R (2012) Global and regional mortality from 235 causes of death for 20 age groups in 1990 and 2010: a systematic analysis for the Global Burden of Disease Study 2010. Lancet 380:2095–2128
3. Boffa JJ, Ying L, Placier S et al (2003) Regression of renal vascular and glomerular fibrosis: role of angiotensin II receptor antagonism and metalloproteinases. J Am Soc Nephrol 14:1132–1144
4. Placier S, Boffa JJ, Dussaule JC et al (2006) Reversal of renal lesions following interruption of nitric oxide synthesis inhibition in transgenic mice. Nephrol Dial Transplant 21:881–888
5. Huby AC, Rastaldi MP, Caron K et al (2009) Restoration of podocyte structure and improvement of chronic renal disease in transgenic mice overexpressing renin. PLoS One 21, e6721

6. Kavvadas P, Weis L, Abed AB et al (2013) Renin inhibition reverses renal disease in transgenic mice by shifting the balance between profibrotic and antifibrotic agents. Hypertension 61:901–907
7. Chatziantoniou C, Dussaule JC (2008) Is kidney injury a reversible process? Curr Opin Nephrol Hypertens 17:76–81
8. Mauer M, Zinman B, Gardiner R et al (2009) Renal and retinal effects of enalapril and losartan in type 1 diabetes. N Engl J Med 361:40–51
9. Hou G, Vogel W, Bendeck MP (2001) The discoidin domain receptor tyrosine kinase DDR1 in arterial wound repair. J Clin Invest 107:727–735
10. Curat AC, Vogel WF (2002) Discoidin domain receptor 1 controls growth and adhesion of mesangial cells. J Am Soc Nephrol 13:2648–2656
11. Gross O, Beirowski B, Harvey SJ et al (2004) DDR1-deficient mice show localized subepithelial GBM thickening with focal loss of slit diaphragms and proteinuria. Kidney Int 66:102–111
12. Gross O, Girgert R, Beirowki B et al (2010) Loss of collagen-receptor DDR1 delays renal fibrosis in hereditary type IV collagen disease. Matrix Biol 29:346–356
13. Flamant M, Placier S, Rodenas A et al (2006) Discoidin domain receptor 1 null mice are protected against hypertension-induced renal disease. J Am Soc Nephrol 17:3374–3381
14. Ruiz-Ortega M, Lorenzo O, Suzuki Y et al (2001) Proinflammatory actions of angiotensins. Curr Opin Nephrol Hypertens 10:321–329
15. Fliser D, Buchholz K, Haller H (2004) EUropean Trial on Olmesartan and Pravastatin in Inflammation and Atherosclerosis (EUTOPIA) investigators. Antiinflammatory effects of angiotensin II subtype 1 receptor blockade in hypertensive patients with microinflammation. Circulation 110:1103–1107
16. Chatziantoniou C, Boffa JJ, Ardaillou R et al (1998) Nitric oxide inhibition induces early activation of type I collagen gene in renal resistance vessels and glomeruli in transgenic mice: role of endothelin. J Clin Invest 101:2780–2789
17. Boffa JJ, Tharaux PL, Placier S et al (1999) Angiotensin II activates collagen type I gene in the renal vasculature of transgenic mice during inhibition of nitric oxide synthesis: evidence for en endothelin-mediated mechanism. Circulation 100:1901–1908
18. Guerrot D, Kerroch M, Placier S et al (2011) Discoidin domain receptor 1 is a major mediator of inflammation and fibrosis induced by unilateral ureteral obstruction. Am J Pathol 179:83–91
19. Chevalier R (2006) Specific molecular targeting of renal injury in obstructive nephropathy. Kidney Int 70:1200–1201
20. Chevalier R, Forbes M, Thornhill B (2009) Ureteral obstruction as a model of renal interstitial fibrosis and obstructive nephropathy. Kidney Int 75:1145–1152
21. Vielhauer V, Anders H, Mack M et al (2001) Obstructive nephropathy in the mouse: progressive fibrosis correlates with tubulointerstitial chemokine expression and accumulation of CC chemokine receptor 2- and 5-positive leukocytes. J Am Soc Nephrol 12:1173–1187
22. Lange-Sperandio B, Cachat F, Thornhill B et al (2002) Selectins mediate macrophage infiltration in obstructive nephropathy in newborn mice. Kidney Int 61:516–524
23. Anders H, Vielhauer V, Frink M et al (2002) A chemokine receptor CCR-1 antagonist reduces renal fibrosis after unilateral ureter ligation. J Clin Invest 109:251–259
24. Lange-Sperandio B, Schimpgen K, Rodenbeck B et al (2006) Distinct roles of Mac-1 and its counter-receptors in neonatal obstructive nephropathy. Kidney Int 69:81–88
25. Franco C, Britto K, Wong E et al (2009) Discoidin domain receptor 1 on bone marrow-derived cells promotes macrophage accumulation during atherogenesis. Circ Res 105:1141–1148
26. Kerroch M, Guerrot D, Vandermeersch S et al (2012) Genetic inhibition of discoidin domain receptor 1 protects mice against crescentic glomerulonephritis. FASEB J 26:4079–4091
27. Franco C, Hou G, Ahmad PJ et al (2008) Discoidin domain receptor 1 deletion decreases atherosclerosis by accelerating cell matrix accumulation and reducing inflammation in low-density lipoprotein receptor-deficient mice. Circ Res 102:1202–1211

28. Avivi-Green C, Singal M, Vogel WF (2006) Discoidin domain receptor 1-deficient mice are resistant to bleomycin-induced lung fibrosis. Am J Respir Crit Care Med 174:420–427
29. Timoshanko JR, Kitching AR, Iwakura Y et al (2004) Leukocyte-derived interleukin-1beta interacts with renal interleukin-1 receptor I to promote renal tumor necrosis factor and glomerular injury in murine crescentic glomerulonephritis. Am J Pathol 164:1967–1977
30. Tipping PG, Timoshanko J (2005) Contributions of intrinsic renal cells to crescentic glomerulonephritis. Nephron Exp Nephrol 101:173–178
31. Traykova-Brauch M, Schönig K, Greiner O et al (2008) An efficient and versatile system for acute and chronic modulation of renal tubular function in transgenic mice. Nat Med 14:979–984
32. Hakroush S, Moeller MJ, Theilig F et al (2009) Effects of increased renal tubular vascular endothelial growth factor (VEGF) on fibrosis, cyst formation, and glomerular disease. Am J Pathol 175:1883–1895
33. Koesters R, Kaissling B, Lehir M et al (2010) Tubular overexpression of transforming growth factor-beta1 induces autophagy and fibrosis but not mesenchymal transition of renal epithelial cells. Am J Pathol 177:632–643
34. Kim HG, Tan L, Weisberg E et al (2013) Discovery of a potent and selective DDR1 receptor tyrosine kinase inhibitor. ACS Chem Biol 8:2145–2150
35. Li Y, Lu X, Ren X et al (2015) Small molecule discoidin domain receptor kinase inhibitors and potential medical applications. J Med Chem 58:3287–3301
36. Murray CW, Berdini V, Buck IM et al (2015) Fragment-based discovery of potent and selective DDR1/2 inhibitors. ACS Med Chem Lett 6:798–803

# Chapter 16
# Discoidin Domain Receptors in Liver Fibrosis

**Elvira Olaso, Joana Marquez, Aitor Benedicto, Iker Badiola, and Beatriz Arteta**

## 16.1 Hepatic Fibrosis

Most chronic liver diseases are associated with fibrosis and are characterized by parenchymal damage and inflammation. Accumulation of ECM in the liver following chronic injury may be initiated by a variety of factors including chronic hepatitis, viral infections, alcohol consumption, and drug abuse [1]. Genetic and environmental factors also play key roles in the development of liver fibrosis, as well as mediating the progression and response to fibrosis-causing agents [2]. Recent clinical and experimental evidence indicates that acute hepatic fibrosis can be reversed if the stimulus is removed [3–5]. However, perpetuation of hepatic fibrosis leads to irreversible cirrhosis, an end-stage lesion, which may lead to liver failure and death.

Despite the progress in the field, the assessment of liver fibrosis remains a difficult task. Experts still have to cope with several challenges. Even though liver biopsy is an invasive technique and might lead to sampling errors and inter-observer variability, it is the most used technique to date. However, the development of powerful experimental methods in recent decades has allowed researchers to dissect important aspects of the cellular and molecular mechanisms underlying hepatic fibrosis. Both in vivo and in vitro models can be used to identify and manipulate important mechanisms and gene products. Findings from this research are providing a new source of clinical markers and target-oriented therapies for the management of patients suffering from this disease.

Experimental models of liver fibrosis, which mimic most of the features of human liver fibrosis, have been used extensively to characterize stellate cell

E. Olaso, Ph.D. • J. Marquez, Ph.D. • A. Benedicto, Ph.D. • I. Badiola, Ph.D.
B. Arteta, Ph.D. (✉)
Department of Cell Biology and Histology, School of Medicine and Enfermery, University of the Basque Country, Leioa 48940, Spain
e-mail: beatriz.arteta@ehu.es



R. Fridman, P.H. Huang (eds.), *Discoidin Domain Receptors in Health and Disease*, DOI 10.1007/978-1-4939-6383-6_16

activation in vivo. For example, the time course of acute animal intoxication with carbon tetrachloride ($CCl_4$) has been described extensively [6]. First, $CCl_4$ is metabolized primarily by the cytochrome P450 2E1, and the resulting radicals initiate fatty acid peroxidation. Necrosis is evident within 12 h and peaks in approximately 1–3 days. Dying hepatocytes trigger the influx of inflammatory cells into necrotic areas, which produce cytokines that cause hepatocyte regeneration as well as stellate cell activation to highly fibrogenic myofibroblasts to stimulate the repair process. While acute injury is reversible and diminishes within approximately 3 days, chronic $CCl_4$ administration leads to larger ECM deposition in the space of Disse around sinusoids, leading to bridging fibrosis.

To further analyze the molecular mechanism of stellate cell activation, parenchymal and nonparenchymal cells can be isolated and cultured in vitro to recapitulate the in vivo situation, allowing for detailed investigation of their interaction under controlled conditions. To both confirm and identify more interacting molecular factors, genetic models in which genes are expressed ectopically or knocked-out can be used to establish the biological relevance of a gene product in the pathological liver.

## 16.2 Sinusoidal Features of Hepatic Fibrosis

Hepatic fibrosis is characterized by an excess of ECM deposition in the subendothelial space of the sinusoids, or liver capillaries (for review, see [2]). Liver injury is often followed by hepatocyte necrosis and/or invasion of inflammatory cells, which distort normal cell–cell and cell–matrix interactions. New interactions are established that trigger a fibrogenic response.

In the fibrotic liver, significant quantitative and qualitative changes occur in the composition of ECM. The accumulation of this collagenous ECM, known as capillarization of the sinusoid, has dramatic effects on surrounding cells, including activation of stellate cells (see below), loss of hepatocyte microvilli, and disappearance of endothelial cell fenestrations. As a result, capillarized sinusoids cannot function properly.

Accumulation of ECM in hepatic fibrosis is due to both increased synthesis of matrix components and abnormal ECM degradation. In the normal liver, the subendothelial space, which separates the epithelial hepatocytes from the sinusoidal endothelium, contains a diffuse basement membrane matrix that is rich in collagen type IV, laminin, and proteoglycans. These subendothelial ECM proteins are critical for maintaining the differentiated functions of resident liver cells. As the normal liver becomes fibrotic, ECM content increases 3–5 fold [7]. Moreover, the subendothelial ECM shifts from a low-density basement membrane to one that is rich in fibrillar collagens types I and III [8] (Fig. 16.1a).

Fibrosis reflects a loss of homeostasis between fibrogenesis and matrix degradation. Matrix metalloproteinases (MMPs) are the main extracellular matrix-degrading

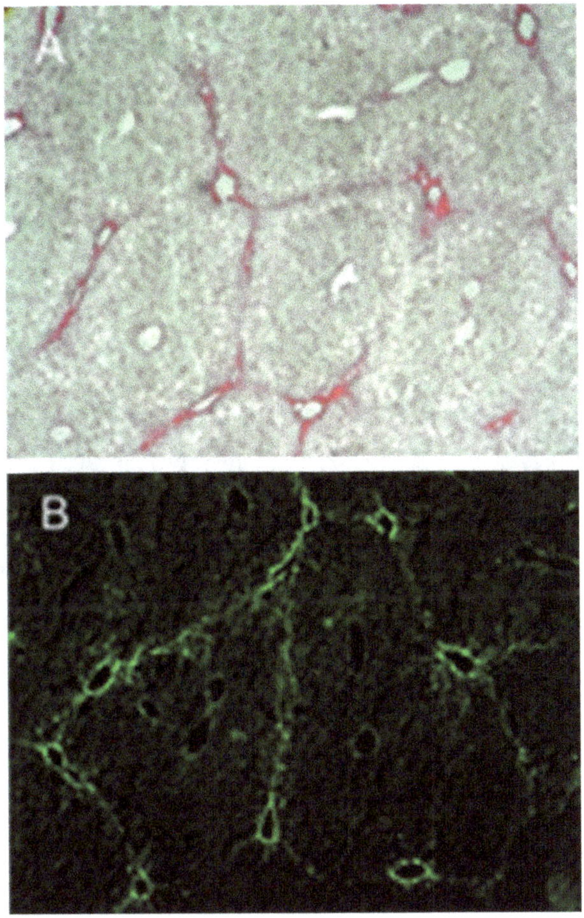


**Fig. 16.1** Mouse model of liver fibrosis. Mice were intoxicated with $CCl_4$, and their livers were analyzed immunohistochemically. (**a**) Sirius red staining of collagenous fibers in fibrotic mouse livers (in red). (**b**) In situ analysis shows gelatinase activity in the fibrous bands (*light green*)

enzymes in the liver [9, 10] (Fig. 16.1b). MMP-1 is downregulated, and MMP-2 and MMP-9 are upregulated in human liver diseases. These latter two enzymes degrade basement membrane collagen, whereas MMP-1 degrades interstitial collagens types I and III [11]. Increased expression of MT1-MMP (membrane-type 1-matrix metalloproteinase) has also been observed in fibrotic human livers; MT1-MMP can activate latent MMP-2.

In experimental liver fibrosis, progressive liver injury is associated with upregulation of the MMP inhibitors, TIMP-1 and TIMP-2, relative to MMP-1, which may explain the decrease in MMP-1 activity that accompanies progressive fibrosis [9]. In rodents, MMP-13 is the major interstitial collagenase that degrades type I/III collagens. Some studies indicate that MMP-2 may also disrupt collagen type I in rodents [12]. There are several hepatic cellular sources of MMPs; the main MMP producers are stellate cells and liver-associated macrophages.

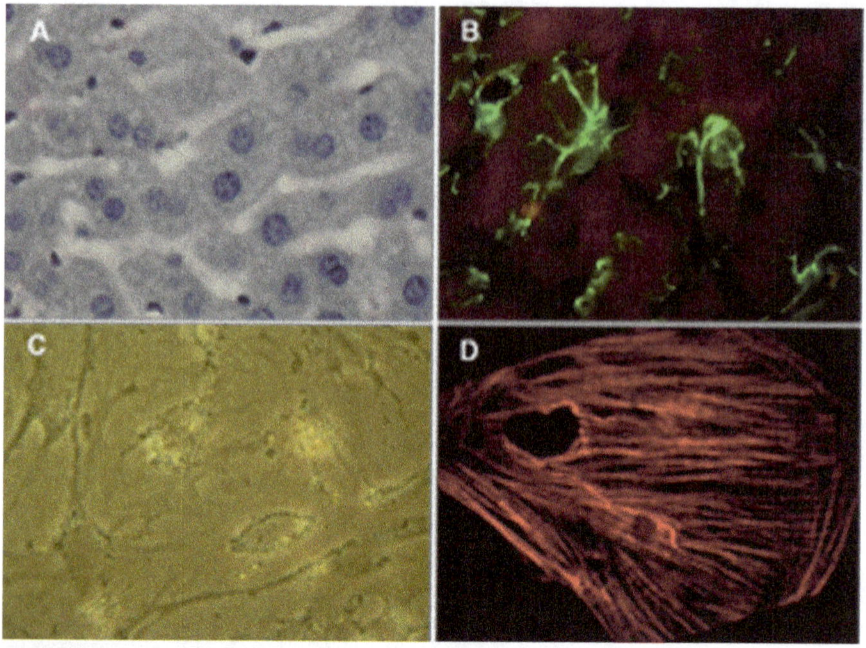


**Fig. 16.2** Stellate cells in healthy and injured mouse livers. (**a**) Hematoxylin/eosin staining of histological liver sections from healthy mice. (**b**) GFAP stellate cells (*green*) in the sinusoids of healthy livers. (**c**) Early culture of mouse stellate cells isolated from healthy livers. (**d**) Alpha-smooth muscle actin expression in activated mouse stellate cells (*red*)

## 16.3 Stellate Cells are the Main Cell Type Responsible for Liver Fibrosis

Stellate cells are the main matrix-producing cells involved in liver fibrosis and hold the key to understanding its pathogenesis. Stellate cells are residential mesenchymal cells that are embedded in the subendothelial space of Disse, where they maintain close and constant interactions with the hepatocytes and endothelial cells [13]. In a healthy liver, stellate cells remain quiescent and contain large amounts of vitamin A droplets [14] (Fig. 16.2). Stellate cells express desmin, a myogenic intermediate filament, and neural markers, such as glial fibrillary acidic protein (GFAP) [15]. Stellate cells are also involved in liver development and regeneration [16].

The response of stellate cells to the ECM is the most essential factor in perpetuating hepatic fibrosis (for review see [17]). In response to injury, activated stellate cells transdifferentiate into myofibroblasts. Other cells that undergo a similar type of activation are the mesangial cells in the kidney, pulmonary mesenchymal cells, and stellate cells in the pancreas. Stellate cell activation is the transition from a quiescent cell to a highly proliferative, fibrogenic, and contractile one. Activated stellate cells also contribute to the amplified inflammatory response [18, 19].

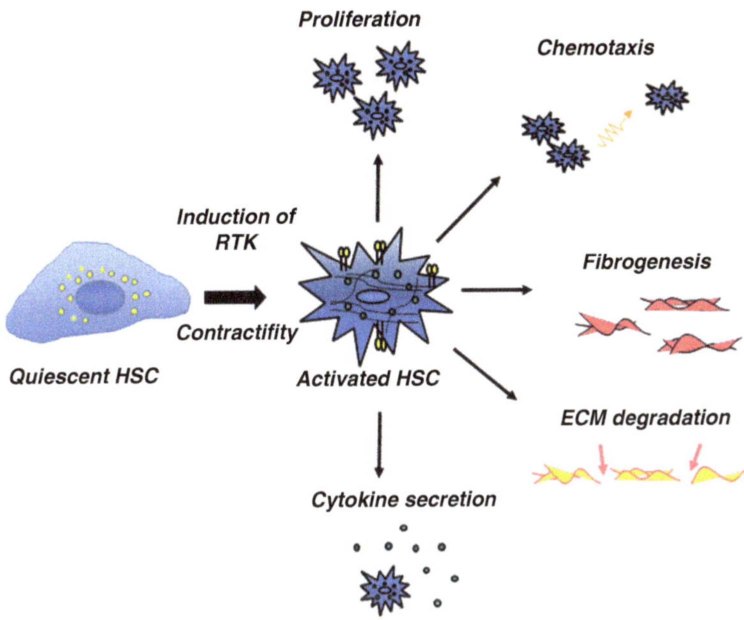


**Fig. 16.3** The two-step model of stellate cell activation

## 16.4 The Two-Step Model of Stellate Cell Activation

Activation of stellate cells involves a drastic change in gene expression that gives rise to a dramatic restructuring of their phenotypic characteristics. This complex phenomenon has been extensively described as a two-phase process, consisting of an initiation phase, followed by perpetuation (see [20] for review) (Fig. 16.3). The *initiation* or preinflammatory phase refers to early changes in gene expression and phenotype, which render the cells responsive to other cytokines and stimuli, whereas *perpetuation* results from the effects of these stimuli on maintaining the activated phenotype and generating fibrosis.

### *16.4.1 Initiation*

Stellate cell activation begins within minutes of liver injury; it involves a coordinated induction of genes, which encode both regulatory and structural proteins, in response to paracrine stimuli from injured hepatocytes and activated liver sinusoidal endothelial cells (LSECs). Prominent CXCL4 expression in the activated LSECs exerts a fibrotic response in adjacent quiescent stellate cells [21]. Infiltration of macrophages (such as resident Kupffer cells) coincides with the appearance of stellate cell activation markers. In vitro coculture with Kupffer cells modulates the fibrogenic response of the stellate cells [22]. Furthermore, the addition of Kupffer cell-conditioned medium during the activation of stellate cells in culture leads to a closer recapitulation of the changes in gene expression that follow chronic $CCl_4$ treatment in vivo [23].

Among the very early changes, upregulation of genes like Collagen I (*COL1A1*), alpha smooth muscle actin (*ASMA*), insulin-like growth factor-binding protein 3 (*IGFP3*), and genes encoding receptor tyrosine kinases (RTKs) have been described [24, 25]. The main phenotypic changes include the sudden expression of *ASMA*, loss of normal retinoid-storing capacity, and increased rough endoplasmic reticulum. These newly acquired phenotypic features render stellate cells susceptible to paracrine and autocrine stimuli.

### 16.4.2 Perpetuation

Perpetuation of stellate cell activation involves discrete changes in cell behavior. The net effect of these changes is to increase accumulation of collagenous ECM. The signals and phenotypic characteristics produced after the first activation process remain in the stellate cell and perpetuate the pathologic conditions of the liver tissue through the continued release of mediators from chronically inflamed and injured tissue. The number of activated stellate cells increases in the injured areas via proliferation and chemotactic recruitment. Once there, they produce fibrotic collagenous matrix, supporting the fibrogenic transformation of the liver.

Activated stellate cells also secrete MMPs, which contribute to ECM remodeling. The increased production of cytokines and growth factors promote the continuous autocrine and paracrine activation of stellate cells. A more detailed description of these changes may be found elsewhere [17]. In brief:

(a) *Proliferation*. Increased numbers of stellate cells in the injured liver arise, in part, from local proliferation in response to polypeptide growth factors. These stimulating factors signal primarily through RTKs that are responsible for cell growth and differentiation. The expression of RTKs in quiescent stellate cells is very low, suggesting that only activated stellate cells are able to respond to these factors. PDGF is the best-characterized mitogenic and proliferative cytokine [26]. A large number of other mitogens can also drive stellate cell proliferation, including endothelin-1, thrombin, and others (see [20] for review).

(b) *Chemotaxis*. Hepatic stellate cells migrate to the sites of inflammation in response to chemotactic factors. In these locations, they amplify the inflammatory signal promoting lymphocyte activation [27], leading to a proinflammatory loop, which recruits more stellate cells. This migratory response is partly mediated by cytokines, which constitute one of the main sources of stimulatory molecules that participate in liver fibrosis. In fact, many cytokines act as chemoattractants by signaling through RTKs mediated pathways.

(c) *Fibrogenesis*. With the dominant role of stellate cells in producing ECM clearly established, attention has focused on fibrogenic mediators, particularly TGF-β1 and its receptors. Autocrine secretion is among the most important sources of TGF-β1. A large number of other factors stimulate fibrogenesis by activated stellate cells, including retinoids, interleukin-1β, tumor necrosis factor, and acetaldehyde, although none is as potent as TGF-β1.

As previously stated, the main component of the scarring liver tissue is collagen. Transcriptional activation of the type I collagen gene may occur through a variety of mechanisms. Increased collagen alpha(I) deposition by stellate cells may also result from posttranscriptional events, such as increased mRNA stabilization (for review see [2]). The ECM also directly participates and contributes in the activation of HSCs via collagen accumulation [28, 29]. In turn, activated stellate cells contribute massively to ECM deposition, therefore creating a fibrogenic feedback loop. Elucidation of the cellular mechanism elicited by the ECM to activate stellate cells is an area of intense research as a new approach to find targeted therapies for liver fibrosis.

(d) *ECM degradation*. As noted above, quantitative and qualitative changes in matrix protease activity play an important role in the extracellular matrix remodeling that accompanies liver fibrosis injury. Stellate cells express virtually all key components required for pathologic matrix degradation. Therefore, these cells not only play a key role in matrix production, but they are also involved in matrix degradation.

In the liver, stellate cells are the main source of MMP-2, they also secrete significant amounts of MMP-13 (the rodent equivalent of MMP-1) and stromelysins (MMP-3,-7,-10, and -11), and produce TIMP-1 and TIMP-2. Recent research in liver fibrosis recovery pinpoints TIMP-1 as an amplifying factor for chronic liver fibrosis by two interrelated mechanisms: first, TIMP-1 inhibits MMP activity, allowing matrix deposition; second, TIMP-1 inhibits stellate cell apoptosis, a main step in liver fibrosis recovery (for review see [30]).

(e) *Cytokine release*. Increased production and/or activity of cytokines is critical for both autocrine and paracrine perpetuation of stellate cell activation. Virtually all features of stellate cell activation can be attributed to autocrine cytokines, including Transforming Growth Factor (TGF)-β1, Platelet Derived Growth Factor (PDGF), Fibroblast Growth Factor (FGF), Hepatocyte Growth Factor (HGF), Platelet Activating Factor (PAF), and endothelin-1, among others. Additionally, HSCs respond to chemotactic factors produce by other cell types by migrating to the sites of inflammation, where they amplify the inflammatory signal and promote lymphocyte activation [27]. This response can lead to a proinflammatory loop, which recruits more stellate cells. Besides, stellate cells can also amplify inflammation during liver injury through the release of neutrophil and monocyte/macrophage chemoattractants, such as chemokines from the CCL and CXCL families. The persistent accumulation of immune cells may prolong tissue inflammation and aggravate immunopathology. On the other hand, anti-inflammatory cytokines, such as IL-10, which is produced by stellate cells, have also been identified.

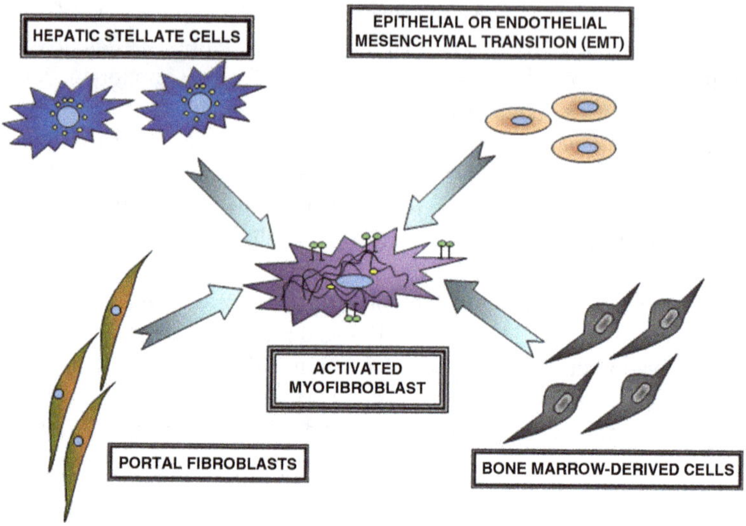


**Fig. 16.4** Cell sources of liver myofibroblasts

## 16.5 Nonstellate Cell Sources of Activated Myofibroblast in the Fibrotic Liver

While increasing data point out stellate cells as the main source of myofibroblasts in the fibrotic liver, there are at least three other potential sources of myofibroblasts in the pathological liver (Fig. 16.4). First, portal fibroblasts contribute to myofibroblasts in chronic cholestatic disorders where fibrosis is initially located around portal tracts. Portal fibroblasts do not express desmin or GFAP, and the lack of both markers differentiates portal myofibroblasts from those derived from activated stellate cells.

Second, recent studies have proposed hepatocytes, cholangiocytes, and endothelial cells as myofibroblast precursors through epithelial or endothelial mesenchymal transition (EMT). EMT is a complex biological process in which endothelial cells lose their specific endothelial cell markers, such as vascular endothelial (VE) cadherin, and acquire a mesenchymal or myofibroblastic phenotype, initiating expression of mesenchymal cell products, including ASMA, vimentin, and types I and III interstitial collagens [31]. These cells include hepatocytes ($CD45^-$ and $albumin^+$), cholangiocytes ($CD45^-$ and $CK19^+$), or endothelial cells ($Tie^{2+}$). However, the contribution of cells undergoing EMT to the myofibroblast population is, as yet, unclear; moreover, recent publications provide strong evidence against EMT as a source of myofibroblasts in liver [32].

Third, bone marrow-derived mesenchymal stem cells (BM-MSCs) may be recruited to the liver and differentiate into hepatic myofibroblasts [33]. BM-derived stem/progenitor cells express MMP-13 and MMP-9 [34]. Whether BM-derived myofibroblasts contribute to ECM deposition in the course of human liver fibrosis is unknown. In experimental liver fibrosis, only a small contribution of BM-derived cells to the myofibroblast population has been detected [34].

## 16.6 Receptor Tyrosine Kinases in Stellate Cell Activation

The identification of RTKs has emerged as a major theme in cell growth and differentiation, apoptosis, and transformation. All RTKs contain a glycosylated extracellular domain to which a ligand binds, a single transmembrane domain, and an intracellular domain with a catalytic tyrosine kinase domain [35]. The tyrosine kinase domain is the most strictly conserved region and is essential for signaling. For most of the RTKs, the ligand is a peptide growth factor, and, in many cases, ligands originate primarily from autocrine or paracrine sources.

RTKs play major roles in stellate cell activation and perpetuation of fibrosis. Autocrine and paracrine ligands signal through RTKs to regulate the main actions of activated stellate cells, such as proliferation, chemotactic migration, and chemokine production. The behavior of the PDGF receptor typifies the role of RTKs during stellate cell activation. Upregulation of the PDGF-β receptor is a very early event in stellate cell activation; within 12 h of injury, the de novo expression of PDGF-β R mRNA is observed in rat stellate cells. A similar induction occurs in stellate cells activated by in vitro culture [36].

Other members of the RTK family, such as the receptors for the FGF, the insulin growth factor (IGF), and the vascular endothelial growth factor (VEGF), have also been identified as mediators of stellate cell growth in activated stellate cells [37]. Because VEGF is the principal effector of angiogenesis, the findings indicate that stellate cells can participate in a classical "angiogenic" response, broadening their potential roles in both wound healing and tumor formation [38]. The DDR subfamily of RTKs, which is composed of two members, DDR1 and DDR2, is unique among RTKs because their ligand is collagen instead of a peptide growth factor (see chapter by Dr. Shimada). The important role of DDRs in stellate cell activation is discussed in detail in the following sections.

## 16.7 Stellate Cell Interactions with the ECM: Integrins and DDRS

Stellate cells interact with the ECM, which directly participates in their activation. Indeed, the accumulation of a collagenous matrix contributes directly to the activation of the quiescent stellate cells into myofibroblast-like cells [39]. However, the exact mechanism that triggers stellate cell activation in response to ECM molecules remains largely unknown.

Integrins are heterodimeric transmembrane receptors that transduce ECM signals in the liver and other tissues. Stellate cells express integrins $\alpha_1\beta_1$, $\alpha_2\beta_1$, $\alpha_v\beta_1$, $\alpha_v\beta_3$, and $\alpha_6\beta_4$; of these, $\beta_1$ mediates adhesion to collagen type I, and $\alpha_1\beta_1$ mediates cellular contraction on collagen type I. While the binding of fibronectin to $\alpha_5\beta_1$-integrin results in an increase the synthesis type I collagen in the early stages, in the later ones, this integrin has been shown to be downregulated [40]. However, HSCs persist in synthesizing type I collagen indicating the existence of additional mechanisms. DDRs, which bind type I collagen, are upregulated in activated HSCs and have been shown to contribute to their proliferation and invasiveness [29]. Therefore, DDRs may represent the elusive transducer of stellate cell activation in response to collagen type I in late stages. With the identification of a DDR in stellate cells (see below), a potentially new mode of cell–matrix interaction emerged that could explain how interstitial collagen type I provokes activation of stellate cells during sinusoidal capillarization.

## 16.8 DDR2 is an Early-Induced RTK in Stellate Cell Activation Following Acute Liver Injury

In order to identify additional RTKs that are induced during stellate cell activation, Friedman and colleagues performed homology polymerase chain reaction using primers derived from the conserved kinase domain of RTKs and a template consisting of activation-specific cDNAs from rat stellate cells [37]. The cloning strategy was validated by the identification of the PDGFβ receptor. A partial fragment of DDR2 cDNA was also recovered. We confirmed DDR2 protein expression and phosphorylation in response to collagen in rat stellate HSC-T6 cells [29] and human stellate LX cells [41]. Furthermore, increased phosphorylated DDR2 protein was observed during in vitro stellate cell activation, which recapitulates most of the stellate cell changes observed in vivo [29].

To explore further the possible implications of DDR2 signaling in the liver, we analyzed rat livers after acute injury by a single dose of $CCl_4$ or after bile duct ligation [29]. Maximal DDR2 mRNA expression was observed in stellate cells that were freshly isolated 48 h after $CCl_4$. Interestingly, this time point is considerably later than the peak for PDGF-β receptor mRNA (12 h). In animals with ligation of the common bile duct, a marked upregulation of DDR2 mRNA is observed after 7 days, in comparison to sham-operated control animals. Increased expression of DDR2 also occurs in experimental alcohol liver fibrosis [42, 43], although the progressive increase in DDR2 mRNA expression was reported only after several weeks of alcohol consumption. Together, these reports show that DDR2 expression is closely associated with fibrotic liver tissues.

## 16.9 Other Sources of DDR2 in the Pathologic Liver

(a) *Biliary epithelial cells*. Primary biliary cirrhosis is characterized by chronic progressive destruction of small intrahepatic bile ducts with portal inflammation and cholestasis, leading to fibrosis. Mao and colleges [44] utilized a novel restriction analysis system to profile the expression of tyrosine kinases (TKs) in primary biliary cirrhosis. Using this technique, DDR2 expression was observed in human mRNA isolated from biliary epithelial cell-enriched cell fractions. In situ hybridization of DDR2 mRNA in primary biliary cirrhosis livers revealed its expression in the small bile duct epithelial regions and in fibroblasts/stromal cells of fibrotic regions. A similar pattern was observed in livers of primary sclerosing cholangitis, although the amount of small ducts that were positively stained was lower than in primary biliary cirrhosis. Additionally, DDR2 transcripts were noted within the fibrotic lesions in cirrhotic livers of patients with other diseases, including alcoholic liver disease and chronic hepatitis C.

(b) *Bone marrow-derived cells*. As reported above, bone marrow-derived cells can be recruited to the injured liver. In vitro maturation of bone marrow-derived cells induced by collagen may be independent of integrins, since it is not inhibited by the presence of a blocking antibody against $\beta_1$-integins. The expression of DDR2 [45] in bone marrow-derived dendritic cells is critical for cell activation and the functional capacity of dendritic cells to regulate immune responses [46]. Indeed, depletion of DDR2 in dendritic cells by specific siRNA attenuated their increased expression of IL-12 and CD86 and decreased their capacity to stimulate T-cell proliferation.

## 16.10 DDR2 Receptor Promotes In Vitro MMP-2-Mediated Proliferation and Chemotactic Invasion by Stellate Cells

The discovery of the dual activities of DDRs as both RTKs and collagen receptors suggested a role for DDRs in proliferation and matrix remodeling [47]. Since RTKs also mediate cell migration, DDR2 was presumed to take part in stellate cell recruitment to the injured tissue. Using both signaling and nonsignaling constructs of DDR2, these activities were tested in cells in vitro. HSC-T6 cells were stably transfected with signaling and nonsignaling forms of DDR2 to investigate the roles and activities of DDR2 in stellate cell proliferation, MMP-2 activity, and chemotactic migration [29]. HSC-T6 overexpressing either wild-type DDR2 (wt-DDR2), a constitutively active chimeric DDR2 receptor that renders the receptor constitutively active (Fc-DDR2), a truncated nonsignaling receptor expressing the extracellular domain (ec-DDR2), or a kinase-dead nonsignaling DDR2 (kd-DDR2) were engineered (Fig. 16.5). Enhanced stellate cell proliferation, MMP-2 release, and chemotactic migration through the reconstituted basal membrane Matrigel in response to Kupffer cell culture supernatants were directly proportional to the increase in signaling pathways activated by DDR2 (Table 16.1). Cells expressing the constitutively Fc-DDR2 construct required higher concentrations of TIMP-2, the natural inhibitor of MMP-2, to achieve comparable levels of inhibition to those of control cells, thus, demonstrating a link between DDR2 signaling and MMP-2 activity in stellate cell proliferation. Furthermore, a correlation between DDR2 expression and HSC-T6 cell proliferation has also been confirmed by means of RNA interference [42].

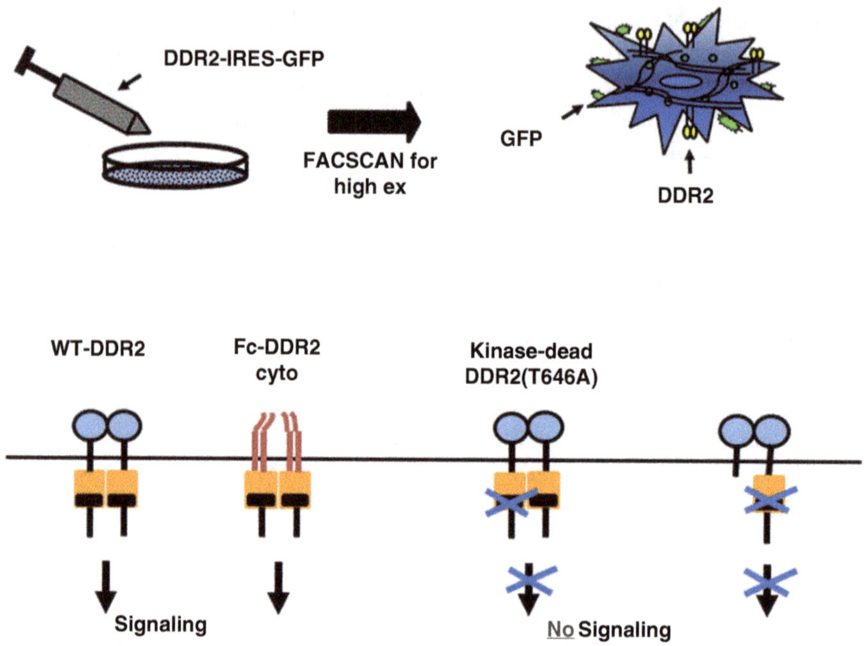


**Fig. 16.5** DDR2-IRES-GFP retroviral expression constructs. HSC-T6 stellate cells were infected with retroviruses expressing several signaling and nonsignaling DDR2-IRES-GFP constructs. High expressers were recovered by FACS can cell sorting for GFP and cultured for a brief period. Wt-DDR2: wild-type DDR2. Fc-DDR2: constitutively active chimera. Kd-DDR2: nonsignaling DDR2 lacking the kinase domain. Ec-DDR2: nonsignaling extracellular domain of DDR2. Control: cells expressing only GFP

**Table 16.1** DDR2 implication in stellate cell profibrogenic functional parameters

| | Chemotaxis | Proliferation | MMP synthesis |
|---|---|---|---|
| Control | +++ | +++ | ++ |
| Ec-DDR2 | + | + | NA |
| Kd-DDR2 | ++ | ++ | + |
| Fc-DDR2 | ++++ | ++++ | +++ |

*NA* not analyzed

## 16.11 Activated Stellate Cells Are the Main DDR2-Expressing Cells in Experimental Chronic Hepatic Fibrosis

Whereas our earlier studies explored the contribution of DDR2 receptors to stellate cell responses after acute injury, their characterization in a chronic model of injury was essential in order to model human liver disease. Therefore, we explored DDR2 expression in livers from mice that were chronically treated with $CCl_4$. Basal DDR2 expression was nearly absent in healthy livers, whereas its protein expression and phosphorylation was readily detectable in mice that had been treated for 2 weeks, which reached maximal levels after 4–6 weeks of treatment. At that time point, DDR2 protein expression is shown following desmin and ASMA expression in cells of the fibrous scar, that is, an expression followed by activated stellate cells in the same areas (Fig. 16.6).

To confirm further the expression of DDR2 and its cellular location in livers collected from mice chronically treated with $CCl_4$ as reported above, we utilized heterozygous DDR2 knock-out/LacZ knock-in mice. In these mice, β-galactosidase expression was used as a surrogate marker for DDR2. In injured livers of these mice, strong β-galactosidase staining appeared in fibroblast-like cells mainly located in the fibrotic septa [38]. As a whole, these in vivo and in vitro data involve stellate cells as the major DDR2-expressing cells in the fibrotic mouse liver following injury with $CCl_4$, correlating with their role as the primary cell type responsible for pericentral fibrosis.

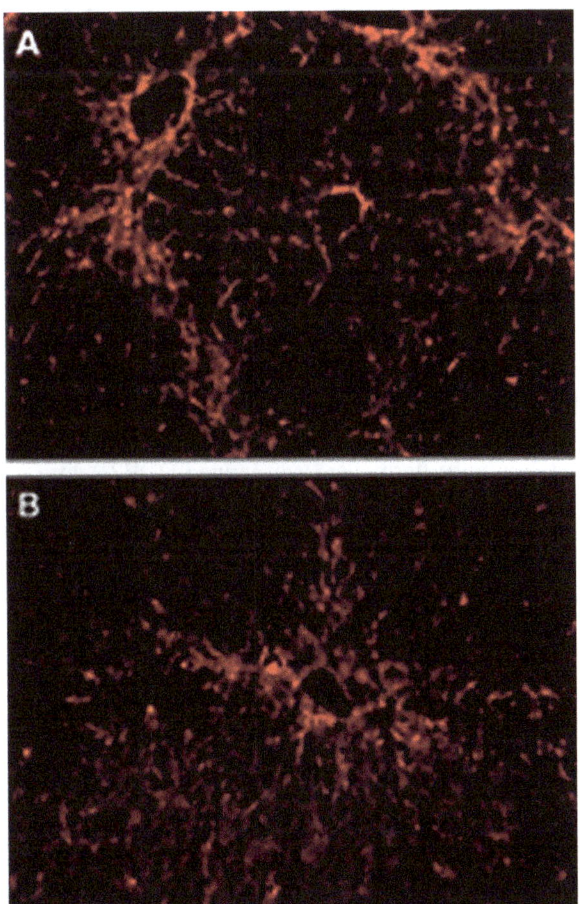


**Fig. 16.6** Expression of desmin and DDR2 in a mouse model of liver fibrosis. (**a**) Desmin expression in $CCl_4$-injured mouse liver (*red*). Desmin is expressed in both quiescent stellate cells in nonaffected sinusoids and in activated ASMA-expressing cells in the fibrous scar. (**b**) DDR2 expression in the $CCl_4$-injured mouse liver (*red*) is mainly located at the fibrous scar

Chronic fibrosis develops, in part, as a result of the chronic and excessive synthesis of collagen by stellate cells in response to autocrine stimuli. As a result, signaling of ECM receptors in the stellate cell is altered, contributing to a vicious cycle of inflammatory and fibrogenic stimulation. Our next goal was to establish the specific role of DDR2 in the contribution of stellate cells to chronic fibrosis. A major hurdle in the study of a single receptor in stellate cell activation is the high degree of cross-signaling pathways; thus, we utilized DDR2 knock-out ($DDR2^{-/-}$) mice to circumvent this difficulty.

## 16.12 Analysis of the Role of DDR2 in Chronic Liver Fibrosis Using DDR2 Knock-Out Mice

DDR2 knock-out mice are viable and fertile, their livers appear grossly normal [48]. Within a few weeks after birth, these mice develop progressive shortening of the axial skeletal bones and the snout, reflecting a defect in osteoblast activity. The skin from DDR2-defficient mice contains reduced MMP activity and shows defective experimental wound healing [49]. In vitro proliferation and MMP-2 promoter activity in skin fibroblasts isolated from $DDR2^{-/-}$ mice were significantly attenuated [50]; these results encouraged us to further study chronic hepatic fibrosis in the DDR2-deficient mice.

### *16.12.1 Loss of DDR2 Promotes Experimental Chronic Hepatic Fibrosis*

In contrast to changes seen in acute injury, chronic $CCl_4$ administration increased the density of activated stellate cells, collagen type I deposition, and gelatinolytic activity in the fibrotic areas of injured $DDR2^{-/-}$ livers compared to wild-type ones. Collagenase activity was similar in all fibrotic mice. We also studied macrophage recruitment (Kupffer cells and monocyte-derived macrophages) because these cells actively participate in ECM remodeling and stellate cell recruitment. We found that fewer macrophages localized to the hepatic scars from injured $DDR2^{-/-}$ mice compared to those of wild-type ones.

Stellate cells and macrophages were isolated from livers of chronically $CCl_4$-treated mice and cultured in vitro. Compared to activated stellate cells from fibrotic wild-type liver, activated stellate cells from fibrotic $DDR2^{-/-}$ livers showed increased proliferation when cultured in the presence of collagen type I, enhanced MMP-dependent chemotactic invasion of reconstituted basal membrane, and reduced collagenase and gelatinase (G/C) activities. Macrophages from fibrotic, $DDR2^{-/-}$ livers exhibited increased collagenase and gelatinase (G/C) activities, higher cytokine mRNA expression, and stronger chemoattractive activity toward $DDR2^{-/-}$ stellate cells, in comparison to macrophages from fibrotic, wild-type livers. These results suggest a combined DDR2-dependent role for stellate cells and macrophages in chronic liver fibrosis.

Finally, we analyzed gene expression of five key cytokines involved in liver fibrogenesis and inflammatory responses in these mice (Osteopontin-OPN-, TNF-α, IL-6, IL-10, and TGF-β). All of these cytokines were overexpressed in fibrotic livers from DDR2$^{-/-}$ mice, along with increased expression of OPN, IL-6, and TGF-β and decreased IL-10 by HSCs. Additionally, their macrophages increased expression of TNF-α, IL-6, *MCP*-1, and TGF-β. Interestingly, basal levels of some of these cytokines were also altered in healthy, untreated DDR2$^{-/-}$ livers and/or in their quiescent stellate cells; thus, it is tempting to speculate that the absence of DDR2 predisposes hepatic tissue to fibrosis. In this regard, we have recently shown that DDR2-deficiency fosters experimental colon carcinoma metastasis [51]. In contrast to stellate cells, quiescent macrophages from untreated DDR2$^{-/-}$ mice do not exhibit significant differences in gene expression compared to macrophages from untreated wild-type livers, indicating that the effect of DDR2 removal on macrophages only occurs following persistent liver injury.

### 16.12.2 Is DDR2 Involved in the Inflammatory and Angiogenic Reaction of Endothelial Sinusoidal Cells to Chronic Liver Fibrosis?

LSECs participate actively in the inflammatory reaction following acute liver injury [52] and exert paracrine effects driving stellate cells activation. LSECs also contribute to the angiogenic response to chronic injury. During progressive injury, vascular disorganization creates a hypoxic milieu that drives activated stellate secretion of VEGF [38] followed by LSEC proliferation and migration to generate new blood vessels. Recently, expression of DDR2 in endothelial cells from injured tissues has been described [53, 54]. Our previous studies did not identify DDR2 expression in LSECs. However, we did not analyze DDRs in LSECs from fibrotic mice. Since DDR2-deficient mice show an altered proangiogenic interaction between stellate cells and LSECs in experimental metastasis, it is also feasible that an induction of DDRs expression on LSEC might dictate the relationship between those two cell types during chronic injury from a different etiology [51]. Ongoing experiments will determine a role for DDRs in LSECs responses to liver injury.

In conclusion, loss of DDR2 promotes chronic liver fibrosis following $CCl_4$ injury. The fibrogenic sinusoidal milieu generated in chronic fibrotic livers of DDR2$^{-/-}$mice recruits more stellate cells to injured regions, which in turn enhances hepatic fibrosis. Taken together, these findings suggest that DDR2 orchestrates gene programs and paracrine interactions between stellate cells and macrophages to attenuate chronic hepatic fibrosis. A preliminary model of the role of DDR2 in liver fibrosis is shown in Fig. 16.7. Ongoing experiments will determine in vitro the molecular mechanism by which expression of DDR2 in stellate cells affects macrophages and liver sinusoidal endothelial cells in acute and chronic liver fibrosis.

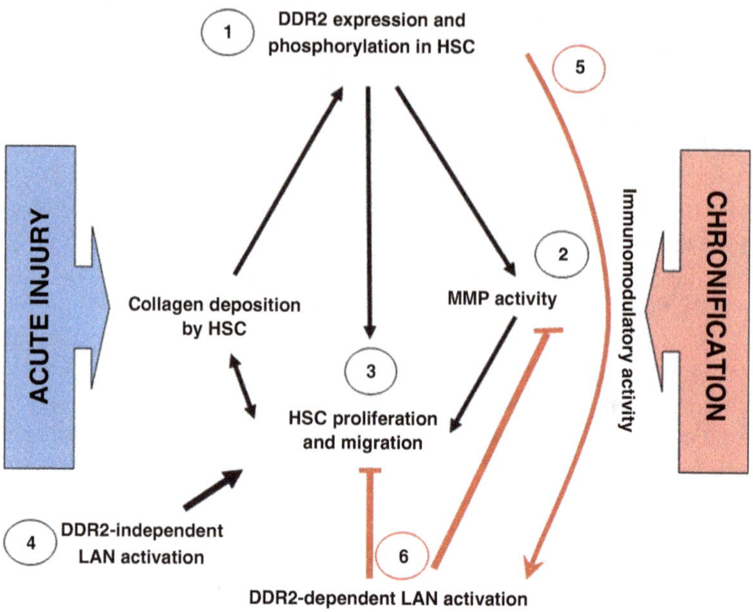


**Fig. 16.7** Role of DDR2 in liver fibrosis. Acute injury induces stellate cell (HSC) transactivation, involving collagen deposition, increased DDR2 expression (1), enhanced MMP activity (2), and stellate cell proliferation and migration to injured areas (3). Activated HSCs respond to collagen by signaling through DDR2, further inducing MMP-2 activity, HSC proliferation, and HSC migration. Increased HSCs results in increased collagen deposits in the injured tissue, thus creating a positive feedback loop. Acute injury also induces DDR2-independent activation of liver-associated macrophages (LAM) (4). Activated LAMs promote HSC proliferation and migration through the secretion of soluble growth and chemotactic factors. Persistent hepatic insult and the consequent injury further increases recruitment of activated HSCs. This profibrotic microenvironment promotes DDR2-dependent synthesis of immunoregulatory soluble factors by activated HSCs (5) and alters LAM release of proliferation and migration factors and MMPs (6)

## 16.13 DDR1 Expression in the Injured Liver

While DDR1 is a subject of intense study in a wide variety of human pathologies, its implication in liver fibrosis remains unclear. Cirrhotic human livers express DDR1 [55, 56]. All DDR1 isoforms are present in human diseased livers, although the relative expression of the different isoforms varies among injury type. In the fibrotic human liver, the DDR1 protein has been immune detected in parenchymal hepatic cells, particularly hepatocytes at the portal-parenchymal interface, and in the bile duct epithelia. Because DDR1 has not been detected in cells expressing FAP (fibroblast activation protein), and a nonfibroblastic origin for hepatic DDR1 is suspected.

We have analyzed DDR1 expression in our experimental models of liver injury; our preliminary studies showed that DDR1 mRNA expression is increased in cultured hepatocytes that were freshly isolated from rats 48 h after acute intoxication

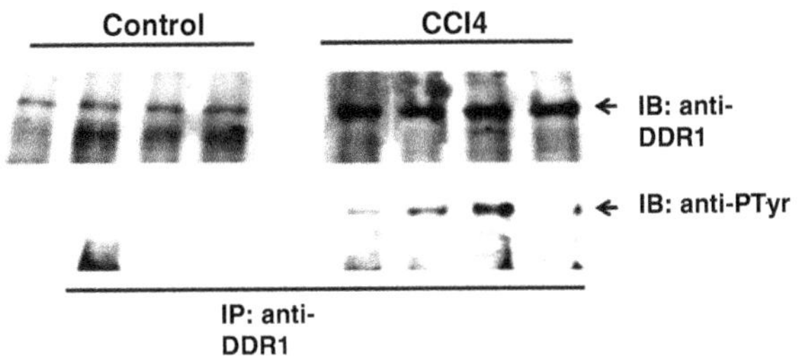


**Fig. 16.8** DDR1 expression in a mouse model of liver fibrosis. Increased total and phosphorylated DDR1 expression in chronic $CCl_4$-injured mouse livers compared to control. *IB* immunoblot, *IP* immunoprecipitation

with alcohol (unpublished data). Similarly, we have observed an increase in DDR1 expression and phosphorylation in fibrotic mice treated with $CCl_4$ (unpublished data) (Fig. 16.8). Immunohistochemical detection of β-galactosidase, a surrogate marker for DDR1 in transgenic heterozygous mice (kindly donated by Regeneron Pharmaceutics), shows DDR1 protein expression restricted to cells within or close to the fibrous septa (unpublished data). Ongoing experiments will determine the exact cell type(s) that express DDR1 and its role in liver injury. Data on DDR1 signaling in other tissues showing that the inhibition of DDR1 reduces cell survival, homing, and colonization in lung cancer bone metastasis might predict a role for DDR1 in liver cell proliferation and MMP-1/MMP-13 synthesis [57, 58].

Expression of DDR1 in nonepithelial cells is currently under debate. Expression of DDR1 in smooth muscle cells and macrophages has been reported in atherogenesis where DDR1 expression was directly related to macrophage accumulation [59]. Guerrot et al. [60] by using DDR1$^{-/-}$ mice have shown the important role of DDR1 in renal disease. In these studies, a reduced accumulation of fibrillar collagen and TGF-β expression was correlated with F4/80$^{+}$ cell count and proinflammatory cytokines. Given the crucial role of myofibroblasts and macrophages in liver fibrosis, a detailed analysis of DDR1 expression in activated macrophages and myofibroblasts is timely and imperative.

## 16.14 Immune Cell Population in Liver Fibrosis: Is There a Role for DDRS?

The inflammatory reaction generated by immune cells is a major regulator of the fibrotic response in liver disease, regardless of the etiology of injury [61]. As the insult becomes chronic, cytokines and chemokines produced by the immune cells promote fibrosis development (for exhaustive reviews see [62–64]).

In turn, the excessive accumulation of collagenous ECM correlates with the recruitment of leukocytes (monocytes, monocyte-derived macrophages, neutrophils, and lymphoid cells) at the sites of injury.

Several models of chronic fibrosis in DDR1-deficient mice such as atherosclerotic plaque formation [59] and renal fibrosis [60, 65] revealed a role for DDR1 in the inflammatory response to tissue injury, since removal of DDR1 gene resulted in reduced leukocyte/macrophage recruitment to scar areas. Macrophages isolated from healthy DDR1-deficient mice show decreased adhesion and motility and MMPs production [59]. However, macrophage expression of DDR1 was not demonstrated. Supporting the proinflammatory role of DDR1 in vivo, in vitro studies suggest that DDR1 signaling in response to collagen regulates leukocyte recruitment, migration, and production of profibrogenic factors. Cultured macrophages, neutrophils, dendritic cells, and T lymphocytes show inducible DDR1 and/or DDR2 expression [66, 67]. The absence of in vivo studies on liver chronic diseases makes it difficult to confirm a role for DDR1 in the inflammatory response to liver fibrosis.

To our knowledge, only one in vivo study has analyzed the role of DDR2 in immune cells during liver fibrosis (see Section 16.12.1). The results obtained by Olaso et al [49] showed ex vivo that liver macrophages isolated from DDR2-deficient mice possess an enhanced ability to attract stellate cells. While robust in vivo data are missing, in vitro data suggest a possible role for DDR2-expressing leukocytes and dendritic cells in the immune response to fibrosis such as TNF-α secretion and stimulation of naïve T cells [45].

In conclusion, the mechanisms that govern the functional roles of immune cell populations during chronic liver fibrosis are far from being established. Besides, the divergent maturation phenotypes make it even more difficult to elucidate any functional consequences of DDRs activation on the different immune cell subsets. It is tempting to speculate that while DDR2 expression on hepatic leukocytes modulates MMPs production favoring ECM degradation and fibrosis resolution [68]. DDR1 regulates profibroinflammatory activities of immune cells [69]. The final outcome would be the net result of the overall factors and activities carried out by stromal and immune cells in the liver microenvironment.

## References

1. Friedman S (1993) Seminars in medicine of the Beth Israel Hospital, Boston. The cellular basis of hepatic fibrosis. Mechanisms and treatment strategies. N Engl J Med 328(25): 1828–1835
2. Bataller R, Brenner D (2005) Liver fibrosis. J Clin Invest 115:209–218
3. Ellis E, Mann D (2012) Clinical evidence for the regression of liver fibrosis. J Hepatol 56(5):1171–1180
4. Kisseleva T, Cong M, Paik Y et al (2012) Myofibroblasts revert to an inactive phenotype during regression of liver fibrosis. Proc Natl Acad Sci U S A 109(24):9448–9453
5. Troeger J, Mederacke I, Gwak G et al (2012) Deactivation of hepatic stellate cells during liver fibrosis resolution in mice. Gastroenterology 143(4):1073–1083, e1022
6. Tsukamoto H, Matsuoka M, French SW (1990) Experimental models of hepatic fibrosis: a review. Semin Liver Dis 10(1):56–65
7. George J, Chandrakasan G (1996) Glycoprotein metabolism in dimethylnitrosamine induced hepatic fibrosis in rats. Int J Biochem Cell Biol 28(3):353–361

8. Rojkind M, Giambone M, Biempica L (1979) Collagen types in normal and cirrhotic liver. Gastroenterology 76(4):710
9. Nagase H, Visse R, Murphy G (2006) Structure and function of matrix metalloproteinases and TIMPs. Cardiovasc Res 69(3):562–573, Review
10. Visse R, Nagase H (2003) Matrix metalloproteinases and tissue inhibitors of metalloproteinases: structure, function, and biochemistry. Circ Res 92(8):827–839, Review
11. Fields G (2013) Interstitial collagen catabolism. J Biol Chem 288(13):8785–8793
12. Matrisian L (1992) The matrix degrading metalloproteinases. Bioessays 14:455–463
13. Yin C, Evason KJ, Asahina K et al (2013) Hepatic stellate cells in liver development, regeneration, and cancer. J Clin Invest 123(5):1902–1910
14. Blaner W, O'Byrne S, Wongsiriroj N et al (2009) Hepatic stellate cell lipid droplets: a specialized lipid droplet for retinoid storage. Biochim Biophys Acta 1791(6):467–473
15. Cassiman D, Libbrecht L, Desmet V et al (2002) Hepatic stellate cell/myofibroblast subpopulations in fibrotic human and rat livers. J Hepatol 36:200–209
16. Roskams T (2008) Relationships among stellate cell activation, progenitor cells, and hepatic regeneration. Clin Liver Dis 12(4):853–860
17. Friedman SL (2008) Hepatic stellate cells: protean, multifunctional, and enigmatic cells of the liver. Physiol Rev 88:125–172
18. Tsukamoto H (1999) Cytokine regulation of hepatic stellate cells in liver fibrosis. Alcohol Clin Exp Res 23(5):911–916
19. Wang S, Ohata M, Schrum L et al (1998) Expression of interleukin-10 by in vitro and in vivo activated hepatic stellate cells. J Biol Chem 273(1):302–308
20. Olaso E, Friedman S (1998) Molecular regulation of hepatic fibrogenesis. J Hepatol 29(5):836–847, Review
21. Ding B, Cao Z, Lis R et al (2014) Divergent angiocrine signals from vascular niche balance liver regeneration and fibrosis. Nature 505(7481):97–102
22. Cubero F, Nieto N (2008) Ethanol and arachidonic acid synergize to activate Kupffer cells and modulate the fibrogenic response via tumor necrosis factor alpha, reduced glutathione, and transforming growth factor β-dependent mechanisms. Hepatology 48(6):2027–2039
23. De Minicis S, Seki E, Uchinami H et al (2007) Gene expression profiles during hepatic stellate cell activation in culture and in vivo. Gastroenterology 132(5):1937–1946
24. Mannaerts I, Schroyen B, Verhulst S, Van Lommel L, Schuit F, Nyssen M, van Grunsven LA(2013) Gene expression profiling of early hepatic stellate cell activation reveals a role for Igfbp3 in cell migration. PLoS One 8(12):e84071
25. Xia Y, Chen R, Song Z et al (2010) Gene expression profiles during activation of cultured rat hepatic stellate cells by tumoral hepatocytes and fetal bovine serum. J Cancer Res Clin Oncol 136(2):309–321
26. Fang L, Zhan S, Huang C et al (2013) TRPM7 channel regulates PDGF-BB-induced proliferation of hepatic stellate cells via PI3K and ERK pathways. Toxicol Appl Pharmacol 272(3):713–725
27. Viñas O, Bataller R, Sancho-Bru P et al (2003) Human hepatic stellate cells show features of antigen-presenting cells and stimulate lymphocyte proliferation. Hepatology 38(4):919–929
28. Davis B, Pratt B, Madri J (1987) Retinol and extracellular collagen matrices modulate hepatic Ito cell collagen phenotype and cellular retinol binding protein levels. J Biol Chem 262(21):10280–10286
29. Olaso E, Ikeda K, Eng FJ et al (2001) DDR2 receptor promotes MMP-2-mediated proliferation and invasion by hepatic stellate cells. J Clin Invest 108(9):1369–1378
30. Iredale J, Thompson A, Henderson N (2013) Extracellular matrix degradation in liver fibrosis: biochemistry and regulation. Biochem Biophys Acta 1832:876–883
31. Piera-Velazquez S, Louneva N, Fertala J et al (2010) Persistent activation of dermal fibroblasts from patients with gadolinium associated nephrogenic systemic fibrosis. Ann Rheum Dis 69(11):2017–2023

32. Chu A, Diaz R, Hui J et al (2011) Lineage tracing demonstrates no evidence of cholangiocyte epithelial-to-mesenchymal transition in murine models of hepatic fibrosis. Hepatology 53(5):1685–1695
33. Russo F, Alison M, Bigger B et al (2006) The bone marrow functionally contributes to liver fibrosis. Gastroenterology 130:1807–1821
34. Higashiyama R, Moro T, Nakao S et al (2009) Negligible contribution of bone marrow-derived cells to collagen production during hepatic fibrogenesis in mice. Gastroenterology 137(4):1459–1466.e1
35. Ullrich A, Schlessinger J (1990) Signal transduction by receptors with tyrosine kinase activity. Cell 61(2):203–212
36. Pinzani M, Milani S, De Franco R et al (1996) Endothelin 1 is overexpressed in human cirrhotic liver and exerts multiple effects on activated hepatic stellate cells. Gastroenterology 110(2):534–548
37. Ankoma-Sey V, Matli M, Chang KB et al (1998) Coordinated induction of VEGF receptors in mesenchymal cell types during rat hepatic wound healing. Oncogene 17(1):115–121
38. Olaso E, Salado C, Egilegor E et al (2003) Proangiogenic role of tumor-activated hepatic stellate cells in experimental melanoma metastasis. Hepatology 37(3):674–685
39. Gaça M, Zhou X, Issa R et al (2003) Basement membrane-like matrix inhibits proliferation and collagen synthesis by activated rat hepatic stellate cells: evidence for matrix-dependent deactivation of stellate cells. Matrix Biol 22(3):229–239
40. Dodig M, Ogunwale B, Dasarathy S, Li M, Wang B, McCullough AJ (2007) Differences in regulation of type I collagen synthesis in primary and passaged hepatic stellate cell cultures: the role of α5β1-integrin. Am J Physiol 293:G154–G164
41. Xu L, Hui A, Albanis E et al (2005) Hepatic stellate cell lines, LX-1 and LX-2: new tools for analysis of hepatic fibrosis. Gut 54:142–151
42. Luo Z, Liu H, Sun X et al (2013) RNA interference against discoidin domain receptor 2 affects ameliorates alcoholic liver disease in rats. PLoS One 8(2), e55860
43. Zhang X, Yan M, Liu L et al (2010) Expression of discoidin domain receptor (DDR2) in alcoholic liver fibrosis in rats. Arch Med Res 41(8):586–592
44. Mao T, Kimura Y, Kenny T et al (2002) Elevated expression of tyrosine kinase DDR2 in primary biliary cirrhosis. Autoimmunity 35(8):521–529
45. Lee J, Kang C, Guan X et al (2007) Discoidin domain receptor 2 is involved in the activation of bone marrow-derived dendritic cells caused by type I collagen. Biochem Biophys Res Commun 352:244–250
46. Barun P, Yoon D, Lee J et al (2012) Collagen I enhances functional activities of human monocyte-derived dendritic cells via discoidin domain receptor 2. Cell Immunol 278:95–102
47. Vogel W, Gish G, Alves F et al (1997) The discoidin domain receptor tyrosine kinases are activated by collagen. Mol Cell 1(1):13–23
48. Labrador J, Azcoitia V, Tuckermann J et al (2001) The collagen receptor DDR2 regulates proliferation and its elimination leads to dwarfism. EMBO Rep 2(5):446–452
49. Olaso E, Lin H, Wang L et al (2011) Impaired dermal wound healing in discoidin domain receptor 2-deficient mice associated with defective extracellular matrix remodeling. Fibrogenesis Tissue Repair 4(1):5
50. Olaso E, Labrador J, Wang L et al (2002) Discoidin domain receptor 2 regulates fibroblast proliferation and migration through the extracellular matrix in association with transcriptional activation of matrix metalloproteinase-2. J Biol Chem 277(5):3606–3613
51. Badiola I, Olaso E, Crende O et al (2012) Discoidin domain receptor 2 deficiency predisposes hepatic tissue to colon carcinoma metastasis. Gut 61(10):1465–1472
52. Knolle PA, Gerken G (2000) Local control of the immune response in the liver. Immunol Rev 174:21–34
53. Goldsmith E, Zhang X, Watson J et al (2010) The collagen receptor DDR2 is expressed during early cardiac development. Anat Rec (Hoboken) 293:762–769
54. Zhang S, Bu X, Zhao H et al (2014) A host deficiency of discoidin domain receptor 2 (DDR2) inhibits both tumour angiogenesis and metastasis. J Pathol 232(4):436–448

55. Shackel N, McGuinness P, Abbott C et al (2002) Insights into the pathobiology of hepatitis C virus-associated cirrhosis: analysis of intrahepatic differential gene expression. Am J Pathol 160(2):641–654
56. Song S, Shackel N, Wang X et al (2011) Discoidin domain receptor 1: isoform expression and potential functions in cirrhotic human liver. Am J Pathol 178(3):1134–1144
57. Curat CA, Vogel WF (2002) Discoidin domain receptor 1 controls growth and adhesion of mesangial cells. J Am Soc Nephrol 13(11):2648–2656
58. Valencia K, Ormazábal C, Zandueta C, Luis-Ravelo D, Antón I, Pajares MJ, Agorreta J, Montuenga LM, Martínez-Canarias S, Leitinger B, Lecanda F (2012) Discoidin domain receptor 1 controls growth and adhesion of mesangial cells. Clin Cancer Res 18(4):969–980. doi:10.1158/1078-0432.CCR-11-1686
59. Franco C, Britto K, Wong E et al (2009) Discoidin domain receptor 1 on bone marrow-derived cells promotes macrophage accumulation during atherogenesis. Circ Res 105(11):1141–1148
60. Guerrot D, Kerroch M, Placier S et al (2011) Discoidin domain receptor 1 is a major mediator of inflammation and fibrosis in obstructive nephropathy. Am J Pathol 179(1):83–91
61. Pellicoro A, Ramachadran P, Iredale JP et al (2014) Liver fibrosis and repair: immune regulation of wound healing in a solid organ. Nat Rev Immunol 14(3):181–194
62. Borthwick LA, Wynn TA, Fisher AJ (2013) Cytokine mediated tissue fibrosis. Biochim Biophys Acta 1832(7):1049–1060
63. Ramadori G, Saile B (2004) Inflammation, damage repair, immune cells, and liver fibrosis: specific or nonspecific, this is the question. Gastroenterology 127(3):997–1000
64. Tanaka H, Leung PS, Kenny TP et al (2012) Immunological orchestration of liver fibrosis. Clin Rev Allergy Immunol 43(3):220–229
65. Flamant M, Placier S, Rodenas A et al (2006) Discoidin domain receptor 1 null mice are protected against hypertension-induced renal disease. J Am Soc Nephrol 17(12):3374–3381
66. Sun X, Phan TN, Jung SH et al (2012) LCB 03–0110, a novel pan-discoidin domain receptor/c-Src family tyrosine kinase inhibitor, suppresses scar formation by inhibiting fibroblast and macrophage activation. J Pharmacol Exp Ther 340(3):510–519
67. Afonso PV, McCann CP, Kapnick SM, Parent CA (2013) Discoidin domain receptor 2 regulates neutrophil chemotaxis in 3D collagen matrices. Blood 121(9):1644–1650
68. Poudel B, Yoon DS, Lee JH, Lee YM, Kim DK (2012) Collagen I enhances functional activities of human monocyte-derived dendritic cells via discoidin domain receptor 2. Cell Immunol 278(1–2):95–102
69. Yoshimura T, Matsuyama W, Kamohara H (2005) Discoidin domain receptor 1: a new class of receptor regulating leukocyte-collagen interaction. Immunol Res 31(3):219–230

# Chapter 17
# The Role of DDRs in Atherosclerosis

**Michelle P. Bendeck**

## 17.1 Introduction

Interactions between cells and the extracellular matrix (ECM) play critical roles in tissue morphogenesis, in normal tissue function, and in the abnormal manifestations of pathology. ECM collagens function as guidance and patterning molecules and convey tissue strength and resistance to mechanical forces. Collagens also activate biochemical pathways in cells, including those regulating cell proliferation, migration, fibrosis, and inflammation. There are several distinct families of collagen receptors which mediate cell contact. These include integrins, immunoglobulin-like receptors, glycoproteins, and discoidin domain receptors (DDRs). The DDRs are receptor tyrosine kinases that bind selectively to collagens, and their roles in the vascular system during the development of atherosclerosis are the focus of this review.

Arteries are composed of a highly ordered structure of cells and ECM. In the intimal layer, endothelial cells sit upon a basement membrane rich in laminin, fibronectin, and types IV and VIII collagen. In the media, individual smooth muscle cells (SMCs) are surrounded by a basement membrane and are in turn embedded in fibrillar types I, III and V collagen, type XVIII collagen, fibronectin, and proteoglycans, and each layer is separated by a fenestrated elastic lamina. The outermost layer, the adventitia, is rich in fibroblasts, collagen types I and III, and elastin. Atherosclerosis initiates with injury to the endothelial cell layer, caused by elevated levels of circulating lipids and/or alterations in hemodynamic forces. As a result, lipids begin to accumulate in the subendothelial layer of the intima (Fig. 17.1). Endothelial cell expression of inflammatory cell adhesion molecules triggers the adherence and diapedesis of leukocytes, and infiltrating macrophages attempt to phagocytose the

M.P. Bendeck (✉)
Department of Laboratory Medicine and Pathobiology, University of Toronto, Ted Rogers Centre for Heart Research, 661 University Ave., 14th Floor, Toronto, ON, Canada, M5G1M1
e-mail: michelle.bendeck@utoronto.ca



R. Fridman, P.H. Huang (eds.), *Discoidin Domain Receptors in Health and Disease*, DOI 10.1007/978-1-4939-6383-6_17

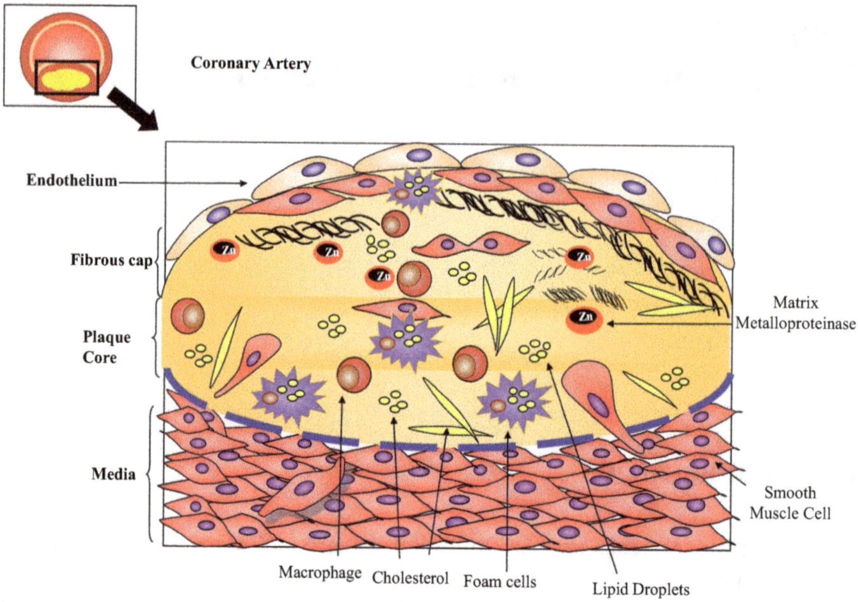


**Fig. 17.1** Schematic illustrating the pathogenic processes involved in atherosclerosis. Cholesterol accumulates in the subendothelial space, eventually giving rise to crystals. Macrophages infiltrate the plaque and phagocytose cholesterol, giving rise to foam cells. Macrophages release cytokines and growth factors which attract more macrophages and stimulate the proliferation and migration of medial smooth muscle cells into the plaque. Smooth muscle cells synthesize collagen and other extracellular matrix molecules, giving rise to the fibrous cap. Both cell types secrete MMPs, which degrade matrix and facilitate cell migration

lipid, giving rise to foam cells. These cells also release cytokines and growth factors which stimulate the proliferation and migration of SMCs from the media. During atherosclerotic plaque development, there are pronounced changes in the composition of the vascular ECM. Coincident with their migration into the intima, SMCs transition from a contractile to a synthetic phenotype, and begin to deposit types I and III collagen, elastin, and fibronectin. They also synthesize an array of different ECM molecules seen only in small quantities in normal vessels, including type VIII collagen, osteopontin, and tenascin. In addition to enhanced and altered ECM synthesis, degradation of the ECM also occurs due to increased activity of proteinases including the matrix metalloproteinase (MMP), cathepsin, and plasminogen activator/plasmin families. ECM remodeling can also leave behind bioactive fragments, including fragments of types I and IV collagen and elastin, each of which may have effects on target cells distinct from that of the intact molecule. This active synthesis of the ECM contributes to plaque expansion and lumenal restriction, which are harmful consequences of atherosclerosis. However, collagen synthesis is also important for the formation of a strong fibrous cap which overlies the plaque and protects it from rupture. Since plaque rupture is the cause of myocardial infarction and sudden cardiac death, it is important to maintain a strong fibrous cap and therefore complete inhibition of collagen synthesis in atherosclerosis is not desirable.

Since collagens are both abundant and bioactive in atherosclerosis, it was logical to propose that they might act through cellular DDRs to mediate disease pathogenesis. In this review, we summarize recent literature reporting on the expression and functions of DDRs in the pathogenesis of atherosclerosis. In Section 17.2, we provide an introduction to DDR1 and DDR2. In Sect. 17.3, we examine the growing literature on the role of these receptors expressed on inflammatory plaque macrophages and vascular smooth muscle cells in atherosclerosis. Finally in Sect. 17.4, we outline the critical roles that DDR1 plays in vascular calcification, an important complication of atherosclerosis and diabetic vascular disease.

## 17.2 DDRs: Unique Collagen Receptors

### *17.2.1 Structure*

The DDRs are a class of receptors containing a collagen-binding extracellular discoidin domain and a cytoplasmic tyrosine kinase domain [1, 2]. There are two members of the family expressed in mammals; DDR1 and DDR2 are expressed from two distinct gene loci, and there are 6 isoforms of DDR1 which are the result of alternative splicing in the juxtamembrane domain of the receptor [3]. The expression and function of these DDR1 isoforms has not been thoroughly investigated, but current knowledge suggests that they may trigger distinct downstream signaling pathways. Only one isoform of DDR2 has been discovered. DDR1 can also be modified by glycosylation [4, 5], and by proteolytic cleavage resulting in shedding of the extracellular domain [6, 7]. Cleavage is mediated by membrane type matrix metalloproteinase (MT-MMP) or disintegrin and metalloproteinase (ADAM) [6]. It is not clear whether cleavage results in alterations in receptor signaling.

Several collagens serve as ligands for the DDRs. DDR1 binds to types I–VI and type VIII collagen, whereas DDR2 not only has higher affinity for the fibrillar collagens types I–III, but also binds nonfibrillar types VIII and X collagen. Both DDRs require triple-helical conformation of the collagen to induce binding and activation; denatured collagens do not bind [3]. Specific binding sites for DDRs within collagens have been identified and include a GVMGFO sequence in types I, II, and III collagen, whereas the specificity of DDR1 adhesion to type IV collagen is determined by five amino acids adjacent to the GVMGFO sequence [8]. Studies have demonstrated DDR dimerization prior to collagen binding, unlike other tyrosine kinase receptors which require ligand for dimerization [9–11].

### *17.2.2 DDR Knockout Mice*

Mice with targeted deletion of DDR1 and DDR2 have been generated, and though the mice are born viable, both receptors play important roles in development. Targeted deletion of the DDR1 gene results in defects in blastocyst implantation and mammary gland development, and a slight but transient dwarfism which resolves postnatally [12]. DDR2-null mice exhibit more severe dwarfism, skeletal malformation, and delayed wound healing [13]. DDR1/2 double knockout mice die during embryogenesis, suggesting that there is compensatory overlap in the functions of these two receptors (W. Vogel and F. Alves; personal communication).

## 17.3 The Roles of DDRs in Arterial Injury and Atherosclerosis

### *17.3.1 DDR Expression and Function After Arterial Injury*

We have discovered that DDR1 mediates critical cellular responses to collagen in both SMCs and macrophages during the development of arterial disease and atherosclerosis. Animal models of mechanical endothelial denudation are commonly used to study cellular and molecular events in restenosis and atherosclerosis. SMCs undergo a phenotypic switch from quiescent to activated after endothelium denuding arterial injury and contribute to the growth of a thickened neointimal layer through proliferation, migration, and ECM synthesis. The ECM is not simply an inert scaffold, but there are dynamic interactions between cells and the ECM that contribute to SMC injury responses. For example, after injury, SMCs synthesize fibrillar type I and III collagens [14], and the short chain type VIII collagen [15–17], all of which can stimulate SMC migration. SMCs also synthesize matrix metalloproteinase (MMP) enzymes which facilitate their migration from medial to intimal layers of the arterial wall by degrading and clearing ECM barriers [18, 19]. Since the DDRs function as collagen receptors and increase MMP production, we hypothesized that DDRs might be important mediators of SMC growth and migration following arterial injury. We subjected $Ddr1^{+/+}$ and $Ddr1^{-/-}$ mice to endothelium denuding wire injury of the carotid artery and observed upregulation of DDR1 after injury, and reduced neointimal thickening in $Ddr1^{-/-}$ mice compared to wild-type mice [20]. Thus, our results supported the hypothesis that DDR1 plays an important role as a collagen receptor mediating intimal thickening after vascular injury. Using cell culture models that mimic critical steps in the vascular response to injury, and comparing SMCs isolated from wild-type mice and DDR1-null mice, we showed that DDR1 plays important roles regulating attachment to collagen types I and VIII, migration, proliferation, and MMP-2 and 9 activity [20, 21]. With these studies, we were the first to study DDRs in the vascular system, and the first to identify important roles for DDR1 in mediating cell growth and migration in any system.

### *17.3.2 DDR1 Signaling*

Collagen binding to the DDR leads to receptor autophosphorylation. However, the kinetics of DDR activation are unusual compared to other receptor tyrosine kinases. DDR1 phosphorylation occurs 1 h after exposure to ligand, peaks at 90 min, and is sustained for 18 h [1]. In studies to determine the mechanisms of DDR1 in mediating SMC migration, we transfected DDR1-null SMCs with either full-length DDR1b, or mutant versions of the receptor which were kinase-dead, or contained a deletion to eliminate the kinase domain [21]. The full-length receptor rescued the deficits in adhesion, migration, and MMP production seen in DDR1-deficient SMCs, while the kinase-deficient mutants restored attachment but not migration or MMP production [21]. Our next studies focused on identifying the signaling pathways activated by DDR1 in SMCs. The nonreceptor tyrosine kinase Src has been implicated in mediating DDR1 phosphorylation [7, 22]. Following collagen stimulation of SMCs, we showed that DDR1 co-immunoprecipitates with Src, and Src is required for phosphorylation of DDR1 [23]. Furthermore, treatment with a Src inhibitor attenuated DDR1-stimulated cell migration. DDR1 also mediated the activation of Erk1/2 in a Src-dependant manner, but though inhibition of Erk1/2 attenuated migration, the effect was similar in DDR1 expressing and nonexpressing cells. Taken together our data indicates that collagen stimulates SMC migration mediated through DDR1 and Src. DDR1 is also required to activate Erk1/2, but this pathway does not lead to cell migration.

Many other studies have defined roles for DDR1 in cell growth, migration, and matrix remodeling in other biologic systems (reviewed in [24]). Intriguingly, there can be substantial variations in DDR1 signaling and cellular outcomes. In mammary epithelial cells, DDR1 stimulation leads to Erk1/2 activation and cell proliferation [25], whereas in mesangial cells DDR1 suppresses Erk1/2 to maintain cell quiescence [26]. DDR1 is capable of crosstalk with other cell adhesion pathways. Some studies have reported DDR1 inhibition of the function of the integrin α2β1 collagen receptor. For example, DDR1 inhibits the phosphorylation of STAT1/3, acting antagonistically to the α2β1 integrin receptor, and attenuating cell migration in response to type I collagen [27]. In kidney epithelial cells, DDR1 inhibits cdc42 mediated cell spreading normally stimulated by α2β1 activation [28]. Conversely, there is evidence that DDR1 activation can increase β1 integrin activation and binding to collagen [29, 30]. Different results may be attributed to differences in cell origin, DDR1 isoform expression, or in the type or physical state of the collagen matrix to which the cell is exposed.

DDR1 interacts with the cadherin cell–cell adhesion molecules. DDR1 signals cooperatively with integrins to mediate the upregulation of N-cadherin in pancreatic tumor cells [31]. There is an emerging body of evidence suggesting that DDR1 physically associates with and stabilizes E-cadherin at cell–cell junctions; in this case DDR1 functions as a scaffolding molecule, since these effects are independent of binding to collagen and do not require receptor activation [32–35]. At present, there is little known about DDR crosstalk with integrins and cadherins in the vascular system, and this will be an important focus for future investigation.

DDR2 is expressed in SMCs [36, 37] and has been proposed as a specific marker of cardiac fibroblasts [38]. A germline deletion of DDR2 in mice led to smaller hearts, disruptions in collagen organization in the heart, and shortened myocyte length with mild cardiac functional impairments [39]. We conducted studies in vitro comparing SMCs isolated from *Ddr2*$^{+/+}$ and *Ddr2*$^{-/-}$ mice and did not find any deficiencies in smooth muscle cell adhesion, migration, or proliferation in response to type I collagen in the *Ddr2*$^{-/-}$ SMCs [40]. However, it is interesting to note that DDR2 has been implicated in mediating SMC migration in response to hypoxia [41], and migration and proliferation in response to hyperbaric oxygen treatment [42]. DDR2 expression was upregulated following cyclic stretching of VSMCs [37]. These groups also found that administration of siRNA to knockdown DDR2 attenuated intimal thickening following balloon catheter injury of the carotid artery in the rat [43]. By contrast, we saw no significant difference in neointimal thickening following wire injury of the carotid arteries of *Ddr2*$^{+/+}$ vs. *Ddr2*$^{-/-}$ mice (unpublished data). The discrepancy in these results may be due to differences in species or experimental procedures.

### 17.3.3 DDR Functions in Atherosclerosis

Mechanical injury models are useful to study the responses of SMCs to injury, but they do not represent the full complexity of the atherosclerotic plaque. The plaque is an active inflammatory, lipid-rich environment which is home to macrophages, and lymphocytes as well as SMCs. Ferri and colleagues have shown that DDR1 and DDR2 are present in the atherosclerotic plaques of fat-fed primates, localized to inflammatory cells in the plaque core and SMCs in the fibrous cap [36]. To gain a better understanding of the roles of DDR1 in atherosclerotic vascular disease, we have studied DDR1 function in a mouse model of atherosclerosis. Genetic deletion of the low density lipoprotein receptor (*Ldlr*) renders mice unable to clear lipoproteins from the circulation, and when these mice are fed a high-cholesterol Western style diet, hypercholesterolemia results [44]. This is followed by the development of fibro-fatty atherosclerotic plaques similar to humans (Fig. 17.1), comprised of extracellular lipids, infiltrating macrophage inflammatory cells which engulf lipids giving rise to foam cells, and SMCs which accumulate to form a protective fibrous cap.

We bred *Ddr1*$^{-/-}$ mice into the *Ldlr*$^{-/-}$ background to generate *Ddr1*$^{+/+}$; *Ldlr*$^{-/-}$ mice and *Ddr1*$^{-/-}$; *Ldlr*$^{-/-}$ mice, and fed the mice a high-cholesterol diet for 12 or 24 weeks, then compared atherosclerotic lesions which formed in the aorta. DDR1 immunostaining was evident in the plaques that formed in *Ddr1*$^{+/+}$; *Ldlr*$^{-/-}$ mice, localizing to both macrophages and SMCs [45]. Targeted deletion of *Ddr1* resulted in decreased atherosclerotic plaque size. However, the plaques in *Ddr1*$^{-/-}$; *Ldlr*$^{-/-}$ mice exhibited accelerated ECM accumulation. The contents of both collagen and elastin were increased, and the mRNAs for procollagen a1(I) and a1(III) and tropoelastin were elevated. Surprisingly, there was no difference in SMC content of the plaques between the genotypes. This was an important difference compared to *Ddr1*$^{-/-}$ mice subject to wire injury where SMC accumulation in the neointimal lesion was reduced (Sects. 17.3.1 and [20]). This emphasizes the importance of studying gene function in the most relevant experimental models.

We were excited to find that there was a significant decrease in macrophage content in the plaques of *Ddr1*$^{-/-}$; *Ldlr*$^{-/-}$ mice compared to *Ddr1*$^{+/+}$; *Ldlr*$^{-/-}$ mice. The mRNA levels for macrophage chemotactic protein 1 (MCP-1), and the vascular cell adhesion molecule 1 (VCAM1) were decreased in plaques from *Ddr1*$^{-/-}$; *Ldlr*$^{-/-}$ mice [45]. These are key mediators of monocyte recruitment from the circulation. In situ gelatinase activity in the plaque, and the expression of mRNA for several matrix metalloproteinases were reduced in the plaques from *Ddr1*$^{-/-}$; *Ldlr*$^{-/-}$ mice. Expression of several MMPs was also reduced in cultured macrophages harvested from the *Ddr1*$^{-/-}$; *Ldlr*$^{-/-}$ mice. Since MMPs are required to degrade matrix barriers and allow inflammatory cell infiltration in the plaque [46], our data suggested a novel role for DDR1 in macrophage invasion via MMP-mediated degradation of the ECM during atherogenesis. Taken together, our studies showed that DDR1 regulates both inflammation and fibrosis early in plaque development, and that systemic deletion of DDR1 attenuated plaque growth and resulted in the formation of stable, matrix-rich plaques [45].

### 17.3.4 DDR1 Functions in Macrophages

To further resolve the functions of DDR1 on macrophages in atherosclerosis, we performed bone marrow transplantation experiments. Female *Ddr1*$^{+/+}$; *Ldlr*$^{-/-}$ hosts were irradiated to destroy bone marrow, then transplanted with bone marrow from *Ddr1*$^{+/+}$; *Ldlr*$^{-/-}$ or *Ddr1*$^{-/-}$; *Ldlr*$^{-/-}$ male donor mice. This gave rise to two groups of chimeric mice which we termed the bone marrow deletion group (*Ddr1*$^{-/-\rightarrow+/+}$) with a deficiency of DDR1 in bone marrow-derived cells, or the control transplant group (*Ddr1*$^{+/+\rightarrow+/+}$) [47]. The mice were fed a high-cholesterol diet for 12 weeks, and atherosclerotic plaques were studied. We observed decreased atherosclerotic plaque size in the bone marrow deletion group compared to the control group. The decrease in plaque size was directly proportional to the decrease in macrophage number, which suggested that bone marrow deficiency of DDR1 limited macrophage accumulation in the plaque, and plaque expansion. By contrast there were no significant differences in SMC number and ECM content of the plaques.

To measure macrophage recruitment, we performed studies to track the infiltration of labeled macrophages into early fatty streak lesions. *Ddr1*$^{+/+}$; *Ldlr*$^{-/-}$ and *Ddr1*$^{-/-}$; *Ldlr*$^{-/-}$ were fed high-cholesterol diet for 2 weeks, then given a single injection of BrdU to label monocytes proliferating in the bone marrow [47]. The recruitment of BrdU-labeled macrophages in the fatty streak lesions was significantly decreased in *Ddr1*$^{-/-}$; *Ldlr*$^{-/-}$ mice, as was lesion size. Potential mechanisms behind the reduction in recruitment were revealed by in vitro experiments showing that DDR1-deficient macrophages exhibited decreased attachment to and invasion of type IV collagen matrix, suggesting that DDR1 is required for macrophage penetration through the subendothelial basement membrane [47].

Our results are consistent with others who have demonstrated proinflammatory roles for DDR1. The absence of DDR1 reduced leukocyte recruitment in inflammatory renal-vascular hypertension induced by long-term angiotensin II infusion [48]. DDR1 deficiency also led to reduced macrophage infiltration and inflammation following partial ureteral obstruction, a model of renal interstitial inflammation [49].

Gross and colleagues demonstrated reduced inflammation in DDR1-deficient mice studied using a model of hereditary glomerulonephritis [50]. In lung fibrosis induced by bleomycin treatment, DDR1 deficiency was associated with decreased MCP-1 and Il-6 expression and decreased numbers of alveolar macrophages present in the lung, and a decrease in fibrosis [51].

There is a growing body of evidence supporting a role for DDRs in regulating inflammatory cell functions. DDR1 has been implicated in the signaling of type I collagen to stimulate monocyte to macrophage differentiation and cytokine production [52, 53]. DDR1 can also regulate dendritic cell antigen presentation [54], and nitric oxide production [55]. In human macrophages, DDR1b upregulates the production of interleukin 8, MIFα, and MCP-1 in a p38 and NFKB-dependent manner [52]. However, p38 activation is dependent upon Shc-A recruitment to DDR1, and since only DDR1b contains the Shc-A-binding domain, other isoforms may not activate this signaling pathway. Indeed, activation of the DDR1a and DDR1b isoforms leads to different responses after transfection and overexpression in THP-1 cells [53]. DDR1 expression is mediated via T-cell receptor activation and enables T-cell migration through 3-D collagen gels by promoting amoeboid migration independent of adhesion [56, 57]. DDR2 also plays roles in inflammation by promoting cytokine and antigen-uptake activity in dendritic cells [58].

At present, our understanding of the role of DDR1 in mediating inflammatory changes in atherosclerosis is far from complete. Is DDR1 involved in dendritic cell surveillance of arterial tissues? To what extent does DDR1 mediate adhesion to the different collagens as monocytes infiltrate deeper into the plaque? Is there crosstalk between DDRs and integrins in macrophages? The phenotype of mice with combined deficiency of the α1 integrin and apoE is similar to DDR1/LDLR-deficient mice (small plaques with decreased macrophages and increased fibrous cap size). This suggests that DDR1 and the collagen-binding α1 integrin may have some functions in common. These are all areas for future investigation.

### 17.3.5 DDR1 Functions in Smooth Muscle Cells

We observed pronounced ECM accumulation after systemic deletion of DDR1 in the atherosclerotic mouse model (Section 17.3.3 and [45]). However, ECM accumulation was not affected by bone marrow deletion of DDR1. Therefore, we hypothesized that the enhanced ECM accumulation that occurred in mice with systemic deletion of DDR1 could be due to increased synthesis by *Ddr1*$^{-/-}$ SMCs. We generated mice with DDR1 deficiency in vessel-wall-derived cells by transplanting bone marrow from *Ddr1*$^{+/+}$; *Ldlr*$^{-/-}$ mice into *Ddr1*$^{-/-}$; *Ldlr*$^{-/-}$ hosts, creating a vessel wall deletion group (*Ddr1*$^{+/+\rightarrow -/-}$), which was compared to a control transplant group (*Ddr1*$^{+/+\rightarrow +/+}$) [59]. Atherosclerotic plaque size was dramatically and significantly increased in the vessel wall deletion group compared to the control group, due to pronounced accumulation of collagen, elastin, proteoglycans, and fibronectin in the plaques. The plaques that formed therefore had large fibrous caps. This was accompanied by an increased number of SMCs in the plaques, whereas macrophage numbers were equivalent between the groups.

We investigated further by performing in vitro studies with SMCs harvested from *Ddr1*$^{+/+}$ and *Ddr1*$^{-/-}$ mice. ECM accumulation is the result of an imbalance between synthesis and degradation. Therefore, we measured expression of several ECM molecules, and a panel of matrix degrading MMPs. *Ddr1*$^{-/-}$ SMCs had higher levels of mRNAs encoding collagen and elastin, and dramatically increased production of type I collagen compared to *Ddr1*$^{+/+}$ SMCs [59]. Paradoxically, the *Ddr1*$^{-/-}$ SMCs also displayed increased expression of mRNA for several MMPs and exhibited increased MMP-dependent migration. However, considering this data together with the results from the in vivo vessel wall deletion transplant studies, we concluded that increased ECM synthesis outpaced the synthesis of MMPs in DDR1-deficient SMCs, and the net effect was increased matrix accumulation. Interestingly, our results may also be consistent with previous studies reporting a role for DDR1 in inhibiting collagen fibrillogenesis [60].

Figure 17.2 presents a graphic summary of our studies investigating DDR1 in atherosclerosis. Compared to *Ddr1*$^{+/+}$; *Ldlr*$^{-/-}$ mice, the plaque formed after systemic deletion of DDR1 (*Ddr1*$^{-/-}$; *Ldlr*$^{-/-}$) (**b**) is smaller, with decreased accumulation of macrophages, whereas matrix accumulation is accelerated due to increased synthesis by SMCs. The plaques formed in mice with bone marrow deletion of DDR1 (**c**) are smaller, with reduced accumulation of macrophages, but comparable numbers of SMCs and equivalent matrix accumulation compared to control transplant mice. Plaques formed in mice with deletion of DDR1 on vessel wall-derived SMCs are large (**d**), due to the dramatically increased accumulation of ECM, and the increased number of SMCs in the plaque compared to control transplant mice. Taken together, these studies suggest important implications for the role of DDR1 in the long-term complications of atherosclerosis. Functional DDR1 leads to the development of large, highly inflammatory, matrix-poor plaques with thin fibrous caps. This type of plaque is unstable and susceptible to rupture. As discussed in Sect. 17.4, DDR1 also plays a role in another late complication of atherosclerosis, calcification.

## 17.4 The Role of DDR1 in Vascular Calcification

Ectopic calcification in the vascular system is remarkably prevalent, occurring in 30–50% of the population over 45, and nearly all people over 70 [61, 62]. Calcification occurring in the aortic arch and coronary arteries causes a two- to eightfold increase in the incidence of cardiovascular and cerebrovascular events, even independent of other risk factors [63, 64]. In the coronary arteries, calcification is correlated with atherosclerotic plaque burden, increased risk of myocardial infarction, postangioplasty dissection, and plaque rupture [65, 66]. Vascular calcification also causes morbidity and mortality in type 2 diabetes, where it manifests with both medial and plaque calcification in central and peripheral arteries [67], and

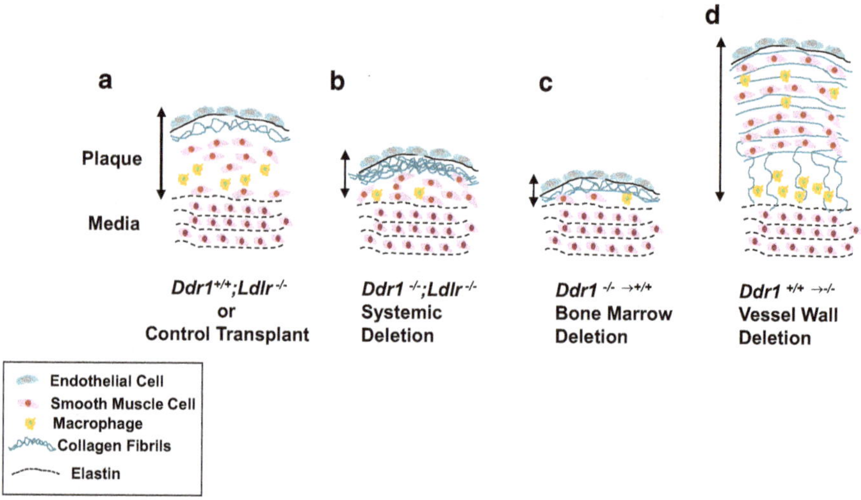


**Fig. 17.2** Graphic summary of studies investigating DDR1 in atherosclerosis. (**a**) Plaque that develops in the *Ldlr*$^{-/-}$ mouse or control transplant mouse. (**b**) Plaque that develops in mice with systemic deletion of DDR1 (*Ddr1*$^{-/-}$; *Ldlr*$^{-/-}$) is smaller with fewer macrophages, but more ECM in the fibrous cap, and similar number of SMCs compared to *Ldlr*$^{-/-}$ mouse. (**c**) Plaque that develops in mice with bone marrow deletion of DDR1 is small, with very few macrophages, but SMC number and ECM content are similar to control transplant mice. (**d**) Plaque that develops in mice with vessel wall deletion of DDR1 is large, with more SMCs, and a dramatically increased accumulation of ECM compared to control transplant mice

peripheral vascular calcification is the leading risk factor for lower limb amputations in diabetes [68]. This is an area of prime interest because at present there are no effective pharmaceutical treatments to reduce vascular calcification, and these patients are refractory to interventions such as angioplasty and stenting.

Interest in this field has heated up since the discovery that vascular calcification is an actively regulated form of calcified tissue metabolism. The causes of vascular calcification are multifactorial and include impaired phosphate metabolism, increased inflammation with consequent oxidative stress, alterations in systemic metabolic factors (hyperglycemia and hyperlipidemia), and alterations in tissue mechanical stress [69]. Despite this diversity of causes, there is a final common pathogenic pathway involving phenotypic transition of resident vascular wall SMCs [70, 71]. This pathway involves the induction of BMPs which activate the osteochondrogenic transcription factors Runx2, Msx2, and Sox-9, leading to cellular transdifferentiation [72]. It was discovered many years ago that collagen provides a permissive matrix for vascular calcification [73]; however, understanding of the mechanisms involved was incomplete and it was not known whether the process was mediated by collagen receptor signaling. Previous work has implicated both DDR1 and DDR2 in mediating chondrogenic and osteogenic differentiation [74, 75], and in playing important roles in bone and cartilage pathologies including arthritis [13, 76, 77], providing support for the notion that DDR1 might play similar roles in the mineralization of vascular tissues.

We hypothesized that the collagen receptor DDR1 might regulate VSMC transdifferentiation and atherosclerotic plaque calcification in LDLR-deficient mice. In plaques from *Ddr1*$^{+/+}$; *Ldlr*$^{-/-}$ mice fed a high-cholesterol diet, we saw large areas of mineralization with morphologic characteristics of endochondral metaplasia, and cells expressing chondrocyte markers [78]. This was significantly attenuated in *Ddr1*$^{-/-}$; *Ldlr*$^{-/-}$ mice, which suggested an important role for DDR1 in mediating calcification. Furthermore in the *Ddr1*$^{-/-}$; *Ldlr*$^{-/-}$ mice, mineralization was reduced even in the face of markedly increased collagen deposition, which is also a positive stimulus for calcification. We also showed that DDR1-deletion reduced inflammatory macrophage accumulation in atherosclerotic plaques [45, 47], and previous studies have implicated cytokines released by macrophages in triggering the osteochondrocytic phenotypic switch in SMCs [79–81]. The reduction in calcification in *Ddr1*$^{-/-}$; *Ldlr*$^{-/-}$ mice was correlated with reduced macrophage accumulation and decreased immunostaining for TNFα in the vessel wall, consistent with the notion that inflammation influences calcification [78]. However, a SMC autonomous pathway was definitely involved, as in vitro assays revealed that *Ddr1*$^{+/+}$ SMCs calcified the extracellular matrix and expressed alkaline phosphatase activity, whereas *Ddr1*$^{-/-}$ SMCs did not. Taken together, our studies suggest that DDR1 upregulates autonomous osteogenic pathways in SMCs and contributes to inflammatory responses modulating calcification. Ongoing research in the lab is focused on determining the possible osteogenic mediators that are regulated by DDR1.

## 17.5 Summary and Conclusions

Through its functions on SMCs and macrophages, we have discovered that DDR1 mediates plaque expansion, inflammation, and the late complications of instability and atherosclerotic vascular calcification. This is summarized in Fig. 17.3. There is a convergence of information suggesting that DDR1 on macrophages functions to enhance inflammatory responses by mediating adhesion, infiltration via MMP-mediated ECM degradation, and cytokine expression. In the vascular system, this has important implications for the development of atherosclerotic plaques and for long-term complications like rupture and calcification, as inflammation is tightly correlated with these phenomena. Moreover, our studies suggest that inhibiting the actions of DDR1 may tip the balance toward ECM deposition, possibly serving to stabilize and protect plaques from rupture. We hope that these studies will eventually lead to the development of improved therapies for cardiovascular disease. Therefore, we suggest that inhibiting DDR1 may lead to smaller, less inflammatory, more stable, and less calcified atherosclerotic plaques.

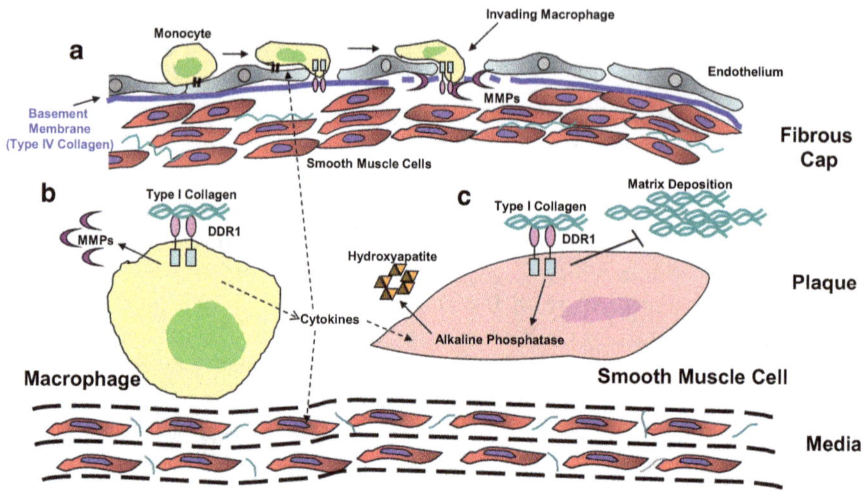


**Fig. 17.3** Functions of DDR1 in the atherosclerotic plaque. (**a**) DDR1 binding to type IV collagen in the subendothelial basement membrane stimulates macrophages to produce MMPs which allow the penetration of the basement membrane and digestion of plaque ECM, promoting macrophage recruitment and invasion of the plaque. (**b**) DDR1 on macrophages stimulates the production of cytokines and growth factors which stimulate the production of adhesion molecules on endothelial cells thereby attracting more macrophages, and stimulate the migration of smooth muscle cells into the plaque. (**c**) DDR1 on SMCs is required for cellular transdifferentiation, and for the deposition of hydroxyapatite in the matrix leading to plaque calcification. In addition, DDR1 suppresses matrix deposition in the plaque, leading to formation of a thin, vulnerable fibrous cap

# References

1. Vogel W, Gish GD, Alves F, Pawson T (1997) The discoidin domain receptor tyrosine kinases are activated by collagen. Mol Cell 1:13–23
2. Shrivastava A, Radziejewski C, Campbell E et al (1997) An orphan receptor tyrosine kinase family whose members serve as nonintegrin collagen receptors. Mol Cell 1:25–34
3. Vogel WF, Abdulhussein R, Ford CE (2006) Sensing extracellular matrix: an update on discoidin domain receptor function. Cell Signal 18:1108–1116
4. Fu HL, Valiathan RR, Payne L et al (2014) Glycosylation at Asn211 regulates the activation state of the discoidin domain receptor 1 (DDR1). J Biol Chem 289:9275–9287
5. Phan TN, Wong EL, Sun X et al (2013) Low stability and a conserved N-glycosylation site are associated with regulation of the discoidin domain receptor family by glucose via post-translational N-glycosylation. Biosci Biotechnol Biochem 77:1907–1916
6. Vogel WF (2002) Ligand-induced shedding of discoidin domain receptor 1. FEBS Lett 514:175–180
7. Slack BE, Siniaia MS, Blusztajn JK (2006) Collagen type I selectively activates ectodomain shedding of the discoidin domain receptor 1: involvement of Src tyrosine kinase. J Cell Biochem 98:672–684
8. Xu H, Raynal N, Stathopoulos S, Myllyharju J, Farndale RW, Leitinger B (2011) Collagen binding specificity of the discoidin domain receptors: binding sites on collagens II and III and molecular determinants for collagen IV recognition by DDR1. Matrix Biol 30:16–26

9. Leitinger B (2003) Molecular analysis of collagen binding by the human discoidin domain receptors, DDR1 and DDR2. Identification of collagen binding sites in DDR2. J Biol Chem 278:16761–16769
10. Mihai C, Chotani M, Elton TS, Agarwal G (2009) Mapping of DDR1 distribution and oligomerization on the cell surface by FRET microscopy. J Mol Biol 385:432–445
11. Xu H, Abe T, Liu JK, Zalivina I, Hohenester E, Leitinger B (2014) Normal activation of discoidin domain receptor 1 mutants with disulfide cross-links, insertions, or deletions in the extracellular juxtamembrane region: mechanistic implications. J Biol Chem 289:13565–13574
12. Vogel WF, Aszodi A, Alves F, Pawson T (2001) Discoidin domain receptor 1 tyrosine kinase has an essential role in mammary gland development. Mol Cell Biol 21:2906–2917
13. Labrador JP, Azcoitia V, Tuckermann J et al (2001) The collagen receptor DDR2 regulates proliferation and its elimination leads to dwarfism. EMBO Rep 2:446–452
14. Strauss BH, Chisholm RJ, Keeley F, Gotlieb AI, Logan RA, Armstrong PW (1994) Extracellular matrix remodeling after balloon angioplasty injury in a rabbit model of restenosis. Circ Res 75:650–658
15. Bendeck MP, Regenass S, Tom WD et al (1996) Differential expression of $\alpha 1$ type VIII collagen in injured, platelet-derived growth factor-BB stimulated rat carotid arteries. Circ Res 79:524–531
16. Sibinga NES, Foster LC, Hsieh C-M et al (1997) Collagen VIII is expressed by vascular smooth muscle cells in response to vascular injury. Circ Res 80:532–541
17. MacBeath JRE, Kielty CM, Shuttleworth CA (1996) Type VIII collagen is a product of vascular smooth-muscle cells in development and disease. Biochem J 319:993–998
18. Bendeck MP, Zempo N, Clowes AW, Galardy RE, Reidy MA (1994) Smooth muscle cell migration and matrix metalloproteinase expression after arterial injury in the rat. Circ Res 75:539–545
19. Cho A, Reidy MA (2002) Matrix metalloproteinase-9 is necessary for the regulation of smooth muscle cell replication and migration after arterial injury. Circ Res 91:845–851
20. Hou G, Vogel W, Bendeck MP (2001) The discoidin domain-receptor tyrosine kinase DDR1 in arterial wound repair. J Clin Invest 107:727–735
21. Hou G, Vogel WF, Bendeck MP (2002) Tyrosine kinase activity of discoidin domain receptor 1 is necessary for smooth muscle cell migration and matrix metalloproteinase expression. Circ Res 90:1147–1149
22. Dejmek J, Dib K, Jonsson M, Andersson T (2003) Wnt-5a and G-protein signaling are required for collagen-induced DDR1 receptor activation and normal mammary cell adhesion. Int J Cancer 103:344–351
23. Lu KK, Trcka D, Bendeck MP (2011) Collagen stimulates discoidin domain receptor 1-mediated migration of smooth muscle cells through Src. Cardiovasc Pathol 20:71–76
24. Borza CM, Pozzi A (2014) Discoidin domain receptors in disease. Matrix Biol 34:185–192
25. Hilton HN, Stanford PM, Harris J et al (1783) KIBRA interacts with discoidin domain receptor 1 to modulate collagen-induced signalling. Biochim Biophys Acta 2008:383–393
26. Curat CA, Vogel WF (2002) Discoidin domain receptor 1 controls growth and adhesion of mesangial cells. J Am Soc Nephrol 13:2648–2656
27. Wang CZ, Su HW, Hsu YC, Shen MR, Tang MJ (2006) A discoidin domain receptor 1/SHP-2 signaling complex inhibits alpha2beta1-integrin-mediated signal transducers and activators of transcription 1/3 activation and cell migration. Mol Biol Cell 17:2839–2852
28. Yeh YC, Wang CZ, Tang MJ (2009) Discoidin domain receptor 1 activation suppresses alpha(2)beta(1) integrin-dependent cell spreading through inhibition of Cdc42 activity. J Cell Physiol 218:146–156
29. Xu H, Bihan D, Chang F, Huang PH, Farndale RW, Leitinger B (2012) Discoidin domain receptors promote alpha1beta1- and alpha2beta1-integrin mediated cell adhesion to collagen by enhancing integrin activation. PLoS One 7, e52209
30. Staudinger LA, Spano SJ, Lee W et al (2013) Interactions between the discoidin domain receptor 1 and $\beta 1$ integrin regulate attachment to collagen. Biol Open 2:1148–1159

31. Shintani Y, Fukumoto Y, Chaika N, Svoboda R, Wheelock MJ, Johnson KR (2008) Collagen I-mediated up-regulation of N-cadherin requires cooperative signals from integrins and discoidin domain receptor 1. J Cell Biol 180:1277–1289
32. Eswaramoorthy R, Wang CK, Chen WC et al (2010) DDR1 regulates the stabilization of cell surface E-cadherin and E-cadherin-mediated cell aggregation. J Cell Physiol 224:387–397
33. Yeh YC, Wu CC, Wang YK, Tang MJ (2011) DDR1 triggers epithelial cell differentiation by promoting cell adhesion through stabilization of E-cadherin. Mol Biol Cell 22:940–953
34. Hidalgo-Carcedo C, Hooper S, Chaudhry SI et al (2011) Collective cell migration requires suppression of actomyosin at cell-cell contacts mediated by DDR1 and the cell polarity regulators Par3 and Par6. Nat Cell Biol 13:49–58
35. Wang CZ, Yeh YC, Tang MJ (2009) DDR1/E-cadherin complex regulates the activation of DDR1 and cell spreading. Am J Physiol Cell Physiol 297:C419–C429
36. Ferri N, Carragher NO, Raines EW (2004) Role of discoidin domain receptors 1 and 2 in human smooth muscle cell-mediated collagen remodeling: potential implications in atherosclerosis and lymphangioleiomyomatosis. Am J Pathol 164:1575–1585
37. Shyu KG, Chao YM, Wang BW, Kuan P (2005) Regulation of discoidin domain receptor 2 by cyclic mechanical stretch in cultured rat vascular smooth muscle cells. Hypertension 46:614–621
38. Morales MO, Price RL, Goldsmith EC (2005) Expression of discoidin domain receptor 2 (DDR2) in the developing heart. Microsc Microanal 11:260–267
39. Cowling RT, Yeo SJ, Kim IJ et al (2014) Discoidin domain receptor 2 germline gene deletion leads to altered heart structure and function in the mouse. Am J Physiol Heart Circ Physiol 307:H773–H781
40. Hou G, Wang D, Bendeck MP (2012) Deletion of discoidin domain receptor 2 does not affect smooth muscle cell adhesion, migration, or proliferation in response to type I collagen. Cardiovasc Pathol 21:214–218
41. Chen SC, Wang BW, Wang DL, Shyu KG (2008) Hypoxia induces discoidin domain receptor-2 expression via the p38 pathway in vascular smooth muscle cells to increase their migration. Biochem Biophys Res Commun 374:662–667
42. Shyu KG, Wang BW, Chang H (2009) Hyperbaric oxygen activates discoidin domain receptor 2 via tumour necrosis factor-alpha and the p38 MAPK pathway to increase vascular smooth muscle cell migration through matrix metalloproteinase 2. Clin Sci 116:575–583
43. Shyu KG, Wang BW, Kuan P, Chang H (2008) RNA interference for discoidin domain receptor 2 attenuates neointimal formation in balloon injured rat carotid artery. Arterioscler Thromb Vasc Biol 28:1447–1453
44. Ishibashi S, Brown MS, Goldstein JL, Gerard RD, Hammer RE, Herz J (1993) Hypercholesterolemia in low density lipoprotein receptor knockout mice and its reversal by adenovirus-mediated gene delivery. J Clin Invest 92:883–893
45. Franco C, Hou G, Ahmad PJ et al (2008) Discoidin domain receptor 1 (Ddr1) deletion decreases atherosclerosis by accelerating matrix accumulation and reducing inflammation in low-density lipoprotein receptor deficient mice. Circ Res 102:1202–1211
46. Deguchi JO, Aikawa M, Tung CH et al (2006) Inflammation in atherosclerosis: visualizing matrix metalloproteinase action in macrophages in vivo. Circulation 114:55–62
47. Franco C, Britto K, Wong E et al (2009) Discoidin domain receptor 1 on bone marrow derived cells promotes macrophage accumulation during atherogenesis. Circ Res 105:1141–1148
48. Flamant M, Placier S, Rodenas A et al (2006) Discoidin domain receptor 1 null mice are protected against hypertension-induced renal disease. J Am Soc Nephrol 17:3374–3381
49. Guerrot D, Kerroch M, Placier S et al (2011) Discoidin domain receptor 1 is a major mediator of inflammation and fibrosis in obstructive nephropathy. Am J Pathol 179:83–91
50. Gross O, Girgert R, Beirowki B et al (2010) Loss of collagen-receptor DDR1 delays renal fibrosis in hereditary type IV collagen disease. Matrix Biol 29:346–356
51. Avivi-Green C, Singal M, Vogel WF (2006) Discoidin domain receptor 1 deficient mice are resistant to bleomycin-induced lung fibrosis. Am J Respir Crit Care Med 174:420–427

52. Matsuyama W, Wang L, Farrar WL, Faure M, Yoshimura T (2004) Activation of discoidin domain receptor 1 isoform b with collagen up-regulates chemokine production in human macrophages: role of p38 mitogen-activated protein kinase and NF-kappa B. J Immunol 172:2332–2340
53. Kamohara H, Yamashiro S, Galligan C, Yoshimura T (2001) Discoidin domain receptor 1 isoform-a (DDR1a) promotes migration of leukocytes in three-dimensional collagen lattices. FASEB J 15:2724–2726
54. Matsuyama W, Faure M, Yoshimura T (2003) Activation of discoidin domain receptor 1 facilitates the maturation of human monocyte-derived dendritic cells through the TNF receptor associated factor 6/TGF-{beta}-activated protein kinase 1 binding protein 1{beta}/p38{alpha} mitogen-activated protein kinase signaling cascade. J Immunol 171:3520–3532
55. Kim SH, Lee S, Suk K et al (2007) Discoidin domain receptor 1 mediates collagen-induced nitric oxide production in J774A.1 murine macrophages. Free Radic Biol Med 42:343–352
56. Hachehouche LN, Chetoui N, Aoudjit F (2010) Implication of discoidin domain receptor 1 in T cell migration in three-dimensional collagen. Mol Immunol 47:1866–1869
57. Chetoui N, El Azreq MA, Boisvert M, Bergeron ME, Aoudjit F (2011) Discoidin domain receptor 1 expression in activated T cells is regulated by the ERK MAP kinase signaling pathway. J Cell Biochem 112:3666–3674
58. Poudel B, Ki HH, Lee YM, Kim DK (2013) Induction of IL-12 production by the activation of discoidin domain receptor 2 via NF-kappaB and JNK pathway. Biochem Biophys Res Commun 434:584–588
59. Franco C, Ahmad PJ, Hou G, Wong E, Bendeck MP (2010) Increased cell and matrix accumulation during atherogenesis in mice with vessel wall-specific deletion of discoidin domain receptor 1. Circ Res 106:1775–1783
60. Flynn LA, Blissett AR, Calomeni EP, Agarwal G (2010) Inhibition of collagen fibrillogenesis by cells expressing soluble extracellular domains of DDR1 and DDR2. J Mol Biol 395:533–543
61. Allison MA, Criqui MH, Wright CM (2004) Patterns and risk factors for systemic calcified atherosclerosis. Arterioscler Thromb Vasc Biol 24:331–336
62. Bild DE, Detrano R, Peterson D et al (2005) Ethnic differences in coronary calcification: the multi-ethnic study of atherosclerosis (MESA). Circulation 111:1313–1320
63. Iribarren C, Sidney S, Sternfeld B, Browner WS (2000) Calcification of the aortic arch: risk factors and association with coronary heart disease, stroke, and peripheral vascular disease. JAMA 283:2810–2815
64. Wayhs R, Zelinger A, Raggi P (2002) High coronary artery calcium scores pose an extremely elevated risk for hard events. J Am Coll Cardiol 39:225–230
65. Rumberger JA, Simons DB, Fitzpatrick LA, Sheedy PF, Schwartz RS (1995) Coronary artery calcium area by electron-beam computed tomography and coronary atherosclerotic plaque area. A histopathologic correlative study. Circulation 92:2157–2162
66. Sangiorgi G, Rumberger JA, Severson A et al (1998) Arterial calcification and not lumen stenosis is highly correlated with atherosclerotic plaque burden in humans: a histologic study of 723 coronary artery segments using nondecalcifying methodology. J Am Coll Cardiol 31:126–133
67. Shao JS, Cheng SL, Sadhu J, Towler DA (2010) Inflammation and the osteogenic regulation of vascular calcification: a review and perspective. Hypertension 55:579–592
68. Snell-Bergeon J, Budoff M, Hokanson J (2013) Vascular calcification in diabetes: mechanisms and implications. Curr Diab Rep 13:391–402
69. Wu M, Rementer C, Giachelli CM (2013) Vascular calcification: an update on mechanisms and challenges in treatment. Calcif Tissue Int 93:365–373
70. Speer MY, Li X, Hiremath PG, Giachelli CM (2010) Runx2/Cbfa1, but not loss of myocardin, is required for smooth muscle cell lineage reprogramming toward osteochondrogenesis. J Cell Biochem 110:935–947

71. Speer MY, Yang HY, Brabb T et al (2009) Smooth muscle cells give rise to osteochondrogenic precursors and chondrocytes in calcifying arteries. Circ Res 104:733–741
72. Sallam T, Cheng H, Demer LL, Tintut Y (2013) Regulatory circuits controlling vascular cell calcification. Cell Mol Life Sci 70:3187–3197
73. Watson KE, Parhami F, Shin V, Demer LL (1998) Fibronectin and collagen I matrixes promote calcification of vascular cells in vitro, whereas collagen IV matrix is inhibitory. Arterioscler Thromb Vasc Biol 18:1964–1971
74. Lund AW, Stegemann JP, Plopper GE (2009) Mesenchymal stem cells sense three dimensional type I collagen through discoidin domain receptor 1. Open Stem Cell J 1:40–53
75. Wu SC, Hsiao HF, Ho ML et al (2015) Suppression of discoidin domain receptor 1 expression enhances the chondrogenesis of adipose-derived stem cells. Am J Physiol Cell Physiol 308:C685–C696
76. Schminke B, Muhammad H, Bode C et al (2013) A discoidin domain receptor 1 knock-out mouse as a novel model for osteoarthritis of the temporomandibular joint. Cell Mol Life Sci 71(6):1081–1096
77. Xu L, Servais J, Polur I et al (2010) Attenuation of osteoarthritis progression by reduction of discoidin domain receptor 2 in mice. Arthritis Rheum 62:2736–2744
78. Ahmad PJ, Trcka D, Xue S et al (2009) Discoidin domain receptor-1 deficiency attenuates atherosclerotic calcification and smooth muscle cell-mediated mineralization. Am J Pathol 175:2686–2696
79. Tintut Y, Patel J, Territo M, Saini T, Parhami F, Demer LL (2002) Monocyte/macrophage regulation of vascular calcification in vitro. Circulation 105:650–655
80. Deuell KA, Callegari A, Giachelli CM, Rosenfeld ME, Scatena M (2012) RANKL enhances macrophage paracrine pro-calcific activity in high phosphate-treated smooth muscle cells: dependence on IL-6 and TNF-alpha. J Vasc Res 49:510–521
81. Byon CH, Sun Y, Chen J et al (2011) Runx2-upregulated receptor activator of nuclear factor {kappa}B ligand in calcifying smooth muscle cells promotes migration and osteoclastic differentiation of macrophages. Arterioscler Thromb Vasc Biol 31:1387–1396

# Chapter 18
# Discoidin Domain Receptors in Cardiac Development

**David E. Justus, Adam Hoffman, Ekaterina Mironova, Alexander Hartman, Jack G. Goldsmith, Jay D. Potts, and Edie C. Goldsmith**

## 18.1 Introduction

Development of the four chambered heart proceeds by a series of stages conserved across species which include formation of paired heart fields, tube formation through fusion of the heart fields, looping of the heart tube, valve formation, and septation. This process requires establishment and maintenance of cell–cell interactions, deposition and remodeling of the cardiac extracellular matrix (ECM), and proper interactions between cardiac cells their extracellular environment [1]. Heart development begins with the formation of bilateral heart fields which fuse into a single, straight heart tube composed of an inner endothelial cell layer and an outer myocardial cell layer separated by an acellular matrix referred to as the "cardiac jelly" [2]. The tubular heart undergoes looping, to properly position the future atria and ventricles, and then must be divided into four chambers to achieve separation of the atria and the ventricles as well as right–left demarcation. The cardiac septa divide the left and right halves of the heart to ensure separation of the pulmonary and systemic circuits and the atrioventricular (AV) valves, which develop from embryonic cardiac cushions, ensure unidirectional blood flow through the heart. Endocardial cells overlying the AV cushions undergo an epithelial to mesenchymal transformation (EMT), wherein the endothelial cells lose their cell–cell contacts and migrate through the ECM. The newly transformed mesenchymal cells proliferate in the cushions, giving rise to fibroblasts within fully mature valve leaflets.

The cardiac ECM is a dynamic structure that changes dramatically during development. Overall, it is comprised from the structural protein collagen, noncollagenous glycoproteins (such as laminin, fibronectin, and tenascin) which provide sites

D.E. Justus • A. Hoffman • E. Mironova • A. Hartman • J.G. Goldsmith
J.D. Potts • E.C. Goldsmith (✉)
Department of Cell Biology and Anatomy, University of South Carolina,
6439 Garners Ferry Rd., Columbia, SC 29209, USA
e-mail: edie.goldsmith@uscmed.sc.edu



R. Fridman, P.H. Huang (eds.), *Discoidin Domain Receptors in Health and Disease*, DOI 10.1007/978-1-4939-6383-6_18

for cellular attachment, and proteoglycans that lend structural support and also provide organizational cues for the ECM [3]. The myocardial ECM is composed largely of collagen types I and III [4–6], with the proportions of types I and III changing during embryonic development [7], in addition to type IV collagen which is located in the basement membrane of myocytes. During development, as load on the fetal/neonatal heart increases, the collagen network develops to preserve the integrity and maintain function of the heart [8]. Collagen interconnects myocytes, connects myocytes to capillaries, and forms a weave network which surrounds and mechanically couples groups of myocytes [9]. Studies have shown that during development the collagen content in the rat heart increases from birth through adulthood [10, 11]. At birth, the collagen weave network in rat hearts is essentially nonexistent but develops rapidly during the early postpartum period reaching an organization indistinguishable from the adult collagen network [12]. Other studies have shown the transient expression of different collagen subtypes during various stages of heart development [13, 14] and specifically within the cardiac cushions and maturing valves [15]. Fibroblasts are the primary producers of collagen in the heart [16, 17], with myocytes producing type IV collagen found in their basement membrane [11].

For decades, our understanding of ECM receptors, including those for collagen, was limited to the integrin protein family. In the late 1990s, two novel receptor tyrosine kinases, Discoidin Domain Receptors (DDRs) 1 and 2, were found to bind collagen [18, 19]. Receptor tyrosine kinases are a large family of transmembrane proteins involved in the conversion of extracellular stimuli into a variety of cellular responses including proliferation, migration, adhesion, differentiation, and survival [20]. All family members have an extracellular ligand binding domain, a single membrane spanning segment, and a conserved kinase domain in their cytoplasmic region. DDR1 and DDR2 bind a variety of different collagens within the ECM [18, 19, 21–24] and one unique feature of these receptors is their protracted activation kinetics after ligand binding [18, 19]. A key consequence of DDR activation is the induction of matrix metalloproteinases (MMPs), a family of zinc proteases which can breakdown collagen and other ECM components and play a significant role in matrix remodeling of the heart [25, 26]. The ability to bind and become activated by intact but not denatured collagen [18] and upregulate MMPs, places DDRs in a position where they can sense the physical state of collagen within the ECM and regulate its turnover, potentially making these receptors initiators of matrix remodeling. Given the dramatic changes in the collagen network that occur during cardiac development, the focus of this chapter is to summarize the work to date examining the role of these two receptors in heart development.

## 18.2 DDR2 Expression in the Early Heart

The heart is comprised of four resident cell types: myocytes, muscle cells responsible for generating the force necessary to pump blood; fibroblasts, connective tissue cells that deposit and remodel the cardiac extracellular matrix; and endothelial and smooth muscle cells which form the vessels within the heart. While there are markers for myocytes, endothelial and smooth muscle cells which allow their identification and lineage tracing, no such marker exists for cardiac fibroblasts. Identification of DDR2 as a novel collagen receptor [18, 19] and characterization of its expression on dermal fibroblasts during development [27] and as a marker for mesenchymal cells [28] prompted examination of DDR2 expression in the heart. Reverse-transcriptase-PCR analysis of RNA isolated from multiple time points during embryonic and neonatal development as well as from adult rat hearts indicated that DDR2 transcripts were present at embryonic day 11.5 (ED11.5) and that the message continued to be expressed throughout development and in the adult heart [29]. Further analysis of isolated neonatal cardiac myocytes and fibroblasts revealed that DDR2 mRNA was only present in the fibroblasts; a finding which was confirmed at the protein level where DDR2 was only detected in cardiac fibroblast lysates, not myocyte, endothelial or smooth muscle cell lysates. This study demonstrated that within the heart, DDR2 expression was restricted to the cardiac fibroblasts although outside of the heart it is expressed on a number of other cell types [30, 31]. With this caveat in mind, DDR2 has become a commonly used marker to identify fibroblasts within the heart not only during development [30, 32] but also in multiple disease models [33–35].

As the vertebrate heart develops, it exists as a tubular structure consisting of two cell layers and predominantly two cell types—cardiac myoctes and endocardial (endothelial) cells. Previous studies have shown that fibroblasts do not appear in the myocardium until later in development when collagen is produced [36] and that they originate from the proepicardium [37, 38], but the use of DDR2 as a fibroblasts marker permitted further temporal investigation of this observation. Rodent hearts were examined by immunofluorescence at ED11, 14, 16, neonatal day four, and adulthood for the distribution of DDR2 [39]. At the earliest time point examined (ED11), DDR2-positive staining was predominantly observed on the epicardial surface of the heart with very little signal detected in the myocardium. Around ED14, DDR2-positive cells displayed a distinct distribution at the apex of the heart or at the junction between atria and ventricles, where some of these cells appeared to be migrating into the myocardium. At this developmental time point, DDR2-positive cells were also detected within the atria and the cardiac cushions. The number of DDR2-positive cells increased at ED16 with staining detected across the ventricular free wall and within the septum in addition to atrial and epicardial expression observed earlier in development. In the neonatal heart, when expression of collagen has increased due to increase in load on the heart [12], DDR2-positive cells were distributed throughout the myocardium, a pattern which persisted in adult heart. These cells appeared to be elongated and situated between myocyte layers, representing fibroblasts within the heart [29].

## 18.3 Role of DDR2 in Epithelial-Mesenchymal Transformation in Cardiac Valve Development

Epithelial to mesenchymal transformation (EMT) is a common process which occurs during normal development and in pathological situations such as cancer. Within the heart, EMT is essential for the development and formation of cardiac valves.

Cardiac cushions, the precursor structures to valves, form in the atrioventricular (AV) canal and outflow tract regions of the developing heart [2, 40]. Prior to EMT, cushions consist mainly of an acellular matrix comprised of a variety of ECM constituents but notably lack collagens [41]. In a temporal and spatially restricted manner, endothelial cells overlying cushions undergo EMT which is characterized by activation of endothelial cells (including loss of cell–cell attachment), mesenchymal cell formation, and invasion of newly formed mesenchyme into the cushion matrix (Fig. 18.1; [42]). It is this population of newly derived mesenchymal cells that give rise to cardiac valvular fibroblasts which produce the collagen matrix within the forming valves [43].

While the studies summarized above revealed that DDR2 is expressed in the heart during the temporal window when EMT occurs, Morales et al. discovered that DDR2-positive cells were found within the cardiac cushions concomitant with the expression of type I collagen [39]. RNA isolated from embryonic chick hearts representing developmental stages spanning EMT (stages 12–20) and subjected to reverse transcriptase-PCR revealed that in the chick heart DDR2 mRNA is present throughout the developmental window encompassing EMT as well (Fig. 18.2a). The developmental stages used in these studies represent pre-EMT (st12), time during which signals initiating EMT are released (st14), endothelial cell activation (st15-16), transformation of endocardial cells (st17-18), and invasion, migration, and proliferation of mesenchymal cells in the cushions (st19-23). While DDR2 message was detected prior to EMT in this experiment, this result is consistent with other studies which detected DDR2 expression early in the rodent heart [29] likely due to DDR2-positive cells on the epicardial surface of the heart [39]. Additional studies examining the expression of DDR2 in isolated AV canals revealed that DDR2 is not only expressed in the AV canal (stage 17) but also that the receptor is specifically expressed on mesenchymal cells isolated from the AV canal region [44]. Furthermore, real-time PCR analysis of isolated AV canals from stages 14–23 demonstrated consistent levels of DDR2 expression during EMT. Western blotting confirmed the presence of DDR2 protein in AV canal extracts and suggested that protein levels increased during the progression of EMT.

To characterize the localization of DDR2 in the developing chick heart, DDR2-specific in situ hybridization probes were developed and the distribution of this receptor examined. Whole mount in situ hybridization of stage 18 chick embryos revealed extensive DDR2 reactivity throughout the embryo (Fig. 18.2b; unpublished) while the control sense probe had essentially no reactivity. To further analyze the localization of DDR2 mRNA, whole mount staining was carried out on embryos from stages 15–20 followed by vibratome sectioning to facilitate visualization of DDR2. As shown in Fig. 18.2c (unpublished), a small amount of DDR2 staining was detected in the cushions and on the endocardium in stage 15 hearts. However, by stage 18, as the number of mesenchymal cells within the cushions increased, so did the degree of DDR2 staining. By stage 20, the cushion area was completely filled with DDR2-positive mesenchymal cells. Immunofluorescence of similar staged chick hearts also showed progressive increases in DDR2-positive cells within the AV cushions [44]. To correlate DDR2 expression with specific

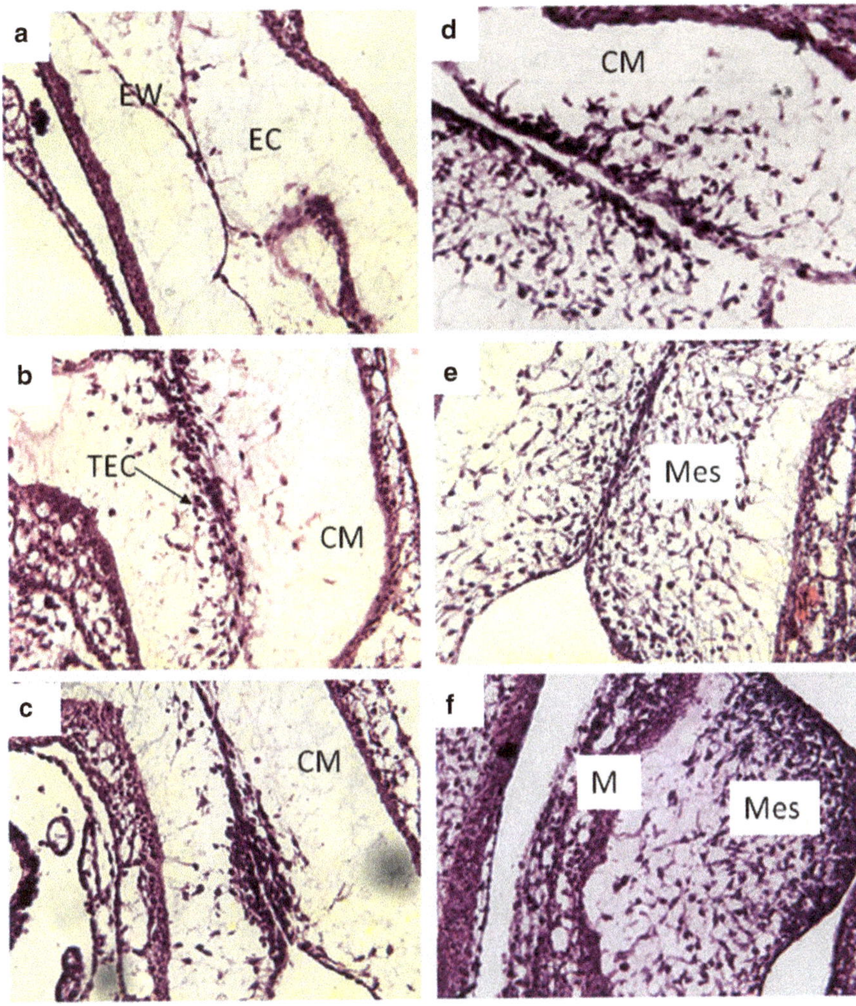


**Fig. 18.1** EMT in cardiac cushions. H&E staining of stage 14–24 chick cardiac cushions clearly illustrates the transformation and invasion of endocardial cushions by mesenchymal cells. (**a**) Endothelial cells maintain a clear border on the endocardial wall (EW) of stage 14 embryos. Note the absence of cells within the endocardial cushions (EC). (**b–d**) Transformed endothelial cells (TEC) assume an elongated spindle shape as they lose cell–cell contacts, differentiate into mesenchymal cells, and migrate into the cushion matrix (CM). (**e**) By stage 20, the CM is densely populated with mesenchymal cells (Mes) which ultimately reach the underlying myocardial (M) surface (**f**)

functional changes at the cellular level during EMT, an in vitro AV canal explant model system was used. This model involves culture of an isolate AV canal on a three-dimensional collagen gels and faithfully recapitulates critical events during the transformation process, including activation of endothelial cells and invasion and migration of mesenchyme within a collagen matrix. In this system, immunofluorescence

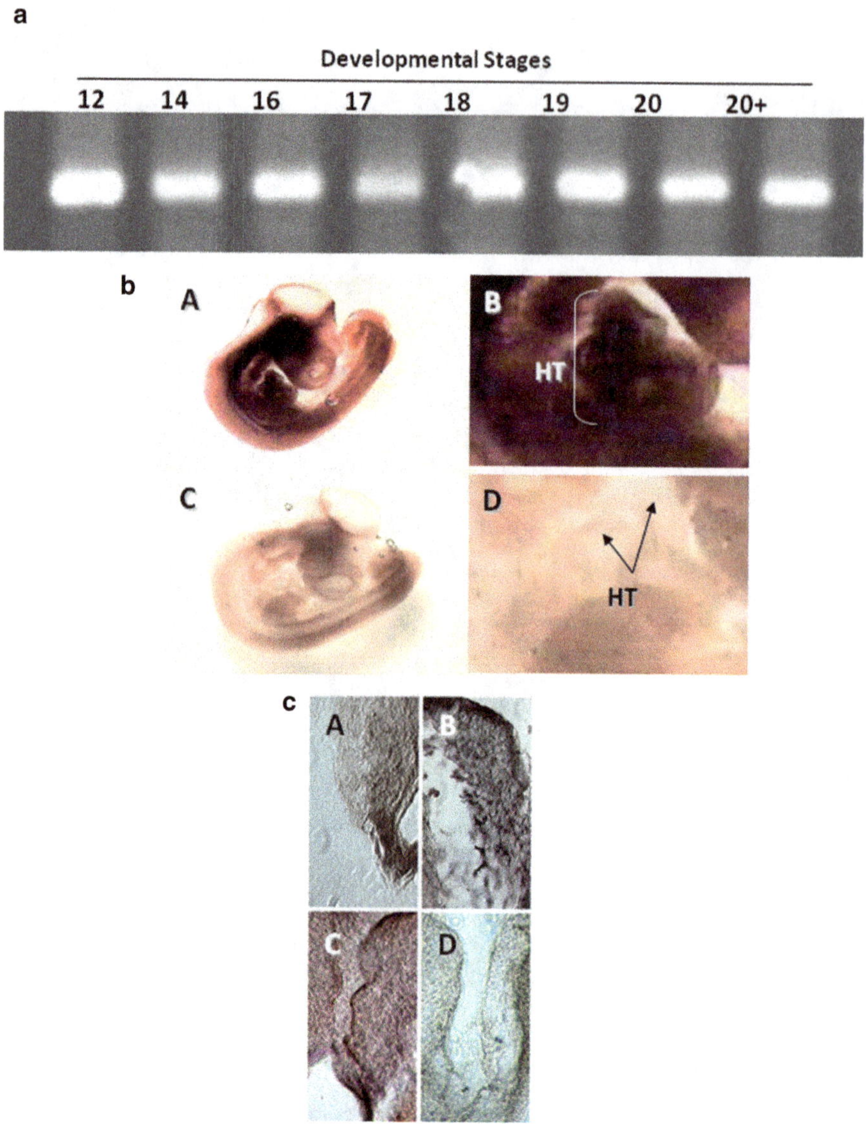


**Fig. 18.2** DDR2 expression in the developing chick heart). (**a**) RNA was isolated from embryonic chick hearts from developmental stages 12–20+. These stages encompass the developmental window during which EMT occurs within the cardiac cushions. Reverse transcriptase-PCR was carried out using primers specific for chick DDR2. DDR2 mRNA was found in all developmental time points examined and not bands were detected in the control no reverse transcriptase sample. (**b**) Whole mount in situ hybridization to characterize DDR2 mRNA distribution in stage 18 embryos. *Panels A* and *C* represent low magnification images of embryos stained with antisense (*top*) and sense (*bottom*) DDR2 probes, whereas *Panels B* and *D* are higher magnification images specifically showing the heart. (**c**) Distribution of DDR2 mRNA in st15(A), st18(B), and st20(C) AV cushions. Note the increase in DDR2-positive staining as EMT progresses. Sense control (**d**) showed no reactivity

failed to detect DDR2 on competent endothelial cells (cells that can undergo transformation but have to yet receive an activation signal), but DDR2 was present on activated endothelial cells and migrating mesenchymal cells within the collagen gel [44]. Combined these studies suggest that DDR2 is first expressed when endothelial cells are stimulated to undergo activation, making this receptor an early marker for activation. However, it is unclear what molecule(s) initiate expression of DDR2 in AV endocardial cells. DDR2 expression on activated endothelial cells and migrating mesenchymal cells may provide a mechanism for these cells to sense collagen within their extracellular environment and respond to that stimulus through proliferation, migration, and production of collagen as has been observed in other systems [45].

Given the robust expression of DDR2 during EMT and its potential to mediate migration of mesenchymal cells into the cushion matrix, isolated chick AV canals were cultured in the presence of an anti-EC-DDR2 antibody [44]. As shown in Fig. 18.3a (unpublished), in the absence of antibody treatment numerous mesenchymal cells are observed migrating away from the cushion explant. In the presence of anti-EC-DDR2 antibody, while an initial layer of endothelial cells migrated away from the explant, these cells never underwent transformation, developed a rounded morphology, and were not capable of migrating into the underlying collagen matrix (Fig. 18.3b; unpublished). This lack of migration was not likely due to cell death, as myocardial cells associated with the explant were still contracting when images were collected. A global DDR2 knockout (DDR2 KO) mouse was developed [46] and although the authors reported no cardiac abnormalities, DDR2$^{-/-}$ mice are viable but exhibit dwarfism due to impaired bone growth. AV cushions were obtained from wild-type and DDR2 KO mice to determine if endothelial cells within KO cushions would undergo activation and transformation into migrating mesenchyme. Endothelial cells in explanted AV cushions from wild-type mice clearly underwent EMT as evidenced by the extensive cellular migration away from the explant (Fig. 18.3c; unpublished). In contrast, AV cushions isolated from DDR2 KO mice were noticeably smaller in size than cushions from wild-type mice (Fig. 18.3d; images c and d in this panel were photographed under identical conditions) and demonstrated a severely impaired capacity to undergo EMT, with limited cell migration away from the explant observed (Fig. 18.3d; unpublished). These in vitro studies suggest that DDR2 plays a significant role in EMT during cardiac valve development; however, examination of hearts from DDR2 KO mice revealed the near normal appearance of AV valves in these mice. Immunohistochemistry of hearts from embryonic day 12 (ED12) wild-type and DDR2 KO embryos revealed no obvious structural differences between the cushions, with numerous mesenchymal cells visible in the cushion matrix and little difference in the distribution of collagen types I and III, and periostin (Fig. 18.4; unpublished). By embryonic day 18 (ED18), the wild-type valve leaflets demonstrated a flattened, elongated morphology, whereas the DDR2 KO leaflets had a rounded appearance (Fig. 18.4). While leaflets from both wild-type and DDR2 KO mice appear to exhibit the same degree of cellularity and similar distribution of collagens, there is a dramatic difference the amount and localization of periostin within the ED18 leaflets. Periostin is a matricellular protein which has been shown to bind a number of ECM proteins, including multiple

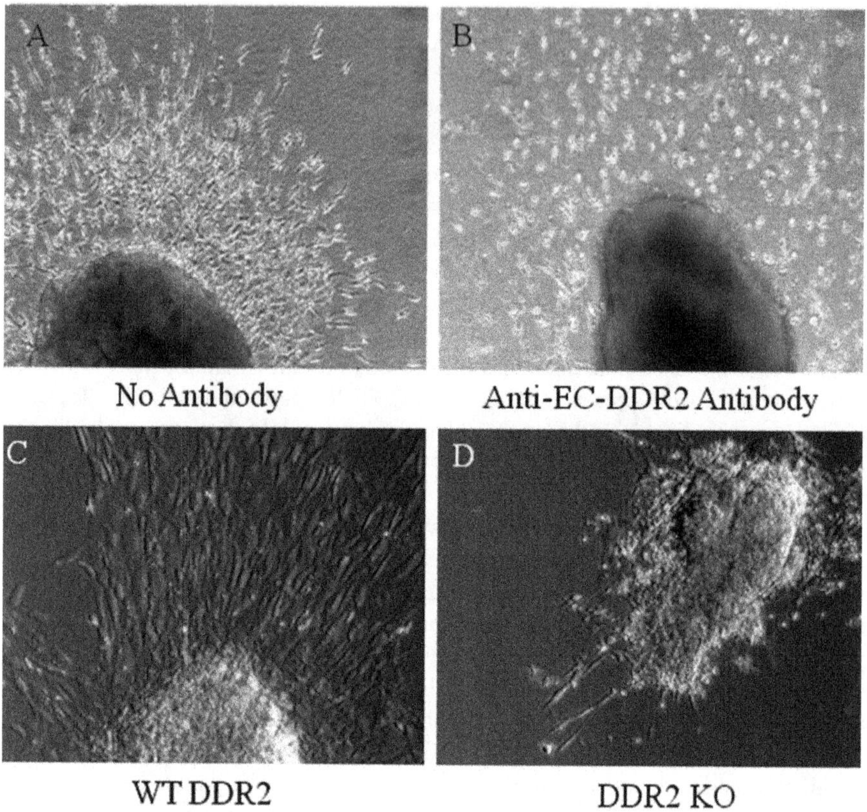


**Fig. 18.3** Inhibition of DDR2 prevents EMT in AV cushion explants. Using an in vitro EMT assay, the role of DDR2 was examined. AV cushions were isolated from st14 chick embryos, placed on three-dimensional collagen gels an either untreated (**a**) or treated with 200 μg/mL anti-EC-DDR2 antibody (**b**). While EMT proceeds normally in the untreated cushion explants (**a**; evidenced by the extensive migration and invasion of mesenchymal cells away from the explant), the addition of anti-EC-DDR2 antibody to the culture media impaired transformation (**b**), resulting in cells with a rounded morphology and limited ability to invade the underlying matrix. Isolation of AV cushions from wild-type (**c**) or ED9.5-10.5 DDR2 KO mice (**d**) revealed a similar finding. Wild-type AV cushions exhibited extensive outgrowth of cells away from the explant (**c**), whereas DDR2 KO cushions demonstration minimal capacity for EMT (**d**), with few migrating mesenchymal cells observed

collagens, as well as many members of the integrin protein family [30]. In addition, periostin has also been shown to regulate collagen fibril assembly and diameter [47]. Further studies are needed to confirm a connection between DDR2 and periostin in valve development, but given that both can interact with collagen, the potential exists for DDR2 to indirectly regulate collagen assembly/organization in the maturing valve leaflet through periostin. The formation of valves in vivo despite in vitro studies which indicate that inhibiting DDR2 should negatively impact valve formation suggests the existence of other compensating mechanisms in vivo.

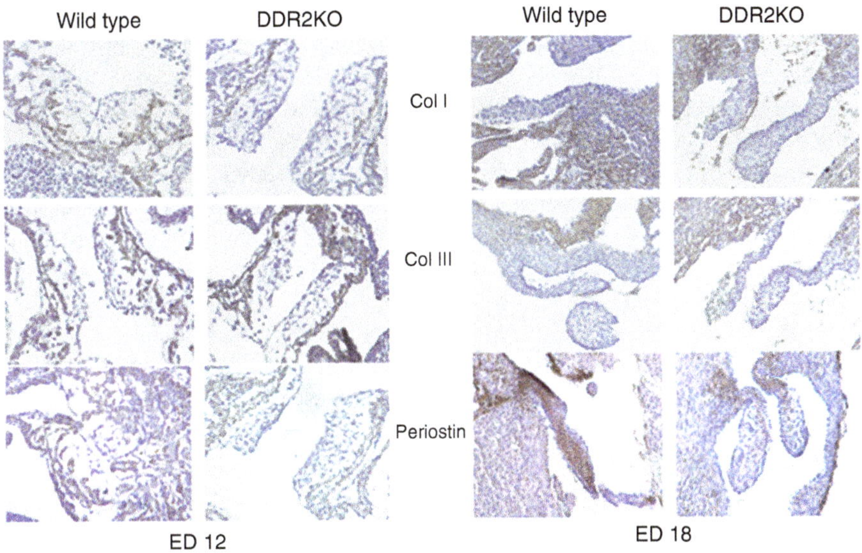


**Fig. 18.4** Immunohistochemical characterization ECM components within wild-type and DDR2 KO valves. Hearts were isolated from ED12 and ED18 wild-type and DDR2 KO mice and stained to examine the expression and distribution of collagen types I and III, and periostin. *Left panels*—ED12 wild-type and DDR2 KO cushions. Cushions from both genotypes demonstrate the presence of mesenchymal cells within the AV cushions and similar distributions of ECM components. *Right panels*—ED18 wild-type and DDR2 KO valve leaflets. Wild-type leaflets possess a flattened, extended morphology compared to DDR2 KO leaflets and DDR2 KO leaflets had significantly reduced levels of periostin expression distributed mainly on the atrial surface of the leaflet

One such mechanism may involve the contribution of other cell types, such as proepicardium/epicardial derived cells, neural crest, or bone marrow derived cells [48], to AV valve formation.

## 18.4 DDR2 and Cardiac Fibroblasts

Cardiac fibroblasts mediate myocardial ECM remodeling not only during heart development but also during cardiovascular disease. These remodeling events are associated with changes in both collagen expression and the production of MMPs [49, 50]. Given that DDR2 binds collagen within the ECM and within the heart is expressed only on fibroblasts, examination of this receptor's role in collagen remodeling is warranted. Most studies that examine cellular remodeling use the well-established collagen gel assay to recapitulate this process in vitro [51]. In this model system, three-dimensional collagen scaffolds are prepared containing the cell type of interest and the change in gel area over time is monitored as an indication of cellular activity. Analysis of RNA extracted from fibroblasts cultured in collagen gels indicated that as contraction increased over time the expression of DDR2 also increased, suggesting a role of DDR2 in fibroblast-mediated remodeling of collagen matrices. To determine the functional role of DDR2 in fibroblast-mediated matrix remodeling, neonatal cardiac fibroblasts were preincubated either without antibody (controls) or with increasing concentrations (50, 100, or 200 μg/mL) of an anti-EC-DDR2 antibody [44], embedded in collagen type I gels and the gels were allowed to polymerize. Over a 72 h period, the anti-EC-DDR2 antibody inhibited collagen gel contraction in a dose-dependent manner (Fig. 18.5; unpublished) and this inhibition was statistically significant ($p<0.05$) from control at all antibody concentrations tested. A maximal inhibition of gel contraction (approximately 37 % compared to control) was observed after 24 h using 200 μg/mL anti-EC-DDR2 antibody. Increasing culture time in the presence of the anti-EC-DDR2 antibody did not further decrease gel contraction. Inhibiting the interaction of DDR2 on cardiac fibroblasts with collagen clearly reduced the ability of these cells to physically contract and remodel their matrix environment. While the level of inhibition was significant, it is not surprising that blocking DDR2 did not completely inhibition collagen gel contraction as fibroblasts express other ECM receptors, namely integrins, which are also involved in this process.

Integrins are heterodimeric cell surface receptors which bind a variety of ECM components, including collagens, and have long been recognized as major players in cell migration, adhesion, differentiation, and remodeling. Cardiac fibroblasts express a multitude of integrins, most notably α1β1 and α2β1 which are capable of binding collagen [31, 52]. Previous work has shown that blocking β1 integrin on cardiac fibroblasts dramatically impairs their ability to remodel collagen gels [53]. There has been much interest regarding crosstalk between integrins and DDRs, and while activation of these receptor families occurs independently [54], there appears to be crosstalk between them and even signaling regulation which can be cell type specific [55]. Treating cardiac fibroblasts with anti-β1 integrin and anti-EC-DDR2 antibodies, singly or in combination, allowed for examination of crosstalk between these two receptor families during matrix remodeling. Exposure of cardiac fibroblasts to an anti-β1 integrin antibody for 24 h significantly reduced gel contraction by almost 60 % compared to control gels while treatment with the anti-DDR2 antibody yielded a significant, but smaller, decrease in contraction (Fig. 18.6; unpublished). When both antibodies were combined, gel contraction was even further reduced, with the resulting effect appearing additive in nature, implying that the receptors do indeed using different pathways to mediate collagen remodeling. However, inhibition of both DDR2 and β1 integrin did not completely abrogate gel contraction, suggesting that additional collagen receptors, perhaps DDR1 or other integrins, are participating in the remodeling process.

The importance of DDRs on fibroblasts in facilitating mechanical deformation and remodeling of collagen matrices has also been examined in an age-related manner by comparing neonatal and adult cardiac fibroblasts. Fibroblast-mediated contraction of collagen matrices depends on alpha smooth muscle actin (α-SMA), a cytoskeletal protein found in fibroblasts [56–58] and is recognized to depend on integrin receptors for attachment to the collagen matrix [53, 59]. The contribution of DDRs in mediating collagen contraction has been examined by Wilson et al. [60].

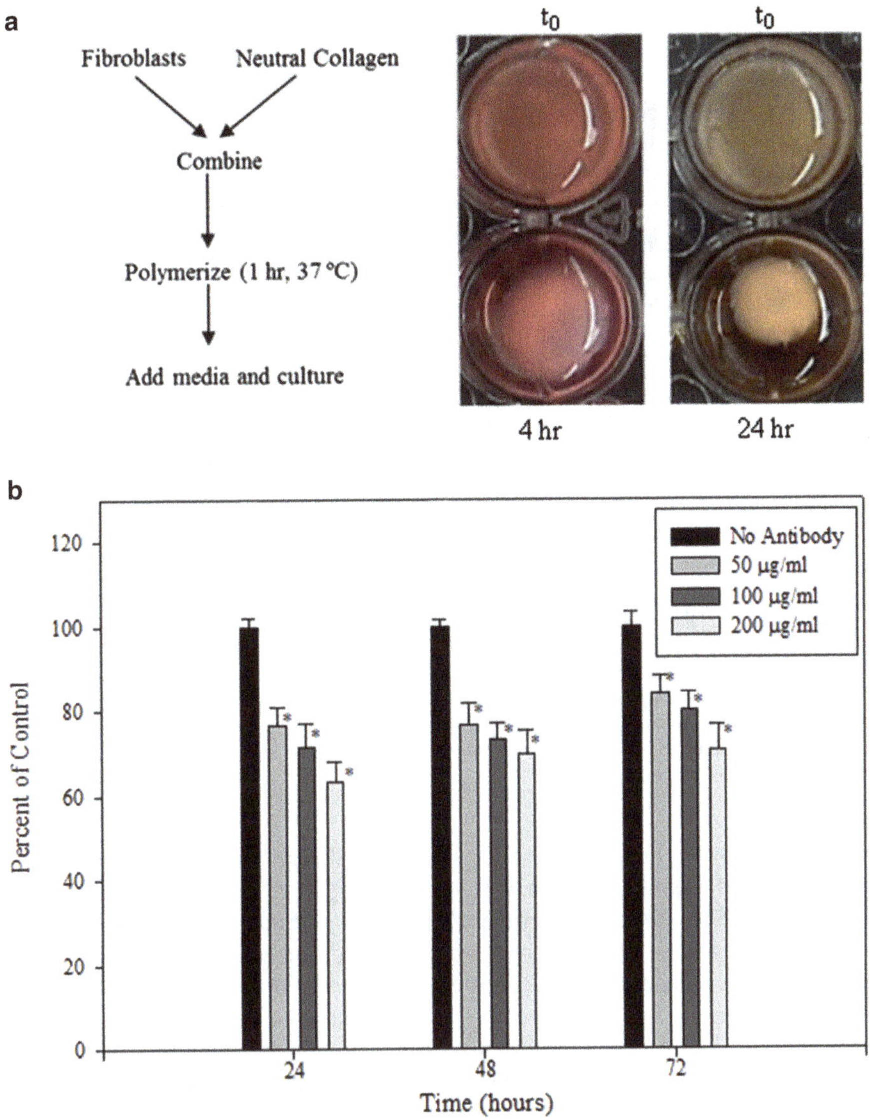


**Fig. 18.5** Blocking DDR2 on cardiac fibroblasts inhibits collagen gel contraction. (**a**) *Left hand*—flow chart for the preparation of collagen gels; *right panel*—images of collagen gels at $t_0$ (*upper panels*) and at 4 and 24 h later. (**b**) Treatment of cardiac fibroblasts with an antibody against the extracellular portion of DDR2 significantly inhibited fibroblast-mediated gel contraction. Furthermore, this response appears to occur in a dose-dependent manner. $*p<0.05$ by Student's $t$-test; $n$ = three independent experiments with three replicates per experiment

Using rat cardiac fibroblasts, Wilson et al. quantified the deformation of monolayer collagen matrices and 3D collagen gels by neonatal and adult cardiac fibroblasts. Experiments using monolayer culture were conducted using the protocol established

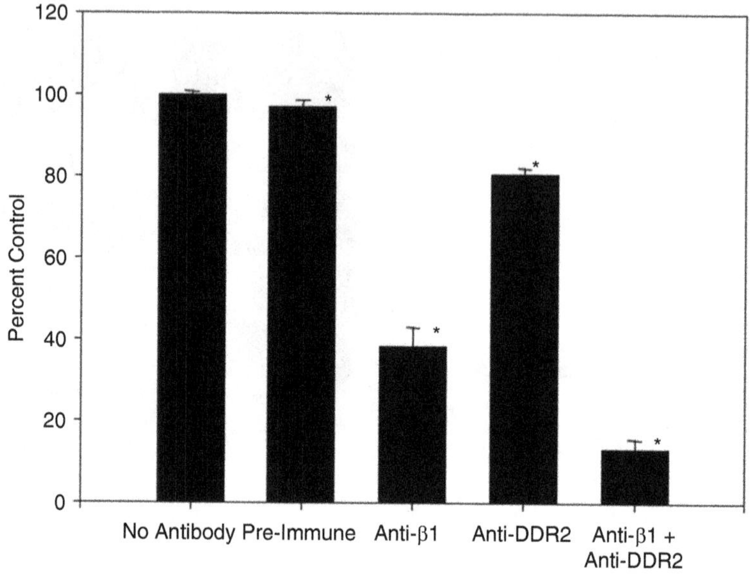


**Fig. 18.6** DDR2 and β1 integrin independently mediate collagen matrix remodeling. Treatment of neonatal cardiac fibroblasts with anti-β1 integrin (50 μg/mL) or anti-DDR2 (200 μg/mL) antibodies individually or in combination significantly inhibited contraction of three-dimensional collagen gels compared to untreated control gels. While addition of preimmune (200 μg/mL) antibodies demonstrated a slight decrease contraction, it was not near the extent observed with either β1 or DDR2 blocking antibodies. $*p<0.05$ by Student's *t*-test; $n$=three independent experiments with three replicates per experiment

by Stone et al. [61] which utilized gold nanorods and darkfield microscopy to calculate 2D strain maps. Real-time PCR and Western blotting were used to quantify DDR1, DDR2, and α-SMA expression for these same cultures. Strain calculations found no statistically significant difference between the neonate and adult fibroblasts in monolayer culture even though RT-PCR showed statistically different levels of DDR1 and DDR2 transcription between the cell types. Western blotting only showed a difference in DDR2 expression, with elevated levels found in the adult cells. There was neither discernible difference in α-SMA transcription nor expression, which likely accounted for the lack of differences in the calculated strain values.

In the 3D culture system, differences between neonate and adult fibroblasts became apparent. Both the age of the fibroblast and the length of time the gels were incubated yielded significant differences. Adult cells had more DDR1 and DDR2 protein expression early, with continued culture time allowing expression in the neonate cells to catch up and then surpass the adult fibroblasts. Regression analysis of expression for DDRs and α-SMA with gel contraction determined that DDR1 and DDR2 expression are similarly regulated and that α-SMA and DDR1/DDR2 expression in neonates are correlated. This is not the case for adult fibroblasts, where no correlation between DDR expression and α-SMA was established, likely accounting for the increased contractile properties of neonatal fibroblasts.

Differences in 2D and 3D culture results have been established by Cukierman et al. [62], with the 3D system providing a more in vivo like environment. The age-dependent expression of DDRs reported by Wilson et al. are consistent with the results of Ieda et al. [63], who reported that cardiac fibroblasts from developing hearts have altered gene expression, including higher levels of ECM protein expression than those from mature hearts.

## 18.5 DDR1 in the Developing Heart

Although several studies have examine the expression or role of DDR1 in the development of other organs [27, 64–68], little is known about its potential role in heart development. Similar to what has been observed with the DDR2 knockout mouse, DDR1 null mice are viable, smaller in size than wild-type littermates, and have reproductive defects [69]. Like the DDR2 knockout, no developmental cardiac abnormalities were noted in DDR1 null mice. Reverse transcriptase-PCR analysis of RNA obtained from fetal hearts revealed that DDR1 is expressed in the heart at ED11.5 and that this expression continues through fetal development (unpublished observation) with DDR1 being expressed by both myocytes (unpublished observation) and fibroblasts [60]. As described by Wilson et al. [60], DDR1 is expressed by both neonatal and adult cardiac fibroblasts but to a higher degree by neonatal fibroblasts cultured on 2D collagen substrates. While this work represents an initial analysis of DDR1 expression in the heart, it is clear that further studies are needed to determine how DDR1 contributes to cardiac development.

## 18.6 Conclusions and Future Directions

While several studies have characterized the expression pattern of DDR2 in the developing heart and examined potential contributions of DDR2 in this process, surprisingly little is known about the contribution of DDR1. As noted, both DDR1 and DDR2 knockout mice are viable, with no reported cardiac defects. The absence of heart defects in these mice may be due to rescue by the other DDR family member or by additional ECM receptors such as integrins. A recent paper by Unsoeld et al. [70] examining axonal growth during development suggests that DDR2 compensates for DDR1 deletion during neuronal development. In addition to the extensive ECM deposition and remodeling that occurs during development, cardiovascular disease is often associated with the aberrant and excessive accumulation of collagen (fibrosis) which impairs left ventricular function. To date, no studies have been published examining the role of DDRs in the progression of cardiovascular disease. Recent work by Cowling et al. [71] using a novel DDR2-null model demonstrated a reduction in adult heart size in the null animals compared to wild type along with slight systolic and diastolic dysfunction in DDR2-null mice subjected to stress.

In addition, analysis of collagen in the DDR2-null hearts revealed a greater collagen volume compared to wild type, suggesting defects in collagen deposition and organization within the DDR2-null hearts potentially due to altered expression of prolyl 4-hydroxylase by cardiac fibroblasts. Given this data and the ability of these receptors to alter expression of matrix metalloproteinases, key enzymes involved in ECM remodeling, it would be interesting to see how the DDR1 and DDR2 knockout mice respond in common cardiovascular disease models (i.e., pressure overload, volume overload, or myocardial infarction). Lastly, given the specific localization of DDR2 on fibroblasts within the heart, the development of a DDR2-promoter to drive gene expression in these cells would be a valuable tool for researchers studying fibroblast lineage and function.

**Acknowledgments** The authors would like to thank Ms. Mary Morales for technical assistance, Regeneron Pharmaceuticals for providing the DDR2 knockout mice, and the National Institutes of Health (HL73937) and the Mid-Atlantic Affiliate of the American Heart Association (0060217U and 13GRNT17070086) for funding.

## References

1. Linask KK, Han M, Cai DH et al (2005) Cardiac morphogenesis: matrix metalloproteinase coordination of cellular mechanisms underlying heart tube formation and directionality of looping. Dev Dyn 233:739–753
2. Markwald RR, Wessels A (2001) Overview of heart development. In: Tomanek RJ, Runyan RB (eds) Formation of the heart and its regulation. Birkhauser, Boston
3. Goldsmith EC, Carver W, Borg TK (2001) The role of the extracellular matrix and its receptors in modulating cardiac development. In: Tomanek RJ, Runyan RB (eds) Formation of the heart and its regulation. Birkhauser, Boston
4. Bishop JE, Greenbaum R, Gibson DG et al (1990) Enhanced deposition of predominantly type I collagen in myocardial disease. J Mol Cell Cardiol 22:1157–165
5. Carver W, Terracio L, Borg TK (1993) Expression and accumulation of interstitial collagen in the neonatal rat heart. Anat Rec 236:511–520
6. Borg TK, Terracio L (1990) Interaction of the extracellular matrix with cardiac myocytes during development and disease. In: Robinson T (ed) Issues in Biomedicine. Karger, Basel
7. Thompson RP, Fitzharris TP, Denslow S et al (1979) Collagen synthesis in the developing chick heart. Tex Rep Biol Med 39:305–319
8. Borg TK, Caulfield JB (1981) The collagen matrix of the heart. Fed Proc 40:2037–2041
9. Caulfield JB, Borg TK (1979) The collagen network of the heart. Lab Invest 40:364–372
10. Hopkins F, McCutcheon EP, Wekstein DR (1973) Postnatal changes in rat ventricular function. Circ Res 32:685–691
11. Eghbali M, Robinson TF, Seifter S et al (1989) Collagen accumulation in heart ventricles as a function of growth and aging. Cardiovasc Res 23:723–729
12. Borg TK, Caulfield JB (1979) Collagen in the heart. Texas Rep Biol Med 39:321–333
13. Cheah KS, Lau ET, Au PK (1991) Expression of the mouse alpha 1 (II) collagen gene is not restricted to cartilage during development. Development 111:945–953
14. Iruela-Arispe ML, Sage EH (1991) Expression of type VIII collagen during morphogenesis of the chicken and mouse heart. Dev Biol 144:107–118
15. Kruithof BPT, Krawitz SA, Gaussin V (2007) Atrioventricular valve development during late embryonic and postnatal stages involves condensation and extracellular matrix remodeling. Dev Biol 302:208–217

16. Eghbali M, Czaja MJ, Zeydel M et al (1988) Collagen chain mRNAs in isolated heart cells from young and adult rats. J Mol Cell Cardiol 20:267–276
17. Weber KT (1989) Cardiac interstitium in health and disease: the fibrillar collagen network. JACC 13:1637–1652
18. Vogel W, Gish GD, Alves F et al (1997) The discoidin domain receptor tyrosine kinases are activated by collagen. Mol Cell 1:13–23
19. Shrivastava A, Radziejewski C, Campbell E et al (1997) An orphan receptor tyrosine kinase family whose members serve as non-integrin collagen receptors. Mol Cell 1:25–34
20. Lemmon MA, Schlessinger J (2010) Cell signaling by receptor tyrosine kinases. Cell 141:1117–1134
21. Leitinger B (2003) Molecular analysis of collagen binding by the human discoidin domain receptors, DDR1 and DDR2. Identification of collagen binding sites in DDR2. J Biol Chem 278:16761–16769
22. Leitinger B, Steplewski A, Fertala A (2004) The D2 period of collagen II contains a specific binding site for the human discoidin domain receptor, DDR2. J Mol Biol 344:993–1003
23. Leitinger B, Kwan AP (2006) The discoidin domain receptor DDR2 is a receptor for type X collagen. Matrix Biol 25:355–364
24. Hou G, Vogel W, Bendeck MP (2001) The discoidin domain receptor tyrosine kinase DDR1 in arterial wound repair. J Clin Invest 107:727–735
25. Brauer PR (2006) MMPs—role in cardiovascular development and disease. Front Biosci 11:447–478
26. Spinale FG (2007) Myocardial matrix remodeling and the matrix metalloproteinases: influence on cardiac form and function. Physiol Rev 87:1285–1384
27. Chin GS, Lee S, Hsu M et al (2001) Discoidin domain receptors and their ligand, collagen, are temporally regulated in fetal rat fibroblasts in vitro. Plast Reconstr Surg 107:769–776
28. Alves F, Vogel W, Mossie K et al (1995) Distinct structural characteristics of discoidin 1 subfamily receptor tyrosine kinases and complementary expression in human cancer. Oncogene 10:609–618
29. Goldsmith EC, Hoffman A, Morales MO et al (2004) Organization of fibroblasts in the heart. Dev Dyn 230:787–794
30. Snider P, Standley KN, Wang J et al (2009) Origin of cardiac fibroblasts and the role of periostin. Circ Res 105:934–947
31. Goldsmith EC, Bradshaw AD, Zile MR, Spinale FG (2014) Myocardial fibroblast-matrix interactions and potential therapeutic targets. J Mol Cell Cardiol 70:92–99
32. Banerjee I, Fuseler JW, Price RL et al (2007) Determinatin of cell types and numbers during cardiac development in the neonatal and adult rat and mouse. Am J Physiol Heart Circ Physiol 293:H1883–H1891
33. Talasila A, Germack R, Dickenson JM (2009) Characterization of P2Y receptor subtypes functionally expressed on neonatal cardiac myofibroblasts. Br J Pharmacol 158:339–353
34. Sun J, Li SH, Liu SM et al (2009) Improvement in cardiac function after bone marrow cell therapy is associated with an increase in myocardial inflammation. Am J Physiol Heart Circ Physiol 296:H43–50
35. Squires CE, Escobar GP, Payne JF et al (2005) Altered fibroblast function following myocardial infarction. J Mol Cell Cardiol 39:699–707
36. Borg TK (1982) Development of the connective tissue network in the neonatal hamster heart. Am J Anat 165:435–443
37. Dettman RW, Denetclaw W, Ordhal CP et al (1998) Common epicardial origin of coronary vascular smooth muscle, perivascular fibroblasts, and intermyocardial fibroblast in the avian heart. Dev Biol 193:169–181
38. Gittenberger-De Groot AC, Vrancken Peeters MPFM, Mentink MMT et al (1998) Epicardium-derived cells contribute a novel population to the myocardial wall and the artioventricular cushions. Circ Res 82:1043–1052
39. Morales MO, Price RL, Goldsmith EC (2005) Expression of discoidin domain receptor 2 (DDR2) in the developing heart. Microsc Microanal 11:260–267

40. Bolender DL, Markwald RR (1979) Epithelial-mesenchymal transformation in chick atrioventricular cushion morphogenesis. Scanning Electron Microsc 13:313–321
41. Bouchey D, Argraves WS, Little CD (1996) Fibulin-1, vitronectin and fibronectin expression during avian cardiac valve and septa development. Anat Rec 244:540–551
42. Ramsdell AF, Markwald RR (1997) Induction of endocardial cushion tissue in the avian heart is regulated, in part, by TGFβ-3-mediated autocrine signaling. Dev Biol 188:64–74
43. Sinning AR, Lepera RC, Markwald RR (1988) Initial expression of type I procollagen in chick cardiac mesenchyme is dependent upon myocardial stimulation. Dev Biol 130:167–174
44. Goldsmith EC, Zhang X, Watson J et al (2010) The collagen receptor DDR2 is expressed during early cardiac development. Anat Rec 293:762–769
45. Olaso E, Labrador JP, Wang L et al (2002) Discoidin domain receptor 2 regulates fibroblast proliferation and migration through the extracellular matrix in association with transcriptional activation of matrix metalloproteinase-2. J Biol Chem 277:3606–3613
46. Labrador JP, Azcoitia V, Tuckermann J et al (2001) The collagen receptor DDR2 regulates proliferation and its elimination leads to dwarfism. EMBO Rep 2:446–452
47. Norris RA, Damon B, Mironov V et al (2007) Periostin regulates collagen fibrillogenesis and the biomechanical properties of connective tissue. J Cell Biochem 101:695–711
48. de Vlaming A, Sauls K, Hajdu Z et al (2012) Atrioventircular valve development: New perspectives on an old theme. Differentiation 84:103–116
49. Bishop JE, Laurent GJ (1995) Collagen turnover and its regulation in the normal and hypertrophying heart. Eur Heart J 16C:38–44
50. Spinale FG, Coker ML, Thomas CV et al (1998) Time-dependent changes in matrix metalloproteinase activity and expression during the progression of congestive heart failure. Circ Res 82:482–495
51. Grinnell F (1994) Fibroblasts, myofibroblasts and wound contraction. J Cell Biol 124:401–404
52. Ross RS, Borg TK (2001) Integrins and the myocardium. Circ Res 88:1112–1119
53. Burgess ML, Carver WE, Terracio L et al (1994) Integrin-mediated collagen gel contraction by cardiac fibroblasts: effects of angiotensin II. Circ Res 74:291–298
54. Vogel W, Brakebusch C, Fassler R et al (2000) Discoidin domain receptor 1 is activated independently of beta(1) integrin. J Biol Chem 275:5779–5784
55. Leitinger B (2014) Discoidin domain receptor functions in physiological and pathological conditions. Int Rev Cell Mol Biol 310:39–87
56. Arora PD, McColloch CA (1994) Dependence of collagen remodeling on alpha-smooth muscle actin expression by fibroblasts. J Cell Physiol 159:161–175
57. Hinz B, Celetta G, Tomasek JJ et al (2001) Alpha-smooth muscle actin expression upregulates fibroblast contractile activity. Mol Cell Biol 12:2730–2741
58. Linjen P, Petrov V, Fagard R (2003) Transforming growth factor-beta 1-mediated collagen contraction by cardiac fibroblasts. Biomech Model Mechanobiol 1:59–67
59. Carver W, Molano I, Reaves TA et al (1995) Role of the alpha 1 beta 1 integrin complex in collagen gel contraction in vitro by fibroblasts. J Cell Physiol 165:425–437
60. Wilson CG, Stone JW, Fowkles V et al (2011) Age-dependent expression of collagen receptors and deformation of type I collagen substrates by rat cardiac fibroblasts. Microsc Microanal 17:555–562
61. Stone JW, Sisco PN, Goldsmith EC et al (2007) Using gold nanorods to probe cell-induced collagen deformation. Nano Lett 7:116–119
62. Cukierman E, Pankov R, Stevens DR et al (2001) Taking cell matrix adhesions to the third dimension. Science 122:1708–1712
63. Ieda M, Tsuchihashi T, Ivey KN et al (2009) Cardiac fibroblasts regulate myocardial proliferation through beta 1 integrin signaling. Dev Cell 16:233–244
64. Lai C, Lemke G (1991) An extended family of protein-tyrosine kinase genes differentially expressed in the vertebrate nervous system. Neuron 6:691–704
65. DiMarco E, Cutuli N, Guerra L et al (1993) Molecular cloning of trkE, a novel trk-related putative tyrosine kinase receptor isolated from normal human keratinocytes and widely expressed by normal human tissues. J Biol Chem 268:24290–24259

66. Sanchez MP, Tapley P, Saini SS et al (1994) Multiple tyrosine protein kinases in rat hippocampal neurons: isolation of Ptk-3, a receptor expressed in proliferative zones of the developing brain. Proc Natl Acad Sci 91:1819–1823
67. Gross O, Beirowski B, Harvey SJ et al (2004) DDR1-deficient mice show localized subepithelial GBM thickening with focal loss of slit diaphragms and proteinuria. Kidney Int 66:102–111
68. Meyer zum Gottesberge AM, Gross O, Becker-Lendzian U, Massing T, Vogel WF (2008) Inner ear defects and hearing loss in mice lacking the collagen receptor DDR1. Lab Invest 88:27–37
69. Vogel WF, Aszodi A, Alves F et al (2001) Discoidin domain receptor 1 tyrosine kinase has an essential role in mammary gland development. Mol Cell Biol 21:2906–2917
70. Unsoeld T, Park JO, Hutter H (2013) Discoidin domain receptors guide axons along longitudinal tracts in *C. elegans*. Dev Biol 374:142–152
71. Cowling RT, Yeo SJ, Kim IJ et al (2014) Discoidin domain receptor 2 germline gene deletion leads to altered heart structure and function in the mouse. Am J Physiol Heart Circ Physiol 307:H773–H781

# Index




R. Fridman, P.H. Huang (eds.), *Discoidin Domain Receptors in Health and Disease*, DOI 10.1007/978-1-4939-6383-6

GPSR Compliance
The European Union's (EU) General Product Safety Regulation (GPSR) is a set of rules that requires consumer products to be safe and our obligations to ensure this.

If you have any concerns about our products, you can contact us on

ProductSafety@springernature.com

In case Publisher is established outside the EU, the EU authorized representative is:

Springer Nature Customer Service Center GmbH
Europaplatz 3
69115 Heidelberg, Germany

www.ingramcontent.com/pod-product-compliance
Lightning Source LLC
Chambersburg PA
CBHW060743150926
51501CB00003B/82

*9781493963812*